T0338250

# MULTISCALE ANALYSIS OF DEFORMATION AND FAILURE OF MATERIALS

# Microsystem and Nanotechnology Series

### Series Editors – Ronald Pethig and Horatio Dante Espinosa

| | | |
|---|---|---|
| Fluid Properties at Nano/Meso Scale | Dyson et al | September 2008 |
| Introduction to Microsystem Technology | Gerlach | March 2008 |
| AC Electrokinetics: Colloids and Nanoparticles | Morgan & Green | January 2003 |
| Microfluidic Technology and Applications | Koch et al | November 2000 |

# MULTISCALE ANALYSIS OF DEFORMATION AND FAILURE OF MATERIALS

**Jinghong Fan**

*Kazuo Inamori School of Engineering,*
*Alfred University, New York, USA*

A John Wiley and Sons, Ltd, Publication

*Registered office*
John Wiley & Sons Ltd, The Atrium, Southern Gate, Chichester, West Sussex, PO19 8SQ, United Kingdom

For details of our global editorial offices, for customer services and for information about how to apply for permission to reuse the copyright material in this book please see our website at www.wiley.com.

*Library of Congress Cataloguing-in-Publication Data*

Fan, Jinghong.
  Multiscale analysis of deformation and failure of materials / Jinghong Fan.
     p. cm.
  Includes bibliographical references and index.
  ISBN 978-0-470-74429-1 (cloth)
1. Deformations (Mechanics) 2. Materials–Analysis–Data processing. 3. Multivariate analysis. I. Title.
TA417.6.F36 2010
620.1'123–dc22                                        2010025737

A catalogue record for this book is available from the British Library.

Print ISBN: 9780470744291
ePDF ISBN: 9780470972274
oBook ISBN: 9780470972281

Set in 9/11pt, Times Roman by Thomson Digital, Noida (India)
Printed and bound in Singapore by Markono Print Media Pte Ltd.

*To my wife, Zheng Ying*
*Daughter Ying Fan and Son Qiang Fan*
*for*
*inspiration and loving support*

# Contents

# About the Author

Dr. Jinghong Fan is a Professor at the Kazuo Inamori School of Engineering at Alfred University, New York, USA. Dr. Fan graduated from the Department of Naval Architecture, Shanghai Jiao Tong University, China and received MS and Ph.D. degrees from the Department of Aerospace Engineering and Engineering Mechanics at the University of Cincinnati, USA. Dr. Fan serves as the Chairman of the Scientific Committee of the Research Center on Materials Mechanics at Chongqing University. He is Co-Chair of the International Conference on Heterogeneous Materials Mechanics (ICHMM) 2004, 2008, and 2011.

Dr. Fan has developed the generalized particle dynamics method by which classical molecular dynamics can be extended to a large material domain. His pioneering work includes showing experimentally the quantitative size effects of layer thickness of microstructure of pearlitic steel on ratcheting (cyclic creep) and then developed a hierarchical multiscale method to describe the discovered size effects by linking variables at micro/meso/macroscopic scales of continuum and the scale of dislocation. Publications include *Foundation of Nonlinear Continuum Mechanics* and the Chinese version of *Multiscale Analysis of Deformation and Failure of Materials* as well as more than 140 papers. His research interests include multiscale modeling and simulation for deformation, defect initiation and evolution under mechanical loading and processing conditions of thin-layer ceramics coating as well as multiscale analysis for interactions between medical implants and bio-cells. His traditional research fields include constitutive laws, plasticity, composite materials, damage, fracture, and fatigue.

# Series Preface

In the past decade, micro- and nano-technology have received unprecedented attention from governments around the world, industry, the press, and the public in the hope to witness revolutionary discoveries, which when translated to products could impact and transform our everyday lives. Following the success of the semiconducting industry, and more recently the information technology industry, the expectation for major nanotechnology breakthroughs, in the early part of the 21st century, is very high. However, micro and nano technologies are less mature and rely on scientific advances to fulfil their promise. In this regard, the development of model capabilities with *predictive power* is essential. Taking advantage of modern supercomputers, *in-silico* modelling of bottom up fabrication of complex 3-D molecular systems, prediction of mechanical, electrical, optical and thermal performance of new nanomaterials (e.g., metallic and semiconducting nanowires and carbon nanotubes), and protein-protein interactions, just to mention a few examples, is now possible. Such advances are poised to impact and accelerate developments in materials, manufacturing, electronics, medicine and healthcare, energy, the environment and world security. Books in this series focus in promoting the dissemination of such advances through scholarly work of the highest quality. The Series is intended to serve researchers and scientists who wish to keep abreast of advances in the expanding field of nano- and micro-technology, and as a resource for teachers and students of specialized undergraduate and post-graduate courses.

The earlier book *Fluid Properties at Nano/Meso Scale*, by Peter Dyson, Rajesh Ransing, Paul Williams and Rhodri Williams, provides a comprehensive numerical treatment of fluidics bridging the nanoscale, where molecular physics is required as a guiding principle, and the microscale where macro continuum laws operate. In this book Jinghong Fan takes us step by step through a wide range of multiscale modeling methods and simulations of the solid state at the atomistic/nano/submicron scales and up through those covering the micro/meso/macroscopic scale. The book is a timely and very useful presentation of modelling approaches and algorithms with a reach to a broad set of problems in nano and biotechnologies. We are introduced to the concept of material-cells that act as links to provide seamless, bottom-up and top-down, transitions between neighbouring sub-scales. This can be used for a progressive understanding of crystal lattice defects at the atomic scale, through to the dynamics of lattice dislocations, and then to macroscopic properties such as plasticity and electrical resistivity. Other examples include a description of how an atom-based continuum theory can be developed to understand hydrogen storage in carbon nanotubes, and how a multiscale analysis of biological cell-surface interactions can aid the development of medical implants.

The pedagogic treatment given by Professor Fan to his book makes it suitable for inclusion in the final year of undergraduate materials science courses in engineering and physical sciences, as well as in computational graduate courses. The book as a whole should be considered as recommended reading for researchers across a wide range of disciplines including materials science, mechanical engineering, applied chemistry and applied physics.

Ronald Pethig                                                                                        Horacio D. Espinosa

# Preface

Experience shows that in-depth understanding of material properties can result in great improvement to products and promote the development of novel ones through synergies with other disciplines, for example design. Therefore it is essential to recognize that materials are inherently of a hierarchical, multiscale character. Properties should not be considered as monolithic quantities only at macroscopic levels, as historically taught. Rather, important material properties can arise at a myriad of length scales ranging from atomic to microscopic to mesoscopic to macroscopic. Computational simulation is also recognized now as an essential element between theory and experimentation. These concepts comprise the foundations of a new interdisciplinary field of study at the interface of engineering and material science, which is referred to in the current literature as multiscale, multi-physics modeling and simulation.

Study of this field necessarily draws from foundations in electronic structure and atomistic-scale phenomena, which are the basic building blocks of materials. Engineers and scientists are increasingly drawn together by this unifying theme to develop multiscale methods to bridge the gaps between lower-scale and macroscopic theory. This amalgam of fields demands a departure from classical solid mechanics curricula in engineering colleges, as well as condensed matter curricula in the fields of physics and chemistry. The need for curricula changes has been accelerated by recent advances in bio- and nanotechnologies.

This book describes the author's research experience in developing multiscale modeling methods across atomistic/nano/submicron scales and micro/meso/macroscopic continuum analysis. Researchers may be interested in how the concept of material neighbor-link cells can seamlessly transform information bottom-up and top-down, how meso-cells link micro- and macroscopic scales, and how their connection to dislocation theory can help investigate, for example, the size effects of cyclic plasticity and failure.

Wide applications of multiscale analysis are introduced in the book, including how atomistic-based continuum theory can be developed for hydrogen storage of carbon nanotubes, how rate effects on dislocation nucleation can be identified by atomistic analysis so its results can be compared with laboratory testing, how new states can be predicted by using the nudged elastic band method to find minimum energy path and saddle point to distinguish the large-scale separation of activation volume which is the physical basis for the distinction between yield and creep and to find the mechanism for the high strength and high ductility of nanostructured metals (e.g., nano-twinned copper), and how multiscale problems can be extracted from biology, such as the multiscale analysis of cell/surface interactions for medical implants.

Students and practitioners interested in these emerging ideas and approaches must develop an appropriate background. This textbook is written with the intention of providing students with the necessary background and advanced knowledge for multiscale modeling and simulation. The enthusiastic feedback provided by undergraduate and graduate students at Alfred University, USA and Shanghai University, China while using this book in a multiscale analysis course has been rewarding and encouraging.

This book not only describes the background, principles, methods, and applications of various atomistic and multiscale analyses, but also emphasizes new concepts and algorithmic developments through various homeworks. Emphasis is placed on the development of simulation skills and use of software for computer atomistic simulations. Associated with Chapter 10 is a Computational Simulation Laboratory Infrastructure (CSLI). CSLI contains computer UNITS with one-to-one correspondence to the sections of Chapter 10, which can be downloaded from the book's website http://multiscale.alfred.edu and used for computational lab practice through courses or self-learning.

My great thanks are due to Prof. D. McDowell of Georgia Institute of Technology, Dr. V. Yamakov of National Institute of Aerospace, Prof. A. Clare of New York State College of Ceramics and Prof. R. Loucks of the Physical Department of Alfred University for constructive suggestions. Thanks are also due to Dr. M. Chinappi, Dr. A. Cao, Dr. Y. Chen, Mr. B. Wang, Mr. D. Parker, Mr. R. Stewert, Mr. H. Lu, and Ms. L. He who have made contributions to various sections of the book. I would also like to express my gratitude to my colleagues, Professors X. Peng, J. Zhang, X. Zeng, and B. Chen in China for their extensive collaboration.

*Jinghong Fan*
Alfred Village, New York

# Abbreviations

| | | | |
|------|------|------|------|
| 1D | One-dimensional | MC | Monte Carlo |
| 2D | Two-dimensional | MD | Molecular dynamics |
| 3D | Three-dimensional | MEAM | Modified embedded atom method |
| ADP | Angular dependent potential | MEP | Minimum energy path |
| BCC | Body-centered cubic | MEMS | Micro electro-mechanical systems |
| CADD | Couple atomistic analysis with discrete dislocation | MO | Molecular orbital |
| | | MS | Molecular statics |
| CNT | Carbon nanotube | NAMD | Nanoscale molecular dynamics |
| CSLI | Computational simulation laboratory infrastructure | NEB | Nudged elastic band |
| | | NEMS | Nano electro-mechanical systems |
| DC | Direct coupling | NLC | Neighbor-link cell |
| DFT | Density function theory | PBC | Periodic boundary condition(s) |
| DT | Deformation twinning | PDB | Protein data bank |
| EAM | Embedded atom method | PES | Potential energy surface |
| ESCM | Embedded statistical coupling method | PSF | Protein structure file |
| | | PN | Peierls-Nabarro |
| FCC | Face-centered cubic | QC | Quasicontinuum method |
| FE | Finite element | QM | Quantum mechanics |
| FEA | Finite element analysis | $R_{cut}$ | Cutoff radius for interatomic potential |
| FEAt | Finite element and atomistic model | | |
| FEM | Finite element method | RT | Rice-Thomson *or* Room temperature |
| GP | Generalized particle dynamics | | |
| GULP | General Utility Lattice Program | RVE | Representative volume element (= Representative unit cell) |
| kMC | Kinetic Monte Carlo | | |
| $k_B$ | Boltzmann constant | SCS | Self-consistent scheme |
| HCP | Hexagonal close-packed cell | SOFC | Solid oxide fuel cells |
| HF | Hartree-Fock | TB | Tight binding |
| LAMMPS | Large-scale atomic/molecular massively parallel simulation | TST | Transition state theory |
| | | $U^{tot}$ | Total system energy |
| LCAO | Linear combination of atomic orbitals | VMD | Visual molecular dynamics |
| | | VV | Velocity Verlet |
| LDA | Local density approximation | XRD | X-ray diffraction |
| LF | leap-frog | YAG | $Y_3Al_5O_{12}$ synthetic garnet |
| LJ | Lennard-Jones | YSZ | Yttria stabilized zirconia |
| MAAD | Macroscopic atomistic *ab initio* dynamics | | |

# 1

# Introduction

The objectives, categories and significance of multiscale modeling and simulation for deformation and failure of materials are described. Applications are classified into two intrinsically related categories. A framework of spatial multiscale analysis covering a large range of scales is proposed. Two examples are given to show how problems can be extracted from practice to formulate models for multiscale simulation.

## 1.1 Material Properties Based on Hierarchy of Material Structure

The aim of this section is to show how material property depends on the hierarchy of material internal structure and why multiscale modeling and simulation are essential to get in-depth knowledge for deformation and failure. This knowledge gain is important since experience tells us that many disciplines and industries would benefit from advances in fundamental understanding of material behavior. This gain covers a wide field, ranging from engineers engaged in the design of ceramics for wrapping the front edges of agricultural mechanical cutters such as the plow or harrow to material scientists involved in the construction of new alloys for turbine blades for jet engines and engineers crafting sub-micron motors. During this kind of development, old theory may be corrected and new theory based on findings related to the deep understanding may be developed.

### 1.1.1 Property-structure Relationship at Fundamental Scale

A fundamental concept of materials science is that the properties of materials follow from their atomic and microscopic structures.[1] The difference in electron structure of atoms results in various material properties as shown by the periodic table of chemical elements. In addition the atomic arrangement and microstructure at higher length scales affect material properties greatly.

Magnesium alloys are less ductile than aluminum alloys because the former alloys possess a close-packed hexagonal structure (HCP) with only three slip systems, while the latter alloys possess a more symmetric face-centered cubic structure (FCC) with 12 slip systems. Figure 1.1 illustrates several nanoscale structures composed of carbon atoms.[2] These various structures have substantially different properties. The graphite in-plane structure has high in-plane strength. The graphite carbon fiber has superior strength as high as about 2.7 GPa along the fiber direction. Moreover, the single-walled carbon nanotube composed of carbon atoms may possess even higher strength.

*Multiscale Analysis of Deformation and Failure of Materials*   Jinghong Fan
© 2011 John Wiley & Sons, Ltd

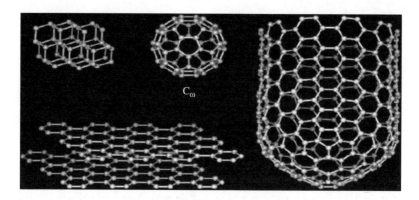

**Figure 1.1**   Morphology of different carbon atom structures (From http://smalley.rice.edu)

For ferroelectric ceramics such as barium titanate ($BaTiO_3$) at temperatures below its Curie temperature ($123.0 \pm 0.6\,°C$), the crystal lattice structure transforms from a cubic lattice into a tetragonal lattice. This phase transformation from paraelectric to ferroelectric results in distinguished positive and negative electric centers of the crystal. Under applied electric or stress field, the crystal may transform to another polarization direction in one of six possible spontaneous polarization directions (see Section 3.7.2 for more detail). These examples elucidate how the atom arrangement substantially affects material properties.

### 1.1.2   Property-structure Relationship at Different Scales

The distribution of atomistic point defects such as lattice vacancies, line defects such as atom dislocations, surface defects such as grain boundaries and surface faults shows different microscopic structures which determine the deformation and failure mechanisms on the fundamental level. Their evolution at the larger microstructural or mesoscopic scale results in void expansion and coalescence, dislocation multiplication and interactions, crystallization and/or phase transformation, micro-crack formation and propagation, impurity segregation to grain boundaries or other defect sites.

While these meso-scale material systems are still controlled by their atomic and microstructures, the clusters of these structures, however, have different morphology and interatomic interactions; thus, the meso-scale structure has distinguishing influence on higher-scale response and properties.

Furthermore, material behavior at macroscopic scale is an averaging of material responses over degrees of freedom in a relatively large volume and time period, thus material properties may exhibit homogeneous, isotropic and linear behavior between stress and strain at small deformation. When deformation is large, deformation induced anisotropy and nonlinear stress-strain behavior such as plasticity, creep, fracture and fatigue can be observed and measured.

### 1.1.3   Upgrading Products Based on Material Structure-property Relationships

The following example demonstrates how one can use the concept that material properties follow from their microscopic structures. Ordinary ceramics, such as alumina ($Al_2O_3$), are opaque because of the pores formed in the sintering and densifying process. The pores result in scattering of incident light rays and local stress concentration. Ceramics become opaque when the porosity reaches 3%. To make traditionally opaque $Al_2O_3$ ceramics into optically transparent materials a fundamental change is required in material structure at microscopic scale. Adding a small quantity of 0.1 wt% MgO causes the high-temperature densification process for the $Al_2O_3$ powder to fully compact. This simple invention makes the ceramic transparent through improving density and decreasing porosity. The ability of anti-erosion in sodium vapor and the strength and life at high temperature are also significantly improved. This microstructure modification enables the ceramic to work persistently in sodium vapor over 1000 °C. Therefore, it can be used as a key component of the sodium vapor bulb which has an intensity of illumination over six times that of common bulbs. Similar examples can be found from many materials including granular materials and bio-inspired composites. However, the relationship between micro-structure and material properties is not well understood. It requires new ideas and methods such as multiscale modeling schemes described in this book to understand the relationship and, in turn, to improve material behavior and products.

## 1.1.4   Exploration of In-depth Mechanisms for Deformation and Failure by Multiscale Modeling and Simulation

From the above description, it is seen that material properties follow from their atomic and microscopic structures and exhibit different properties at different scales. This fact tells us that if one would like to investigate the deformation and failure mechanisms in-depth, one should not just stay at the macroscopic or phenomenological scale, as historically taught. Rather, important properties and material responses can be manifested at a myriad of length scales ranging from atomic to microscopic to mesoscopic to macroscopic.

Since the time mankind discovered iron, material knowledge was, and still is, mostly empirical in the form of well-guarded recipes of how to make a sword or a tin can. Thus almost all of our knowledge so far in material science and engineering is empirical, and the current ongoing effort is to connect it to the fundamental physical knowledge. For example, many physical or constitutive relations of materials such as Hooke's stress-strain law, the creep equations of deformation versus time, the Hall-Petch relation between grain size and yield stress are empirically derived from test data and practice.

Of course, there are many attempts to justify them using physics models (and whether these models are reliable is sometimes questionable) but that is about where it ends. They do not follow directly from the laws of physics. But without this connection to basic physics, we cannot make predictions or design materials, except by wild guesses based on our empirical laws. Understanding of the fundamental physical mechanisms behind material properties is still a major scientific task and it is the reason why multiscale analysis is so important.

This is based on the judgment that multiscale analysis investigates the effects of a hierarchy of internal structures on material behavior, thus making it realistic to understand fundamental mechanisms and to avoid misunderstandings based on macroscopic observations and resulting assumptions. This becomes possible due to the fact that the rapid development of computer and computational algorithms makes modeling and simulation an increasingly powerful tool.

Traditionally, physics takes theory and experiment as its two fundamental bricks. In many cases, theory is developed based on simplified assumptions and can only be verified with limited tests. Multiscale simulation sits in between these two, which makes the development of theory more realistic and even allows "thought test" or tests which cannot be done by experiment. Furthermore, multiscale simulation makes the content of experiment rich and deep, reducing its uncertainty and making it easier to be intrinsically connected to or be explained by the theory.

Because it starts from the fundamental scale and can be calculated quantitatively, multiscale analysis can bring new concepts, methods, mechanisms and phenomena for, say, upgrading or designing materials. These may not be possible or at least need a longer time to develop when one stays only at the phenomenological stage. The significance of multiscale analysis in science and technology is thus defined and illustrated.

Multiscale analysis is an interdisciplinary approach. It combines solid mechanics, material science, physics of condensed matter, chemistry, biology, etc. In addition, multiscale analysis shows prospects for the deep development of solid mechanics and computational mechanics that integrates our understanding of multiple sciences such as chemistry and physics. It therefore turns out that multiscale analysis is an intermediate means to link the fundamental knowledge one learns in chemistry and physics to their advanced applications, say in nanotechnology and biotechnology.

## 1.2  Overview of Multiscale Analysis

Direct atomistic simulation is an impractical means of capturing macroscopic behavior due to strong inherent limitations in size and time scales. Thus, multiscale modeling schemes are needed for the purpose of effectively linking different scales and minimizing computational requirements.

### 1.2.1  Objectives, Contents and Significance of Multiscale Analysis

The objectives and contents of multiscale analysis serve to develop various effective methods so that intrinsic relationships can be established qualitatively and quantitatively to integrate related physical and geometrical variables at different scales.

These relationships in small scales are important in finding various mechanisms of material deformation and failure, while at the macroscopic scale they provide foundations for developing physical or constitutive laws for materials. These bottom-scale up based constitutive laws are different from those established only by phenomenological testing and analysis. They provide intrinsic connections with material deformation and failure mechanism at different scales, thus may be used to develop new approaches and foundations for improving and designing materials to satisfy specific property requirements at macroscopic scale.[3]

The classification of multiscale modeling methods is still under discussion and different categories can be defined based on the methodology, feature and usage of the methods. Just like the history of development of constitutive laws of materials from the 1950s to the 1980s, it is quite natural that there are various models and classifications at the primary stage. It is expected that the same trend will happen to the development of multiscale modeling; each method proposed will serve as a brick to build a truly effective framework of multiscale analyses in the coming decades.

### 1.2.2  Classification Based on Multiscale Modeling Schemes

From a methodological point of view, multiscale analysis can be classified into hierarchical (sequential) and concurrent (parallel) methods. The former is basically a one-way coupling from a lower scale to an adjacent upper scale. The latter requires full coupling among different scales.

#### 1.2.2.1  Hierarchical (Sequential) Methods

There are a large number of physical phenomena in nature in which the coupling of variables at different scales is not so strong, thus investigation limited to a single scale would still be effective. When the coupling becomes weak, the effect of lower scales on the upper scale may be represented by their effect on a few variables of the upper scale. There are two approaches dealing with this weak coupling problem. One is a phenomenological approach which determines parameters by fitting test data. This approach limits its accuracy to the method of data fitting.

The other is a hierarchical (sequential) method which determines several key variables of the upper scale based on a lower-scale simulation using appropriate methods that obey physical laws. Taking the analysis on Buzzards Bay tide circumfluence as an example,[4] Clementi and Reddaway[5] derived atomistic potential based on primary data obtained from quantum mechanics on water molecules. They then used the obtained potential parameters for molecular dynamics (MD) simulation to determine viscosity and density of the fluid. Finally, the MD-derived viscosity and density were effectively used for the bay fluid dynamics simulations.

The hierarchical approach should be applied where only weak coupling exists between small-scale and large-scale variables. In other words, the change in large scale is uniform or quasi-static when viewed from the small scale, so that the small-scale simulation can be used in the upper scale through a statistical averaging process.

### 1.2.2.2 Concurrent (Parallel) Methods

The concurrent approach performs calculations simultaneously to consider strong coupling among scales. It is suitable for cases where a strong coupling exists between different scales, such as turbulence and elastoplastic crack propagation in a finite medium. In the latter case, the crack tip is controlled by atomic bonding through electronic motion and its interaction with atomic nuclei described by quantum mechanics (QM), and the crack tip is constrained by atoms described by MD and further constrained by continuum region. The latter connects to external boundary, transforms the applied force and is usually simulated by the so-called finite element (FE) method. This widely used method in the analysis of engineering structures divides the material continuum domain into many small elements by FE nodes and the lines in between. The stress and strain field can be obtained by the solution of simultaneous algebraic equations (see Section 6.3). The connection and coupling between domains of different scales will be described in Section 5.3 and 7.1.

## 1.2.3   Classification Based on the Linkage Feature at the Interface Between Different Scales

Recently, a new classification of multiscale methods is proposed[6] based on how the atoms in the atomistic domain are connected to the continuum domain. In this classification, approaches that relate atoms and finite element nodes in a one-to-one manner, or through a form of interpolation, will be referred to as direct coupling (DC) methods. According to this definition, most of the existing multiscale methods belong to DC methods.

While DC methods are straightforward, there is a difficulty for truly seamless transition along the boundaries between the atomistic and the continuum domain.[7] This is caused by the intrinsic incompatibility between materials at the two sides of the interface (or boundary) which causes non-physical phenomena, including the so-called ghost forces (see Section 6.4.8).

This incompatibility arises because the constitutive behavior of the continuum on one side is local in nature but that of atoms on the other side is non-local. Here, non-local constitutive behavior indicates that the force at any atom depends also on atoms which do not directly connect with the atom but operate in its neighborhood through interatomic forces; local behavior indicates that the force (stress) of a material point depends only on the deformation gradient (strain) at the same point of the continuum. Due to this local behavior, atoms in the continuum region cannot feel interactions from other atoms nearby as their counterparts in the atomistic region can.

The other shortcoming of DC methods is that in most cases the finite element size has to be nearly the interatomistic distance at the atomistic/continuum interface to perform well. This feature causes substantial difficulty when the model sizes are increased, as one can see below.

In many cases one needs a large atomistic region embedded in a continuum of microns dimension for materials modeling. Moreover, due to rapid CPU development (with quad-core processors on the market and even hundreds-core processors in the next five years), multimillion atom simulations will be easily done even on one computer soon. Because DC methods need to have an extremely small mesh size at the interface of FEM domain, they will have substantial difficulty in meeting this new challenge.

An alternative to DC methods called ESCM (embedded statistical coupling method) is proposed recently using statistical averaging over selected time interval and volume in atomistic subdomains at the MD/FE interface to determine nodal displacement for the continuum FE domain.[6] Another non-DC method called GP (generalized particle dynamics) method is proposed using constant material neighbor-link cells at the interface region to mutually transfer information from bottom-scale up or from top-scale down to quantitatively link variables at different scales.[8] Because the GP method conducts calculation of all scales of generalized particle in each corresponding atomistic scale with the same potential function and numerical algorithm as the atomistic scale, this method is also called extended molecular dynamics. The GP and ESCM non-DC methods will be introduced, respectively, in Chapter 5 and Section 5.10.6.

## 1.3 Framework of Multiscale Analysis Covering a Large Range of Spatial Scales

In this section we propose two classes of multiscale modeling problems in applications. This type of classification is important to form a framework of multiscale analysis to cover a large range of scales.

### 1.3.1 Two Classes of Spatial Multiscale Analysis

Although material properties follow from the atomic structure and the micro/mesoscopic structure, the effects of the characteristics of the two kinds of structures on deformation and failure of materials are different. Considering the difference, spatial multiscale analyses can be separated into two correlative classes.

The first class of multiscale analysis covers the range from atomic scale (angstrom) to microscopic continuum scale (sub-micrometer), covering the nanometer scale. A great application domain of the first class of multiscale analysis lies in applications of micro electro-mechanical systems (MEMS) and nano electro-mechanical systems (NEMS) and bio-technology systems. The key task of this class of multiscale analysis is to develop effective methods smoothly connecting QM/atomic-scale analysis with continuum-scale microscopic analysis to raise the efficiency and accuracy of simulations at small scales.

The second class of multiscale analysis ranges from microscopic to mesoscopic and to macroscopic scale in the continuum domain. In many practical problems, chemical compositions of materials have already been fixed and the effects of electronic and atomic structure is clear, thus the property of the materials depends further on their microstructure in the continuum domain. The main purpose of this class of multiscale analysis is to derive the relationship between property and structure hierarchy, and it usually spans microscopic, mesoscopic and macroscopic scales of continuum. This category is an extension of the existing mesomechanics or two-scale analysis to three or more scales, thus it involves a great many practical engineering problems.

### 1.3.2 Links Between the Two Classes of Multiscale Analysis

The joint region which links the two classes of multiscale analysis is the microscopic domain of continuum. Its dimension is about 10–100 nm. If these two classes of problems can be respectively solved well, nanoscale simulations are expected to link these two categories. If so, the multiscale analysis may span several scales from angstrom to mm or centimeters. This link through nanoscale can be realized through QM-supported atomistic analysis to obtain physical parameters such as frictional coefficient, viscosity and dislocation density for the microscopic analysis of continuum and then connect them to meso- and macroscopic scale.

### 1.3.3 Different Characteristics of Two Classes of Multiscale Analysis

There are some substantial differences in physical nature between the two classes of multiscale analysis, and one may expect the corresponding methodologies to have some differences. The first difference is that the dimension of the atomic structure is fixed but the definition of micro-, meso- and macroscopic structure of continuum in the second class of multiscale analysis is relatively defined. For example, the radius of an aluminum atom is 0.143 nm, its crystal lattice is FCC and the lattice constant is 0.404 nm, i.e., $\sqrt{8}$ times the radius. At this scale, QM and atomistic analysis such as MD should be employed for the analysis.

In the second class of multiscale analysis, however, the definition of microstructure is relative to that of mesostructure and macrostructure. This relativity is the reason why the definition of characteristic length at microscopic scale can be so different in different disciplines. Even in the same discipline, there are different definitions for different problems. Although the relativity of these definitions brings some complexities to the analysis, it enlarges the usage and capability of the second multiscale analysis in practical applications. In fact, according to the demands of given practical problems, three or more distinctly different scales including micro-, sub-micro, meso- and sub-mesoscopic scales can be defined to make the result more meaningful. The relative definition of these scales can directly aim at the target, e.g., grain boundaries or voids, etc.

### 1.3.4 Minimum Size of Continuum

A logical question in dividing two classes of spatial multiscale modeling problems is: what is the minimum size of material that can be considered as continuum? At this moment, there are different opinions as to the answer. It is reported recently that the Hall-Petch relationship is valid when the material is over 20 nm.[9] This relationship describes continuum behavior whereby yield stress is inversely proportional to the square root of grain size, and its underlying mechanism is dislocation pile-up near the grain boundary. This experimental result implicitly indicates that plasticity may be used for high decimal nanometers.

On nanoscale, materials exhibit quite different mechanical properties from micron-scale. An important reason is that the ratio of the surface area to the volume at the nanoscale can be much larger than that at the micron scale. This is so because one can easily prove that the area/volume ratio is $3/r$ for a sphere and $6/a$ for a cube, where $r$ is the radius of the sphere and $a$ the length of the cube edge. It can be seen that the surface/volume ratio of a 1 nm sphere (cube) is 1000 times that of a 1 μm sphere (cube). Therefore, surface energy is very important at the nanoscale so that if the material size is less than a characteristic size, the classical continuum constitutive equation should be revised to consider the effects of surface energy and surface tension.

Recent work shows that continuum theory may be extended to the problems with size of several nanometers if the elastic theory of continuum takes surface elasticity theory into consideration.[10] Smalley, the winner of the Nobel Prize, and his colleagues pointed out that continuum mechanics laws are incredibly reliable,[11] so that they can be applied to the analysis of the behavior of intrinsically discrete solids with the size of a few nanometers. However, the valid minimum size for plasticity analysis remains unsolved.

## 1.4 Examples in Formulating Multiscale Models from Practice

In order to understand how multiscale models can be extracted from problems in practice, two examples, one for inelastic deformation and the other for failure of materials, are introduced in this section. The one for failure or fracture analysis belongs to the first category of application within the atomistic submicron domain and uses both hierarchical and concurrent multiscale methods. The one for deformation analysis belongs to the second category within micro/meso/macroscale continuum domain and uses a hierarchical multiscale method. Emphases are put to explain concepts introduced. Solution schemes will be described in the following chapters.

## 1.4.1  Cyclic Creep (Ratcheting) Analysis of Pearlitic Steel Across Micro/meso/macroscopic Scales

This section introduces an example for inelastic deformation. Rail steel exhibits distinct cyclic creep (ratcheting). This is the phenomenon that for fixed peak and valley values of cyclic stress the inelastic strain accumulates in the direction of mean stress with the increase of cyclic number. This phenomenon occurs when the cyclic loading is asymmetric so that mean stress is non-zero, and it plays an important role in deformation behavior. This kind of cyclic creep is greatly detrimental to railway wear resistance and has been an intensive research area investigated phenomenologically. These single-scale type models have met difficulties in matching experimental data for a long time. In this section, we start from the experimental observations to show how multiscale analysis is necessary.

### 1.4.1.1  Structure Hierarchy at Multiple Scales for High Carbon Rail Steel

A careful experimental study was conducted to investigate size effects of microstructure of high-carbon steel on its mechanical behavior. The detail for how these specimens with designed microstructure of different sizes were fabricated and how the size effects were characterized and confirmed by experiments will be described in Section 8.8. Figure 1.2 shows the microstructure and mesoscopic structure of high-carbon rail steel which is used for railways with heavy loading. The specimens were heat-treated to 860 °C and kept at that temperature unchanged for a certain time interval, then cooled down at the rate of 2 °C/s to room temperature. It can be seen in Figure 1.2(a) that the material is composed of numerous pearlitic colonies with random orientations, and in Figure 1.2(b) that each pearlitic colony is composed of alternately cementite and ferrite lamellae. The average size of a pearlitic colony is about 20 μm.

The cementite lamella is hard and thin with chemical composition of $Fe_3C$ and the thickness about 28∼35 nm. The ferrite lamella is softer with body-centered cubic (BCC) lattice at room temperature and a thickness about 8.1 times that of the cementite lamella. The sizes of carbon and iron atoms are respectively 0.077 and 0.124 nm with average about 0.1 nm. If the thickness of a cementite lamella and 1 mm dimension are taken as the characteristic length, respectively, for microstructure and macroscopic

(a)                                            (b)

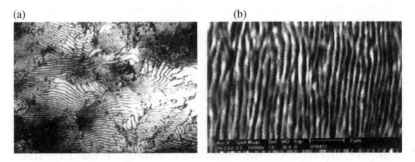

**Figure 1.2**  Hierarchy of material structures of high-carbon rail steel.[12] (a) Meso- and macroscopic structures are composed of randomly oriented pearlitic colonies (magnification x = 1000). (b) Microstructure is composed of alternately arranged cementite (black) and ferrite (white) lamellae (x = 10,000) (Reproduced from Fan, J., Gao, Z. and Zeng, X. (2004) Cyclic plasticity across micro/meso/macroscopic scales. *Proceedings of Royal Society of London A*, 460(2045), 1477, Royal Society of London)

structure, the length ratio of the macroscopic to mesoscopic to microscopic to atomistic characteristic length is approximately 10,000,000:200,000:300:1 and the order ratio is roughly 7:5:2:1.

This fact means that the characteristic length of macroscopic scale is about $10^5$ times that of microscopic scale, and $10^7$ times that of atomic scale. Because the characteristic length 30 nm of microscopic scale is about 300 times that of atomic scale, it is appropriate to conduct microscopic analysis using continuum mechanics. Each lamellae unit is composed of one cementite layer and one ferrite layer, whose thickness is approximately taken as interlaminar spacing.

### 1.4.1.2 Experimental Finding for Size Effects on Ratcheting

A significant finding by experiments is that cyclic creep (ratcheting) behavior of high-carbon rail steel exhibits strong size effects. Experimental results show that the specimens elongated along the longitudinal direction with the number of cycles, indicating extensive cyclic creep (ratcheting). It shows that the macroscopic cyclic stress-strain curve depends on interlaminar spacing of the material. Specifically, cyclic creep increases distinctly with the increase of interlaminar spacing as shown in Figure 1.3; it is seen that when the interlaminar spacing decreases from 319 nm to 287 nm (94%) and 267 nm (85%), the cyclic creep is reduced to 70% and 47%, respectively.

This finding is consistent with other known experimental observations that the strength, fatigue life and wear-resistance of pearlitic steel depend greatly on interlaminar spacing. For instance, there is no evidence for gross plastic deformation of the cementite plates thicker than 0.1 μm, and there is no evidence for extensive brittle fragmentation of cementite plates thinner than 0.01 μm.[13] Experiments also revealed an inverse square root relation between the yield strength and the interlaminar spacing of the material, i.e., the thinner the lamella, the higher the yield strength.

### 1.4.1.3 An Introduction of Multiscale Modeling Scheme for Ratcheting

Now, it is clear that the ratcheting problem involves material structure and property at micro-, meso- and macroscopic scales. If one neglects, say, cementite and ferrite lamellae in the microscopic scale by assuming homogeneous media in the pearlitic colony, as some two-scale models do, one will be misled and unable to find the underlying reasons why lamella thickness is dominant in the deformation and failure mechanisms.

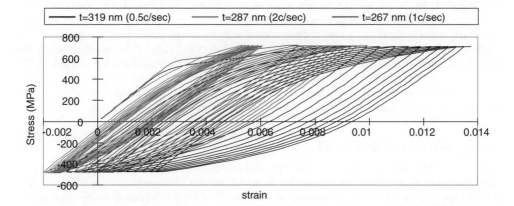

**Figure 1.3** Cyclic creep (ratcheting) of the first seven cycles for layer thickness of 319, 287 and 267 nm of high-carbon rail steel under applied asymmetric loading of $115 \pm 605$ MPa[12] (Reproduced from Fan, J., Gao, Z. and Zeng, X. (2004) Cyclic plasticity across micro/meso/macroscopic scales. *Proceedings of Royal Society of London A*, 460 (2045),1477, Royal Society of London)

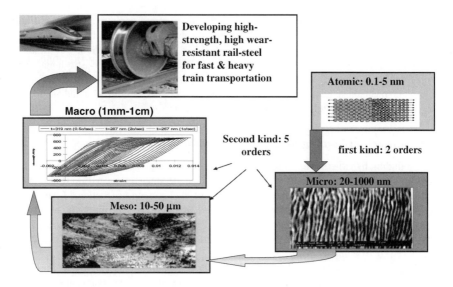

**Figure 1.4** A sketch of the multiscale analysis for pearlitic rail steel with high-strength and high-wear resistance (Reproduced from Fan, J., Gao, Z. and Zeng, X. (2004) Cyclic plasticity across micro/meso/macroscopic scales. *Proceedings of Royal Society of London A*, 460 (2045),1477, Royal Society of London)

This example provides a convincing proof for the importance of multiscale analysis in investigating the deformation and failure of materials. An illustration of the two classes of multiscale modeling of rail steel is provided in Figure 1.4. The multiscale analysis across micro/meso/macroscopic scales of continuum was reported.[12] In this analysis, a generic pearlitic colony and a unit cell consisting of a cementite and a ferrite lamella are taken as the representative unit cell (RVE) for, respectively, mesoscopic and microscopic scale. The basic method proposed in bridging the three scales of this hierarchical multiscale analysis was to take the RVE at the intermediate scale to connect its lower and upper scales. Specifically, an effective constitutive equation of the pearlitic colony (meso-cell) was developed by microstructure analysis, assuming elasticity for cementite and elastoplasticity for ferrite phase and dislocation model for plasticity constants.

The developed constitutive equation was then used to quantitatively link macroscopic analysis through a self-consistent scheme. As a result, the quantitative relationships between microscopic variables (interlaminar spacing and plasticity parameters of ferrite layer) and macroscopic variables (stress and strain) are established so that size effects can be simulated by this across three-scale simulation. The numerical results obtained show a good comparison with the experimental results. The detailed approach in taking the unit cell of the intermediate scale to link the lower scale and upper scale and the comparison between the simulation and experimental data will be discussed in detail in Chapter 8.

## 1.4.2 Multiscale Analysis for Brittle-ductile Transition of Material Failure

This section introduces an example for failure analysis of materials. A key issue in fracture failure analysis is to develop a criterion for predicting whether there will be brittle cleavage induced by micro-crack or plastic passivation in the crack tip induced by dislocations. These two failure mechanisms show greatly different energy levels. Energy consumed in the latter by ductility is far larger than in brittle fracture. On the other hand, many materials show transition between brittle and ductile behavior. It depends on temperature and strain rate. In the first two sections following we will introduce the single-scale models within continuum and in the last two sections show verification and corrections of multiscale analysis, respectively, by the hierarchical and concurrent methods.

### 1.4.2.1 Rice-Thomson (RT) Model

The first criterion for the brittle-ductile transition was proposed by the Rice-Thomson (RT) model.[14] RT model focuses on the competition between two mechanisms: one is loss of atomic cohesion bonding force which causes brittle fracture; the other is the emission of crack tip dislocations which causes ductile failure. The plasticity behavior depends on the competition between the maximum energy for resisting unstable slipping (or critical energy release rate $G_{disl}$) and cleavage surface energy $R_f$ (or energy release rate for brittle fracture). If the former is larger than the latter then brittle fracture will appear, otherwise, there will appear plastic deformation by slipping. Because of the difficulty in accurately calculating $G_{disl}$, the RT model sometimes provides acceptable prediction, but sometimes it does not.

### 1.4.2.2 Rice's Revised Version Based on Peierls-Nabarro (PN) Model

Since the 1990s some work has been done to improve the RT model. Rice set up a new framework within the context of continuum mechanics to investigate the initiation of dislocation from the crack tip.[15] The key idea is to provide a constitutive law $\tau(\delta)$ between shear stress and displacement on the slip plane ahead of the crack tip as a relative boundary condition to investigate dislocation emission.

In the continuum model, the distribution of shear stress on the slip plane ahead of the crack tip follows Frenkel's sinusoidal distribution [16, 17] proposed originally by Peierls [18] and Nabarro. Therefore, this model is called the Peierls-Nabarro (PN) model. In addition, a rigid body slip model between the top and the bottom half crystals was used to develop the constitutive equation $\tau(\delta)$.

Based on this equation, Rice then derived the energy release rate $G_{disl}$ by J integral. The advantage of this approach is that it does not assume the pre-existence of dislocation but traces dislocation evolution from its initiation to final release at the crack tip. Furthermore, this approach is capable of estimating the required energy $G_{disl}$ so that a comparison with the cleavage surface energy $R_f$ can be made to predict whether material exhibits brittle or ductile property.

However, it was found recently that the stress distribution approach based on the PN model and the shear displacement analysis based on rigid body movement does not estimate the value of $G_{disl}$ correctly.[19] The reason is that it is very difficult for the PN model to transform the discrete atomic structure into a continuum model. Furthermore, the rigid displacement model is too rough so that the constitutive equation for $\tau(\delta)$ is not developed on a solid foundation.

### 1.4.2.3 Bridging Atomistic-continuum Scale by Hierarchical (Sequential) Multiscale Method

It is proved that the information required for establishing the constitutive equation $\tau(\delta)$ can be systematically obtained through atomistic calculation.[20, 21] The $\tau(\delta)$ obtained can then be put into Rice's continuum framework to investigate the brittle-ductile transition. In fact, Cleri et al. carried on MD simulation and obtained a more reliable $\tau(\delta)$ curve. This curve is greatly different from the sinusoidal curve of the PN model. The differences include that the curve $\tau(\sigma)$ is asymmetric about the original point and the positive maximum shear stress is not equal to the negative one. In addition, the shear stress is not zero when the ratio of displacement $\delta_r$ is half of the Burgers vector.

A more reliable $G_{disl}$ can be obtained by applying the MD-determined $\tau(\sigma)$ curve to Rice's continuum model. This example shows how a hierarchical modeling method from bottom-scale simulation can offer information to up-scale through establishing the constitutive equation of the intermediate slip zone.[20, 21] Here, the slip zone consists of atoms above and below the slip plane and it is the connection between the atomistic scale and the continuum scale. This example also shows that multiscale analysis can improve models existing for a long time.

### 1.4.2.4  Bridging Atomistic-continuum Scale by Concurrent (Parallel) Multiscale Method

Recently, Miller and Rodney applied the quasicontinuum (QC) concurrent method for investigating the dislocation nucleation during nanoindentation.[22] This work is looking for a general criterion for dislocation nucleation and it is meaningful for brittle-ductile transition. QC method is the one with the longest history in multiscale analysis and will be introduced in Sections 6.4–6.7. Using this concurrent multiscale method, a numerical simulation model with a larger size was used. Consequently plentiful and detailed information was obtained, including:

- In two-dimensional study (2D), it is seen that the dislocation nucleation event is the instantaneous appearance of a dislocation dipole of finite size, corresponding to the collective motion of about 10 atoms on either side of the slip plane. One row of 10 atoms moves about 0.6 Å along the slip direction, while the next row moves about the same distance in the other direction. Other than these 20 or so atoms, there is very little movement. Although the final configuration involves two fully formed dislocations with a large separation between them, the true nucleus of plasticity is a dipole with about half the full Burgers vector.
- In three-dimensional study (3D), this process becomes the spontaneous formation of a loop, again of finite size depending only on the size of the indenter. The diameter of this critical nucleating disk (or dipole in 2D) is defined as the "nucleation diameter", $d_{nuc}$.
- Atomistic analysis in the multiscale model reveals that dislocation nucleation involves the collective motion of a finite disk of atoms over two adjacent slip planes. This nucleation mechanism highlights that the traditional model based on a Peierls-Nabarro model such as Rice's method is not accurate. Instead, a nonlocal consideration of the crystal orientation relative to the indentation in the development of a nucleation criterion is necessary.

Based on the detailed information obtained from this concurrent multiscale analysis, a new theory for dislocation nucleation is obtained which can more accurately predict the location, type and size of the dislocation disk. This example shows the power of multiscale analysis in verification or correction of the old theory, in finding new phenomena and mechanisms, and in developing a new theory.

## 1.5  Concluding Remarks

The importance of multiscale analysis in science and engineering has been recognized for some time. However, due to lack of corresponding models and tools, multiscale analysis has attracted wide attention only in recent years. The needs of both developing technologies such as nanotechnology and biotechnology, and developing interdisciplines in science such as theory and methods in heterogeneous medium (e.g., heterogeneous materials mechanics) have provided the impetus. Progress has been made possible by the emerging computational modeling and simulation tools, high-speed computers, and increasingly high resolution and rapid characterization instruments that have laid down the foundation for a new generation of multiscale analysis.

The following example illustrates the unprecedented opportunity for investigating new concepts, mechanisms, and materials as explained in detail in this chapter. Recently, experimental studies[23–24] with high-resolution TEM have confirmed that by introducing coherent nanotwins of copper, an unusual combination of ultrahigh strength of about 1 GPa (10 times higher than that of conventional copper wire) and high ductility (14% elongation to failure) was obtained while retaining an electrical conductivity comparable to that of pure copper. This is a breakthrough in the concept of strengthening mechanism because traditionally the high electrical conductivity and ultrahigh strength are contradictory; the same contradiction is true for fulfilling the dual requirements of ultrahigh strength and high ductility.

Traditional approaches for strengthening materials are based on strategies that control the generation of various defects described in Section 1.1.2. By producing solute atoms, precipitates and dispersed particles, forest dislocations, and grain boundary, this kind of strengthening forms obstacles to motions of lattice dislocations. It causes the material to harden but reduces ductility because plasticity deformation is limited by constrained dislocation motions. Furthermore, it increases the scattering of conducting electrons at these defects, thus increasing the electrical resistivity of the metal. On the other hand, gliding of dislocations along twin boundaries (TB) is feasible because twins form mirror symmetric coherent boundary structures. Thus, it opens a new avenue to strengthen materials without losing high ductility and conductivity.

Intensive atomistic analyses have been performed to investigate why TBs can substantially strengthen materials. By finding saddle points in the minimum energy path for the dislocation passing of coherent TB (see Sections 9.4 and 9.5.4), an atomistic-mechanistic framework for elucidating the origin of ductility in terms of interactions of dislocations with TB interfaces has been developed.[25] Recently, a large-scale molecular dynamics simulation was performed to investigate the effects of TB spacing on strengthening.[26] It is found that there occurs a transition in deformation mechanism at a critical TB spacing for which strength is maximized. When the TB spacing is larger than the critical value, the classical Hall-Petch type of strengthening due to dislocation pileups (see Sections 8.6.2 and 8.11) and the cutting through twin planes are the strengthening mechanisms. However, if the spacing is smaller than the critical value, the deformation mechanism switches to a dislocation-nucleation-controlled softening mechanism along with the TB migration mechanism. The fact that the result of atomistic simulation is qualitatively consistent with the experimental result of this newly discovered phenomenon [27] signifies the importance of atomistic simulation that will be introduced in the next two chapters. On the other hand, the fact that this TB simulation incorporated as many as 18,500,000 atoms and had a total computation time of 22.8 central processing unit years in the Kraken Cray XT5 supercomputer system signifies the need of multiscale simulation. This will be introduced in Chapters 5–8.

# References

[1] Shackelford, F. (2004) *Introduction to Materials Science for Engineers*, 6th edn., Prentice Hall.

[2] http://smalley.rice.edu/

[3] McDowell, D. L. and Olson, G. B. (2008) Concurrent design of hierarchical materials and structures. *Scientific Modeling and Simulation*, **15**(1), 207 (DOI 10.1007/s10820-008-9100-6).

[4] Rudd, R. E. and Broughton, J. Q. (2000) Concurrent coupling of length scales in solid state systems. *Phys. Stat. Sol.* (b), **217**(1), 251.

[5] Clementi, E. and Reddaway, S. F. (1988) Global scientific and engineering simulations on scalar, vector and parallel LCAP-type supercomputers. *Phil. Trans. of the Royal Society of London* A, **326**(1591), 445.

[6] Saether, E., Yamakov, V., and Glaessgen, E. H. (2009) An embedded statistical method for coupling molecular dynamics and finite elements analysis. *Int. J. Numer. Meth. Eng.*, **78**, 1292.

[7] Curtin, W. A. and Miller, R. E. (2003) Atomistic/continuum coupling in computational materials science, *Modelling Simul. Mater. Sci. Eng.*, **11**, R33.

[8] Fan, J. (2009) Multiscale analysis across atoms/continuum by a generalized particle dynamics method. *Multiscale Modeling and Simulation*, **8**(1), 228.

[9] Hansen, N. (2004) Hall-Petch relation and grain boundary strengthening. *Scripta Materialia*, **51**, 801.

[10] Duan, H. L., Wang, J., Huang, Z. P., *et al.* (2005) Size-dependent effective elastic constants of solids containing nano-inhomogeneities with interface stress. *J. Mech. Phys. Solids*, **53**, 1574.

[11] Yakobean, B. and Smalley, R. E. (1997) Fullerene nanotubes: C-1000000 and beyond. *American Scientist*, **85**, 324.

[12] Fan, J., Gao, Z., and Zeng, X. (2004) Cyclic plasticity across micro/meso/macroscopic scales. *Proceedings of Royal Society of London* A, **460**(2045), 1477.

[13] Langford, G. (1977) Deformation of pearlite. *Metall. Trans.* A, **8**, 861.

[14] Rice, J. R. and Thomson, R. (1974) Ductile versus brittle behavior of crystals. *Philos. Mag.*, **29**(10), 59.

[15] Rice, J. R. (1992) Dislocation nucleation from a crack tip: An analysis based on Peierls concept. *J. Mech. Phys. Solids*, **40**(2), 239.

[16] Rice, J. R. and Beltz, G. E. (1994) The activation energy for dislocation nucleation at a crack. *J. Mech. Phys. Solids*, **42**, 333.

[17] Beeltz, G. E. and Rice, J. R. (1992) Dislocation nucleation at metal-ceramic interfaces. *Acta Metall. Mater.*, **40**, S321.

[18] Peierls, R. E. (1940) The size of dislocation. *Proc. Phys. Soc.*, **52**, 34.

[19] Cleri, F., Yip, S., Wolf, D., *et al.* (1997) Atomistic-scale mechanism of crack-tip plasticity: Dislocation nucleation and crack-tip shielding. *Phys. Rev. Lett.*, **79**(7), 1309.

[20] Cleri, F., Wolf, D., Yip, S., *et al.* (1997) Simulation of dislocation nucleation and motion from a crack tip. *Acta Mater.*, **45**(12), 4993.

[21] Cleri, F., Phillpot, S. R., Wolf, D., *et al.* (1998) Atomistic simulations of material fracture and the link between atomic and continuum length scales. *J. Am. Ceram. Soc.*, **81**(3), 501.

[22] Miller, R. E. and Rodney, D. (2008) On the nonlocal nature of dislocation nucleation during nanoindentation. *J. Mech. Phys. Solids*, **56**, 1203.

[23] Lu, L., Shen, Y., Chen, X., Qian, L., and Lu, K. (2004) Ultrahigh strength and high electrical conductivity in copper. *Science*, **304**, 422.

[24] Lu, K, Lu, L., and Suresh, S. (2009) Strengthening materials by engineering coherent internal boundaries at the nanoscale. *Science*, **324**, 349.

[25] Zhu, T., Li, J., Samanta, A., Kim, H. G., and Suresh, S. (2007) Interfacial plasticity governs strain rate sensitivity and ductility in nanostructured metals. *PNAS*, **104**(9), 3031.

[26] Li, X., Wei, Y., Lu, L., Lu, K., and Gao, H. (2010) Dislocation nucleation governed softening and maximum strength in nano-twinned metals. *Nature*, **464**, 2010.

[27] Lu, L., Chen, X., Huang, X., and Lu, K. (2009) Revealing the maximum strength in nanotwined copper. *Science*, **323**, 607.

# 2

# Basics of Atomistic Simulation

Atomistic simulation, including molecular statics (MS), molecular dynamics (MD), and Monte Carlo (MC) statistical methods, is a powerful tool in investigating material properties and underlying mechanisms of deformation and failure. It is the foundation of bottom-scale up analyses and has been applied to simulate material systems in nanotechnology and biotechnology. In this chapter, the background, basic principles, numerical algorithms, boundary conditions, statistical ensembles, energy minimization, data processing of atomistic analysis are introduced. Detailed procedures, programming, and examples of atomistic simulations are introduced in Chapter 10 and can be enhanced through Computational Simulation Laboratory Infrastructure (CSLI) practice which can be downloaded from the website http://multiscale.alfred.edu.

## 2.1 The Role of Atomistic Simulation

Material property is dependent on atomistic structure and its evolution. An in-depth investigation of the evolution of atomistic configuration with loading and time is crucial to understand the mechanisms of material deformation and failure. Since the atoms are on the angstrom scale, it would be difficult to conduct direct experimental observation. Therefore, numerical simulation of materials at the atomistic scale becomes more essential.

### 2.1.1 Characteristics, History and Trends

Molecular dynamics (MD), one method of atomistic simulation, studies the atomic dynamic behaviour at the atomic nanoscale. The kinetic energy of the system can be ignored when the temperature is at zero Kelvin. Molecular statics (MS) or static lattice calculation is thus adopted in that case and considered a special case of MD. Atomistic analysis can be conducted statistically by Monte Carlo (MC) methods in which random process will be investigated for deformation and failure. Other methods of atomistic simulation determine minimum energy path and saddle point on that path, which are effective to determine the evolution of atomistic configuration and activation energy and activation volume for the initiation of phase transition process such as dislocation nucleation in the atomistic system.

Atomistic simulation has a long and successful history in science and engineering. Molecular dynamics can be dated back to the 1950s. Due to the limitation of speed and memory of computers at that time, the application was limited to studies of the collective or averaged thermomechanical properties of thermodynamics and physicochemistry for certain physical systems.[1–4]

*Multiscale Analysis of Deformation and Failure of Materials*   Jinghong Fan
© 2011 John Wiley & Sons, Ltd

Since the 1980s, MD methods have become widely accepted and improved because of rapid computer developments and innovations in computational algorithms. In the early 1990s, MD textbooks and conference proceedings with systematic explanations appeared. An article entitled "Dynamics and conformational energetic of a peptide hormone: Vasopressin" was published in 1985.[5] Using MD simulation, the work explored the dynamic structure of the tissue, found six low energy states and determined trajectories of atoms. This was the first time that MD was applied to the area of biology and is a validation of MD's reliability and effectiveness. These developments show that MD has gradually become an effective atomistic simulation method.[5-11]

MD simulation captures the overall features and attractive behaviour of the simulated system and makes it possible to visualize and observe the phenomena in detail at the atomic scale. Valuable information which can not be obtained in experiments is provided. This uniqueness makes MD attractive especially in the fields of physics, chemistry, biology, mechanics and material science. As a result of MD simulation, new phenomena have been proved or discovered including the reverse Hall-Petch effect which quantitatively describes the relation between yield stress and grain size at the nanoscale.[12] The value of atomistic simulation to explore natural phenomena and to develop technology is apparent.

## 2.1.2  Application Areas of Atomistic Simulation

Atomistic simulations have applications in some of the following areas.

### 2.1.2.1  Material Defects and Structure

Crystal defects and stable structures of non-crystal materials have been subjects of atomistic analysis for many years.[13-15] The basic principle is that the stable structure or defect state has the lowest total potential energy. The subject is widely studied because of its importance in technology. For instance, zirconium-based ceramic materials are used to store nuclear wastes, therefore the structural stability of the zirconium is an important safety issue. Du et al.[16, 17] used MD simulation for this problem and found polymerization structures of silicon-oxygen network in ceramic materials.

Atomistic simulation is also used to study the ceramic materials in solid oxide fuel cells (SOFC).[18] In SOFC, oxygen ions are conducted by oxides and directly react with fuels. Specifically, oxygen enters from one side (cathode) as negative ions traveling through the electrolyte and then reacts with fuels such as hydrogen on the other side (anode). The continuous flow of charged ions generates electrical power. Defects such as vacancies help oxygen move rapidly inside the solid oxide electrolyte. Compared to conventional power plants, SOFC can achieve much higher efficiency with lower pollution. Atomistic simulations of SOFC electrolytes have produced key understandings of their defect structures and provided insights for improving their oxygen ion conductivity (see Section 3.7.1 for details). This is expected to help lower the SOFC working temperature and expand application potential.

### 2.1.2.2  Fracture

Fracture happens in different ways and at different speeds depending on several parameters. MD simulation is used to obtain in-depth understanding of fracture mechanisms. Obviously, this investigation is technically important. Sections 3.11.3, 6.62–6.65, 6.10.3–6.10.7 and Section 9.4 introduce fracture mechanisms and applications.

### 2.1.2.3  Surface and Interface

Surface physics has gained ground since the 1980s and now attracts attention as a result of the development of high resolution microscopes and high resolution electronic and nanoprobing techniques. MD simulation plays an important part in the understanding of surface reconstruction, surface melting, surface diffusion and surface roughening. Surface effects are more important in nanowires and nanotubes in which the ratio of surface area over volume is high. Under this condition how the surface tension affects the material property, such as Young's modulus, dislocation nucleation, and the constitutive behaviour, is one of the important topics.

### 2.1.2.4 Friction and Adhesion

Friction and adhesion are closely related to the design of surface. When designing fasteners for cars and airplanes the surface must have a certain friction coefficient. Otherwise, fasteners would loosen due to relaxation or require larger mounting torque. The latter may cause bolt strength problems. MD analysis of friction and adhesion between two objects is unfolding, driven directly by the emergence of the atomic force microscope (AFM). Sections 9.7.2 and 9.7.3 introduce the adhesion problem of bio-cells on medical implants.

### 2.1.2.5 Biology

MD can be used in the study of macromolecules and bio-films. The three most abundant biological macromolecules are proteins, nucleic acids (DNA, RNA) and polysaccharides. They are polymers composed of multiple covalently linked identical or nearly identical small molecules or monomers. The latter are amino acid for protein, nucleotide for nucleic acid, and monosaccharide for polysaccharide. The atomic dynamic phenomena and events observed in MD simulation may play a key role in controlling the biochemical and thermomechanical processes that affect the properties and functions of biological cells. Examples using MD for this purpose are illustrated in Section 9.6.3. MD can also be helpful in analyzing the characters in transportation processes of pharmaceutical molecules which is important in the early stage of drug design. Numerical simulation may reduce the research and development costs of new drugs.

### 2.1.2.6 Molecular Cluster Aggregate

The study of aggregates made up of different kinds of atoms is of great technical significance. An example is related to alloy design in which some atoms (e.g., copper atoms in aluminum) gather to form aggregates and gain different mechanical properties. Thus cluster aggregates of metals are essential. The aggregates vary in size from several to thousands of atoms and may perform a link between large molecular systems and solids. Interestingly, different large aggregates have very similar energy. It is difficult to determine their structure by applying the principle of minimum energy as used for finding stable structures in non-crystal materials. Because the aggregates have a large ratio of surface area over volume and are anisotropic, some traits such as melting character are quite different from those of the solids. The large surface strength can intensify the catalytic role in chemical reactions. For automobile emission systems, scientists developed materials with aggregates and cell structure to transform harmful carbon monoxide into carbon dioxide. This innovation effectively reduces exhaust gas pollution.

## 2.1.3 An Outline of Atomistic Simulation Process

The idea of atomistic simulation is to simulate an atomic system from atomic interactions. These interactions can be determined for a certain atomic configuration with interatomic potentials. The latter can be obtained from experience, experimental data or calculated with quantum mechanics (QM). Based on the principle of total potential energy, that energy can be minimized to explore the stable structure of the system (static lattice calculation), the force on each atom can be calculated to generate atom motion (MD), the energy required for a certain process can be associated with probabilities to statistically study the process (MC method), the minimum energy path evolved from a given initial configuration to the final configurations, and the activation energy/activation volume for phase transition can be determined (MS of

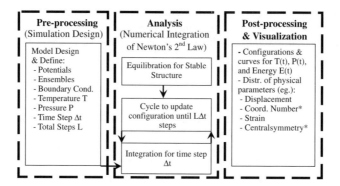

**Figure 2.1** Flowchart of atomistic simulation process. *The coordination number CN and central symmetry parameter Ci are used, respectively, to find the possible dislocation cores and stacking faults by software AtomEye.[19] For normal FCC structure CN=12, if at some locations CN is very large (e.g., CN=14), it may form the core of dislocation (see Section 5.7 and 10.4.3)

infrequent events). Other properties of the system can also be calculated from the total energy or analyzed through a simulated process. Since MD is a commonly used method in multiscale analysis and has wide applications, it can be used as an example to explain the simulation process.

Figure 2.1 shows a flowchart of the MD simulation. The left dash box shows pre-processing with simulation design and the choices of potentials and ensembles (see Sections 2.2.2 and 2.7). After a configuration of atoms is constructed, the potentials are used to determine the force on each atom. Then the atoms are allowed to move for a short time $\Delta t$ with the initial velocities and calculated accelerations, following Newton's second law of particle motion. Then the forces are calculated again with the new atom positions and the step is repeated. The middle box shows this numerical analysis process with integration for a time step $\Delta t$ and cycling to update configurations for L time steps.

L is the loading steps to complete the designed simulation time $t$ (i.e., $t = L \Delta t$) and is usually about $10^4$–$10^7$. To follow the fast atomic vibrations, of the order of $(1–10)$ THz $(10^{12}–10^{13}$ Hz$)$, the step-by-step integration process requires very small time steps, typically of $10^{-15}$ s $(=1$ fs$)$ for metals and ceramics, 2 fs for biological materials. This numerical analysis is not only used for the analysis under given loading but can also be used first for obtaining the equilibrium status and checking whether the used potential functions are correct. The latter can be done, for instance, to check whether the obtained lattice constant is consistent with the observed one. In Sections 10.7 and 10.8 there are examples to show how to carry on numerical simulation with thermodynamic ensemble to get different equilibrium status. The right box is for data processing which uses different softwares such as AtomEye, vmd, gnuplot, etc. to visualize the obtained configuration and obtain other useful information.

As is seen from the above description the MD simulation is not complicated. It is simply a cycling process that "finds forces at atoms, update their positions, and repeats to determine the atom forces at the new positions". However, the obtained results can be quite different for different users. The following three factors are essential to get high accuracy for the simulation:

– A realistic initial condition for both position and velocity vectors of the simulation system is absolutely important. Obtaining the equilibrium status of the system before any loading or excitation is essential for getting reliable results. This is a prerequisite condition to carry on non-equilibrium simulation such as system deformation under external loading. To reach that equilibrium status, the system needs to undertake a relaxation process under a certain thermodynamic ensemble such as the so-called "npt" or "nve" esemble (see Section 2.7).

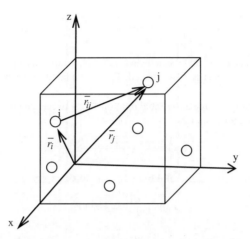

**Figure 2.2**   Coordination of atoms. $\bar{r}_{ij}$ connects particles i and j in Cartesian coordinate system

– A rational design of the simulation model and its boundary condition (see Sections 2.5 and 2.6 respectively).
– An accurate potential function to calculate the correct interatomic force. This is a core topic which will be presented in several sections of this chapter.

These factors will be discussed in the detailed introduction of simulation method in the following sections.

### 2.1.4   An Expression of Atomistic System

Suppose the total number of the atoms in the simulation system is N. The position of atom i ($i = 1, 2, \ldots N$) can be defined by the radius vector $\bar{r}_i$ or by its components $x_i$, $y_i$, and $z_i$ of that atom along the three axes X, Y, Z. Figure 2.2 shows the vector $\bar{r}_{ij}$ which connects atom i with atom j. The first subscript represents the starting atom and the second indicates the ending atom. In atomistic simulation, the length of the vector $\bar{r}_{ij}$, namely the scalar $r_{ij}$ is frequently used to describe the distance and relative position between atoms i and j. Note that the subscripts i and j in $r_{ij}$ are not the same as the expressions of coordinate components of tensors and vectors which frequently appear in books on mechanics. The same symbol in various disciplines has different meanings. To make symbols consistent in several disciplines, symbol definition is flexible. When there is no need to clarify a specific component of vectors and tensors abstract notations are adopted. Subscripts are sometimes used to denote Cartesian tensors and sometimes used to denote interacting atoms.

## 2.2   Interatomic Force and Potential Function

### 2.2.1   The Relation Between Interatomic Force and Potential Function

From mechanics principle, we know that the force $\bar{F}_i$ at atom i applied by other atoms in the simulation system can be obtained through the derivatives of potential energy $U$ of the system with respect to its position vector $\bar{r}_i$ as follows:

$$\bar{F}_i = -\frac{\partial U(\bar{r}_1, \bar{r}_2 \ldots \bar{r}_N)}{\partial \bar{r}_i}, \quad i = 1, 2, \ldots, N \tag{2.1a}$$

The negative sign signifies the fact that when the work, $\bar{F}_i d\bar{r}_i$, is done during the displacement $d\bar{r}_i$ by the force $\bar{F}_i$, the system potential energy $U$ is reduced. This fact can be seen when both sides of (2.1a) are multiplied by $d\bar{r}_i$.

The following is an example of the differential relationship between energy and force via the harmonic potential energy between atom i and j:

$$U = \frac{1}{2}k(r_{ij})^2 = \frac{1}{2}k(\bar{r}_{ij} \cdot \bar{r}_{ij}) = \frac{1}{2}k(\bar{r}_j - \bar{r}_i)^2 \tag{2.1b}$$

Substituting the last vector form of (2.1b) into (2.1a), the force expression can be obtained:

$$\bar{F}_i = -\frac{\partial U}{\partial \bar{r}_i} = k(\bar{r}_j - \bar{r}_i) \tag{2.1c}$$

$$\bar{F}_j = -\frac{\partial U}{\partial \bar{r}_j} = -k(\bar{r}_j - \bar{r}_i) \tag{2.1d}$$

Examples (2.1c) and (2.1d) show correct values and directions of the interatomic force between atoms i and j. The direction of the force applied at atom i points from atom i to j. The force at atom j is opposite in direction but equal in value to the force at atom i. In traditional MD, however, formulations are usually similar in form to the first or second equation of (2.1b), i.e., $r_{ij}$ or $\bar{r}_{ij}$ is directly used.

As seen in (2.1c) and (2.1d), if the potential energy $U$ is known the interaction forces between atoms can be determined. Through integrations of Newton's second law, evolution of dynamics variables such as position vector, velocity and acceleration of each atom in the system with time can be determined. Therefore, the models for determination of the potential function U and its related force field of the atomistic system is the key for atomistic simulation. These models are referred to as "interatomic potentials" or "force fields" models. The first term is used in the physics community while "force field" is frequently used in the chemistry community.

## 2.2.2   Physical Background and Classifications of Potential Functions

Atoms are composed of electrons, protons and neutrons. In classical molecular dynamics, the atom structure is not considered and is replaced by a single mass point as illustrated in Figure 2.2. However, the interatomic potential and forces have their origin at the subatomic level and therefore the atom structure must be considered.

In the three-dimensional (3D) structure of an atom, the negatively charged electron cloud is attracted to, and thus orbits around, its positively charged nucleus. Within the quantum mechanics (QM) description of electron motion, a probabilistic approach is employed to evaluate the probability densities which describe the probability that electrons occupy particular spatial locations. The term "electron cloud" is related to spatial distributions of these electron densities.

The negatively charged electron cloud is also increasingly attracted by the nuclei of its neighbor atoms as the distance between the neighbor nuclei decreases. On reaching the particular distance called the equilibrium bond length, the attraction is equilibrated by a repulsive force between the positively charged nuclei. A further decrease in the inter-nuclei distance results in a continued growth of the repulsive force.

A brief description of the physical background of interatomic forces should be used when only two atoms interact. In reality, interactions will occur between electron clouds and nuclei of multiple atoms. They mainly depend on the bonding type and the surrounding atom arrangement and can be classified as follows:

### 2.2.2.1 Pair Potentials

These interactions are considered along the centers of two atoms. The effects of other atoms on the pair interaction are not considered. This is the earliest empirical potential and has produced considerable data for the parameters. It is used for gas, liquid, metals, biomaterials, etc. It will be discussed in the next section.

### 2.2.2.2 Ionic Bond Potentials

This is related to ionic bond in which electrostatic (Coulomb) interactions often appear, for instance, in ceramics such as oxides (e.g., $Al_2O_3$ and $Ce_2O_3$) and will be introduced in Sections 3.2–3.4.

### 2.2.2.3 Covalent Bond Potentials

This is related to covalent bonds. They are caused by the overlap of electron orbits of multiple atoms. These potentials are important and will be introduced in several sections including Section 3.8 for silicon and carbons.

### 2.2.2.4 Metal Bond Potentials

In metals, bonding is primarily non-directional and can be characterized by positive ions embedded in an electron cloud. It can be better described by the embedded atom method (EAM) and the modified EAM (MEAM) which will be described in Sections 3.10–3.11.

### 2.2.2.5 Constructing Binary and Higher Order Potentials from Monoatomic Potentials

This is a very important issue in current potential development efforts and will be discussed in Section 3.12.

### 2.2.2.6 QM-Based Potential Development

All the potentials are intimately linked to the interactions of electrons which are fundamentally governed by the QM laws. In Chapter 4, the energy link between potential functions in atomistic analysis and those in QM are given, and can be used to develop more accurate potentials based on the methods introduced there.

## 2.3  Pair Potential

Pair potentials are the simplest interatomic interactions, and are dependent on the distance $r_{ij}$ between two atoms. The total pair potential of a generic atom i with other near atoms in the simulation system can be expressed as follows:

$$U_i = \frac{1}{2} \sum_{j=1(j\neq i)}^{Ne} V_{ij}(r_{ij}) \tag{2.2a}$$

where the factor $\frac{1}{2}$ is introduced due to the fact that the pair potential $V_{ij}$ is equally shared by both atom i and j, so atom i should only account for a half. Similar to (2.1b), the summation symbol over a generic atom j ($j = 1, \ldots Ne$) covers all atoms within its neighborhood sphere defined with atom i as the center and

$r_{cut}$ as the radius. The latter is also called cutoff radius indicating if $r_{ij} > r_{cut}$ the interatomic potential is too small and can be neglected. The total potential energy $U^{tot}$ of the atomistic system can be expressed as the sum of pair potentials of all atoms as follows:

$$U^{tot} = \sum_{i=1}^{N} U_i \qquad (2.2b)$$

Substituting (2.2b) into (2.2a), the total potential can be expressed by the following two summations: the first covers all atoms j, ranged from 1 to Ne, inside the neighborhood sphere of atom i, and the second covers all the atoms, ranged from 1 to N, in the system:

$$U^{tot} = \frac{1}{2} \sum_{i=1}^{N} \sum_{j=1(j\neq i)}^{Ne} V_{ij}(r_{ij}) \qquad (2.2c)$$

## 2.3.1    Lennard-Jones (LJ) Potential

A variety of mathematical models exist that describe the interatomic interaction. In 1924, Jones proposed the following potential function[20] (currently known as the Lennard-Jones[21] potential or LJ potential) to describe pairwise atomic interactions:

$$V_{LJ}(r) = 4\varepsilon\left[\left(\frac{\sigma}{r_{ij}}\right)^{12} - \left(\frac{\sigma}{r_{ij}}\right)^{6}\right] \qquad (2.3a)$$

$$r_{ij} = |\bar{r}_i - \bar{r}_j| \qquad (2.3b)$$

where $r_{ij}$ denotes the distance between atom i and j, and the term $1/r^{12}$ simulates the repulsive force between two atoms. The introduction of the $1/r^{12}$ term, from the Pauli exclusion principle, states that when the electron clouds of two atoms begin to overlap, the energy of the system increases rapidly, implying that two electrons cannot have the same quantum state (see Postulate 4 of Section 4.3). The $1/r_{ij}^{6}$ term describes the attractive force between atoms simulating the van der Waals force. Van der Waals force is weaker than repulsive force; therefore the corresponding exponent is 6, much smaller than that of the repulsion term.

There are two parameters, $\sigma$ and $\varepsilon$, in LJ potential (2.3a). In Figure 2.3, $\sigma$ is the collision diameter, the distance at which the potential $V(r)$ is zero. The parameter $\varepsilon$ is the bond energy at the equilibrium position $r = r_0$ where the force $F(r) = 0$ by which $r_0$ is also called bond length. $\varepsilon$ is with the negative sign and is the

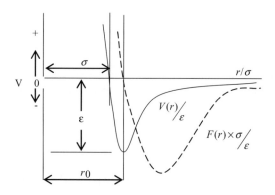

**Figure 2.3**    The variation of potential and force with the dimensionless interatomic distance

minimum energy for an atomic pair. It represents the work needed to move coupled atoms from the equilibrium position $r_0$ to infinity. This is the reason why $\varepsilon$ is called bond or dissociation energy, it signifies the maximum energy that the pair of atoms can absorb before debonding (brittle) or slip (ductile) deformation occurs.

In Figure 2.3, the solid and dotted lines represent the variation of dimensionless energy ($V(r)/\varepsilon$) and force ($F(r)\sigma/\varepsilon$) to the change of the dimensionless distance $\frac{r}{\sigma}$ respectively. The equilibrium interatomic distance $r_0$ is related to the collision diameter $\sigma$ as

$$r_0 = 2^{1/6}\sigma \tag{2.4}$$

The values of $\sigma$ and $r_0$ are usually several angstrom, $1\,\text{Å} = 10^{-10}\,\text{m}$. Since the term in the square bracket of (2.3a) is dimensionless, the unit of potential energy V depends on $\varepsilon$. The value of $\varepsilon$ is very small, therefore a small energy unit such as electron volt (eV), rather than joules, is commonly used. One eV represents the work done by a unit electric charge under a unit voltage, $eV = 1.60219 \times 10^{-19}\,\text{J}$. The interactive force between two atoms can be obtained from equation (2.3a) by taking the derivative with respect to $r$, i.e.,

$$f_{ij}(r) = -\frac{\partial V(r)}{\partial r} = 24\frac{\varepsilon}{\sigma}\left[2\left(\frac{\sigma}{r}\right)^{13} - \left(\frac{\sigma}{r}\right)^{7}\right] \tag{2.5}$$

In the more than 80 years since LJ potential was proposed, a significant amount of data for parameters $\varepsilon$ and $\sigma$ has been accumulated, examples of data are given in Table 2.1. It is worth noting that these values are obtained under special conditions. If conditions change, values may also change.

LJ pair potential has been used to simulate a wide variety of materials and phases, with the latter including rare gas, solid, and metallic liquid.[22–24] The latest applications include crystal growth on liquid surface,[25–27] diffusion dynamics in liquids and solids,[28] equilibrium structure of the over-cooling liquid,[29–30] material free surface[31] and grain boundaries,[32] fracture process in two-dimensional solid,[33] and simulations of fracture process of quasi-crystals.[34]

### 2.3.2   The 6-12 Pair Potential

The potential in (2.3a) with exponents 6–12 can be rewritten as:

$$V_{ij}(r) = \frac{K_R}{r^{12}} - \frac{K_A}{r^{6}} \tag{2.6}$$

where $K_R = 4\varepsilon\sigma^{12}$ and $K_A = 4\varepsilon\sigma^6$, this type of potential is called 6–12, with 6 and 12 denoting the power index in the formula. $K_R$ and $K_A$ are the constants of the repulsive and attractive term. Since the net force is

**Table 2.1**   LJ pair potential parameters for a single chemistry element

| Chemical element | $\varepsilon$ [eV] | $\sigma$ (Å) | Chemical element | $\varepsilon$ [eV] | $\sigma$ (Å) |
|---|---|---|---|---|---|
| Carbon, C | 0.0065 | 4.0000 | Iron, Fe | 0.0007 | 4.1400 |
| Sulfur, S | 0.0087 | 4.0000 | Magnesium, Mg | 0.0012 | 3.0030 |
| Hydrogen, H | 0.0009 | 2.0000 | Calcium, Ca | 0.009 | 1.705 |
| Nitrogen, N | 0.0069 | 3.5000 | Phosphorus, P | 0.0020 | 1.446 |
| Oxygen, O | 0.0086 | 3.2000 | Silicon, Si | 0.0022 | 2.357 |
| Gold, Au | 0.0375 | 1.8700 | Sodium, Na | 0.0056 | 2.35 |
| Chlorine, Cl | 0.0051 | 4.4000 | Potassium, K | 0.003 | 1.333 |
| Aluminum, Al | 0.0050 | 2.8500 | Fluorine, F | 0.0016 | 1.41 |
| Copper, Cu | 0.0041 | 2.3380 | | | |

zero at the equilibrium state, from (2.1a) the equilibrium bond length $r_0$ can be determined by

$$\frac{dV_{ij}}{dr} = 0 \tag{2.7}$$

Substituting $V(r)$ of (2.6) into (2.7) we have

$$r_0 = 2\left(\frac{K_R}{K_A}\right)^{1/6} \tag{2.8}$$

It is easy to verify that the relation (2.4) between $\sigma$ and $r_0$ in Lennard-Jones potential agrees with the above equation.

The potential function can also be used to estimate the energy stored in the material. Taking argon as an example, assume $K_A = 10.37 \times 10^{-78}$ J·m$^6$ and $K_R = 16.16 \times 10^{-135}$ J·m$^{12}$, then (2.8) gives the equilibrium bond length $r_0 = 0.382$ nm $= 3.82$ Å. Substituting $r_0$ into formula (2.6) gives $V(0.382) = -1.66 \times 10^{-21}$ J, i.e., the bond energy. Once this atomic potential energy is known, it is easy to calculate the total bonding energy in a given volume by times the pair potential energy and the total pairs of particles. 1 mol argon has the atom pairs of Avogadro number $N = 0.602 \times 10^{24}$ thus the total bonding energy in 1 mol can be estimated as Vbond $= -1.66 \times 10^{-21} \times 0.602 \times 10^{24} = -0.999$ (kJ/mol).

In Section 10.2 and Homework (10.6), there are detailed descriptions for how to make a FORTRAN code for using LJ potential to calculate the acceleration components along the x, y, and z directions.

### 2.3.3 Morse Potential

As we mentioned, the LJ potential increases rapidly when the distance, r, between two atoms is towards 0 because the repulsive term is inversely proportional to $r^{12}$. In reality, the distance between two atoms can seldom be less than $0.9\sigma$, thus the LJ potential may be too hard to describe the repulsive force for some materials such as copper. For those materials, the introduction of a soft pair potential function may be suitable. Morse potential discussed in this section, and Born-Mayer potential and Buckingham potential to be discussed in Section 3.2.2, all belong to this kind of "soft" potential function in the sense that they use an exponential function $\exp(-kr)$ to decay the atomic interaction when the distance is small, especially for the repulsive force.

Morse potential is expressed as follows:

$$V_{ij}(r_{ij}) = D[\exp\{-2\alpha(r_{ij}-r_o)\} - 2\exp\{-\alpha(r_{ij}-r_o)\}] \tag{2.9a}$$

where the first term on the right is the repulsive potential and the second is attractive. It shows that the decaying of the repulsive term with separation distance is much faster than the attractive term because its exponential index is double the attractive one. Formula (2.9a) can be rewritten as the form:

$$V_{ij}(r_{ij}) = D\{\exp(-2\alpha_0\eta_{ij}) - 2\exp(-\alpha_0\eta_{ij})\} \tag{2.9b}$$

where $r_0$ is the equilibrium bond length and $D$ is the dissociation energy parameter. In (2.9a), $\alpha$ is an inverse length scaling factor whose unit is 1/Å. The parameters $\eta_{ij}$, $\alpha_0$ in (2.9b) are dimensionless as:

$$\eta_{ij} = \frac{r_{ij}-r_0}{r_0}, \quad \alpha_0 = \alpha r_0 \tag{2.10a, b}$$

The interatomic force between atom i and j can be determined by taking the derivative of equation (2.9b) as:

$$f_{ij} = -\frac{dV_{ij}(r_{ij})}{dr_{ij}} \tag{2.10c}$$

**Table 2.2** Parameters related to the Morse potential (Reproduced from Komanduri, R., Chandrasekaran, N., and Raff, L. M. (2001) Molecular dynamics (MD) simulation of uniaxial tension of some single-crystal cubic metals at nanolevel. *International Journal of Mechanical Sciences*, 43, 2237, Elsevier)

| Material | Crystal structure | Dissociation energy, $D$ (eV) | Equilibrium radius, $r_0$ (Å) | $\alpha$ (Å$^{-1}$) | Lattice constant (Å) |
|---|---|---|---|---|---|
| Aluminum | FCC | 0.2703 | 3.253 | 1.1650 | 4.05 |
| Copper | FCC | 0.3429 | 2.866 | 1.3590 | 3.62 |
| Nickel | FCC | 0.4205 | 2.780 | 1.4199 | 3.52 |
| Iron | BCC | 0.4172 | 2.845 | 1.3890 | 2.87 |
| Chromium | BCC | 0.4414 | 2.754 | 1.5721 | 2.89 |
| Tungsten | BCC | 0.9906 | 3.032 | 1.4116 | 3.17 |

Substituting (2.9b) into (2.10c) gives the force on atom i applied by atom j:

$$f_{ij} = \frac{2D}{r_0}\alpha_0\{\exp(-2\alpha_0\eta_{ij})-\exp(-\alpha_0\eta_{ij})\} \tag{2.10d}$$

Parameters of Morse potential used in MD, such as $D$, $\alpha$, equilibrium bond length $r_0$, and the lattice constant of aluminum, nickel, iron, chromium, and tungsten are listed in Table 2.2. The influence of $\alpha$ and $D$ on the curves of interatomic force versus interatomistic distance are seen in Figure 2.4. If other conditions do not change, the potential well depth increases with the value of $\alpha$ or $D$. This indicates that the dissociation energy increases and more debonding force is required to break the bond. On the other hand, the influence of $D$ and $\alpha$ on the material behavior is different. When $D$ increases, the range of effective interatomic distance of the Morse potential increases; while $\alpha$ increases, the effective range decreases. Morse potential and Lennard-Jones potential are both widely used for simulation in chemistry, physics, and engineering.

## 2.3.4 Units for Atomistic Analysis and Atomic Units (au)

In atomistic analysis, the quantity is usually very small, small units are used. Some commonly used units are introduced below:

- Time: $1 \times 10^{-12}$ seconds, i.e., picoseconds (ps)
- Length: $1 \times 10^{-10}$ meters, i.e., angstroms (Å)
- Atomic mass unit (amu): $1.6605402 \times 10^{-24}$ gram, or $1.6605402 \times 10^{-27}$ kilograms)
- Electron volt (eV): $1.60219 \times 10^{-12}$ erg, or $1.60219 \times 10^{-19}$ J.
- Avogadro's number: $N = 6.02217 \times 10^{23}$ mol$^{-1}$
- Boltzmann constant: $k_B = 1.38062 \times 10^{-16}$ erg K$^{-1}$ or $1.38062 \times 10^{-23}$ J K$^{-1}$

More detailed applications of these units can be seen in Section 10.5.5. Here, we will also introduce Atomic Units (au). These units form a system of units convenient for atomic physics, electromagnetism, and quantum electrodynamics, especially when the focus is on the properties of electrons (see Chapter 4). There are different kinds of atomic units, Hartree units and Rydberg units. Here, Hartree atomic units given by http://en.wikipedia.org/wiki/atomic_units are introduced in Table 2.3.

Unit transformation is important. Examples of units from eV to energy (kCal or kJ) per mole are given in Section 10.5.5; examples from au units of length and energy to Å and eV are given in Section 4.1.

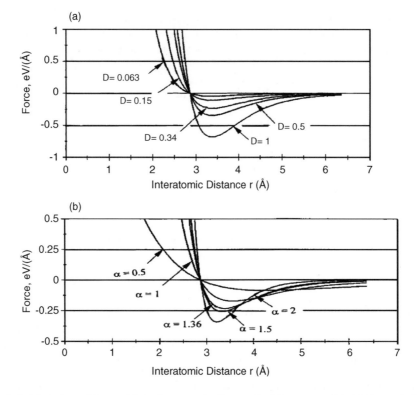

**Figure 2.4** Variations of force with the interatomic distance for Morse potential. (a) The variation of $D$ ($\alpha$ and $r_0$ kept unchanged). (b) The variation of $\alpha$ ($D$ and $r_0$ kept unchanged) (Reproduced from Komanduri, R., Chandrasekaran, N., and Raff, L. M. (2001) Molecular dynamics (MD) simulation of uniaxial tension of some single-crystal cubic metals at nanolevel. *International Journal of Mechanical Sciences*, 43, 2237, Elsevier)

**Table 2.3** Hartree atomic units (au) from http://en.wikipedia.org/wiki/atomic_units

| | | | Fundamental Atomic Units | |
|---|---|---|---|---|
| Quantity | Name | Symbol | SI value | Planck unit scale |
| mass | electron rest mass | $m_e$ | $9.109\,3826(16) \times 10^{-31}\,\text{kg}$ | $10^{-8}\,\text{kg}$ |
| length | Bohr radius | $a_0 = \hbar/(m_e c \alpha)$ | $5.291\,772\,108(18) \times 10^{-11}\,\text{m}$ | $10^{-35}\,\text{m}$ |
| charge | elementary charge | $e$ | $1.602\,176\,53(14) \times 10^{-19}\,\text{C}$ | $10^{-18}\,\text{C}$ |
| angular momentum | Reduced Planck's constant | $\hbar = h/(2\pi)$ | $1.054\,571\,68(18) \times 10^{-34}\,\text{J s}$ | (same) |
| energy | Hartree energy | $E_h = m_e c^2 \alpha^2$ | $4.359\,744\,17(75) \times 10^{-18}\,\text{J}$ | $10^9\,\text{J}$ |
| electrostatic force constant | Coulomb's constant | $1/(4\pi\in_0)$ | $8.9875516 \times 10^9\,\text{C}^{-2}\,\text{N m}^2$ | |

**Homework**

(2.1) The latent heat of vaporization (the energy released during vaporization) of liquid oxygen is 213 kJ/kg and the average bond energy of O-H in water is 458.9 kJ/mol.

(a) Change the oxygen latent heat of 213 kJ/kg to eV per oxygen ($O_2$) molecule.

(b) Build a LJ potential model which can approximately represent the interaction between $O_2$ molecules in liquid oxygen. The bond length in liquid oxygen is 3.374 Å. For calculation, assume that the latent heat per $O_2$ is the equilibrium potential energy of the interaction, but discuss if this assumption is reasonable and whether this assumption can result in inaccurate estimation of the potential energy.

(c) Change the O-H bond energy of 458.9 kJ/mol to eV per bond.

Hint: (1) For $O_2$, 32 g mass is a mole which includes $6.02217 \times 10^{23}$ $O_2$ molecules. The latter number is the Avogadro's number.

(2) For (b) discussion: should the common way using Avogadro's number as bond number in a mole be an accurate bond calculation in this case?

(2.2) There are two types of interatomic potential, one is a long-ranged potential such as Coulomb potential, the other is a short-ranged potential which is enacted within a small range of interatomic spacing such as pair potential. Do you think there is a short-ranged potential between the two H+ in water molecules? How about the long-ranged potential between them?

(2.3) A Morse potential model for O-H bond in water molecules found in literature has the following form:

$$\Phi(r) = D \times (1-\exp[-\alpha(r-r_0)])^2 - D]$$

Explain the physical meaning of D and $r_0$. Plot the potential versus interatomic distance with $D = 6.204$ eV, $\alpha = 2.22$ Å$^{-1}$, $r_0 = 0.924$ Å.

(2.4) The simulation of water molecules is important in many fields. Do you think an O-H potential model enough to simulate the water molecule adequately? (Hint: think about the structure of the water molecule)

## 2.4 Numerical Algorithms for Integration and Error Estimation

### 2.4.1 Motion Equation of Particles

In MD, the configuration evolution with time for an atomic system can be obtained by integration of differential equations expressed by Newton's second law for all atoms. The common form used in MD to calculate the velocity, acceleration and position of a generic atom i under force $f_{ij}$ applied by its neighborhood atom j is:

$$f_{ij} = m_i \frac{d^2 r_{ij}}{dt^2} \tag{2.11a}$$

Or

$$\frac{d^2 r_{ij}}{dt^2} = \frac{f_{ij}}{m_i} \tag{2.11b}$$

where $m_i$ denotes the mass of atom i. The left side of (2.11b) gives the acceleration, its first and second integrations give velocity and position of atom i which are caused by the force $f_{ij}$. The direction of force and acceleration point from atom i to atom j. To get the total force and acceleration of atom i, a summation for all forces on atom i should be performed. This is clearly shown by the vector form of Newton's second law for a generic atom i:

$$\bar{F}_i = m_i \frac{d^2 \bar{r}_i}{dt^2} \quad (i = 1, 2 \ldots N) \tag{2.12}$$

$$\bar{F}_i = \sum_{j=1(j \neq i)}^{Ne} \bar{f}_{ij}(\overline{r_{ij}}) \quad (i = 1, 2 \ldots N) \tag{2.13a}$$

Here $\bar{F}_i$ denotes the total interatomic force subjected by atom i. It is applied by other atoms in the atomistic system. Subscript j denotes a generic atom of these atoms ($j \neq i$). In reality, the interatomic force drops quickly with the distance $r_{ij}$. Only atom j (j = 1, 2, ... Ne) within the neighborhood sphere of atom i is considered in the summation. As introduced in Section 2.3, the sphere is defined with atom i as the center and $r_{cut}$ the radius. Here, $r_{cut}$ is the cutoff radius beyond which the interatomic force on atom i is assumed to be negligible and Ne is the total number of atoms within the neighborhood sphere.

Substituting (2.13a) into (2.12) the following expressions can be obtained:

$$m_i \frac{d^2 \bar{r}_i}{dt^2} = \sum_{j=1(j \neq i)}^{Ne} \bar{f}_{ij}(\overline{r_{ij}}) \tag{2.13b}$$

Motion equations of atomistic systems can be described by forces as is done by Newton's second law. They can also be described by energy description, which will be explained here briefly. For energy description, total energy E, kinetic energy K, potential energy $U(\bar{r})$ and Hamiltonian H of the atomistic system are introduced:

$$E = K + U \tag{2.14a}$$

$$K = \sum_{i=1}^{N} k_i = \frac{1}{2} \sum_{i=1}^{N} m_i (\bar{v}_i)^2 = \sum_{i=1}^{N} \frac{(\bar{p}_i)^2}{2m_i} \tag{2.14b}$$

$$U(\bar{r}) = \sum_{i=1}^{N} U_i(\bar{r}) \tag{2.14c}$$

where the summation covers all atoms, $\bar{v}_i$ is the velocity of particle i and $\bar{p}_i$ the momentum (i.e., $\bar{p}_i = m_i \bar{v}_i$).

Hamiltonian mechanics works with generalized position vectors $\bar{r}_i$ and linear momentum vectors $\bar{p}_i$ which uniquely defines the system by the combination of pairs ($\bar{p}_i, \bar{r}_i$) (i = 1, 2, ..., N) as follows:

$$H(\bar{r}_1, \bar{r}_2 \cdots \bar{r}_N, \bar{p}_1, \bar{p}_2 \cdots \bar{p}_N) = \sum_{i=1}^{N} \frac{\bar{p}_i^2}{2m_i} + U(\bar{r}_1, \bar{r}_2 \cdots \bar{r}_N) \tag{2.14d}$$

$$\dot{\bar{r}}_i = \frac{\partial H}{\partial \bar{p}_i} \quad (i = 1 \ldots N) \tag{2.14e}$$

$$\dot{\bar{p}}_i = -\frac{\partial H}{\partial \bar{r}_i} \quad (i = 1 \ldots N) \tag{2.14f}$$

Hamiltonian mechanics leads to a system of coupled first order differential equations as shown in (2.14e) and (2.14f). The Hamiltonian expression will be useful to describe quantum mechanics in Chapter 4.

Before different types of energy potentials mentioned in Section 2.2 can be discussed in Chapter 3 and Chapter 4, several basic topics in atomistic simulation will be reviewed. These topics include numerical algorithms in this section, model development and boundary conditions in Sections 2.5 and 2.6, statistical ensembles in Section 2.7, energy minimization and statistical data analysis in Section 2.8, and statistical Monte Carlo method in Section 2.9.

Combining (2.1a) and (2.12) we have

$$m_i \frac{d^2 \bar{r}_i}{dt^2} = -\frac{\partial U(\bar{r}_1, \bar{r}_2 \dots \bar{r}_N)}{\partial \bar{r}_i}, \quad i = 1, 2 \dots \dots N \tag{2.15}$$

These equations are the governing equations for atomistic simulation. Because $U$ corresponds to nonlinear interatomic force fields, (2.15) is a set of coupled second order nonlinear differential equations for the N-body atomistic system. They can be solved by discretizing the simultaneous equations in time. The following are different numerical algorithms for the time discretization.

## 2.4.2   Verlet Numerical Algorithm

In order to integrate the differential equations of atom motion on a computer, finite difference schemes are used to discretize the time. From a mathematical point of view, this is an initial value problem under a given boundary condition. It will be carried on by numerous time steps. Each step will have finite time difference $\Delta t$ instead of infinitesimal time $dt$ used in the differential equation. The following Taylor series expansions for a generic function, $u(t)$, can be used to derive explicit finite difference equations:

$$u(t + \Delta t) = u(t) + \frac{\Delta t \, du(t)}{1! \, dt} + \frac{\Delta t^2 \, d^2 u(t)}{2! \, dt^2} + \frac{\Delta t^3 \, d^3 u(t)}{3! \, dt^3} + \cdots \frac{\Delta t^{(n-1)} \, d^{(n-1)} u(t)}{(n-1)! \, dt^{(n-1)}} + R_n(\Delta t^n) \tag{2.16a}$$

This expansion indicates that using the value and derivatives at time $t$ and the time increment $\Delta t$ to express the function at time $t + \Delta t$, the error is the remaining part $R_n$, at the order of $(\Delta t)^n$. The symbol $n$ denotes the first n terms on the right of the equation in which the highest derivative order is (n − 1). $R_n$ will be dropped in (2.16a) for numerical algorithm calculation. This means that error is inevitable by replacing $dt$ with $\Delta t$. It, however, can be very small if $\Delta t$ is small and n is large. The error order, however, depends on the numerical algorithm. In principle, the accuracy can be adequate and is only limited by the speed and memory of the computer.

Using $u(t)$ as position vector $\bar{r}_i(t)$ and n = 4, one can use (2.16a) to express $\bar{r}_i(t + \Delta t)$, $\bar{r}_i(t - \Delta t)$ explicitly as follows:

$$\bar{r}_i(t + \Delta t) = \bar{r}_i(t) + \frac{\Delta t \, d\bar{r}_i(t)}{1! \, dt} + \frac{\Delta t^2 \, d^2 \bar{r}_i(t)}{2! \, dt^2} + \frac{\Delta t^3 \, d^3 \bar{r}_i(t)}{3! \, dt^3} + R_n(\Delta t^4) \tag{2.16b}$$

$$\bar{r}_i(t - \Delta t) = \bar{r}_i(t) - \frac{\Delta t \, d\bar{r}_i(t)}{1! \, dt} + \frac{\Delta t^2 \, d^2 \bar{r}_i(t)}{2! \, dt^2} - \frac{\Delta t^3 \, d^3 \bar{r}_i(t)}{3! \, dt^3} + R'_n(\Delta t^4) \tag{2.16c}$$

Their sum by (2.16b) and (2.16c) will cancel out the first and third derivatives, leaving:

$$\bar{r}_i(t + \Delta t) + \bar{r}_i(t - \Delta t) = 2\bar{r}_i(t) + \frac{d^2 \bar{r}_i}{dt^2} (\Delta t)^2 + O((\Delta t)^4) \tag{2.16d}$$

where $O((\Delta t)^4)$ is the sum of the remaining part of $R_n$ and $R'_n$. The error is in the order of $(\Delta t)^4$. Following Newton's second law (2.2) we have

$$\frac{d^2 \bar{r}_i}{dt^2} = \frac{\bar{F}_i(t)}{m_i} \tag{2.16e}$$

Substituting (2.16e) into (2.16d) to replace the second derivative and dropping $O((\Delta t)^4)$, the desired recursion equation of position vector at time $(t + \Delta t)$ is obtained as follows:

$$\bar{r}_i(t+\Delta t) = -\bar{r}_i(t-\Delta t) + 2\bar{r}_i(t) + \left(\frac{\bar{F}_i(t)}{m_i}\right)(\Delta t)^2 \tag{2.16f}$$

This Verlet algorithm is a three-step method which involves the parameters at three different times, $t - \Delta t$, $t$, and $t + \Delta t$. The position vector at the last and current steps, i.e., at $t - \Delta t$ and $t$ can be used to calculate the position vector $\bar{r}_i(t+\Delta t)$ at the next step. Using this recursion equation, one can successively calculate $\bar{r}_i(t_0 + \Delta t)$, $\bar{r}_i(t_0 + 2\Delta t) \cdots \bar{r}_i(t_0 + L\Delta t)$ with $t_0$ being the initial time. Applying this process to all atoms ($i = 1, \ldots, N$), one can obtain the configuration of the system at any specific time. In this method $\bar{V}_i(t)$ is not explicit in the recursion procedure. It can be approximated by Taylor expansion to the first derivatives in (2.16b) and (2.16c) and by dropping the second order error $O((\Delta t)^2)$

$$\bar{V}_i(t) = \frac{d\bar{r}_i}{dt} = \frac{1}{2(\Delta t)}[\bar{r}_i(t+\Delta t) - \bar{r}_i(t-\Delta t)] \tag{2.16g}$$

The part expressed by $O((\Delta t)^n)$ in (2.16d) denotes the local truncation error. This is a part of the global error because, to reach the prescribed final time $t_0 + t$, more steps are needed which causes more error. For the position vector, if the local truncation error is $\sim(\Delta t)^{k+1}$, usually the global error for position vector is $\sim(\Delta t)^k$. In this case, the algorithm is called a k-th order method (for the position vector). The error is one order higher than the local order due to the accumulation of the position error with the integration steps. The Verlet algorithm is third order in position and potential energy, and second order in velocity and kinetic energy.

## 2.4.3   Velocity Verlet (VV) Algorithm

Swope et al.[36] proposed the Velocity Verlet algorithm which starts from the position $\bar{r}_i(t)$ and velocity $\bar{v}_i(t)$ at the current step. From Taylor expansion of (2.16a) and n = 3 we have:

$$\bar{r}_i(t+\Delta t) = \bar{r}_i(t) + \bar{V}_i(t)\Delta t + \frac{1}{2}\left(\frac{\bar{F}_i(t)}{m_i}\right)(\Delta t)^2 \tag{2.16h}$$

where $\bar{F}_i(t)$ is the force at the current time. After calculating the force $\bar{F}_i(t+\Delta t)$ at the next step, one can get the average acceleration from the $\bar{F}_i(t)$ and $\bar{F}_i(t+\Delta t)$. This acceleration expressed in the following parenthesis can be used to determine the velocity at the next time step $t + \Delta t$ as follows:

$$\bar{V}_i(t+\Delta t) = \bar{V}_i(t) + \frac{1}{2}\left[\frac{\bar{F}_i(t)}{m_i} + \frac{\bar{F}_i(t+\Delta t)}{m_i}\right]\Delta t \tag{2.16i}$$

The remaining terms in both (2.16h) and (2.16i) are $O(\Delta t)^3$, thus the local error is third order, and the method is a second order one because it is with second order global error. The integration process of Velocity Verlet algorithm is as follows:

$$\bar{r}_i(t_0) \rightarrow \bar{r}_i(t_0 + \Delta t) \rightarrow \bar{r}_i(t_0 + 2\Delta t) \rightarrow \cdots \bar{r}_i(t_0 + L\Delta t) \tag{2.16j}$$

This is different from Verlet algorithm because it does not start from $t_0 - \Delta t$. Velocity Verlet algorithm is widely used in MD because one can obtain the position vector $\bar{r}_i(t)$ and the velocity $\bar{V}_i(t)$ at the same time. In Section 10.2.1.2, we have used this numerical algorithm to develop a Fortran 90 code in which a subroutine is used to calculate the force of $\bar{F}_i(t+\Delta t)$. Substituting it to (2.16i) the velocity at $(t + \Delta t)$ can be determined. Readers who hope to understand this numerical algorithm deeply are recommended to look at that section and the code in detail.

It is seen from (2.16h) that the initial conditions for solving the differential equation of Newton's second law by Velocity Verlet algorithm require both initial position vectors and initial velocity vectors. In other words, before the first time step velocity must be assigned to the system in addition to the positions. Several programs automatically generate velocity corresponding to the assigned temperature.

In Sections 10.2.1 and 10.8, there are detailed descriptions and practice for how to make the FORTRAN code for the VV numerical algorithm.

### 2.4.4   Other Algorithms

Leap-frog (LF) algorithm[37] is a second-order method. It starts from $\bar{V}_i(t-\Delta t/2)$ and $\bar{r}_i(t)$ and is computed using the following recursion equations:

$$\bar{V}_i\left(t+\frac{\Delta t}{2}\right) = \bar{V}_i\left(t-\frac{\Delta t}{2}\right) + \left(\frac{\bar{F}_i(t)}{m_i}\right)\Delta t \qquad (2.17a)$$

$$\bar{r}_i(t+\Delta t) = \bar{r}_i(t) + \bar{V}_i\left(t+\frac{\Delta t}{2}\right)\Delta t \qquad (2.17b)$$

The initial velocity can be taken as:

$$\bar{V}_i(t_0) = \frac{1}{2}\left[\bar{V}_i\left(t_0-\frac{\Delta t}{2}\right) + \bar{V}_i\left(t_0+\frac{\Delta t}{2}\right)\right] \qquad (2.17c)$$

The Leap-frog (LP) and Velocity Verlet (VV) method are used widely in MD code such as in DL_POLY software. It is up to the individual which method to use when developing code.

Beeman's algorithm[38] is the third-order method and is computed using the following recursion equations:

$$\bar{r}_i(t+\Delta t) = \bar{r}_i(t) + \bar{V}_i(t)\Delta t + \left[\frac{4\bar{F}_i(t)-\bar{F}_i(t-\Delta t)}{m_i}\right]\frac{(\Delta t)^2}{6} \qquad (2.17d)$$

$$\bar{V}_i(t+\Delta t) = \bar{V}_i(t) + \left[\frac{2\bar{F}_i(t+\Delta t)+5\bar{F}_i(t)-\bar{F}_i(t-\Delta t)}{m_i}\right]\frac{(\Delta t)}{6} \qquad (2.17e)$$

Gear algorithm[39] is one of the predictor-corrector integration algorithms with high accuracy. It includes three steps. First, one predicts a position, velocity and acceleration of atoms by Taylor expansions. Second, one can find the force at the new position $\bar{r}_i(t+\Delta t)$ and acceleration $\bar{a}(t+\Delta t)$. Third, one can compare the obtained acceleration with the predicted one, and use the difference between them as a corrector. Because the Gear algorithm needs to store derivatives of position vectors at several time steps, the memory requirement is relatively high.[40] However, recent practice shows that if one uses the proper correcting coefficients in the corrector step then only one force calculation is needed. Actually, one of the advantages of Gear-type prediction-correction method is that one can calculate the force only once and does not have to store it for later use. This is unlike the Leap-frog (LF) or the simple Verlet scheme, where one needs forces at two moments of time, and have to store them. In addition, prediction-correction methods are an order of magnitude more correct and more stable than the rest, which means that one may use a several times larger time step, and thus actually save time in force calculations.[41]

## 2.5   Geometric Model Development of Atomistic System

The prerequisite for performing MD simulations is to determine atomistic or molecule structures of various materials in the model. The structure at the atomistic scale is measured by X-rays, and is generally determined by the following three factors. The first is geometric dimension and shape of the primary cell, the second is affiliated symmetry groups and configurations of the primary cell, and the third is the relative coordinates of each type of atom in the cell. As long as the atom coordinates within the cells are

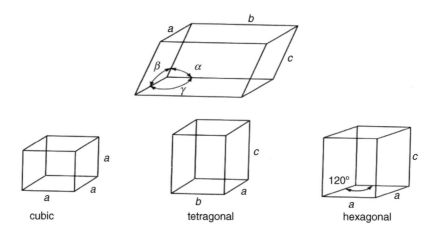

**Figure 2.5**   Examples of primary crystal units

determined based on these three aspects, all the atom coordinates of this model can be obtained from repeated extension of crystal lattice axes within a given model frame.

Geometry dimensions and shape of a primary cell are commonly given by six variables that describe the cell parameters. They are $a, b, c, \alpha, \beta, \gamma$, where $a, b, c$ are crystal lattice constants with units Å, along the x, y, and z axis, respectively. Variables $\alpha$, $\beta$, and $\gamma$ are the angles between the $b$ and $c$, the $a$ and $c$, the $a$ and $b$ axis, respectively as shown in Figure 2.5. Note that the coordinates of atoms within cells are relative to the corresponding values of $a$, $b$, and $c$. Suppose, the coordinates of an atom are $(\frac{1}{3}, \frac{1}{3}, \frac{1}{3})$, then the real lengths of the three coordinates of the atom should be $\frac{a}{3}, \frac{b}{3}, \frac{c}{3}$. If values of $a$, $b$, and $c$ differ between each other, then lengths of this unit cell along different axis are different.

Now, we introduce how to determine atom structures of the primary cell. Take calcium phosphate fluoride for example, which has a structure that is quite important for investigating bioactive materials. The chemical composition of calcium phosphate fluoride is $Ca_{10}(PO_4)_6F_2$. It contains 42 atoms within each cell. They are 10 calcium atoms, 6 phosphorus atoms, 24 oxygen and two fluorine atoms. According to the X-ray databases,[42] the main information obtained for this cell is as follows:

$$a = 9.3718\,\text{Å}, \quad b = 9.3718\,\text{Å}, \quad c = 6.8876\,\text{Å}$$
$$\alpha = 90°, \ \beta = 90°, \ \gamma = 120°$$

Space array group : P63/m

SG array number : 176

(2.18a–f)

Table 2.4 gives detailed information of each atom in the unit cell. It includes the number and position of each type of atom such as 4f, 6h, 12i and 2a. Position symbol f, h, i and a present Wykoff marks, whose first number represents the number of atoms. The atom positions represented by these symbols are shown by X-ray crystallography international diagrams.[43] In the diagrams, symmetry group number and SG array number are given in the database. In the case of calcium phosphate fluoride, it has P63/m space groups and 176 SG. The crystallography table has specified x, y and z coordinates for each atom position as given below:

$$4f:\begin{cases} \dfrac{1}{3},\ \dfrac{2}{3},\ z;\ \dfrac{2}{3},\ \dfrac{1}{3},\ \bar{z} \\[2mm] \dfrac{2}{3},\ \dfrac{1}{3},\ \dfrac{1}{2}+z;\ \dfrac{1}{3},\ \dfrac{2}{3},\ \dfrac{1}{2}-z \end{cases}$$

(2.19a)

**Table 2.4**   Atom position distribution and coordinates of $Ca_{10}(PO_4)_6F_2$ (After Inorganic Crystal Structure Database[42])

| Atom | Charge/e | Number | Position | x/Å | y/Å | z/Å |
|------|----------|--------|----------|-----|-----|-----|
| Ca | 2 | 4 | f | 0.3333 | 0.6667 | 0.00106 |
| Ca | 2 | 6 | h | 0.24155 | 0.99290 | 0.25 |
| P | 5 | 6 | h | 0.39809 | 0.36879 | 0.25 |
| O | −2 | 6 | h | 0.32629 | 0.48435 | 0.25 |
| O | −2 | 6 | h | 0.5780 | 0.46664 | 0.25 |
| O | −2 | 12 | i | 0.34067 | 0.2564 | 0.07089 |
| F | −1 | 2 | a | 0 | 0 | 0.25 |

$$6h : \begin{cases} x,\ y,\ \dfrac{1}{4};\ \bar{y},\ x{-}y,\ \dfrac{1}{4};\ y{-}x,\ \bar{x},\ \dfrac{1}{4} \\[2mm] \bar{x},\ \bar{y},\ \dfrac{3}{4};\ y,\ y{-}x,\ \dfrac{3}{4};\ x{-}y,\ x,\ \dfrac{3}{4} \end{cases} \tag{2.19b}$$

$$12i : \begin{cases} x,\ y,\ z;\ \bar{y},\ x{-}y,\ z;\ y{-}x,\ \bar{x},\ z \\ \bar{x},\ \bar{y},\ \bar{z};\ y,\ y{-}x,\ \bar{z};\ y{-}x,\ x,\ \bar{z} \\ \bar{x},\ \bar{y},\ \dfrac{1}{2}+z;\ y,\ y{-}x,\ \bar{z}+z;\ x{-}y,\ x,\ \dfrac{1}{2}+z \\ x,\ y,\ \dfrac{1}{2}-z;\ \bar{y},\ x{-}y,\ \dfrac{1}{2}-\bar{z};\ y{-}x,\ \bar{x},\ \dfrac{1}{2}-z \end{cases} \tag{2.19c}$$

$$2a : \begin{cases} 0,\ 0,\ \dfrac{1}{4} \\[2mm] 0,\ 0,\ \dfrac{3}{4} \end{cases} \tag{2.19d}$$

In (2.19a–c), a bar on the top of a letter, such as $\bar{z}$, indicates that the coordinate (e.g., z) has a negative value. The x, y, z coordinates of the first atom in any group such as 4f, 6h, 12i are given in Table 2.4. Thus the corresponding values of x, y and z of any group can be found in the last three columns of Table 2.4. This means, in order to obtain coordinates of other atoms in the group, one just needs to put the relevant x, y and z value of Table 2.4 into each coordinate expression of (2.19a–d).

For instance, the positions of four calcium atoms of 4f can be determined by substituting z = 0.00106 given in Table 2.4 into four groups of coordinates in (2.19a). Similarly, the positions of six calcium atoms can be determined by the six groups of coordinates in (2.19b) with x = 0.24155, y = 0.99290 and z = 0.25; the same for the six phosphorus atoms but with x = 0.39809, y = 0.36979 and z = 0.25 as given in Table 2.4.

As an example, Figure 2.6 shows atomistic structure in the calcium phosphate fluoride/water/α-quartz MD system. Atom distribution expressed by the top part of Figure 2.6 is of calcium phosphate fluoride, where the smallest atom is an oxygen atom, and the biggest atom is a calcium atom. The central part of the figure represents distributions of hydrogen and oxygen atoms of water molecules, while the lower part gives distributions of silicon (the larger) and oxygen atoms of α-quartz ($SiO_2$).

Let us take aluminum oxide $Al_2O_3$ as an example to show how to determine the coordinates of atoms Al and O in its primary unit cell. From Ref. [42] or other crystal database one can find the SG number of $Al_2O_3$ which is SG 167. One can also find from the same database that the unit cell consists of 12 Al atoms whose arrangement follows the pattern of "12c", and 18 O atoms whose arrangements follow the pattern

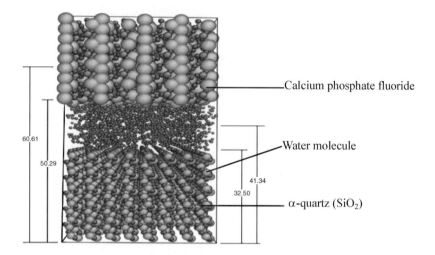

**Figure 2.6**   Atom structure in calcium phosphate fluoride/water/α-quartz MD system

of "18e". To determine the coordinates of "12c" and "18e" patterns, respectively, for the 12 Al and 18 O atoms one needs to use the tables for X-ray crystallography.[43] From that one can find a page which corresponds to SG 167. There, all the coordinates of the atoms are directly given one by one or through an additional process. For the case of $Al_2O_3$ an additional process is necessary. Specifically, in the top of that page for SG 167, there is a row which shows three sets of x, y, z coordinates with a " + " symbol at the end:

$$\text{Hexagonal axes}: \ (0,0,0; \ 1/3,2/3,2/3; \ 2/3,1/3,1/3) +$$

The " + " symbol indicates that the three coordinate sets in the above row should be added to the coordinates, respectively, shown in the list after the words "18e" and "12c" on that page. They read:

$$18e \quad x,0,\frac{1}{4} \quad 0,x,\frac{1}{4} \quad \bar{x},\bar{x},\frac{1}{4} \qquad \bar{x},0,\frac{3}{4} \quad 0,\bar{x},\frac{3}{4} \quad x,x,\frac{3}{4}$$

$$12c \quad 0,0,z \quad 0,0,\bar{z} \quad 0,0,\frac{1}{2}+z \quad 0,0,\frac{1}{2}-z$$

Now take the "12c" coordinates as an example to see how the three sets in the top row combine with the four sets of coordinates in "12c" to produce the 12 sets of coordinates of the 12 Al atoms. The method is simple: each (x, y, z) value listed in the top row can sum the x, y, z values of the four sets in "12c" to produce four new coordinate sets; there are three sets in the top row so they produce in total 12 coordinate sets for the 12 Al atoms. As the first top-row set is zero, let us use the second top-row set (1/3,2/3,2/3) as an example to show the addition as follows:

$$\left(0+\frac{1}{3} \quad 0+\frac{2}{3} \quad z+\frac{2}{3}\right), \left(0+\frac{1}{3} \quad 0+\frac{2}{3} \quad \bar{z}+\frac{2}{3}\right), \left(0+\frac{1}{3} \quad 0+\frac{2}{3} \quad \frac{1}{2}+z+\frac{2}{3}\right),$$

$$\left(0+\frac{1}{3} \quad 0+\frac{2}{3} \quad \frac{1}{2}-z+\frac{2}{3}\right)$$

in which each parenthesis represents the x, y, z coordinates of one atom. Thus, the coordinate sets of four atoms are obtained. For the other eight Al atoms, repeat the process using the remaining sets of (0,0,0) and (2/3,1/3,1/3) of the top row.

The same procedure can be used to determine the O atoms, where each addition of one top-row set to the pattern of 18e will now produce six coordinate sets, so that the additions of the three top-row sets will produce 18 coordinate sets for the 18 O atoms.

After one has the coordinate data of atoms in the primary cell, one needs either to use existing software or to develop computer code to generate coordinates for all atoms in the model. In CSLI, there are two simple codes for BCC and FCC structures. Their names are crystal_M_Simple.f90 and crystal_Structure. f90, respectively, located in the directory of UNITS/UNIT2/MD_CODE (see Section 10.2) and UNITS/UNIT6/NVE_Ar (see Section 10.6). There is also MaterNew_2010_Comb.f90 in UNITS/UNIT14/INI_CONF/DLPOLY_Model for complicated material structures as shown above (see Appendix 10.F).

## 2.6   Boundary Conditions

The treatment of boundary conditions of atomistic modeling is very important. Since the size of atomistic models is small, usually at the nanoscale, a large percentage of atoms are located on the surfaces. These surface atoms have completely different surrounding conditions and forces from the atoms inside of the bulk. These surface effects are usually undesired when simulating a bulk material. Furthermore, surface atoms may vaporize to vacuum if there is no other medium surrounding the model, which may cause instability of the simulation system.

The problem of surface effects can be overcome by implementing periodic boundary conditions. In some cases, nonperiodic and mixed boundary conditions are also used. Each of them has its own advantages and disadvantages. For example, periodic boundary conditions can eliminate boundary effects, but have limitations in the simulation of nonequilibrium state. Free or mixed boundaries can help study nonequilibrium state, but usually bring about boundary effects.

### 2.6.1   Periodic Boundary Conditions (PBC)

In a simulation model with PBC, a given number of atoms move within a supercell and interact with each other. The supercell is surrounded by a periodically repeated environment made up of an infinite number of its own images. Thus, the atoms in a supercell not only interact with atoms in the same cell, but with image atoms in adjacent mapped cells. The supercell can be viewed as a rectangular box, and the images of this simulation box are aligned periodically in all directions. For two dimensional (2D) cases, each cell has eight neighbors, and for three dimensional (3D) cases, there are 26 neighbors.

The coordinates of the atoms in the images can be obtained by adding or subtracting the integer multiple of the cell's edge length. During the simulation, when an atom leaves the unit cell, its image will enter the cell from the other side. Therefore, the number of atoms within the unit cell is kept unchanged.

Figure 2.7 gives PBC a schematic in a 2D problem. The central basic cubic box or supercell is replicated throughout space to form an infinite body. Take atom 1 in the center box for example; when it moves to box C, its image in box G enters the center box from the other side. And all the images of atom 1 move in the exactly same way. By this means the surfaces are eliminated.

PBC enables us to study the property of materials through the simulation of a small number of atoms. Usually if the treatment is appropriate, PBC with the short-ranged interactions is a good approximation on the equilibrium properties apart from phase transitions. In other words, if there is a phase transition in the unit cell the PBC may cause a serious error because the repeat images may not be realistic. However PBC needs to be treated with caution, since it sometimes introduces unrealistic periodic structure into the system. If the potential function (or interaction force) is long-ranged compared to the box size, substantial interactions between an atom and its own image in the neighboring boxes will appear. These artificial interactions generate additional constraints on the system and will affect the simulation results. Therefore for each simulation, it is important to check if the box size is large enough to represent the property of the simulated material system. If the resources are available it should be a

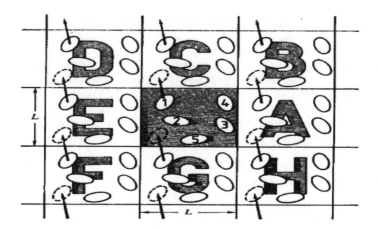

**Figure 2.7** A two-dimensional periodic system (Reproduced from Allen, M. P. and Tildesley, D. J. (1987) *Computer Simulation of Liquids*, Clarendon Press, Oxford, reprinted 1990, Oxford University Press)

normal step to verify whether there is a box size effect on the results by comparing results from different box sizes.

Another effect of PBC is that atoms near a box boundary will be subject to the interactions of image atoms on the other side of the boundary. For instance, in the basic box, the distance between atoms 1 and 3 may be beyond the cutoff radius and so they will not have direct interactions. However, the image of atom 1 in cell A and atom 3 in the basic box have a shorter distance between them, thus their interactions may need to be considered.

This situation should be noted when investigating defects such as vacancy, interstitials and dislocations, because if they are too close to the boundary their interactions with image defects may not be realistic. To avoid this situation, one way is to enlarge the model and distribute the defects far from the basic box boundary.

Other restrictions in using PBC include simulating a wave with a long wavelength. For a basic cube box with edge length $a$, the periodicity of cells will suppress any waves with the wavelength greater than $a$. If the wave generated in the MD simulation cannot pass through the PBC, it will reflect back and the energy will transform to thermal energy in the form of local lattice vibrations. The same limitations apply to the simulation of phonons (lattice thermal vibration) of long wavelength. PBC may also have effects on the rate of rapid phase transformation from liquid to solid or gas.[6]

## 2.6.2 Non-PBC and Mixed Boundary Conditions

For some simulations, it is not appropriate to adopt PBC in all directions. For example, when studying the absorption of surface atoms, PBC cannot be used in the z-direction perpendicular to the surface, but it can be used in the direction parallel to it. If there is external loading such as tensile loading or shearing, then either it must have the same periodicity as the basic box, or the PBC must be abandoned.

Another example is the simulation of nano-indentation. In the nano-indentation test, the typical indenter size is a few dozen nanometers. In order to reduce the boundary effects, the scale of MD simulation must be at least one order larger than that of the indenter. Such a model would easily exceed the capacity of a computer. In order to reduce the calculation requirement, a potential function is customarily introduced to simulate the corresponding indenter. The effective area of the potential may be much smaller than the size of the actual indenter, thus rigid boundary conditions often appear in the PBC form. Such boundary conditions artificially harden the material, and therefore reduce the occurrence of dislocations.

Some systems that have inherent surfaces, such as liquid drops and atom cluster aggregates, require the use of nonperiodic boundary conditions. Other systems like nonuniform or nonequilibrium systems may also need the use of nonperiodic boundary conditions.

Free boundary conditions are used for a system where the surface is the main interest, as compared to PBC that focuses on the inner system. Fixed boundary conditions are used for uniaxial loading simulations or shearing problems in which one boundary should be fixed to investigate the relative deformation.

## 2.7 Statistical Ensembles

Depending on the needs of simulation tasks, the atomistic system can be taken as different thermodynamic ensembles so a certain thermodynamic state can be controlled for the system during the simulation process. These ensembles are introduced as follows.

### 2.7.1 Nve Ensemble

This thermodynamic ensemble is also called microcanonical ensemble NVE. It means that the atom number N, the system volume V, and the total energy E of the system remain constant throughout the simulation.

### 2.7.2 Nvt Ensemble

This ensemble is called canonical ensemble NVT. It means that the atom number N, the system volume V, and the temperature T of the system remain constant throughout the simulation. However, the total energy E is not kept constant. To keep temperature constant, the simulation system should be connected to a thermostat (or thermal bath) to allow the thermostat to provide or absorb heat to maintain the constant system temperature if the system is lower or higher than the desired temperature $T_{req}$, respectively. This can be done using the velocity-scaling method, mimicking the effect of a heat bath. The principle is based on the following important relationship between the temperature and the kinetic energy of the system:

$$T = \frac{2K}{k_B(3N-N_c)} \tag{2.20a}$$

where $k_B$ is Boltzmann constant $1.38062 \times 10^{-23}\,\mathrm{J\,K^{-1}}$. $N_c$ is the constraint degree of freedom so that $(3N - N_c)$ denotes the total degrees of freedom of the 3D atomistic model with the total N atoms. Therefore, if one hopes to increase the temperature of the system, kinetic energy should be increased.

The temperature setting is conducted by the scaled atom velocity method. In practice, the new velocity is obtained by multiplying a scale factor $\lambda$ to the current velocity $\bar{V}_i(t)$ as follows:

$$\bar{V}_{i,\,scaled} = \lambda \bar{V}_i(t) \tag{2.20b}$$

$$\lambda = \sqrt{\frac{T_{req}}{T(t)}} \tag{2.20c}$$

Using (2.20c), it is easy to find that:

$$\Delta T = T_{req} - T(t) = (\lambda^2 - 1)T(t) \tag{2.20d}$$

If the system has higher temperature $T(t)$ than the prescribed temperature $T_{req}$, from (2.20c) $\lambda < 1$, the system's atoms will reduce in velocity, and in turn, reduce the kinetic energy. As a result, the system temperature will be reduced. Conversely, if the kinetic energy is increased, the temperature will increase.

The thermostats used for NVT include:

Berendsen thermostat:[44] the idea for controlling the temperature is the same as the velocity-scaling method. It uses velocity scaling to keep the temperature change $\Delta T$ proportional to the temperature difference between the thermal bath and the system, i.e.,:

$$\Delta T = \frac{\Delta t}{\tau}(T_{bath}-T(t)) \tag{2.20e}$$

Here, $T_{bath}$ is the required temperature for the system. Thus, the velocity scaling factor is given as follows:

$$\lambda = \sqrt{1+\frac{\Delta t}{\tau}\left(\frac{T_{bath}}{T(t)}-1\right)} \tag{2.20f}$$

where $\Delta t$ is the time step for numerical integration, and $\tau$ is a coupling factor. If $\tau = \Delta t$, this method returns to the above simple velocity-scaling method.

Nose-Hoover scheme:[45] the way proposed for adjusting the system temperature is to increase a force term to Newton's law as follows:

$$\bar{F}_i(t)-m_i\chi(t)\bar{v}(t) = m_i\frac{d^2\bar{r}_i}{dt^2} \tag{2.20g}$$

The force expressed by the second term is a frictional force which is proportional to the atom velocity. If the system temperature is higher than the setting temperature, the corresponding kinetic energy and velocity is also higher (see (2.20a), thus more frictional force will be produced to reduce the acceleration and, in turn, to reduce the velocity and the system temperature. (2.20g) can be rewritten as:

$$\frac{d^2\bar{r}_i}{dt^2} = \frac{\bar{F}_i(t)}{m_i}-\chi(t)\bar{v}(t) \tag{2.20h}$$

where $\chi(t)$ is the frictional coefficient which is controlled based on the temperature difference between the system and the thermal bath as follows:

$$\frac{d\chi(t)}{dt} = \frac{N_f k_B}{Q}(T(t)-T_{bath}) \tag{2.20i}$$

where:

$$Q = N_f \kappa_B T_{bath}\tau_T^2 \tag{2.20j}$$

is the effective "mass" of the thermostat, $\tau_T$ is the thermostat relaxation time constant (normally in the range 0.5 to 2 ps), $N_f$ is the number of degrees of freedom in the system and $k_B$ is Boltzmann constant.

## 2.7.3   Npt Ensemble

This is the isobaric-isothermal ensemble, therefore the atom number N, the system pressure P, and the temperature T of the system remain constant throughout the simulation. To control the system pressure, a barostat is used. Specifically, pressure is controlled using a piston, mimicking the volume-pressure relationship. The volume changes with time and, in turn, changes the "instantaneous" pressure so that the average pressure will converge approximately to the requested value. The types of NPT include:

NPT ensemble Hoover: select Hoover NPT with f1 (or $\tau_T$) and f2 (or $\tau_P$) as the thermostat and barostat relaxation times (ps).

NPT ensemble Berendsen: select Berendsen NPT with f1 (or $\tau_T$) and f2 (or $\tau_P$) as the thermostat and barostat relaxation times (ps).

## 2.8    Energy Minimization for Preprocessing and Statistical Mechanics Data Analyses

Before an atomistic simulation is conducted, preprocessing through energy minimization to the produced geometric model is necessary.

### 2.8.1    Energy Minimization

Energy minimization is an approach conducted at absolute zero temperature (0 K) so that kinetic energy is zero. Since the velocity and acceleration is zero, which mimics a quasistatic mechanics experiment for material behavior, it is also called molecular statics, or static lattice calculation. This method is one kind of atomistic analysis for deformation and failure behavior. During this minimization process, defect distribution in the material as mentioned in Section 2.1.2 can be found. It can also be used to study dislocation nucleation from crack tips and deformation for carbon nanowires.

Here, we mainly use the energy minimization process to obtain a realistic initial model configuration for MD. Through this minimization process an atomistic structure state with minimum potential energy will be obtained. Physically, the atomistic structure obtained will be a stable atomistic structure following the minimum principle of the total potential energy.

By this means, the artificial atom arrangements, defects and unstable structural factors will be avoided or reduced. There are different algorithms to perform energy minimization [46] The minimization process is conducted with fast evolution initially, but slows down after several hundred or thousand cycles where the lowest stable total potential energy is established. There are also several types of software that can carry on this minimization process, including GULP[47] and DL_POLY_2.[48] Specifically, in Section 10.3, GULP, short for General Utility Lattice Program, is used to to introduce the static lattice calculations which include optimizations of structure parameters, determinations of potential parameters, carrying on defect analysis and shell model simulations.

### 2.8.2    Data Analysis Based on Statistical Mechanics

Data processing and visualization is very important for atomistic analysis, by which useful information can be extracted and explained. Note that instantaneous information such as position and velocity may not be significant because there is no direct relationship with observable physical variables. To convert the atomistic data to observable physical parameters such as pressure, temperature, force, energy and heat capacity, data averaging methods based on statistical mechanics must be used.

One of the most useful methods for this purpose is to adopt the Ergodic hypothesis. This hypothesis states that the average properties of the simulation system equal the time average of those properties over a certain time period. Taking the determination of temperature, pressure and potential energy of the atomistic system as examples, these variables are not determined by a single time step but by the averages of these variables, say, over several hundred time steps (see Homework (10.9) at end of Section 10.2). This is reasonable because in thermodynamics the system is in equilibrium or quasi-equilibrium in the sense that some observable (thermodynamic) quantities do not change with time. Therefore, if we have an isolated system and we have the time evolution of the position and velocity of each atom we can obtain thermodynamic quantities through the time-averaging process after a transient period.

In general, suppose an arbitrary variable $P$ must be determined in the atomistic system, then one should first get the variable $P(i), P(i + 1), \ldots, P(i + m)$ at time $t_i, t_{i+1} \ldots t_{i+m}$. The effective variable $P$ which is the time average over these time steps can then be obtained by

$$P_{eff} = \langle P(t) \rangle = \frac{1}{m} \sum_{n=i}^{n=i+m} P(t_n) \qquad (2.21a)$$

where $m$ denotes the number of time steps for averaging. The start time step i is not necessarily to take the first time step. It can be determined based on the required information and considered whether the system at these steps is near equilibrium state. It is obvious that if the system is still in the transition state, the average value may not be good to represent the status of a thermodynamic system in equilibrium.

In Section 10.2 and Homework (10.9) there is description and practice on how to develop a simple FORTRAN code to carry on the time averaging described by (2.21a). Formula (2.21a) can also be used in a time averaging over a certain domain, a part of the atomistic system. In Chapter 1, non-DC methods such as ESCM are briefly introduced. In ESCM, bridging between the atomistic and continuum scale is accomplished by inputting the atomistic displacement to the FEM node of the continuum. In the ESCM method, the displacement is not a point displacement at a given time instant, but rather the displacement is determined by statistical mechanics. More specifically, it is obtained through time averaging shown in (2.21a) of $m$ data steps of the average displacement. The latter is a spatial average over a small volume defined by the interface volume cell at the handshaking interface domain.

In MD postprocessing, there is a radial distribution function to show spatial distributions of different types of atoms. The function denotes the density and/or atom numbers of the second type of atoms within a certain radius $r$ surrounding an atom of the first type. A table with various radiuses $r$ is given so that the spatial average can be conducted over that range (see Section 10.5.7.1 for more information).

While the Ergodic hypothesis is effective, the more general formulas of statistical mechanics for ensemble average of the variable $P$ should be briefly introduced as follows:

$$\langle A \rangle_{Ensemble} = \int_p \int_r A(p,r)\rho(p,r)dpdr \qquad (2.21b)$$

where $p$ denotes the set of linear momentum vector $p = \{\bar{p}_i\}$ ($i = 1, 2, \ldots N$); and $r$ is the set of position vectors $r = \{\bar{r}_i\}$ ($i = 1, 2, \ldots N$). From the concept and (2.14d) introduced for Hamiltonian mechanics (see Section 2.4.1), it is seen that the double integration covers the Hamiltonian p-r phase space. In (2.21b) $\rho(p, r)$ is the probability density distribution function. For a canonical system it can be expressed through Hamiltonian as follows:

$$\rho(p,r) = \frac{1}{Q}\exp\left(-\frac{H(r,p)}{k_B T}\right) \qquad (2.21c)$$

$$Q = \int_p \int_r \exp\left(-\frac{H(r,p)}{k_B T}\right)dpdr \qquad (2.21d)$$

where Hamiltonian $H(r, p)$ is given by the sum of kinetic and potential energy as

$$H(r,p) = \sum_{i=1}^{N}\frac{\bar{p}_i^2}{2m_i} + U(\bar{r}_1, \bar{r}_2 \cdots \bar{r}_N) \qquad (2.21e)$$

## 2.9  Statistical Simulation Using Monte Carlo Methods

The frequent used methods in atomistic analysis include MS, MD and MC. Most contents described in this chapter so far besides Section 2.4 for integration of Newton's second law of motion are suitable for both MS and MD. MS can be considered as a special case of MD and it will be discussed in more detail in Chapters 6 and 7 when we introduce multiscale methods such as quasicontinuum method. In this section we will introduce MC method which is a statistical simulation method. This method can be used for both statics and dynamics problems of atomistic systems. The uniqueness of this method is that it uses a statistical method instead of the deterministic method used in MS and MD.

## 2.9.1   *Introduction of Statistical Method*

Monte Carlo method was developed by von Neumann, Ulam and Metropolis. The name Monte Carlo was chosen because of the extensive use of random numbers in calculation; it was coined by Metropolis in 1949 and used in the title of a paper describing the early work at Los Alamos National Laboratory.[49]

Statisticians had used model sampling experiments to investigate problems long before this time. The novel contribution of von Neumann and Ulam (1945)[50] was to realize that determinate mathematical problems could be treated by finding a probabilistic analogue which is then solved by a statistical sampling experiment. Both MS and MD are determinate mathematical problems in which MS is concerned with developing governing equilibrium equations to satisfy the principle of minimum potential energy (see Section 6.2) and MD is for integration of differential motion laws of atoms. Thus there are ways for MS and MD to be analyzed by a statistical method. Taking MS as an example, one needs to find a probabilistic analogue for MS to have the configuration with minimum potential energy, and to find the realistic configuration by comparing different sampling configurations and then choose the one with minimum potential energy.

In 1901, the Italian mathematician Lazzerini performed a simulation by dropping a needle 3407 times on a plane. From the random number of times that the needle fell inside the circle of the plane out of the total dropping number, he estimated $\pi$ to be 3.1415929. This example is given here, following Allen and Tildesley,[6] to illustrate the concept of the statistical method.

Let us conduct the so called "hit and miss" statistical test. The first quadrant of a circle with radius of a unit length, centered at the origin and inscribed in a square OABC, is used for the analysis (see Figure 2.8). In the figure, O is both the center of the circle and the left bottom corner of the square. Points A and C are right and top tangential points of the circle to the lines AB and BC of the unit squares. Then the area with shaded lines, encompassed by radius OA, the circle arc length $\widetilde{AC}$ and radius OC, belongs to the circle, and the area encompassed by the arc length $\widetilde{AC}$ and lines AB and BC of the square is out of the circle.

Note the coordinates $x_p$, $y_p$ of any generic point p inside the square are between 0 and 1. To guarantee that the shot hits a point inside the square at each trial two independent random numbers are chosen from a

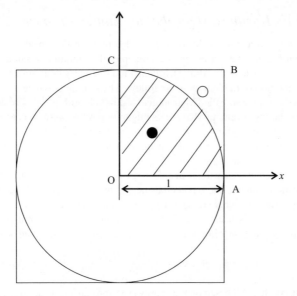

**Figure 2.8**   A hit and miss test to estimate $\pi$ (Reproduced from Allen, M. P. and Tildesley, D. J. (1987) *Computer Simulation of Liquids*, Clarendon Press, Oxford, reprinted 1990, Oxford University Press)

uniform distribution on $(0,1)$. Each pair of these numbers are used as the coordinates of a point. It may fall into the circle (shaded) region as the solid ball or out of the circle (but inside the square) as the open ball shown in Figure 2.8. Specifically, the distance from the random point to the origin O is calculated based on the x and y coordinates. If the distance is less than or equal to one, the shot has landed in the shaded region and a hit is scored, otherwise, it is a "miss" and no score is given. If $\tau_{shot}$ denotes the total number of shots to hit the unit square OABC and $\tau_{hit}$ denotes the random number falling into the circle, then the $\pi$ value can be estimated from the ratio of $\tau_{hit}$ over $\tau_{shot}$ as follows. By the definition of $\pi$ in a circle with unit radius, we have

$$\pi = \frac{Circle\ Area}{r^2} = \frac{Circle\ Area}{1^2} = \frac{4 \times Area\ under\ the\ curve\ CA}{Area\ of\ the\ square\ OABC} \approx \frac{4\tau_{hit}}{\tau_{shot}} \qquad (2.22)$$

where the last equation is an approximate one. It assumes that the area is proportional to the number of shot points for a homogeneous random process. It is obvious that the higher the number $\tau_{shot}$, the higher the accuracy of this assumption. After $10^7$ shots the statistical estimate is 3.14173 correct to four figures. To calculate another decimal place would require an order of magnitude increase in the number of shots. That Lazzerini only used 3407 times to get his 3.1415929 estimation is certainly a lucky case. The key to this statistical Monte Carlo method is the generation of the totally $2\tau_{shot}$ random numbers of the coordinates x, y from a uniform random distribution. Random number generators are simple programs which are included in several frequently used computer languages and code. In Fortran 90, the following commands can be used:

```
Call random_number(x)
Call random_number(y)
Call init_random_seed()
```

Interested readers may refer to the codes developed in Homework (2.5) and (2.6) below and Appendix G in Ref. 6 for more detail.

### 2.9.2   Metropolis-Hastings Algorithm for Statics Problem

Let us return to a statics problem for determination of a realistic deformed configuration. Here, no simultaneous governing equations based on the principle of minimum potential energy need to be determined. What we need to do in the Monte Carlo method is to carry on routine sampling experiments which involve the generation of random numbers followed by a limited number of arithmetic and logic operations. The procedure of the so-called Metropolis-Hastings algorithm[51] is simple: first draw random numbers and calculate the system energy H(A) for configuration A; second advance the system by random numbers and calculate the new system energy H(B) for configuration B; third accept or reject the new configuration according to an energy criterion; then repeat this process until a certain number of trials is reached.

As for the energy criterion, it is easy to see that if energy H(B) is less than H(A), configuration B may be closer to the configuration with minimum potential energy, thus configuration B should be accepted, and then go to the next step. In the case where H(B) is larger than H(A) but not too much, there is a possibility that the H(B) contribution will still be acceptable. To further accept or reject this configuration, draw random number p within $0 < p < 1$ and if the following inequality is valid then it is accepted otherwise, it is rejected:

$$p < \exp(-\frac{H(B)-H(A)}{k_B T}) \qquad (2.23)$$

The set of configurations obtained based on the energy criterion can also be used to calculate a statistical ensemble of thermodynamical properties based on the concepts described in the last section, especially using Eq. (2.21a).

How to move from a previous to a new state is arbitrary, which makes this method widely applicable. However, one should have additional knowledge of the system behavior so it can be used easily for the generation of new configurations. This is quite different from MD or MS where the trajectories or the deformation pattern of each atom can be determined by establishing and solving the governing equations. Here, only the knowledge to guide the motion of the system is needed.

## 2.9.3    Dynamical Monte Carlo Simulations

Although Monte Carlo method has been largely associated with obtaining static or equilibrium properties of model systems, it can also be utilized to study dynamical phenomena. Often, the traditional dynamics leading to certain structural or configurational properties of matter are not completely amenable to a macroscopic continuum description. On the other hand, MD is not computationally capable of probing large systems of interacting particles at long times. Thus dynamical MC method or kinetic MC (kMC) method are capable of bridging the large gap existing between these two well-established dynamical approaches. This is because the dynamics of individual particles are modeled in this technique, but usually only in a coarse-grained way representing average features which would arise from a lower-scale result.

Quite different from the statics problem, dynamic MC or kMC method is sensitive to the manner in which the time series of events, characterizing the evolution of a system, is constructed. Studies have also underscored the importance of utilizing a Monte Carlo sampling procedure in which transition probabilities from one system state to other state are based on a reasonable physical dynamical model. It was stated that if the following three criteria are satisfied then the Monte Carlo method may be utilized to simulate effectively a Poisson process. The latter is a stochastic process in which events occur continuously and independently of one another. While a Poisson process is a continuous time process, its discrete-time counterpart is the Bernoulli process. The latter is frequently used in numerical statistical simulation and the mutual connections between these two common processes can be seen in the next section. The three criteria are:

- First, transition probabilities reflect a "dynamical hierarchy" in addition to satisfying the detailed-balance criterion;
- Second, time increments upon successful events are formulated correctly in terms of the microscopic kinetics of the system; and
- Third, the effective independence of various events can be achieved.

To illustrate these points and the criteria for successful applications of Monte Carlo method for dynamical problems we discuss the approach to and the attainment of Langmuir adsorption-desorption equilibrium in the following section based on the work of Fichthorn and Weinberg.[52]

## 2.9.4    Adsorption-desorption Equilibrium

Let us consider the dynamic process of adsorption/desorption of gas-phase species A with a solid-single-crystalline surface. From a kinetic point of view, adsorption equilibrium (steady state) is established when the net rate of chemisorption of gas-phase A is equal to the net rate of desorption of chemisorbed A to the gas phase. It is assumed that chemisorbed molecules do not interact appreciably with one another and gas-phase molecules arrive independently to a surface containing a uniform and periodic array of adsites. The arrivals continuously occur at random, uncorrelated times; this process may be considered a Poisson process and it can be characterized by an average rate $r_A$ for the adsorption. A similar scenario is applicable to molecules chemisorbed on the surface;

desorption events occur with an average rate $r_D$. The appropriate kinetic expression for this balance is:

$$\frac{d\theta}{dt} = r_A(1-\theta) - r_D\theta \tag{2.24}$$

Here, $\theta$ is an important parameter which denotes the fractional surface coverge of A. With the initiation condition $\theta(t=0)=0$, it is easy to get or verify the following solution for (2.24):

$$\theta(t) = \frac{r_A}{r_A + r_D}(1 - \exp[-(r_A + r_D)t] \tag{2.25}$$

And, in the limits as $t \rightarrow \infty$ the fractional surface coverage is

$$\theta_e = \frac{r_A}{r_A + r_D} \tag{2.26}$$

where the subscript "e" of $\theta$ denotes the equilibrium fractional surface coverage of A. Eqs. (2.25) and (2.26) are exact solutions for the continuous Poisson adsorption/desorption process and will be used to check the validity of MC method.

The MC algorithm for simulating the adsorption equilibrium of a gas-phase species A with a 2D lattice containing N sites can be briefly described as follows. As shown in Figure 2.9 a trial begins when one of the N sites in the surface is selected randomly. If the site is vacant, adsorption may occur with probability WA; and desorption may occur with probability WD if the site is occupied. In the formal case, the path in the flowchart will go right since the answer to the question for occupied is no, so the site is empty; otherwise, it will go left since the answer is yes and the site is occupied. In both cases, a random number $r$ will be produced by a uniform random number generator. If the path goes right and the probability WA is larger than the random number, i.e., $r < WA$ then the absorption process occurs. Similarly, if the path goes left and $r < WD$ then the desorption process occurs.

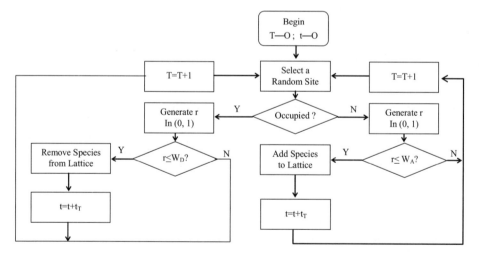

**Figure 2.9**   A flow diagram for simulating as a Poisson process the approach to and the attainment of Langmuirian adsorption equilibrium. T is the (integral) number of trials, t represents real time, r is a uniform random number between 0 and 1, $W_i$ is the transition probability for event i (i = A (adsorption) or D (desorption)), and $\tau_T$ is the real time increment at trial T (Reproduced with permission from Fichthorn, K. A. and Weinberg, W. H. (1991) Theoretical foundations of dynamical Monte Carlo simulations. *The Journal of Chemical Physics*, 95 (2), 1090. Copyright 1991 by the American Physical Society)

If any of this process occurs it is a successful trial. There is no way that the adsorption and desorption can occur at the same time, which satisfies the third criterion that the event is independent. Time is advanced by an increment $\tau_i$ upon successful realization of an event at trial i and the capital letter T is used to count the overall number of trials (including unsuccessful trials), which accumulate over repetition of the algorithm. Based on this MC simulation, one can get the fractional surface coverage A in terms of the real time $t$ during this adsorption/adsorption process. As a result of the calculation, the surface occupation fraction value, $\theta$, can be obtained versus time $t$ and can then be used to compare with the exact solution of Eq. (2.25). The comparison is shown in Figure 2.10.

Before discussing the comparison, it will be shown that through a proper definition of $W_A$ and $W_D$ and the correct utilization of an appropriate $\tau_i$ and the random selection process, the Monte Carlo algorithm can provide a correct solution to (2.24) for the N-site ensemble through a simulation of the Poisson process. Let us discuss this issue by reference to the three criteria introduced in the last Section.

### Criterion 1: The transition probabilities $W_A$ and $W_D$ must be chosen so that MC simulation obeys detailed balance condition, e.g., Eq. (2.26)

This criterion requires that the MC simulation is consistent with a physical model of the simulated phenomenon. To demonstrate the manner in which this criterion of physical consistency is achieved, let us consider the discrete stochastic process of the MC algorithm mentioned. By performing this algorithm, we simulate a sequence of independent Bernoulli trials. This is the discrete-time counterpart of the Poisson process; the probability per trial of a successful adsorption event is $W_A (1 - \theta_i)$ because the probability for adsorption is related to the fraction $(1 - \theta_i)$ of empty sites. On the other hand, the probability per trial of a successful desorption event is $W_D \theta_i$ because the desorption is proportional to the fraction of occupied sites in the lattice structure. The total probability of success per trial is

$$W_A (1-\theta_i) + W_D \theta_i \leq 1 \tag{2.27}$$

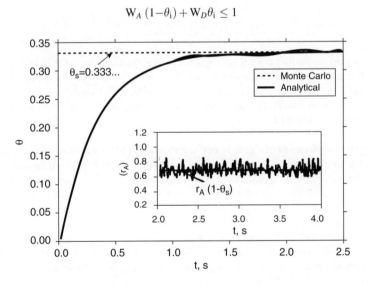

**Figure 2.10** The transient solution of Eq. (2.24) for the initial condition of an empty surface $[\theta(t=0)=0]$ provided by both analytical form of Eq. (2.25) and the Monte Carlo algorithm with $r_A = 1.0$ (site s)$^{-1}$ and $r_D = 2.0$ (site s)$^{-1}$ ($W_A = 1/2$ and $W_D = 1$). The insert depicts the rate of the adsorption measured from the MC simulation at steady-state and the continuum steady-state rate (Reproduced with permission from Fichthorn, K. A. and Weinberg, W. H. (1991) Theoretical foundations of dynamical Monte Carlo simulations. *The Journal of Chemical Physics*, 95 (2), 1090. Copyright 1991 by the American Physical Society)

Here, $\theta_i$ is the fractional coverage of $A$ at trial i. When the system has reached steady state, $\theta_i$ is towards $\theta_S$ which is the simulated equilibrium fractional surface coverage of $A$. In addition, from probability analysis, the expected number of adsorption events in T trials is given by

$$\langle N_{A,T} \rangle = W_A(1-\theta_S)T \tag{2.28}$$

Similar results may be obtained for desorption events as

$$\langle N_{DT} \rangle = W_D \theta_S T \tag{2.29}$$

It is easy to see that when a steady state has occurred in the simulation, the average rate of adsorption obtained from (2.28) is equal to the average rate of desorption from (2.29), i.e.,

$$W_A(1-\theta_S) = W_D \theta_S \tag{2.30}$$

Detailed balance is satisfied at equilibrium (steady state) if $W_A$ and $W_D$ are defined in a way which can meet the physical model represented by Eq. (2.26). For the present case, the transition probabilities could be constructed through normalization of the rates of adsorption and desorption, relative to the larger one of these two. If $r_A > r_D$, then the transition probabilities could be defined

$$W_A = 1 \tag{2.31a}$$

and

$$W_D = \frac{r_D}{r_A} \tag{2.31b}$$

Substituting these values to Eq. (2.30) results in

$$\theta_S = \frac{r_A}{r_A + r_D} \tag{2.32}$$

Note the difference between this equation and (2.26) is on the left hand, one is the saturated fraction $\theta_S$ and the other is the equilibrium surface fraction $\theta_e$. It is seen, however, in systems which are sufficient large, that the saturated value $\theta_S$ should approach $\theta_e$ because the large Bernoulli trials are more close to the saturation value of the continuous Poisson process. Thus, the detailed balance condition, e.g., (2.26) based on a physical model is satisfied.

**Criterion 2: The requirement of a proper correspondence of MC time to real time must be satisfied**
This criterion requires that the time sequence in MC simulation is consistent with the real time sequence. To understand this criterion, it should be noted again that the Poisson process actually is a continuous-time variation of the discrete Bernoulli process which is simulated by the MC algorithm when time is measured in terms of trials. In fact, by replacing the discrete interevent times with appropriate continuous values, the MC algorithm produces a chain of events which is a Poisson process. The continuous interevent times are constructed through the following probability analysis.

The most significant feature attributable to the Poisson process for the purpose at hand is the following exponential characterization of the probability density at time $t_e$ between successive events:[53]

$$f_{t_e}(t) = re^{-rt} \tag{2.33a}$$

From this probability density function $f_{t_e}(t)$, it is easy to prove the mean time period $\langle t_e \rangle$ between successive events is $1/r$. The approval can be shown after obtaining the average probability $\langle p \rangle$ between the two events with time interval $\Delta t$ as follows. By definition

$$\langle p \rangle \equiv \frac{1}{\Delta t} \int_0^{\Delta t} f(t)dt \tag{2.33b}$$

Substituting (2.33a) into the above equation we have the average probability between the two events as

$$\langle p \rangle = \frac{1}{\Delta t} \int_0^{\Delta t} r e^{-rt} dt = r \qquad (2.33c)$$

thus the mean time period between successive events is calculated as the inverse of the rate $r$ as follows:

$$\langle t_e \rangle = 1/r \qquad (2.34)$$

A particularly useful feature of the Poisson process is based on the fact that an ensemble of independent Poisson processes will behave as one large Poisson process such that statistical properties of the ensemble can be formulated in terms of the dynamics of individual processes.[52] The adsorption-desorption equilibrium of an entire system of independent molecules thus can be considered as an N-independent forward-reverse Poisson process, each with some arbitrary but finite rate $r_i$.

Upon each trial i at which an adsorption or desorption event is realized, time should be advanced with an increment $\tau_i$ selected from an exponential distribution (2.33a) with parameter $r_i$ as

$$r_i = (N-M_i)r_A + M_i r_D \qquad (2.35)$$

Here, N is the total number of sites in the ensemble, $M_i$ is the number of sites occupied at trial i (i.e., $\theta_i = M_i/N$). Based on (2.34) and (2.35) for the event I, the mean time between two successful events may be written as

$$\langle t_i \rangle = \frac{f_i}{r_i} = \frac{f_i}{(N-M_i)r_A + M_i r_D} \qquad (2.36)$$

Here, $f_i$ is the fraction of successful trials at which $M_i$ sites are occupied. The reason to introduce $f_i$ is that only a successful trial can be used for accounting the physical time $t$, thus the time calculated by the nominal parameter $r_i$ or $r_A$ and $r_D$ should be reduced by the fraction coefficient $f_i$. Furthermore, over many successful trials at steady state, the average time between successful events is the average of different mean times, i.e.,

$$\langle t_0 \rangle = \sum_i \langle t_i \rangle = \sum_i \frac{f_i}{(N-M_i)r_A + M_i r_D} \qquad (2.37)$$

This equation represents a time weighting of various configurations of simulated N-site ensemble by the MC method. This is consistent with the realistic process that is dictated by the detailed-balance criterion for the equilibrium ensemble. Therefore, the second criterion is satisfied. In general, the ensemble property simulated by MC method at a particular point in time, say, $\theta_s(t)$, could fluctuate about the true continuum ensemble property, say, $\theta(t)$, at that time $t$. In other words, the value is simulated without ever achieving exactly the continuum value. Thus, time-weighted averages must be computed to estimate the true continuum ensemble at any point in time to keep the desired degree of accuracy. By the way, here the time-weighted averages correspond to thermodynamic averages at equilibrium if the detailed–balance criterion 1 for thermal equilibrium is fulfilled.

### Criterion 3: The requirement of the independence of events comprising the time sequence of the process is satisfied
This criterion requires that the event simulated by MC method is independent. Strictly speaking, the formalism for the MC method presented for dynamics here is valid only when independent events are simulated. In the single-site adsorption-desorption process introduced here, successive adsorptions and desorptions are correlated. Nevertheless, a Poisson process can be constructed consisting of one of the events, e.g., adsorption occurring with a rate $r_A(1-\theta)$.

### 2.9.4.1 Results

Under the design of an independent event with the MC algorithm, choosing transition probabilities by normalization (i.e., $W_A = 1/2$, $W_D = 1.0$) to satisfy the detailed-balance criterion and with time incremented in a procedure analogous to that described in (2.35) to (2.37), the time series can be interpreted in terms of a Poisson process. The MC simulation of the gas/lattice dynamic interaction on $128 \times 128$ square lattices was conducted by Fichthorn and Weinberg[54] under the condition $r_A = 1.0$ (site s)$^{-1}$ and $r_D = 2.0$ (site s)$^{-1}$. Figure 2.10 shows the fractional surface coverage of adsorbate as a function of time for an initially empty surface for both the transient analytical (exact) solution of (2.25) and the MC simulation.

The curve in Figure 2.10 is the result of one run only. The insert of Figure 2.10 shows both the analytical (exact) and MC steady-state rate of adsorption (desorption). It is seen that the agreement between the two solutions in the approach to equilibrium and the steady-state fractional surface coverage is excellent.

MC technique is effective in solving time scale problems. Its application in atomistic and multiscale simulations for vapor deposition of forming thin films and deformation and failure can be found in Refs. [53–58]. The important part of the MC modeling code is to form an evenly distributed stochastic data set.

Basic knowledge for Fortran 90 can be found in Section 10.2.1 and Appendix 10.B "Introduction to Fortran 90". This code with some modification and the "absorb" code in homework (2.6) are stored in UNITS/UNIT2/F90_CODES/Monte_Carlo of the CSLI which can be downloaded from the book's website: http://multiscale.alfred.edu.

---

### Homework

(2.5) The following Fortran 90* "Program Monte" can be used for statistical trials of "hit and miss" test to approximately determine the $\pi$ value expressed by the formula (2.22). First, compile and run that code and then draw a curve of the approximate $\pi$ value in terms of the average values obtained by different trial numbers, such as 3407, 10,000, 50,000 and 100,000 (3407 is the number the Italian mathematician Lazzerini used in 1901 to get the approximate value of 3.1415929 for $\pi$). Hint: (1) In the code, the symbol "success" is $\tau_{hit}$, "tot" is $\tau_{shot}$. From (2.22), the term "$4\tau_{hit}/\tau_{shot}$" gives the approximate value of $\pi$. (2) Due to the random characteristics, the approximate value obtained may change from one calculation to another. To get reasonable results for a given number, one may need to carry on the trial more times, for averaging.

```
"program monte
implicit none
integer:: current
real(kind=8):: x, y, success, tot

call init_random_seed()

do current=1,3407
   call random_number(x)
   call random_number(y)
   if (sqrt(x**2+y**2).le.1) then
      success=success+1.0
   endif
   tot = tot+1.0
enddo
write(*,*), 'Estimation of Pi: ', 4*success/tot
end program monte
```

```
 subroutine init_random_seed()
   integer:: i, n, clock
   integer, dimension(:), allocatable:: seed

   call random_seed(size = n)
   allocate(seed(n))

   call system_clock(count=clock)
   seed = clock + 37 * (/(i-1, i = 1, n)/)
   call random_seed(put = seed)
   deallocate(seed)
 end subroutine"
```

(2.6) The Fortran 90 code "absorb" below is developed for use of kMC method to investigate a gas adsorption and desorption process on a crystal plate with $128 \times 128$ square lattices. This code and code "monte.f90" was developed by David Parker, a junior student of mechanical engineering at Alfred University after he learned Section 2.9 in the multiscale course of 2010 by using this book. The rate of $r_A = 1.0$ (site s)$^{-1}$, $r_D = 2.0$ (site s)$^{-1}$ and the probability $W_A = 1/2$ and $W_D = 1$, respectively, for adsorption and desorption are the same as Figure 2.10. Requirements:

(a) Show reasons why this code is consistent with the flow chart 2.9; specifically show key variables and line numbers which describe, respectively, the adsorption and desorption process.

(b) Explain the meaning of the variables including "numetries" and "real(numatoms)/real (128**2)", grid(i,j).

(c) Explain why the time expression shown in lines 29 and 40 is consistent with (2.36).

(d) Compile and run this code and draw the curves in the absorb.out and make comparison with the curve of Figure 2.10, and then discuss the comparison and the results.

```
"Program absorb
implicit none
integer:: numatoms = 0, i, j, current, numtries=100000
real(kind=8):: x, y, chance, WD=1.0, WA=.5, time=0.0, rA=1.0, rD=2.0
logical, dimension(128,128):: grid

open(15,file='absorb.out')

do i=1,128
   do j=1,128
      grid(i,j)=.false.
   enddo
enddo

call init_random_seed()

do current=1,numtries
   call random_number(x)
   call random_number(y)
   call random_number(chance)
   i = x*128+1
```

```fortran
    j = y*128+1
    if (grid(i,j).EQV..false.) then
      if (chance.le.WA) then
        grid(i,j)=.true.
        numatoms=numatoms + 1
        time = time + 1/((128*128-numatoms)*rA+numatoms*rD)
        write(15,'(f13.10,1x,f13.10)'), time, (numatoms/128.0**2)
      endif
    else
      if(chance.le.WD) then
        grid(i,j)=.false.
        numatoms=numatoms-1
        time = time + 1/((128*128-numatoms)*rA+numatoms*rD)
        write(15,'(f13.10,1x,f13.10)'), time, (numatoms/128.0**2)
      endif
    endif
  enddo

  end program absorb

  subroutine init_random_seed() !from
  http://gcc.gnu.org/onlinedocs/gfortran/RANDOM_005fSEED.html
    integer:: i, n, clock
    integer, dimension(:), allocatable:: seed

    call random_seed(size = n)
    allocate(seed(n))

    call system_clock(count=clock)
    seed = clock + 37 * (/(i-1, i = 1, n)/)
    call random_seed(put = seed)
    deallocate(seed)
  end subroutine"
```

## 2.10   Concluding Remarks

Determination of potential function is most important in atomistic analysis. Following the plan to introduce different categories of potentials as described in Section 2.2.2, this chapter introduces pair potentials. This is important with its simplicity and history, its basic principle, and its concise form that describe interatomic interactions. Important potentials such as Tersoff–Brenner potentials for Si and C (see Section 3.8) take advantage of the form of pair potentials. One part of the embedded potential function (EAM) important for metals consists of the pair potential. The latter is introduced in detail in Section 3.11 with the recent developments incorporating not only the crystal data but also the amorphous, liquid and X-ray diffraction data for developing potentials of Al, Cu, and Zr. Note, as shown in these studies for EAM potentials[59–61], the factor "1/2" in (2.2a) and (2.2c) of the pair potential part is taken out if the summation of interactive pair energy between a generic atom $i$ and its neighbor atom $j$ limits only to the neighbor atoms with $j > i$. Thus, double accounting of the pair energy will be avoided. Pair potential discussion will be continued in the next chapters with the first emphasis on parameter determination by experiments and QM (see Sections 3.4, 3.14, and 4.1), and data collection through the Internet (Appendix in Chapter 3) for

Buckingham potential. How to develop combined potentials for two or more chemical elements from monoatomic potentials is also discussed in Section 3.12.

# References

[1] Alder, B. J. and Wainwright, T. E. (1957) Phase transition for a hard sphere system. *J. Chem. Phys.*, **26**, 1208.

[2] Alder, B. J. and Wainwright, T. E. (1959) Studies in molecular dynamics. I. General method. *J. Chem. Phys.*, **31**, 459.

[3] Rahman, A. (1964) Correlations in the motion of atoms in liquid argon. *Phys. Rev.*, **136**, A405.

[4] Verlet, L. (1967) Computer experiments on classical fluids. I. Thermodynamical properties of Lennard-Jones molecules. *Phys. Rev.*, **159**, 98.

[5] Hagler, A. T., Osguthorpe, D. J., Dauber-Osguthorpe, P., and Hempel, J. C. (1985) Dynamics and conformational energetic of a peptide hormone: Vasopressin. *Science*, **227**(4692), 1309.

[6] Allen, M. P. and Tildesley, D. J. (1987) *Computer Simulation of Liquids*, Clarendon Press, Oxford, reprinted 1990.

[7] Frenkel, D. and Smit, B. (1996) *Understanding Molecular Simulation – From Algorithms to Applications*, Academic Press, San Diego.

[8] Ciccotti, G. and Hoover, W. G. (eds.) (1986) *Molecular Dynamics Simulations of Statistical Systems*. Proceedings of the 97th International School of Physics Enrico Fermi. North-Holland, Amsterdam.

[9] Allen, M. P. and Tildesley, D. J. (eds.) (1993) *Computer Simulation in Chemical Physics*. NATO ASI Series 397. Kluwer, Dordrecht.

[10] Leach, A. R. (2001) *Molecular Modelling: Principles and Applications*, Prentice Hall.

[11] Schlick, T. (1993) *Molecular Modeling and Simulation*, Springer, New York.

[12] Schiøtz, J. S., Di Tolla, F. D., and Jacobsen, K. W. (1998) Softening of nanocrystalline metals at very small grain sizes. *Nature*, **391**(5), 561.

[13] Amakov, V., Wolf, D., Phillpot, S. R., *et al.* (2002) Dislocation processes in the deformation of nanocrystalline aluminum by molecular dynamics simulation. *Nature Materials*, **1**, 1.

[14] Catlow, C. R. A. (1977) Point defect and electronic properties of uranium dioxides. *Proceedings of the Royal Society of London A*, **353**, 533.

[15] Williford, R. E., Weber, W. J., Devanathan, R., *et al.* (1999) Native vacancy migrations in zircon. *J. Nucl. Mater.*, **273**, 164.

[16] Du, J. and Corrales, L. R. (2005) First sharp diffraction peak in silicate glasses: structure and scattering length dependence. *Phys. Rev. B*, **72**, 092201.

[17] Du J., Devanathan, R., Corrales, L. R., *et al.* (2006) Short- and medium-range structure of amorphous zircon from molecular dynamics simulations, *Phys. Rev. B*, **74**, 214204.

[18] Devanathan, R., Weber, W. J., Singhal, S. C., and Gade, J. D. (2006) Computer simulation of defects and oxygen transport in yttria-stabilized zirconia, *Solid State Ionics*, **177**, 1251.

[19] Li, J. (2003) AtomEye: An efficient atomistic configuration viewer. *Modelling Simul. Mater. Sci. Eng.*, **11**, 173.

[20] Jones, J. E. (1924) On the determination of molecular fields. II. From the equation of state of a gas. *Proceedings of the Royal Society of London A*, **106**, 463.

[21] Baskes, M. I. (1999) Multi-body effects in fcc metals: A Lennard-Jones embedded-atom potential. *Phys. Rev. Lett.*, **83**(13), 2592.

[22] Jensen, E. J., Kristensen, W. D., and Cotterill, R. M. J. (1973) Molecular dynamics studies of melting: 1. Dislocation density and the pair distribution function. *Philos. Mag.* **27**, 623.

[23] Broughton, J. Q. and Gilmer, G. H. (1983) Molecular dynamics investigation of the crystal-fluid interface. I. Bulk properties. *J. Chem. Phys.*, **79**, 5095.

[24] Swope, W. C. and Anderson, H. C. (1990) 106-particle molecular-dynamics study of homogeneous nucleation of crystals in a supercooled atomic liquid. *Phys. Rev. B*, **41**, 7042.

[25] Waal, B. W. van de (1991) Can the Lennard-Jones solid be expected to be fcc? *Phys. Rev. Lett.*, **67**, 3263.

[26] Wolde, P. R. ten, Ruiz-Montero, M. J., and Frenkel, D. (1995) Numerical evidence for bcc ordering at the surface of a critical fcc nucleus. *Phys. Rev. Lett.*, **75**, 2714.

[27] Shen, Y. C. and Oxtoby, D. W. (1996) bcc symmetry in the crystal-melt interface of Lennard-Jones fluids examined through density functional theory. *Phys. Rev. Lett.*, **77**, 3585.

[28] Dzugutov, M. (1996) Universal scaling law for atomic diffusion in condensed matter. *Nature*, **381**, 137.

[29] Kob, W., Donati, C., Plimpton, S. J., *et al.* (1997) Dynamical heterogeneities in a supercooled Lennard-Jones liquid. *Phys. Rev. Lett.*, **79**, 2827.

[30] Donati, C., Douglas, J. F., Kob, W., *et al.* (1998) Stringlike cooperative motion in a supercooled liquid. *Phys. Rev. Lett.*, **80**, 2338.

[31] Broughton, J. Q. and Gilmer, G. H. (1983) Molecular dynamics investigation of the crystal-fluid interface. II. Structures of the fcc. (111), (100), and (110) crystal-vapor systems. *J. Chem. Phys.*, **9**, 5105.

[32] Cotterill, R. M. J., Leffers, T., and Liltholt, H. (1974) Molecular dynamics approach to grain boundary structure and migration. *Philos. Mag.*, **81**, 265.

[33] Abraham, F. F., Brodbeck, D., Rudge, W. E., and Xu, X. P. J. (1997) A molecular dynamics investigation of rapid fracture mechanics. *J. Mech. Phys. Solids*, **45**, 1595.

[34] Mikulla, R., Stadler, J., Krul, F., *et al.* (1998) Crack propagation in quasicrystals. *Phys. Rev. Lett.*, **81**, 3163.

[35] Komanduri, R., Chandrasekaran, N., and Raff, L. M. (2001) Molecular dynamics (MD) simulation of uniaxial tension of some single-crystal cubic metals at nanolevel. *Int. J. Mech. Sci.*, **43**, 2237.

[36] Swope, W. C. and Anderson, H. C. (1990) 106-particle molecular-dynamics study of homogeneous nucleation of crystals in a supercooled atomic liquid. *Phys. Rev. B*, **41**, 7042.

[37] Honeycutt, R. W. (1970) The potential calculation and some applications. *Methods in Computational Physics*, **9**, 136.

[38] Beeman, D. (1976) Some multistep methods for use in molecular dynamics calculations. *J. Comput. Phys.*, **20**, 130.

[39] Gear, C. W. (1971) *Numerical Initial Value Problems in Ordinary Differential Equations*, Prentice Hall, Englewood Cliffs, NJ.

[40] Toxvaerd, S. A. (1982) New algorithm for molecular dynamics calculations. *J. Comput. Phys.*, **47**, 444.

[41] Personal communication with Vesselin Yamakov, National Institute of Aerospace, USA, 2009.

[42] Inorganic Crystal Structure Database, Fachinformationszentrum Karlsruhe, Germany, 2006.

[43] *International Tables for X-Ray Crystallography*. The Kynoch Press, Birmingham, 1965.

[44] Berendsen, H. J. C., Postma, J. P. M., van Gunsteren, W., *et al.* (1984) Molecular dynamics with coupling to an external bath. *J. Chem. Phys.*, **81**, 3864.

[45] Hoover, W. G. (1991) *Computational Statistical Mechanics*, Elsevier, New York, 121.

[46] Press, W. H., Teukolsky, S. A., Vetterling, W. T., and Flannery, B. P. (1992) *Numerical Recipes*, 2nd edn., Cambridge University Press.

[47] Gale, J. D. (1997) GULP: A computer program for the symmetry-adapted simulation of solids. *J. Chem. Soc. Faraday Trans.*, **93**(4), 629.

[48] Smith, W., Forester, T. R., Todoror, L. T., and Leslie, M. (2006) *The DL-Poly-2 User Manual*, CCLC Daresbury Laboratory, Cheshire, UK, Version 2.17.

[49] Metropolis, N. and Ulam, S. (1949) The Monte Carlo method, *J. Am. Stat. Ass.*, **44**, 335.

[50] von Neumann, J. and Ulam, S. (1945) Random ergodic theorems. *Bull. Am. Math. Soc.*, **51**(9), 660.

[51] Buehler, M. J. (2008) *Atomistic Modeling of Materials Failure*, Springer.

[52] Fichthorn, K. A. and Weinberg, W. H. (1991) Theoretical foundations of dynamical Monte Carlo simulations. *J. Chem. Phys.*, **95**(2), 1090.

[53] Cinlar, E. (1975) *An Introduction to Stochastic Process*, Prentice-Hall, Englewood Cliffs, NJ.

[54] Gill, P. A., Spencer, P. E., and Cocks, A. C. F. (2005) Mixed KMC-continuum models for the evolution of rough surfaces. *Int. J. Multiscale Comput. Eng.*, **3**(2), 239.

[55] Fan, Jing, Boyd, I. D., and Shelton, C. (2001) Monte Carlo modeling of YBCO vapor deposition, CP585, in *Rarefied Gas Dynamics* (eds. T. J. Bartel and M. A. Gallis), American Institute of Physics.

[56] Fan, Jing, Boyd, I. D., and Shelton, C. (2000) Monte Carlo modeling of electron beam physical vapor deposition of yttrium. *J. Vacuum Sci. Techn. A*, **18**(6), 2937.

[57] Sun, Q. and Boyd, I. D. (2005) Theoretical development of the information preservation method for strongly nonequilibrium gas flows. 38th AIAA Thermophysics Conference, Toronto, AIAA paper 4828.

[58] Yang, Yougen (2000) The Monte Carlo simulation of physical vapor decomposition, Ph. D. Dissertation, University of Virginia.

[59] Mendelev, M.I., Kramer, M. J., Becker, C. A., and Asta, M. (2008) Analysis of semi-empirical interatomic potentials appropriate for simulation of crystalline and liquid Al and Cu. *Phil. Mag.*, **88**(12), 1723.

[60] Mendelev, M.I., Kramer, M. J., Ott, R. T., Sordelet, D. J., Yogodin, D., and Paper, P. (2009) Development of suitable interatomic potentials for simulation of liquid and amorphous Cu-Zr alloys. *Phil. Mag.*, **89**(11), 967.

[61] Mendelev, M. I., Sordelet, D. J., and Kramer, M. J. (2007) Using atomistic computer simulations to analyze X-ray diffraction data from metallic glasses, *J. Appl. Phys.*, **102**(4), 3501.

# 3

# Applications of Atomistic Simulation in Ceramics and Metals

In this chapter, applications of atomistic simulation in ceramics and metals are discussed, respectively, in Part 3.1 and Part 3.2. Several models including Born and Shell model, Buckingham and Tersoff potentials of ionic and covalent bonding materials, as well as potential parameter determination of oxides are discussed in Part 3.1, followed by the introduction of Embedded Atom Method (EAM), MEAM and binary potentials of unlike atoms for metals in Part 3.2. Application examples include hydrogen embrittlement simulation, determination of defect structure of doped material and mechanisms of nonstoichiometry by atomistic statics lattice calculation, simulations of conductivity of oxide fuel electrolyte, domain switching of ferroelectric ceramics, and yield mechanism of metallic nanowires by molecular dynamics. Calculations of atomistic stress are discussed. Parameter data collection of potential functions from the internet based on a special technical requirement is exemplified through the simulation needs of nano-ceramics film coating on steel substrate. For beginners, sections with symbol* may not be read in the first time.

## Part 3.1 Applications in Ceramics and Materials with Ionic and Covalent Bonds

## 3.1 Covalent and Ionic Potentials and Atomistic Simulation for Ceramics

### 3.1.1 Applications of High-performance Ceramics

Traditionally, ceramics are often not considered in component design because of their flaw-dependant mechanical properties. Specifically, its fracture occurs suddenly rather than being "warned" of potential failure by plastic deformation as metals. Recently, high-performance engineering ceramics have been developed, that can be substituted for metals in many applications. The tools fabricated from silica-alumina and aluminum oxide can cut metal quicker with longer tool life than the best metallic tools. Engineering ceramics which are highly wear-resistant are used to wrap the front edges of agriculture mechanical cutters such as the plow harrow, which can increase the cutter life more than 10 times.

*Multiscale Analysis of Deformation and Failure of Materials*  Jinghong Fan
© 2011 John Wiley & Sons, Ltd

There are some other examples. SiAlON ceramics are significant in applications because of their high strength, good thermal shock resistance and exceptional resistance to wetting or corrosion. Ferroelectric ceramics such as $BaTiO_3$ is widely used as a dielectric in capacitors and sensors for smart materials and devices. Indium tin oxide, a transparent conductor, is widely used in flat panel display screen and solar panel on the market. Atomistic simulation is important for ceramics and will be discussed in this section.

### 3.1.2    Ceramic Atomic Bonds in terms of Electronegativity

Ceramics are usually referred to as inorganic crystalline solids including ionic materials such as metal chlorides and metal oxides, and covalent materials like silicon nitride. Most bonds, however, are mixtures of ionic and covalent. In the simulations of ceramics, structure construction is usually easy. The periodical boundary condition can be readily satisfied because of the translational symmetry of crystal structures. However, the interatomic potential model has to be carefully constructed with knowledge of the nature of the atomic bonds. Another challenge in the simulation of ceramics is that these materials usually consist of several chemical elements, thus developing force fields consisting of multiple consistent potentials for binary, ternary or more complex systems are essential. These potentials are not limited to the interactions between the same kind of atoms such as Mg-Mg and O-O, but also the interatomic potentials for the so-called unlike elements such as Mg-O. This character will be emphasized in detail in Section 3.4 and 3.12.

The electronegativity property of an atom describes the capability for the atom to receive electrons, and is closely related to the atom's position in the Periodic Table of the Elements. Figure 3.1 shows the electronegativity of several chemical elements. The first row in the horizontal axis shows their positions in the Periodic Table. The second row shows the structure and electronic numbers in the outermost shells, i.e. the "s" and "p" (or sp) shells in the electronic structure of the atom.

The electronegativity as being related to the requirement of atoms stability can be used to determine if the material is more ionic or more covalent. Chemical element Na only has one valence electron and Cl has 7 valence electrons in the outmost sp orbitals. Na can easily transfer the electron to Cl so both can satisfy the "octet rule"–an atom is stable when its valance shell is fully filled with 8 electrons ($s^2p^6$). As a result,

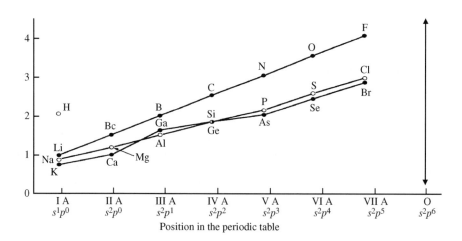

**Figure 3.1**   Electronegativity for some chemical elements (Re-plotted based on Askeland, D. R. (1989) *The Science and Engineering of Materials*, PWS Publishing Company, Boston. Copyright PWS Publishing Company)

Na and Cl become the positive ion (cation) and negative ion (anion), respectively, and strong ionic bonding is formed via Coulomb attraction. On the other hand, Si and C are in the middle of the Periodic Table and both have 4 valence electrons; therefore a covalent bond would be formed by sharing electrons among them, due to their close electronegativities. It is seen obviously that most bonding occurs to try to "help" the atoms achieve the above stability requirement of electronic shell structure.

## 3.2 Born Solid Model for Ionic-bonding Materials

Apart from metals and covalent materials, the atomic bonds in ceramic materials are ionic bonds in most cases. These ionic bonds are bonds between ions with positive and negative charges. As outlined above, ionic bonds exist between atoms with very different electronegativity and metal oxides such as MgO, ZnO, $UO_2$ and many others discussed later in the tables of this section.

### 3.2.1 Born Model

Atomistic simulations of ionic-bonding materials are based on the Born solid model,[2] which assumes crystal atoms interact with each other through long-range electrostatic and short-range forces. The potential energy $U_i$ for any ion or atom i can be written as

$$U_i = \frac{1}{2} \sum_{j=1(\neq i)} \frac{q_i q_j}{4\pi\varepsilon_0 r_{ij}} + \frac{1}{2} \sum_{j=1(\neq i)} V_{ij}(r_{ij}) + \text{multibody potentials} \qquad (3.1)$$

where the second term on the right side is the pair potential and the third term is multi-body potential including the angular effects. This model usually considers only 3-body and sometimes 4-body potentials to avoid the complexity of mathematical expressions. The first term on the right side is the Coulomb potential, $\varepsilon_0$ is the permittivity of the medium, for vacuum $\varepsilon_0 = 0.0552635$ eV·V$^{-2}$·nm$^{-1}$. $q_i$ and $q_j$ are electric charges for ion particles i and j. Positive charges will have a positive sign, while negative charges have a negative sign. Variable $r_{ij}$ is the separation distance between ionic i and j. In some literatures, the constant $1/4\pi\varepsilon_0$ is normalized as 1 for simplicity.

Two points related to the Coulomb potential are as follows. First, Coulomb potential is a long-ranged force in comparison with the short-range forces of pair potential and multi-body potentials. As described in Section 2.2.2, the short-range forces are caused by interactions of electronic clouds. On the other hand, long-ranged Coulomb force is caused by electrostatic forces. Second, the separation distance $r_{ij}$ between ions i and j occurs in the denominator of (3.1), leading to an extremely slow process for convergence of summation. Thus a mathematic transformation technique was developed by Ewald, called Ewald method, which can effectively improve the converging speed.

### 3.2.2 Born-Mayer and Buckingham Potentials

The Coulomb potential, which determines the long-range force among ionic bonds, is explicitly expressed by the first right term of (3.1). Thus the key of the analysis for ionic bond solids and semi-ionic bond solids is to know the pair potential $V_{ij}$ and multi-body potential. The short-range pair potential for most oxides adopts the Born-Mayer potential or the Buckingham model. The Born-Mayer model is expressed by Lewis and Catlow:[3]

$$V_{ij}(r) = A\exp(-r_{ij}/\rho) \qquad (3.2a)$$

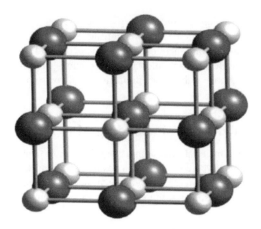

**Figure 3.2** MgO with the rocksalt structure. Small (or yellow) atom is $Mg^{2+}$, large (or red) atom is $O^{2-}$

An additional term $r_{ij}^{-6}$ can be added to account for the attraction force, by which the following Buckingham potential is produced:

$$V_{ij}(r_{ij}) = A \exp(-r_{ij}/\rho) - C r_{ij}^{-6} \qquad (3.2b)$$

Again, the subscript letters $ij$ denote the atoms $i$ and $j$ and are sometimes dropped for simplification. The parameters $A$, $\rho$ and $c$ of the above potential function are generally obtained by fitting the relevant experimental data using the least squares method. Their determinations are sometimes difficult because for many oxides, only structural parameters are known; there is a lack of experimental data on electricity, elasticity, polarization and dynamics behavior that can be used for parameter fitting.

---

**Homework**

MgO has a rocksalt structure (Figure 3.2). The interaction between $Mg^{2+}$ and $O^{2-}$ can be evaluated

using Buckingham potential. The parameters are $A = 1280$, $\rho = 0.3$, $C = 0$,

when r is in angstrom, and E in eV.

(1.1) Plot the short-range potential between $Mg^{2+}$ and $O^{2-}$ and the corresponding short-range force with respect to interatomic distance. In which range is the interaction attraction? In which range is it repulsion?

(1.2) How is an equilibrium distance between $Mg^{2+}$ and $O^{2-}$ achieved? What kind of potential balances the short-range potential? Calculate the equilibrium distance between $Mg^{2+}$ and $O^{2-}$.

(1.3) If the distance between $Mg^{2+}$ and $O^{2-}$ in MgO is the equilibrium distance, what is the lattice parameter of MgO? Comparing it with the value in literature, where does the discrepancy come from?

---

## 3.3 Shell Model

In simulations of ionic materials, predicting how ions respond to the electrostatic field can be important. For example, the simulations of the polarization caused by charged defects or a distortion in the lattice

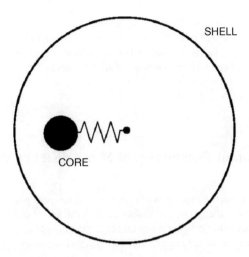

**Figure 3.3**   Shell model: shell and core connected by spring

structure all require an adequate reproduction of the polarizability of the material. In such situations, the electronic polarizability, which is caused by electron displacement under electrostatic fields, cannot be neglected. As the rigid ion model is inappropriate, the shell model was proposed by Dick and Over-hauser[4] utilizing a simple mechanical description to simulate this characteristic. This model treats an ion as a core and an outer shell connected to each other. The core collects all of the ion masses, while the shell takes the polarizable charge. The coupling between nucleus and shell is done by a spring as shown in Figure 3.3. In an electrostatic field, shell will displace from the core but keep its spherical charge distribution. The displacement between shell and core results in a dipole moment.

In the shell model, the shell possesses electric charge $Y_-$, which is usually very different from the total charges of all electrons in the ion. This is because only the outer electrons are effectively affected by the electrostatic field. Note that the charges of shell derived from experimental data are not necessarily negative because of environmental effects on the complex electronic behavior. The short-ranged potential of the Born model in (3.1) is used to simulate the interaction of overlapping electrons among the shells. The core in the shell model represents the nucleus and the inner electrons, with a total electric charge $Y_+$. $Y_+ + Y_-$ equals the total ion charge. The spring with a constant $k$ is used in the coupling between shell and the core. After determining the force F which is imposed on this model in the solving process, the distance between shell and nucleus can be determined by F/k. The polarity, $\alpha$, of the unbound ion electron can be derived from the formula:

$$\alpha = \frac{Y_-^{\,2}}{k} \tag{3.3}$$

where the parameters $Y_-$, $k$ and the distance between the core and shell can be determined by fitting the crystal data, usually the dielectric constants. Parameters of shell models such as spring constants, charges of shells and cores, and the corresponding Buckingham potential parameters can be found in Ref. [5]. The 15 chemical elements involved with these include Mg, Ca, Sr, Ba, Fe, Ti, Al, La, Pr, Nd, Gd, Eu, Tb, Yb and O.

The shell model has been extensively used in static lattice calculations of ceramic materials and provided good results. However it cannot directly fit into the classic MD scheme. As the shell has no mass, it cannot be treated by Newton's law. The shell model has been introduced into MD by two methods:

adiabatic shell model and relaxed shell model. In adiabatic shell model, a small amount of mass is assigned to the shell so it can be treated by Newton's law. In the relaxed shell model, only the cores move according to Newton's law. When a new set of core coordinates are generated, a relaxation will be performed to obtain the local equilibrium positions of all the cores and shells. Section 10.3.5 introduces the use of GULP software for the input file of MgO shell model, and its calculation can be conducted in the CSLI directory UNITS/UNIT2/shell_model.

## 3.4    Determination of Parameters of Short-distance Potential for Oxides

As mentioned, one challenge in the simulation of ceramics is developing potentials of binary, ternary and high-order potentials. This challenge will be discussed in detail for oxides. Lewis and Catlow[3] have developed a systematic method to develop potential functions of oxide ceramics through extrapolation, using the already available potential functions of other types of ceramics.

### 3.4.1    Basic Assumptions

(1)  This method is based on the following assumptions: Assume that the interactions between $O^{2-}$ and $O^{2-}$ ions for any oxide are equivalent, the following Buckingham potential parameters proposed by Catlow for $UO_2$ can then be used for any other oxides:[6]

$$A = 22764\,eV, \quad \rho = 0.41\,\text{Å}, \quad C = 20.37\,eV/\text{Å}^6 \tag{3.4a}$$

(2)  Assume that the long distance potential for interactions between cations and cations is purely determined by the Coulomb potential function. A consequence of this assumption is that the short-range potential effect among cations can be neglected. Catlow thought that the distances among cations are quite large based on the structure of oxide molecules. Consequently, the interactions are relatively smaller than those among oxygen ions, which are already tiny in the space range between balanced crystal structures. Therefore, this assumption is rational.

(3)  Assume that the short-range pairwise potential for interactions between cations and anions is given as the Born-Mayer form. This ignores the contributions of the additional attractive potential term in the Buckingham potential (3.2b), i.e., assumes that the contribution of the term is small and can be taken into account by adjusting the Born-Mayer parameters.

In addition, to include the ion polarizability, the shell model is also used. The parameters for a shell model of oxygen ion are given as follows:[6]

$$Y_- = -3.06\,e, \quad k = 80.21\,eV\,\text{Å}^{-2} \tag{3.4b}$$

For certain oxides, the experimental results, which can determine the parameters, are inadequate. In this circumstance, the essential properties of these potentials must receive a further assumption; one must use the common properties of the same oxide type. This is based on the assumption that some conjunct parameters exist for the same type of oxides, thus the parameters of any oxide can be obtained from experiments on other oxides of the same type.

In the following sections three methods to find oxide potential parameters will be introduced.[3] The first two methods determine parameters by the use of crystal structural information. The third method is only suitable for cases where a great deal of experimental data is available, thus one can systematically derive parameters for both short-range potential functions and the shell model.

## 3.4.2 General Methods in Determining Potential Parameters

There are two basic types of methods for determining potential parameters. One is by fitting the calculation data of quantum mechanics (QM), the other is by fitting experimental data. Mixed methods in fitting both QM and experimental data are superior.

### 3.4.2.1 By Fitting Quantum Mechanics Data

The potential parameters can be determined from QM calculation. A good example is the parameter determination for the interaction of $O^{2-}$ and $O^{2-}$ ions of Buckingham potential shown in (3.4a). These parameters are determined from the results of a QM code called "Atmol2 Gaussians" by the Hartree-Fock (HF) method.[6] HF method is a first-principle method of QM and will be introduced in detail in Section 4.9.

The basic idea is to use HF for calculation of the coupling energy between $O^{2-}$ and $O^{2-}$ ions in terms of the internuclear separation distance r between them. For instance, for a separation distance of 2.8 au (atomic unit, see Table 2.3 in Section 2.3.4) which equals 1.4817 Å, the HF-obtained coupling energy is $-149.5670759$ au. This value should subtract the energy of two isolated $O^-$ ions which are calculated by QM as $-149.5399674$ au. This subtraction results in the interaction energy of $-0.027109$ au. By the relationship of the unit au with the unit eV listed in Section 2.3.4 we obtain the interaction energy for $r = 1.4817$ Å as $-0.737654$ eV. By applying HF for different r, one can then obtain a series of data on energy pairs V(r). (Remark: the energy discussed here is potential energy, which can be negative, as shown in Figure 2.3.)

These pairs can be used to determine the parameters of A, $\rho$ and C in Buckingham potential of (3.2b) by, for instance, the least square fitting method. It is interesting to see how large is the difference between the HF-obtained energy $-0.737654$ eV and the data calculated by Buckingham potential with the obtained parameters. Substituting the parameters in (3.4a) into (3.2b) results in

$$V(r) = 22764 \exp(-r/0.41) - 20.37 r^{-6} \qquad (3.5)$$

which with $r = 1.4817$ Å gives the Buckingham potential as $-0.78962$ eV. The relative error is 7%.

With an increase in computational power, it is also possible to calculate the energy of a periodic structure for a specific material with QM. By varying the structure, a group of energies can be obtained and then a so-called "energy hypersurface" can be generated, which is the function of energy versus structural parameters. This hypersurface can be fitted to the short-ranged potential with the aid of simulation software. As a result, it gives more precise parameters for the studied material.

Fitting with QM data provides the ability to simulate materials without sufficient experimental data and avoiding the experimental uncertainty. However, one should keep in mind that potential parameters obtained via QM are not necessarily the best unless they have been tested and compared to the experiments. Taking the QM-determined three parameters in (3.4a) as an example, in many applications, the values of A and $\rho$ are not changed, while the value of C can be one to four times greater than the listed value to match with experimental data.

The other shortcoming of the QM method is its time/power consumption, i.e., the model size is limited by the computer source. Details about QM concepts and calculation can be found in Chapter 4. Specifically, a more detailed introduction to the determination of parameters in (3.4a) is given in Section 4.1.

### 3.4.2.2 By Fitting Experimental Data

An alternative method to obtain the parameters is by fitting to the experimental data. Material properties, such as lattice parameter and elastic constant, can be measured in experiments and can also be calculated from the derivatives of the total energy. For example lattice parameter is related to the first derivative of the energy; and elastic constant is its second derivative with respect to strain (see (6.9k)). Since the total energy of a given structure can be calculated from the interatomic potential, fitting can be done by finding the potential parameters to produce the material properties closest to the experimental data. The resulting parameters can usually reproduce the material properties very well, but the uncertainty of the experimental data should be considered, especially when using the high-order derivatives.

Actually, energy integration and fitting have been coded into several simulation softwares like GULP[7] and can be executed very fast. Several examples using GULP can be found in Section 10.3.3; its calculation can be conducted in the CSLI directory UNITS/UNIT2/structure.

### 3.4.3 Three Basic Methods for Potential Parameter Determination by Experiments

This section will discuss three types of material properties commonly used in the experimental fitting method for oxides. They provide values of Born-Mayer pair potential parameters A and $\rho$ for 39 oxides, shown in Tables 3.1 to 3.3. These tables are useful for investigations of interactions between oxide cations and anions. However, only the information about the cations is listed in the tables because all anions are oxygen atoms which are known.

#### 3.4.3.1 The First Method

For the case of oxide ceramics with only a small amount of experimental data, the extrapolation-interpolation method is adopted. This method assumes that oxide structures and chemical compositions are similar and that the parameter $\rho$ of Born-Mayer potential for all oxide ceramics is identical. Based on this assumption, a constant $\rho$ can be used to describe the interactions between cations and anions for the first transition series of oxides in the chemical Periodic Table. The parameter A in (3.2b), however, is obtained from fitting crystal structural parameters through equilibrium conditions. Table 3.1 gives the corresponding constant values of A and $\rho$ for the second and third types of transition oxides and for lanthanide and actinide oxides.

#### 3.4.3.2 The Second Method

This method relies on crystal structure with a low symmetry. In this case, the structural data contains a great deal of information. Specifically, and different from the case of cubic crystals, this method applies

**Table 3.1** Parameters A and $\rho$ of Born-Mayer oxide potential function (Reproduced from Lewis, G. V. and Catlow, C. R. A. (1985) Potential models for ionic oxides. *Journal of Physics C*, 18, 1149. Copyright © IOP)

| Cation | Charge | A/eV | $\rho$/Å | Cation | Charge | A/eV | $\rho$/Å |
|--------|--------|------|------|--------|--------|------|------|
| Ca | 2 | 1227.7 | 0.3372 | Zn | 2 | 700.3 | 0.3372 |
| Sc | 2 | 838.6 | 0.3372 | Zr | 4 | 1453.8 | 0.3500 |
| Ti | 2 | 633.3 | 0.3372 | Cd | 2 | 868.3 | 0.3500 |
| V | 2 | 557.8 | 0.3372 | Hf | 4 | 1454.6 | 0.3500 |
| Cr | 2 | 619.8 | 0.3372 | Ce | 4 | 1017.4 | 0.3949 |
| Mn | 2 | 832.7 | 0.3372 | Eu | 2 | 665.2 | 0.3949 |
| Fe | 2 | 725.7 | 0.3372 | Tb | 4 | 905.3 | 0.3949 |
| Co | 2 | 684.9 | 0.3372 | Th | 4 | 1144.6 | 0.3949 |
| Ni | 2 | 641.2 | 0.3372 | U | 4 | 1055.0 | 0.3949 |

**Table 3.2** Determination of potential functions A and $\rho$ with the second method (Reproduced from Lewis, G. V. and Catlow, C. R. A. (1985) Potential models for ionic oxides. *Journal of Physics C*, 18, 1149. Copyright © IOP)

| Cation | Charge | A/eV | $\rho$/Å | Cation | Charge | A/eV | $\rho$/Å |
|---|---|---|---|---|---|---|---|
| Sc | 3 | 1299.4 | 0.3312 | Gd | 3 | 1336.8 | 0.3551 |
| Mn | 3 | 1257.9 | 0.3214 | Ho | 3 | 1350.2 | 0.3487 |
| Y | 3 | 1345.1 | 0.3491 | Yb | 3 | 1309.6 | 0.3462 |
| La | 3 | 1439.7 | 0.3651 | Lu | 3 | 1347.1 | 0.3430 |
| Nd | 3 | 1379.9 | 0.3601 | Pu | 3 | 1376.2 | 0.3593 |
| Eu | 3 | 1358.0 | 0.3556 | | | | |

equilibrium simulations to fit the data of the cell size and atomic coordinates to obtain relevant parameters for given oxides. Table 3.2 lists some obtained values of parameters A and $\rho$ for certain oxide ceramics using this method.

### 3.4.3.3 The Third Method

For certain oxide ceramics, the elastic constants, dielectric constants, cohesion energy and configurations have been determined by experiment. This experimental data is not only good for determining the short-range potential parameters, but can also be used to determine the polarization parameters of the corresponding shell model. Note that the dielectric constant has a close relation to shell parameters, therefore it is simultaneously affected by the short-range potential to a certain degree. Therefore, for a given oxide, the parameters of potential functions for interactions between positive and negative ions will be affected by the choice of methods to determine the parameters of the shell model.

Literature introduced two types of systematic methods to obtain the parameters of shell models for several types of ceramic oxides where a great deal of experimental data is available. The results given by the first method are shown in Table 3 of that reference paper.[3] It gives the $A$ and $\rho$ values of potential functions between certain cations and anions, whose valences are two, three or four; as well as values of $Y_+$ and $Y_-$ for the corresponding shell model, which, respectively, represent the charges of core and shell. The total oxygen ion charge is $-2e$, thus if the core charges are $1.06e$, then the shell will have a negative charge of $-3.06e$ to satisfy the requirement that the whole atom has a charge of $-2e$.

The results obtained from the second method are shown in Table 3.3 (a) and (b). Table 3.3 (a) gives short-range potential parameters (A and $\rho$) for ten oxide ceramics, including ordinary elements such as Mg, Ca, Sr, Ba, Mn, Fe, Co, Ni, Al, U. Their cation parameters of the shell model are presented in Table 3.3 (b), where the corresponding oxygen of the shell model can take $-3e$, and the elastic parameter is referred to the values given by Catlow.[6]

More self-consistent interatomic potentials for the simulation of binary and ternary oxides including parameters of Buckingham potential and shell model are given in Table 3.4. The parameters of structural lattice and elastic stiffness are also given in the related paper.[5]

## 3.5 Applications in Ceramics: Defect Structure in Scandium Doped Ceria Using Static Lattice Calculation

Atomistic simulations have been used to study a wide range of problems related to ceramic materials. Relative to the nature of the problem, different techniques are applied; such as the static lattice calculation technique, using the energy minimization method. The energy of the crystal lattice is calculated and minimized relative to atomic coordinates, to produce an equilibrium structure that utilizes periodical boundary conditions. Using this method, perfect crystal structures with corresponding lattice energy can

**Table 3.3**   Ten oxide short-range potential parameters A and $\rho$ with the third method (Reproduced from Lewis, G. V. and Catlow, C. R. A. (1985) Potential models for ionic oxides. *Journal of Physics C*, 18, 1149. Copyright © IOP)

| (a) Short-range potential parameters between cations and anions | | | | | | | |
|---|---|---|---|---|---|---|---|
| Cation | Charge | A/eV | $\rho$/Å | Cation | Charge | A/eV | $\rho$/Å |
| Mg | 2 | 1428.5 | 0.2945 | Fe | 2 | 1207.6 | 0.3084 |
| Ca | 2 | 1092.4 | 0.3437 | Co | 2 | 1491.7 | 0.2951 |
| Sr | 2 | 959.1 | 0.3721 | Ni | 2 | 1582.5 | 0.2882 |
| Ba | 2 | 905.7 | 0.3976 | Al | 3 | 1474.4 | 0.3006 |
| Mn | 2 | 1007.4 | 0.3262 | U | 4 | 1014.3 | 0.3976 |

| (b) Shell model parameters and ion polarization parameter $\alpha$ | | | | |
|---|---|---|---|---|
| Oxide | $A + /\text{Å}^3$ | $\alpha\text{-}/\text{Å}^3$ | $Y_+/e$ | $K_+/\text{eV Å}^{-2}$ |
| MgO | 0.094 | 2.366 | 1.585 | 361.6 |
| CaO | 1.284 | 2.702 | 3.135 | 110.2 |
| SrO | 2.122 | 2.899 | 3.251 | 71.7 |
| BaO | 2.656 | 3.102 | 9.203 | 459.2 |
| MnO | 1.773 | 2.496 | 3.420 | 95.0 |
| FeO | 2.056 | 2.421 | 2.997 | 62.9 |
| CoO | 1.599 | 2.396 | 3.503 | 110.5 |
| NiO | 1.781 | 2.341 | 3.344 | 93.7 |
| $Al_2O_3$ | 0.052 | 2.132 | 1.458 | 1732 |
| $UO_2$ | 3.759 | 2.618 | 5.350 | 109.7 |

be calculated and a wide range of important material properties can be obtained from the involved energy dependences on different parameters. In addition, interatomic potential models can be fitted from experimental data using static lattice calculation software like GULP.[7]

An important application of the static lattice calculation is the simulation of point defects in crystals; which include vacancies, impurities and interstitial ions. Introducing defects into materials by means of doping has long been a fundamental way of modifying and improving material properties, and has set the basis of life-changing technologies such as semiconductor technology. Therefore studying the effects of defects on material structures and properties carries an essential importance for material science. A way to study these defects is to use static lattice calculation to simulate the defect structure and energy.

As periodical boundary conditions are usually inappropriate (see Section 2.6.1) when simulating isolated defects or defect clusters, particularly when there is a lack of charge neutrality, Mott-Littleton approximation[8] is commonly used for defect simulations. Mott-Littleton approximation treats an infinite crystal lattice with a defect or defect cluster as three spherical regions centered at the defect or defect cluster, as shown in Figure 3.4. Different levels of approximations are applied on the three regions. The inner spherical region (Region I), which contains defects and is mostly perturbed by defects, is explicitly simulated at the atomistic level, with all short-range potentials and Coulomb potential being considered. The outer shell-like region (Region II), which is far away from defects and extends to infinity, is treated as a dielectric continuum in the electric field caused only by the net charge of defects. An interfacial region (Region IIa) is introduced between the two regions to ensure a smooth transition. In Region IIa, ions are approximated as harmonically displaced from their original lattice sites, under the force from ions in Region I, or in a more common approximation, only the defects in Region I.

The total energy of the system and the detailed structure of the inner region can then be obtained through energy minimization. This method has been incorporated in several computer simulation packages such

**Table 3.4** Derived Buckingham potential parameters (Reproduced from Bush, T. S., Gale, J. D., Catlow, C. R. A., and Battle, P. D. (1994) Self-consistent interatomic potentials for the simulation of binary and ternary oxides. *Journal of Materials Chemistry*, 4, 831, by permission of the Royal Society of Chemistry)

| | $A/\mathrm{eV}$ | $\rho/\text{Å}$ | $C/\mathrm{eV}\,\text{Å}^{-6}$ | $M^{x+}$ | | $O^{2-}$ | | | |
|---|---|---|---|---|---|---|---|---|---|
| | | | | $q(\text{core})$ $e$ | $q(\text{shell})$ $e$ | $q(\text{core})$ $e$ | $q(\text{shell})$ $e$ | $k_{+}/\mathrm{eV}\,\text{Å}^{-2}$ | $k/\mathrm{eV}\,\text{Å}^{-2}$ |
| $Li^{+}-O^{2-}$ | 426.480 | 0.3000 | 0.0 | 1.0 | — | 0.513 | −2.513 | — | 20.53 |
| $Na^{+}-O^{2-}$ | 1271.504 | 0.3000 | 0.0 | 1.0 | — | 0.513 | −2.513 | — | 20.53 |
| $K^{+}-O^{2-}$ | 3587.570 | 0.3000 | 0.0 | 1.0 | — | 0.513 | −2.513 | — | 20.53 |
| $Mg^{2+}-O^{2-}$ | 2457.243 | 0.2610 | 0.0 | 1.580 | 0.420 | 0.513 | −2.513 | 349.95 | 20.53 |
| $Ca^{2+}-O^{2-}$ | 2272.741 | 0.2986 | 0.0 | 0.719 | 1.281 | 0.513 | −2.513 | 34.05 | 20.53 |
| $Sr^{2+}-O^{2-}$ | 1956.702 | 0.3252 | 0.0 | 0.169 | 1.831 | 0.513 | −2.513 | 21.53 | 20.53 |
| $Ba^{2+}-O^{2-}$ | 4818.416 | 0.3067 | 0.0 | 0.169 | 1.831 | 0.513 | −2.513 | 34.05 | 20.53 |
| $Fe^{3+}-O^{2-}$ | 3219.335 | 0.2641 | 0.0 | 1.971 | 1.029 | 0.513 | −2.513 | 179.58 | 20.53 |
| $Ti^{4+}-O^{2-}$ | 2088.107 | 0.2888 | 0.0 | 2.332 | 1.678 | 0.513 | −2.513 | 253.60 | 20.53 |
| $Al^{3+}-O^{2-}$ | 2409.505 | 0.2649 | 0.0 | 0.043 | 2.957 | 0.513 | −2.513 | 403.98 | 20.53 |
| $Ca^{3+}-O^{2-}$ | 2339.776 | 0.2742 | 0.0 | 3.0 | — | 0.513 | −2.513 | — | 20.53 |
| $Y^{3+}-O^{2-}$ | 1519.279 | 0.3291 | 0.0 | 3.0 | — | 0.513 | −2.513 | — | 20.53 |
| $La^{3+}-O^{2-}$ | 5436.827 | 0.2939 | 0.0 | 5.149 | −2.149 | 0.513 | −2.513 | 173.90 | 20.53 |
| $Pr^{3+}-O^{2-}$ | 13431.118 | 0.2557 | 0.0 | 1.678 | 1.322 | 0.513 | −2.513 | 302.36 | 20.53 |
| $Nd^{3+}-O^{2-}$ | 13084.217 | 0.2550 | 0.0 | 1.678 | 1.322 | 0.513 | −2.513 | 302.35 | 20.53 |
| $Gd^{3+}-O^{2-}$ | 866.319 | 0.3770 | 0.0 | −0.973 | 3.973 | 0.513 | −2.513 | 299.96 | 20.53 |
| $Eu^{3+}-O^{2-}$ | 847.868 | 0.3791 | 0.0 | −0.991 | 3.991 | 0.513 | −2.513 | 304.92 | 20.53 |
| $Tb^{3+}-O^{2-}$ | 845.137 | 0.3750 | 0.0 | −0.972 | 3.972 | 0.513 | −2.513 | 299.98 | 20.53 |
| $Yb^{3+}-O^{2-}$ | −991029 | 0.3515 | 0.0 | −0.278 | −3.278 | 0.513 | −2.513 | 308.91 | 20.53 |
| $O^{2+}-O^{2-}$ | 25.41 | 0.6937 | 32.32 | — | — | 0.513 | −2.513 | — | 20.53 |

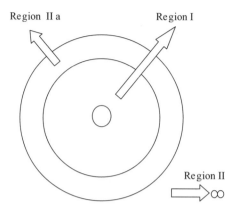

**Figure 3.4** Two-region strategy for defect energy calculations (Reproduced from Dwivedi, A. and Cormack, A. N. (1990) A computer simulation study of the defect structure of calcia-stabilized zirconia. *Philosophia Magazine* A, 61 (1), 1, Taylor & Francis Ltd.)

as GULP for defect analysis. High accuracy can be achieved when sizes of inner regions are reasonably large compared to the size of the defect cluster. Defect energies can be compared for different defect configurations to determine the most likely defect structure. The defect structure can also be used as the basis of other analysis, such as calculating the electronic band structure of the doped material by quantum mechanics.

Studies using this method have provided unique understanding of defects in materials and fundamental topics of material engineering such as doping. Application 1 below is a static lattice simulation of Scandium ($Sc^{3+}$) doped ceria ($CeO_2$), where the structure around a single $Sc^{3+}$ dopant is calculated.

**Application 1: Defect calculation of Scandium doped ceria[9]**
In this example, a static lattice study of the defect structure of Scandium ($Sc^{3+}$) doped ceria ($CeO_2$) is discussed. The simulation used Buckingham potentials and shell model developed in earlier studies. A two-region strategy introduced above was used to perform the defect calculation, with about 150 ions included in Region I. The results showed that the oxygen ions around the Scandium dopant can relax to a lower symmetry configuration with a slightly lower energy than the regular cubic structure. Such subtle structure deformation explained the energy loss peak, with low activation energy observed in the inelastic and dielectric relaxation experiments of the material. The simulated structure was confirmed by X-ray absorption spectroscopy experiments two years later.[10] Also, the simulations showed this relaxation mode happens only for small dopants. It is important to understand the different modes of dopant-vacancy associations in trivalent oxide doped ceria electrolytes, in order to improve the oxygen conductivity of the electrolytes.[11]

## 3.6 Applications in Ceramics: Combined Study of Atomistic Simulation with XRD for Nonstoichiometry Mechanisms in $Y_3Al_5O_{12}$ (YAG) Garnets

### 3.6.1 Background

YAG, short for $Y_3Al_5O_{12}$ garnet, has industrial importance for lasers and scintillators. The reason that a large number of studies have focused on the identification of defects of aluminate garnets is that defects such as cation antisites take a dominant role in its behavior. Specifically, for laser applications, atomic-scale point defects may significantly limit its performance, while for scintillators they may trap electrons that may recombine radiatively to change the functionality.

Studies have found that YAG single crystal with exact stoichiometry is difficult to form.[12] From the phase diagram one can usually determine the quantitative value of each composition (i.e., make it stoichiometric) through a given phase rule.[13] The fact that YAG cannot use the phase diagram to quantitatively determine its microstructure indicates it is nonstoichiometric. The nonstoichiometry can be formed by a variety of defect mechanisms, and different defect mechanisms impact the material properties differently. For example, although both are cation antisite defects (a cation taking a lattice site which should be occupied by a cation of another type), $Y_{Al}$ defect ($Y^{3+}$ takes $Al^{3+}$ site) has been proposed as an electron trap which limits the performance of YAG scintillators, while $Al_Y$ defect ($Al^{3+}$ takes $Y^{3+}$ site) has not. Therefore, understanding the nonstoichiometric mechanism in YAG, can then be used to control the defect formation.[14]

Recently, Patel, Levy, Grimes, *et al.*[15] reported that the use of atomistic simulation can predict the defect structure associated with the deviation from the phase diagram. The crystal lattice was simulated using GULP (see Section 10.3 for detail) which treats the crystal as a Born-like lattice of point charges (see Section 3.2). By comparing the experimental variation in the lattice parameter with atomistic simulation results they predict that nonstoichiometry proceeds via cation antisite defects. Here, the experimental results of lattice constant as a function of derivation from stoichiometry are obtained from XRD; the result of atomistic simulation used here is the defect volume changes. In this section, their work for this interesting nonstoichiometric phenomenon of YAG garnets will be introduced to show the application of the atomistic analysis.

### 3.6.2 Structure and Defect Mechanisms of YAG Garnets

The YAG crystal belongs to the cubic crystal system, with $a = b = c$, and $\alpha = \beta = \gamma = 90°$, therefore, only one lattice parameter is needed. The space group is Ia3d.[16] There are 160 atoms in a unit cell, in which 96 $O^{2-}$ anions form the dodecahedral, octahedral, and tetrahedral interstices, which are filled with $Y^{3+}$, $Al_{(a)}^{3+}$, and $Al_{(d)}^{3+}$, respectively. The crystalline structure of YAG is shown in Figure 3.5.

There are a variety of defect mechanisms that conserve site balance, charge, and mass while facilitating deviations from the stoichiometric YAG composition. Several researchers reported that the antisite would be the dominant defect of YAG, which can affect the crystal performance through electronic trapping and can be easily formed because of the low energy required.[17,18] In addition,

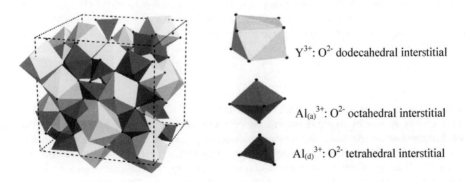

$Y^{3+}$: $O^{2-}$ dodecahedral interstitial

$Al_{(a)}^{3+}$: $O^{2-}$ octahedral interstitial

$Al_{(d)}^{3+}$: $O^{2-}$ tetrahedral interstitial

**Figure 3.5**   The crystalline structure of YAG (Curtsey from Xingguo Zeng, 2010)

**Table 3.5** Buckingham potential parameters for YAG

| Interaction pair | $A$(eV) | $\rho$(Å) | $C$(eV/Å$^6$) |
|---|---|---|---|
| $O^{2-}-O^{2-}$ [20] | 22764.0 | 0.149 | 27.88 |
| $Al^{3+}-O^{2-}$ [21] | 741.9007 | 0.3566 | 0 |
| $Y^{3+}-O^{2-}$ [21] | 2036.838 | 0.3103 | 0 |

**Table 3.6** Comparison between atomistic simulation of stoichiometry YAG and experiments for the lattice constant and elastic constants (Reproduced from Kuklja, M. M. and Pandey, R. (1999) Atomistic modeling of native point defects in yttrium aluminum garnet crystal. *Journal of the American Ceramic Society*, 82 (10), 2881, John Wiley & Sons, Inc.)

| Parameters | Atomic simulation | Experiment results | Error % |
|---|---|---|---|
| Lattice parameter a (Å) | 11.988 | 12 | 0.1 |
| Elastic constant $C_{11}$ ($\times 10^{11}$dyne/cm$^2$) | 34.0 | 33.3 | 2.1 |
| Elastic constant $C_{12}$ ($\times 10^{11}$dyne/cm$^2$) | 12.7 | 11.3 | 12.4 |
| Elastic constant $C_{44}$ ($\times 10^{11}$dyne/cm$^2$) | 11.2 | 11.5 | 2.6 |

earlier study by Kuklja *et al.*[19] reveals that the formation of defect complexes, such as an interstitial and the corresponding vacancy, requires much more energy compared to that for antisite formation. Actually, cation antisites are identified as an intrinsic point defect even in stoichiometric YAG crystal.

### 3.6.3 Simulation Method and Results

Following the basic assumption of Section 3.4.1 the pairwise potentials between cations are neglected. The Buckingham potential of (3.2b) is used for short-distance interactions between $O^{2-}-O^{2-}$, $Al^{3+}-O^{2-}$, and $Y^{3+}-O^{2-}$, and Coulomb potential for long-distance interactions between ions i and j in YAG atomistic simulations. Summing these two together, we have (see (3.1))

$$V_{ij}(r_{ij}) = \frac{q_i q_j}{4\pi\varepsilon_0 r_{ij}} + A\exp(-r_{ij}/\rho) - Cr_{ij}^{-6} \qquad (3.6)$$

The Buckingham potential parameters are listed in Table 3.5.

The simulation starts by checking whether the potential constants are good by checking the obtained crystal lattice constant and the elastic constants for stoichiometric YAG. Table 3.6 shows the comparison between YAG atomistic simulation and experiments for the lattice constant, a, and elastic constants $C_{11}$, $C_{12}$ and $C_{14}$. The agreement between simulation results and experimental data is satisfactory.

The atomic simulation can be employed to calculate the energy for different kinds of defect mechanisms. Also, the lattice parameter and how it varies with different point defects can also be calculated.

Specifically, the work calculated the variation in the YAG lattice parameter with respect to deviations from nonstoichiometry.

### Application 2: Nonstoichiometry explored by the combined study of XRD and atomistic simulation*

Twelve possible defect mechanisms for $Y_2O_3$ excess and six for $Al_2O_3$ excess that can lead to nonstoichiometry in YAG were simulated using GULP.[15] It was found that forming YAl(a) antisite defects ($Y^{3+}$ takes $Al(a)^{3+}$ sites) requires the least energy for $Y_2O_3$ excess nonstoichiometric YAG, and forming AlY antisites requires the least energy for $Al_2O_3$ excess. The lowest energy for $Y_2O_3$ is 1.35 eV and for $Al_2O_3$ is 2.41 eV. The corresponding two mechanisms and reactions produce the simplest defect structure and the nonstoichiometry is compensated only by cation antisites. This finding is not surprising since cation antisites are also YAG's predominant intrinsic defect. Note that, however, the energy difference between the excess of $Y_2O_3$ and $Al_2O_3$ suggests that $Y_2O_3$ excess nonstoichiometry occurs more readily than $Al_2O_3$. In other words, the YAG phase field should extend further in the $Y_2O^3$ direction than $Al_2O_3$.

Furthermore, forming different types of defects has different effects on lattice parameters, which were also calculated as functions of nonstoichiometry. Calculated lattice parameters were then compared with those obtained from XRD experiments of nonstoichiometric YAG samples, as shown in Figure 3.6. It is confirmed that for $Y_2O_3$ excess, the nonstoichiometry proceeds exclusively via YAl antisites. For low $Al_2O_3$ concentrations (<0.2 mol %), the nonstoichiometry is compensated by AlY antisites. However, at higher $Al_2O_3$ concentrations (>0.2 mol %), the experimental data does not correlate with this mechanism. This could be caused by $Al_2O_3$ inclusions that leave $Y^{3+}$ rich YAG but cannot be observed by XRD.

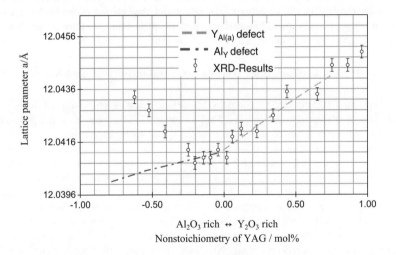

**Figure 3.6** The variation in YAG lattice parameters with respect to deviations from stoichiometry, determined by either atomistic simulation or XRD. The simulation results correspond to the nonstoichiometry caused by cation antisite defects (Reprinted from Patel, A. P., Levy, M. R., Grimes, R. W., *et al.* (2008) Mechanisms of nonstoichiometry in $Y_3Al_5O_{12}$ *Applied Physics Letters*, 93, 191902-1. Copyright © IEEE)

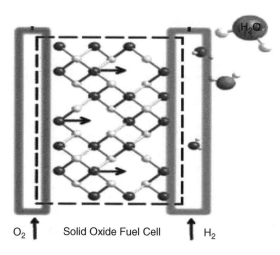

$O_2$ ↑     Solid Oxide Fuel Cell     ↑ $H_2$

**Figure 3.7**   Schematic of solid oxide fuel cell (SOFC)

## 3.7 Applications in Ceramics: Conductivity of the YSZ Oxide Fuel Electrolyte and Domain Switching of Ferroelectric Ceramics Using MD

### 3.7.1   MD Simulation of the Motion of Oxygen Ions in SOFC

MD simulation is a powerful tool when simulating dynamic processes. Application 3 is an MD simulation of the oxygen diffusion in Yttria stabilized zirconia (YSZ). YSZ is a typical material used as the electrolyte SOFC, which conducts oxygen ions from the cathode to the anode. Oxygen ions react with the fuel at the anode, causing the extra electrons to travel through the external circuit back to the cathode to produce electric power as shown in Figure 3.7. The performance and working temperature of SOFC largely depends on the oxygen conductivity of its electrolyte.

---

**Application 3: MD simulation for oxygen ion conductivity of an YSZ oxide fuel cell electrolyte [22]**

In this example, an MD simulation for a SOFC electrolyte made of YSZ, with 8% mole Yttrium is discussed. The simulation cell consisted of $5 \times 5 \times 5$ unit cells with periodic boundary conditions. This potential model uses formal charges for the ions, and short-range Buckingham potentials in addition to the Coulomb interactions. The simulation runs for 2 to 5 ns after the initial equilibration which lasts for 0.05 ns; with an npt ensemble (T = 2000 K). During the simulation, a superposition of projections along the y-axis of oxygen ion positions was recorded from 4000 configurations at intervals of 0.025 ps. As shown in Figure 3.8, the area ($5 \times 5$ unit cells) was divided into $100 \times 100$ pixels, with each pixel assigned to a gray scale based on the number of oxygen ions within that pixel. White and black represent, respectively, the extremes of high and low probability of finding an oxygen ion. Similar plots were obtained for projections along the x- and z-axes. The results indicate that the oxygen ion transportation occurs along [001], even though the tetragonal zirconia has been fully stabilized into a cubic structure. The activation energy and diffusion coefficient were also obtained from the simulation. Figure 3.9 shows the diffusion coefficient $D_0$ as a function of reciprocal temperature (1/T) for the following Arrhenius equation of the conductivity rate:

$$\dot{p} = dp/dt = D_0 \exp(-\Delta H/k_B T) \qquad (3.7)$$

Simulations show that at 1273 °C, a typical working temperature of SOFC, the diffusion coefficient $D_0$ reaches a maximum at around 8% doping concentration.

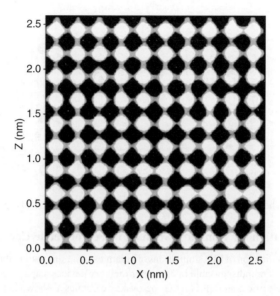

**Figure 3.8** Projection of oxygen ion positions along [010] direction from 4000 molecular dynamics configurations in YSZ at 2000 K. Gray levels represent the probability of finding an oxygen ion from 0 (black) to 1 (white) (Reproduced from Devanathan, R., Weber, W. J., Singhal, S. C., and Gale, J. D. (2006) Computer simulation of defects and oxygen transport in yttria-stabilized zirconia. *Solid State Ionics*, 177, 1251, Elsevier)

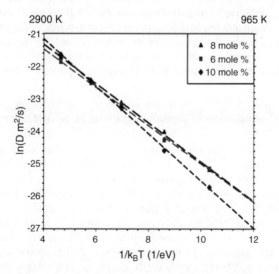

**Figure 3.9** Arrhenius plot of the diffusion coefficient as a function of reciprocal temperature (1/T) for 6 (■), 8(▲), and 10(♦) mol% YSZ. Linear fits to the data are also shown (Reproduced from Devanathan, R.,Weber,W. J., Singhal, S. C., and Gale, J. D. (2006) Computer simulation of defects and oxygen transport in yttria-stabilized zirconia. *Solid State Ionics*, 177, 1251, Elsevier)

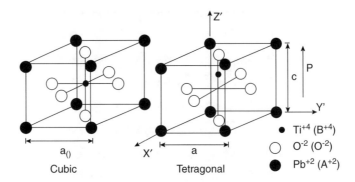

**Figure 3.10**   The cubic and tetragonal structure of an $ABO_3$ ($PbTiO_3$) perovskite oxide.

### 3.7.1.1  MD Simulation for Domain Switching of Ferroelectric Ceramics

Another interesting application of MD simulations of ceramics is the simulation of ferroelectric materials. Ferroelectric materials are indispensable in computer memory devices, capacitors, etc. One of the most commonly used ferroelectric materials is $ABO_3$ perovskite ceramics as shown in Figure 3.10, where A is Pb, B is Ti and the material is $PbTiO_3$. For these ceramics, when the temperature is below the relative Curie temperature, the crystal lattice structure transforms from a cubic lattice ($a = b = c$, $\theta_{ab} = \theta_{bc} = \theta_{ac} = 90°$) into a tetragonal structure ($c > a = b$, $\theta_{ab} = \theta_{bc} = \theta_{ac} = 90°$) with an elongated c axis. In this case, the center of the positive charges does not coincide with the center of the negative charges and therefore polarization occurs along the c axis.

In the crystal, each unit cell can have a polarization along any of the six directions ($\pm x$, $\pm y$, $\pm z$). A region containing unit cells polarized along the same direction is called a domain. The polarization direction of a domain can be switched by an external field such as an electric field or applied stress. This property creates a wide range of applications for these materials. High resolution experiments like neutron diffraction and multiscale simulations have been carried out to understand this phenomenon.[23] In MD simulations, an external field exerts additional forces on the ions, and the ion motion can be calculated accordingly. Therefore the simulation of dynamic behavior of the ions in an alternating electric field can be conducted to study the ferroelectric domain switching. Application 4 below is an example of such simulation.

### Application 4: MD simulation of domain switching in $KNbO_3$ ferroelectric material [24]

Domain switching is an important attribute of ferroelectric materials which has been extensively studied using various techniques. The example introduced here is an MD simulation of polarization reversal in a single domain of $KNbO_3$, a representative perovskite ferroelectric material. Buckingham potentials and shell model were used, which enabled the polarizations of ions to reproduce the tetragonal phase. The material was first equilibrated at 600 K to tetragonal phase with the polarization along the positive z-direction ($Pz > 0$). An external electric field of $5 \times 105$ V/cm was applied along the negative z direction to produce 180° domain switching, monitored by changes of lattice parameter and the angles between crystal lines. During this switching, it will pass the stage of 90° switching. Figure 3.11 shows the polarization reversal over the time range of $8 \sim 17$ ps. The lattice parameter c decreased at first, then returned to its normal value, which indicates that the switching process had occurred. During the switching, the reduction in the $\theta ac$ from 90° to 89.6° and the convergence of lattice parameters a and c indicated that the symmetry changed from a tetragonal structure to orthorhombic, and then returned to tetragonal.

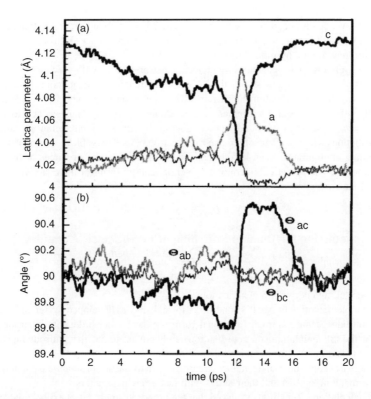

**Figure 3.11** Time evolution of (a) the three Cartesian lattice parameters and (b) the simulation-cell angles during the polarization reversal (Reproduced from Li, J. Y., Rogan, R. C., Ustundag, E., and Bhattacharya, K. (2005) Domain switching in polycrystalline ferroelectric ceramics. *Nature Materials*, 4, 776. Copyright Nature Publishing Group)

## 3.8 Tersoff and Brenner Potentials for Covalent Materials

In this section, a method is introduced mainly for covalent bonding of solid/complex molecular structures. To simulate such materials, three-body potential is needed to take account of the effects of angles between bonds on local electron distributions. That is to say, apart from the interatomic distance $r_{ij}$, the angle $\theta_{ijk}$ between vectors $\bar{r}_{ij}$ and $\bar{r}_{ik}$ is also considered. This is important because the angle has an influence on the molecular shape and the geometry of atomic bonding.

### 3.8.1  Introduction of the Abell-Tersoff Bonder-order Approach

The potential formalism introduced in this section was originally introduced by Abell.[25] This formalism models the local attractive electronic contribution to the binding energy $E_i^{el}$ of an atom i. The involved method is to use an interatomic bonder-order that modulates a two-center interaction as follows:

$$E_i^{el} \equiv - \sum_{j=1(j\neq i)} b_{ij} V_A(r_{ij}) \tag{3.8}$$

where the sum is over nearest neighbors j surrounding atom i, $b_{ij}$ is the bond-order function between atom i and atom j, and $V_A$ (r) is the pair term which represents bonding from valence electronics and is assumed to be valid for different atomic hybridizations.

All multi-body effects such as changes in the local electron density with varying topologies of local bonding are included in the bond-order function. Abell suggested that the major contribution to the bond-order function is the local coordination number z (or CN). This gives the number of the first nearest atoms which directly affect the local environment or electronic condition. Thus he developed a simple relationship $b = z^{-1/2}$ between the bonder-order function b with z. By balancing the attractive local bonding contributions with a pair sum of repulsive interactions, Abell was able to show that the wide range of stable bonding configurations can be rationalized by different ratios of slopes of the repulsive to attractive pair terms. The analytical interatomic potential energy form for the cohesive energy $E_{coh}$ of a collection of atoms becomes:

$$U_{coh} = \sum_i E_i \qquad (3.8a)$$

$$E_i \equiv \sum_{j=1(j>i)} V_{ij} = \sum_{j=1(j>i)} [Ae^{-\alpha r_{ij}} - b_{ij} Be^{-\beta r_{ij}}] \qquad (3.8b)$$

where $E_i$ is the binding energy of a generic atom i.

A practical implementation of Abell's bond-order formalism was developed by Tersoff in 1986[26] and 1989.[27] He introduced an empirical functional form for the bonder-order that incorporates angular interactions, while still maintaining the coordination number of an atom with its surrounding environment as the dominant feature. With his empirical modification of Abell's functional form, he stabilized the diamond lattice against shear, and obtained a reasonable fit to elastic constants and phonon frequencies for silicon, germanium, carbon and their alloys with just a few parameters.

Using QM calculations, Tersoff also showed that the proposed single set of exponential functions for two-center attractive and repulsive terms can provide a reasonable quantitative fit to bond lengths and energies for silicon. A unique feature of Tersoff's bond-order function is that it did not assume different forms for the angular terms for different hybridizations. Instead, it uses an angular function that is determined by a global fit to structures with various coordinations. This feature, together with a physically-motivated functional form that will be introduced below, provides the proposed function with an extraordinary degree of transferability. Here, the transferability means that the potential developed in one material structure can be transferred to be used in more structures, the higher the transferability, the wider usage of the potential function.

## 3.8.2   Tersoff and Brenner Potential

Following Tersoff's pioneering work in materials with covalent bonds (such as carbon, silicon, and germanium), in 1990 Brenner reported an empirical bond-order expression that is appropriate for describing both solid-state carbon and hydrocarbon molecules on an equal footing.[28] The form of its expression allows for bond formation and breaking with appropriate changes in atomic hybridization. Originally, Brenner's potential was developed to model the chemical vapor deposition of diamond films. It has been found later that the expression can be used in simulating a wide range of other chemical processes. These include collision and subsequent reaction of $C_{60}$ with diamond surface;[29] compression of $C_{60}$ between graphite sheets;[30] the pick-up of a module from a surface by another gas-phase molecule;[31] surface pattering via reactive force microscopy;[32] and compression, indentation, reaction, and friction at diamond surfaces.[31–34]

Brenner summarized his experience in developing a potential which does not only require science-based analysis but also a combination of trial-and-error and a certain degree of chemical insight. He reviewed the development process of a potential function for the "second generation" hydrocarbon potential energy.[35] He then thought that this process can be considered as an example of the "art" of effective empirical potential development.

In the following, the detailed expressions of both Tersoff and Brenner potential are given with the unified form:

$$V_{ij} = V_R(r_{ij}) - B_{ij}V_A(r_{ij}), \quad r_{ij} = |\overline{r_{ij}}| \tag{3.9a}$$

where $V_R$ and $V_A$ are pair potential terms that respectively represent the repulsive and attractive forces between atoms i and j. The bond order function $B_{ij}$ is introduced to represent multibody effects. It considers the spatial configuration of the bonds between atom i, j, and the spatial configuration of surrounding bonds.

The silicon potential function model below proposed by Tersoff is an example that considers the local environment:[36]

$$V_{ij} = f_c(r_{ij})\left(Ae^{-\lambda_1 r_{ij}} - B_{ij}e^{-\lambda_2 r_{ij}}\right) \tag{3.9b}$$

$$B_{ij} = \left(1 + \beta^n \varsigma_{ij}^n\right)^{-1/2n} \tag{3.9c}$$

$$\varsigma_{ij} = \sum_{k \neq i,j} f_c(r_{ik}) \, g(\theta_{ijk}) \, e^{\lambda_2^3 (r_{ij} - r_{ik})^3} \tag{3.9d}$$

where the angular effect is introduced by the function $\varsigma_{ij}$ in equations (3.9c) and (3.9d). From (3.9d) it is seen that $\varsigma_{ij}$ depends on the number of pair potential bonds represented by the summation symbol, bond strength $\lambda_2$, and the angle $\theta_{ijk}$ between bonds ij and ik (it is the $\theta$ shown in Figure 3.12).

Function $g(\theta_{ijk})$ is given by:

$$g(\theta) = 1 + {c^2}/{d^2} - {c^2}/{[d^2 + (h - \cos\theta)^2]} \tag{3.9e}$$

The parameters shown in the above equations such as $\beta$, n, $\lambda_1$, $\lambda_2$, c, d, h can be found in the related literature.[26] Variable $f_c$ ($r_{ij}$) is a cut-off radius function by which the cutoff radius $r_{cut}$ is expressed as follows:

$$r_{cut} = R^* f_c(r_{ij}) \tag{3.9f}$$

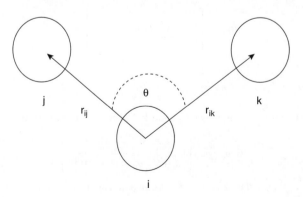

**Figure 3.12**    A schematic for angular effects formulated by three atoms i, j and k

The piecewise smooth function $f_c(r)$ is analytically given as follows:

$$f_c(r) = \frac{1}{2} \begin{cases} 2 & r < R - D_h \\ 1 - \sin\left(\pi(r - R)\big/ 2D_h\right) & R - D_h < r < R + D_h \\ 0 & r > R + D_h \end{cases} \qquad (3.9g)$$

where $D_h$ is called the layer thickness of the potential. It results in the smooth transition in the function value from $r = R - D_h$ (when $f_c(r) = 1$) to $r = R + D_h$ (when $f_c(r) = 0$).

If local bonding is neglected we have $g(\theta_{ijk}) = 0$ and $\varsigma_{ij} = 0$, then B is a constant. Furthermore, assume $B = 2A = \text{constant}$ and $\lambda_1 = 2\lambda_2$, then the potential (3.9b) returns to Morse potential (2.9a) in the range of $r < R - D_h$. In other words, certain properties of the material can be described through the dependence of function $B_{ij}$ on the local atomic environment.

The potential function proposed by Brenner and his co-workers[36] is an expansion of the Tersoff model. It is more accurate but time-consuming. Brenner potential has more detailed terms such as $V_A$, $V_R$ and $B_{ij}$ to take account of different chemical bonds in diamond, graphite phase of carbon, and carbon-hydrogen molecules. The Brenner potential function can be written as

$$V_R(r) = \frac{D^{(e)}}{S-1} e^{-\sqrt{2S}\beta(r - R^{(e)})} f_C(r) \qquad (3.10a)$$

$$V_A(r) = \frac{D^{(e)}S}{S-1} e^{-\sqrt{2/S}\beta(r - R^{(e)})} f_C(r) \qquad (3.10b)$$

where $D^{(e)} = 6.000\,ev$, $S = 1.22$, $B = 21\,nm^{-1}$, $R^{(e)} = 0.1390\,nm$, and the function $f_c$ is a piece-smooth cut-off radius function with the following form:

$$f_c(r) = \begin{cases} 1 & r < 0.17\,nm \\ \frac{1}{2}\{1 + \cos[\pi(r - 0.17)/0.03]\} & 0.17\,nm < r < 0.2\,nm, \\ 0 & r > 0.2\,nm \end{cases} \qquad (3.10c)$$

The multibody coupling factor $B_{ij}$ in (3.9a) is given by:

$$B_{ij} = \frac{1}{2}(B_{ij0} + B_{ji0}) \qquad (3.10d)$$

$$B_{ij_0} = \left[1 + \sum_{k(\neq i, j)} G(\theta_{ijk}) f_C(r_{ik})\right]^{-1/2} \qquad (3.10e)$$

where:

$$\theta_{ijk} = \cos^{-1}(r_{ij}^2 + r_{ik}^2 - r_{jk}^2)/2r_{ij}r_{ik} \qquad (3.10f)$$

It defines the angle between carbon bonds i-j and i-k. Function G is expressed as:

$$G(\theta) = a_0\left[1 + \frac{c_0^2}{d_0^2} - \frac{c_0^2}{d_0^2 + (1 + \cos\theta)^2}\right] \qquad (3.10g)$$

where $a_0 = 0.00020813$, $C_0 = 330$, and $d_0 = 3.5$. More completed parameters for carbon of the Brenner potential (1990) can be found in Appendix 3.A of Ref. [37]. Brenner also developed the second-generation potential for carbon atoms, and the corresponding $V_R$, $V_A$ and $B_{ij}$ can be found in Ref. [38]. This potential will be used in Section 6.10 for developing an atomistic-based continuum model of hydrogen storage with carbon nanotubes.

---

**Homework**

(3.4) Based on the information given in (A) below, prove the relationships given in Ref. [39]:

    (a) The equations (A.1), (A.2) and (A.3) listed in (B) are correct.

    (b) The equilibrium distance $r_0 = 0.145$ nm (hint: the angle is 120° and the potential energy is minimum when it is in equilibrium).

(A)

The Brenner potential (1990) takes the form $V = V_R(r_{ij}) - B_{ij} V_A(r_{ij})$, where $V_R(r) = (D^{(e)}/(S-1)) e^{-\sqrt{2S}\beta(r-R^{(e)})} \times f_c(r)$ and $V_A(r) = (D^{(e)}S/(S-1)) e^{-\sqrt{2/S}\beta(r-R^{(e)})} f_c(r)$ are the repulsive and attractive pair terms (depending only on $r$), $D^{(e)} = 6.00$ eV, $R^{(e)} = 0.1390$ nm, $S = 1.22$, $\beta = 21$ nm$^{-1}$, $f_c$ is the cutoff function, and the multi-body coupling term $B_{ij}$ is given by:

$$B_{ij} = \left[ 1 + \sum_{k(\neq i,j)} G(\theta_{ijk}) \right]^{-\delta}$$

Here

$$G(\theta) = a_0 \left[ 1 + \frac{c_0^2}{d_0^2} - \frac{c_0^2}{d_0^2 + (1+\cos\theta)^2} \right], \quad \delta = 0.5, \quad a_0 = 0.00020813, \quad c_0 = 330, \quad d_0 = 3.5$$

(B)

$$\left. \frac{\partial V}{\partial r_{ij}} \right|_{r_{ij}=r_0, \theta_{ijk}=120°} = 0$$

As

$$r_0 = R^{(e)} - \frac{1}{\beta} \frac{\sqrt{S/2}}{(S-1)} \ln B_0 \tag{A.1}$$

Where $B_0$ is the multi-body coupling term $B_{ij}$ evaluated at $\theta_{ijk} = 120°$ and $B_0 = 0.96$, the equilibrium bond length is $r_0 = 0.145$ nm. The other derivatives can be obtained analytically as

$$\left( \frac{\partial V}{\partial \cos\theta_{ijk}} \right)_0 = \frac{D^{(e)}S}{S-1} \delta a_0 c_0^2 \frac{1}{(d_0^2 + 1/4)^2} B_0^{(\delta+1/\delta)+(1/S-1)} \tag{A.2}$$

$$\left( \frac{\partial^2 V}{\partial r_{ij}^2} \right)_0 = 2D^{(e)} \beta^2 B_0^{(S/(S-1))} \tag{A.3}$$

---

## 3.9 The Atomistic Stress and Atomistic-based Stress Measure

It is interesting to calculate the stress field in an atomistic simulation to investigate the material property and defect effects in the light of continuum mechanical calculations. The commonly used definition of atomistic stress is the virial stress. This is the stress derived from a virial theorem of Clausius in thermodynamics which originally was used to calculate the system pressure.

### 3.9.1   The Virial Stress Measure

The virial stress provides important concepts to couple deformation behaviors of atomistic systems with continuum theories and concepts. For a generic atom $\alpha$ whose virial stress in a tensor index form is written as

$$\sigma_{kl}^\alpha = \frac{1}{\Omega^\alpha}\left[-m^\alpha(\bar{v}^\alpha)_l(\bar{v}^\alpha)_k + \frac{1}{2}\sum_{\beta(\neq\alpha)}\left(\bar{f}^{\alpha\beta}\right)_k\left(\bar{r}^{\alpha\beta}\right)_l\right] \tag{3.10h}$$

where $\Omega^\alpha$ is the volume of an atom. The summation variable is $\beta$ which covers all atoms in the neighborhood of atom $\alpha$. These neighbors apply forces on atom $\alpha$, thus their contributions are summed. The index k, l are various from 1 to 3 and denote the coordinate components of vector and tensor.

Frequently, in conceptual development, the second-order stress tensor, $\underline{\sigma}^\alpha$ is written in an abstract form as

$$\underline{\sigma}^\alpha = \frac{1}{\Omega^\alpha}\left[-m^\alpha\bar{v}^\alpha\bar{v}^\alpha + \frac{1}{2}\sum_{\beta(\neq\alpha)}\left(\bar{f}^{\alpha\beta}\right)\left(\bar{r}^{\alpha\beta}\right)\right] \tag{3.10i}$$

In this book, a line under a letter denotes a second-order tensor which consists of two vectors. In continuum mechanics, the first is the normal vector of the plane which subjects the force, the second is the force vector which applies on the plane. In the first term on the right side in (3.10i) the two vectors are velocity which can have different components; in the second term, the two vectors are force vector $\bar{f}^{\alpha\beta}$ and position vector $\bar{r}^{\alpha\beta}$. Their components along the $k$ and $l$ axis consist of the stress tensor component $\sigma_{kl}^\alpha$ shown in (3.10h).

The first term in (3.10i) is a kinetic part. It was recently shown that the stress including the kinetic contribution is not equivalent to the mechanical Cauchy stress[39] and for low temperature the contribution from the first part may not be significant; for simplicity in the following we mainly consider the second part so that the atomistic stress can be written as

$$\sigma_{kl}^\alpha = \frac{1}{2\Omega^\alpha}\sum_{\beta(\neq\alpha)}\left(\bar{f}^{\alpha\beta}\right)_k\left(\bar{r}^{\alpha\beta}\right)_l \tag{3.10j}$$

The force $\bar{f}^{\alpha\beta}$ is parallel to $\bar{r}^{\alpha\beta}$ and can be expressed through the total potential emergy $\phi$ of the atomistic system as follows:

$$\bar{f}^{\alpha\beta} = -\frac{\partial\phi}{\partial r^{\alpha\beta}}\frac{\bar{r}^{\alpha\beta}}{r^{\alpha\beta}} \tag{3.10k}$$

where $\frac{\partial\phi}{\partial r^{\alpha\beta}}$ gives the force value on atom $\alpha$; the remainder of the right side gives the force direction with

$$\bar{r}^{\alpha\beta} = \bar{r}_\alpha - \bar{r}_\beta \tag{3.10l}$$

We then have

$$\sigma_{kl}^\alpha = \frac{-1}{2\Omega^\alpha}\sum_{\beta(\neq\alpha)}\frac{1}{r^{\alpha\beta}}\left(\frac{\partial\phi}{\partial r^{\alpha\beta}}\right)(\bar{r}^{\alpha\beta})_k(\bar{r}^{\alpha\beta})_l \tag{3.10m}$$

### 3.9.2   The Computation Form for the Virial Stress

To get more understanding about the atomistic stress concept, we list how the atomistic code LAMMPS calculates the stress components. (LAMMPS stands for large-scale atomic/molecular massively parallel simulator, developed at Sandia National Laboratories in the USA, see Section 10.13.) Here, we only list its force part (not its kinetic part). To make the notation simple, the parenthesis before the subscript component k, l will be dropped. For instance the $x_l$ component $(\bar{r}_1)_l$ of the vector $\bar{r}_1$ is simply written as $\bar{r}_{1l}$, thus the six symmetric components of stress tensor for any generic atom $\alpha$ can be expressed as follows:[40]

$$
\sigma_{kl}^{\alpha} = -\left[ mv_k v_l + \frac{1}{2}\sum_{n=1}^{Np}(\bar{r}_{1k}\bar{f}_{1l}+\bar{r}_{2k}\bar{f}_{2l}) + \frac{1}{2}\sum_{n=1}^{Nb}(\bar{r}_{1k}\bar{f}_{1l}+\bar{r}_{2k}\bar{f}_{2l}) + \frac{1}{3}\sum_{n=1}^{Na}(\bar{r}_{1k}\bar{f}_{1l}+\bar{r}_{2k}\bar{f}_{2l}+\bar{r}_{3k}\bar{f}_{3l}) \right.
$$

$$
+ \frac{1}{4}\sum_{n=1}^{Nd}(\bar{r}_{1k}\bar{f}_{1l}+\bar{r}_{2k}\bar{f}_{2l}+\bar{r}_{3k}\bar{f}_{3l}+\bar{r}_{4k}\bar{f}_{4l}) + \frac{1}{4}\sum_{n=1}^{Ni}(\bar{r}_{1k}\bar{f}_{1l}+\bar{r}_{2k}\bar{f}_{2l}+\bar{r}_{3k}\bar{f}_{3l}+\bar{r}_{4k}\bar{f}_{4l})
$$

$$
\left. + \sum_{n=1}^{Nf}\bar{r}_{ik}\bar{F}_{il} \right] \qquad (kl = 11, 22, 33, 12, 13, 23)
$$

$$(3.10n)$$

In the right part of the equation, the subscript "1" denotes quantities belonging to the atom $\alpha$, and other subscripts such as 2, 3 and 4 denote atoms which interact with atom $\alpha$ through interatomic potentials. Each summation symbol denotes a computation loop to make the computation of the corresponding parenthesis conducted one-by-one for all neighbors of atom $\alpha$. In other words, unlike a general mathematical summation symbol, the summation range for n is not explicitly written inside the parenthesis, but implicitly indicates that the summation covers all the pairs with atom $\alpha$. The atoms involved in this summation actually include $(N_p + 1)$ atoms including atom $\alpha$ itself.

Taking the first right term as an example, this term is a pairwise potential energy contribution. Its summation symbol has the same meaning as that in (3.10i), it requires calculation loops with variable n starting from 1 to $N_p$ ($n = 1, 2, \ldots, Np$) to cover all $N_p$ neighbors. $\bar{r}_1$ and $\bar{r}_2$ are the position vectors of the two pair atoms and $\bar{f}_1$ and $\bar{f}_2$ are their corresponding forces which apply, respectively, at atom 1 and 2. The subscript 2 in $\bar{r}_{2k}$ and $\bar{f}_{2l}$ should indicate all position vectors and force vectors in the neighbourhood of atom $\alpha$. This means that the total number of the parenthesis in the first term should be $N_p$ to cover all its neighboring atoms. During the loop $\bar{r}_1$ is stationary but $\bar{f}_1$ is various since the pair atom will change from atom 2 to atom $N_{p+1}$, giving different force on atom 1 (or atom $\alpha$).

One can easily prove that not only the summation symbol of the first term but also the content in the parenthesis is the same as those in (3.10i). In fact, the content in the parenthesis of (3.10i) can be written using (3.10l) as:

$$
\bar{f}^{\alpha\beta}(\bar{r}^{\alpha\beta}) = \bar{f}^{\alpha\beta}(\bar{r}^{\alpha}-\bar{r}^{\beta}) = \bar{f}^{\alpha\beta}\bar{r}^{\alpha}-\bar{f}^{\alpha\beta}\bar{r}^{\beta} = \bar{f}^{\alpha\beta}\bar{r}^{\alpha}+\bar{f}^{\beta\alpha}\bar{r}^{\beta} \qquad (3.10o)
$$

The last equation is obtained because the force, $\bar{f}^{\beta\alpha}$ applied by atom $\alpha$ on atom $\beta$ equals in value but is opposite in direction to the force $\bar{f}^{\alpha\beta}$, i.e., $\bar{f}^{\beta\alpha} = -\bar{f}^{\alpha\beta}$, the latter is the force applied by atom $\beta$ on atom $\alpha$. If we replace $\alpha$ by 1, $\beta$ by 2 in (3.10o) and then write it in a component form, it is exactly the same as the first right term in (3.10n) as:

$$
\frac{1}{2\Omega^{\alpha}}\sum_{\beta(\neq\alpha)}(\bar{f}^{\alpha\beta})_l(\bar{r}^{\alpha\beta})_k = \frac{1}{2\Omega^{\alpha}}\sum_{n=1}^{NP}(\bar{f}^{\alpha\beta})_l(\bar{r}^{\alpha})_k + (\bar{f}^{\beta\alpha})_l(\bar{r}^{\beta})_k =
$$

$$
\frac{1}{2\Omega^{\alpha}}\sum_{n=1}^{NP}(\bar{f}_{1l}\bar{r}_{1k}+\bar{f}_{2l}\bar{r}_{2k}) = \frac{1}{2\Omega^{\alpha}}\sum_{n=1}^{NP}(\bar{r}_{1k}\bar{f}_{1l}+\bar{r}_{2k}\bar{f}_{2l})
$$

$$(3.10p)$$

Similar to the pair potential terms, the right third term in (3.10n) is a bond contribution for the $N_b$ bonds of which atom $\alpha$ is a part. The fourth term is a Tersoff-type potential for angle effects in which each bond involves three atoms. The fifth term relates to the dihedral interaction of two planes and the sixth term belongs to the improper interactions, both of which involve four atoms in the bond. The last term relates to fixes that apply material constraint forces to atom $\alpha$. From the coefficient values of the fourth to sixth terms it is seen that each atom in the set is assumed to have an equal contribution to virial stress. This equal contribution is measured by its average value realized by these coefficients. Specifically, in (3.10n) coefficient 1/4 is used for a dihedral interaction in which four atoms are involved, and coefficient 1/3 for angle effects in which three atoms in the Tersoff 3-body interaction are involved.

### 3.9.3   The Atomistic-based Stress Measure for Continuum

In the stress expression, the force term expressed in (3.10m) needs to be divided by $\Omega^{\alpha}$, which is the volume per atom, to have units of stress (or pressure). However, an individual atom's volume is not easy to compute in a deformed solid. The manual of LAMMPS mentions a way to use the pressure concept to choose volume V of the body to calculate the average atomistic stress.

In fact, per-atom stress is the negative of the per-atom pressure tensor, thus, if the diagonal components of the per-atom stress tensor are summed for all atoms in the system and the sum is divided by 3V, the result should be p, where p is the total pressure of the system. In practice, people use the volume occupied by the atoms as the volume to divide the sum of all atom stress shown in (3.10n) in that domain to get the average stress, which can be considered an atomistic-based stress measure at the continuum scale.

It is important to note that the per-atom stress does not include contributions due to long-range Coulombic interactions. So far it is not clear how this contribution can be easily computed.

---

**Homework**

(3.5)  Suppose the spatial distribution of hydrogen molecules is Close-Packed Hexagonal (AB stacking, see Figure 3.13), and the average spacing between two closest hydrogen molecules is 0.28 nm. Not considering the kinetic part of the virial stress, and only taking into account the van der Waals interactions by LJ potential between the closest molecules, calculate the average inner pressure in the hydrogen molecules with LJ parameters as $\varepsilon_{H_2-H_2} = 3.11 \times 10^{-3}$ eV and $\sigma_{H_2-H_2} = 0.296$ nm.

Hint: (1) $p = -1/3(\sigma_{xx} + \sigma_{yy} + \sigma_{zz})$; (2) For Close-Packed Hexagonal (ABAB stacking), each molecule has 12 closest neighboring molecules. Suppose the coordinate of the central molecule is (0,0,0) and the closest distance is $d_{H_2-H_2}$. The vectors to the closest neighboring molecules and the average volume per each atom can be determined. (3) The average volume per an atom can be obtained as $\Omega = \frac{\sqrt{2}}{2} d^3_{H-H}$.

---

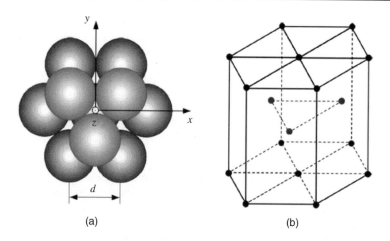

**Figure 3.13**   The Close-Packed Hexagonal (ABAB stacking) for hydrogen molecules

## Part 3.2 Applications in Metallic Materials and Alloys

### 3.10 Metallic Potentials and Atomistic Simulation for Metals

Needless to say, metallic materials and their alloys are still dominant in different engineering structures and our daily life. Frames of airplanes and shuttles are made of light-weight materials such as aluminum alloys. Buildings and ships are primarily made of low carbon steels, while cutting tools and rails are made of high carbon steels. New metal alloys are being developed for ultra-high strength steel to sustain high-explosive and other types of intensive loadings. Nickel based super-alloys and oxide dispersion strengthened (ODS) steel is being developed to sustain extremely high temperatures for jet engines and other components. Lightweight metal-magnesium alloys are being developed for transportation vehicles by solving core issues such as corrosion. To meet these challenges, it is important that mechanisms of diffusion, hardening, anti-corrosion, as well as formation and evolution of dislocations and defects at the nanoscale are investigated by atomistic simulation. Some basic methods of metallic atomistic simulation will be discussed in this and next sections.

The structures of pure metals are simpler than ceramics. They usually have body-centered cubic (BCC), face-centered cubic (FCC) or hexagonal close-packed (HCP) crystal structures. These popular structures are, respectively, shown in Figure 3.14a–c.

On the right of the Figure is shown how the atom number per unit cell of 2, 4, and 2, respectively, is calculated for BCC, FCC, and HCP crystals. The key to understand these calculations is that atoms on the surfaces and corners of the primary cell are shared with their neighboring cells, thus only part of these atoms belongs to the primary unit cell.

Similar to the simulations of ceramics, the periodic boundary condition is readily satisfied. Metallic elements usually have a low electronegativity. They give up their valence electrons to form a "cloud" (or "sea") of electrons surrounding the atom nuclei. The valence electrons move freely within the electron cloud and become associated with several atom nuclei. The positively charged nuclei are held together by mutual attraction to the electrons. This attraction forms a strong metallic bond. The metallic bonding mechanism is quite different from both ionic bond and covalent bond, thus a different type of atomistic potential called Embedded Atom Method (EAM) has been developed.

Before EAM was proposed in 1983, all the studies of structures and basic properties of metallic systems adopted exclusively the pair potentials. In the formulation, the cohesive energy of metal consists of two parts: one is the summation of all the pairwise bonding energy, and the second is the energy which depends on the volume of metal after deformation. Compared to pair potentials, the latter part is the energy necessary to keep the metal in equilibrium, and corresponds to the elasticity energy, which is the energy based on the density of the background electron cloud. In other words, the need to have the additional potential part indicates that pair potentials cannot consider multi-atom effects resulting from the background electron cloud of metal bonding.

In the early 1980, several similar methods to overcome the shortcomings of the pairwise model were proposed. They include EAM,[41, 42] adhesion model,[43] Finnis-Sinclair model,[44] effective medium theory,[45] and the modified effective medium theory.[46] Interestingly, all four models have almost the same mathematical form, which has an energy potential $F_i$ for a generic atom i. $F_i$ is a function of a local parameter $\rho_i$ determined by contributions of atoms within the neighborhood of atom i. The difference between the models depends on the physical explanation of the function $F_i(\rho_i)$, the parameter $\rho_i(r)$ and the pair potential $V(r)$. They are often referred to collectively as "glue model" potentials.

### 3.11 Embedded Atom Methods EAM and MEAM

#### 3.11.1 Basic EAM Formulation

EAM assumes that $F_i$ is the embedded energy, which is the required energy to embed the atom i into the background of a local electron cloud at the atom position i. Mathematically, the EAM formulation can be given as follows:

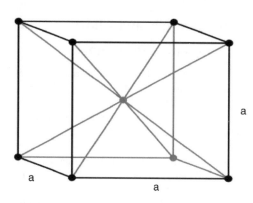

(a) Body-centered cubic (BCC),

Unit cell: $1 + 8 \times \dfrac{1}{8} = 2$

Examples: $\alpha - Fe$ V, Cr, Mo and Mn

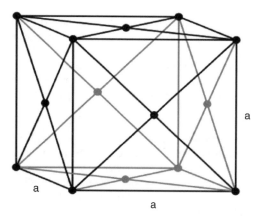

(b) Face-centered cubic (FCC)

Unit cell: $6 \times \dfrac{1}{2} + 8 \times \dfrac{1}{8} = 4$

Examples: $\gamma - Fe$, Al, Ni, Cu, Ag, Pt and Au

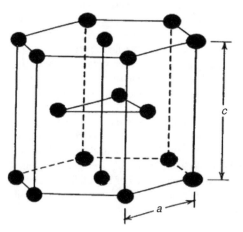

(c) Hexagonal close-packed (HCP). Its unit cell is only 1/3 of the HCP cell and the atom number per unit cell is:

$$1 + 4 \times \frac{1}{6} + 4 \times \frac{1}{12} = 2$$

$$C = 1.633a$$

Examples: Be, Mg, Zn, Zr and $\alpha - Ti$

**Figure 3.14**    BCC, FCC and HCP crystal structures

$$V_{(i)} = F_i(\rho_i) + \text{pair terms} \tag{3.11a}$$

where $\rho_i$ is the electronic density at the atom position i. This density is assumed to be the sum of the electronic density $\rho_j(r_{ij})$ as follows:

$$\rho_i = \sum_{j=1(\neq i)} \rho_j(r_{ij}) \tag{3.11b}$$

where $\rho_j(r_{ij})$ is the electron density distribution of atom j at the position of atom i, before atom i is embedded into the electron cloud.

Substituting (3.11b) into (3.11a) and using (2.2a) to express a pair potential, the potential energy $V_i$ of atom i and the total potential energy $U^{tot}$ of the atomistic system can be expressed as follows:

$$V_i = F_i(\rho_i) + \frac{1}{2}\sum_{j=1(\neq i)} V_{ij}(r_{ij}) = F_i\left(\sum_{j=1(\neq i)} \rho_j(r_{ij})\right) + \frac{1}{2}\sum_{j=1(\neq i)} V_{ij}(r_{ij}) \tag{3.11c}$$

$$U^{tot} = \sum_{i=1}^{N}\left[F_i(\rho_i) + \frac{1}{2}\sum_{j=1(\neq i)} V_{ij}(r_{ij})\right] = \sum_{i=1}^{N}\left[F_i\left(\sum_{j=1(\neq i)} \rho_j(r_{ij})\right) + \frac{1}{2}\sum_{j=1(\neq i)} V_{ij}(r_{ij})\right] \tag{3.11d}$$

In some literatures, (3.9d) is written in a more concise form:

$$U^{tot} = \sum_{i} F(\rho_i) + \frac{1}{2}\sum_{i,j} V_{ij}(r_{ij}) \tag{3.11e}$$

From the summation operation in (3.11b), it can be seen that the electron cloud density $\rho_i$ is approximated by the summation of all the contributions of neighbor atoms. Specifically, atom j's contribution to the cloud at atom i is a function of the distance $r_{ij}$ between atoms j and i.

From (3.11d) we know that to obtain the total energy of the system, we need to determine the pair potential $V_{ij}$, the embedded function $F_i$, and the electron cloud density function $\rho_i$. These functions are not calculated from QM such as the HF method or the electron density function theory (DFT), but fitted from experimental data and therefore remain in the category of empirical potential. Fitting these functions usually requires the following experimental data: lattice parameter, bonding energy, vacancy formation energy, and an elastic constant. EAM is reliable for FCC metals including copper, gold, silver, nickel, palladium, and platinum; but is less reliable for materials in which the direction of atomic bond is important.

## 3.11.2   EAM Physical Background

As discussed in Section 3.10, if only the pair potential is used in the potential function, the body cannot keep its elastic property. It will produce a Cauchy relationship $C_{12} = C_{44}$ that contradicts the real solid's property. Thus the necessary additional potential term must be a volume-dependent term or $F(\rho_i)$ in the EAM modeling, which keeps the simulated system as a solid. In other words, this term offers the cohesive energy to keep the metal atoms bonded together.

The simple way to determine approximately the cohesive energy is not to pull out an atom from the atom system but to put an atom into the atomistic system. Because metal atoms are surrounded by electron background clouds, it is natural to consider the additional energy potential as the embedded energy of the atom in the electron cloud.

A similar idea was proposed by the jellium theory of the effective-medium method, in which the embedded atom is just like jellium. By embedding the hydrogen atom into the effective medium, consisting of the electron cloud, its chemical adhesion heat was successfully calculated.[46] In addition, the quasi-atom method[47] also considers a quasi-atom embedded into a homogeneous or near-homogeneous electron cloud, therefore producing a function similar to the framework of the EAM.

The embedded energy term is a functional which is independent of the actual substrate lattice. Its accurate form is not clear and could be very complex. A simple approximation is to assume that the embedded atom energy only depends on the density of the immediate neighboring electron cloud of the embedded atom.

This assumption is equivalent to the assumption that the embedded atom is located in an electron cloud with locally uniform density. It can further be simplified through the assumption that the electron cloud density at the location of the embedded atom can be approximated by the summation of the electron cloud densities of its neighborhood atoms, so that the embedded energy becomes a simple function of the atom's position as shown in (3.11b).

### 3.11.3   EAM Application for Hydrogen Embrittlement

In their early paper, Daw and Baskes studied hydrogen embrittlement on a nickel (Ni) substrate. The potential function parameters were determined through experimental data, shown in Table 3.7.[41] They include lattice parameter $a_0$, elastic constant $C_{11}$, $C_{12}$, $C_{44}$, sublimation energy $E_s$, vacancy formation energy $E_{iV}^F$, the energy difference of BCC and FCC phase for Ni, and the hydrogen heat of solution and migration energy in Ni.

The function $F(\rho)$ of hydrogen is derived from calculations of electron cloud density; while the function $V(r)$ of H-Ni is determined by the melting heat and migration energy of hydrogen in nickel, assuming that $V_{H-Ni}$ is the geometric mean of $V_{H-H}$ and $V_{Ni-Ni}$ (see next section). Using (3.11d), a model of half infinite nickel plate was developed and simulated. The plate has 17 layers in the direction of [111] with the two ends in this direction free to move. The plate has periodic boundary conditions in the directions of $[1\,\bar{1}\,0]$ and $[1\,1\,\bar{2}]$. Figure 3.15 shows the cross-section of the plate. The hollow circles represent the nickel atoms.

As shown in Figure 3.15(a), four nickel atoms were removed from the central plane to simulate a small crack. External force was imposed on the surface atoms along the [111] direction. As we know, the slip plane is along the [111] plane for FCC crystal so that this design with the stress perpendicular to the slip and direction makes the fracture brittle. When the stress reached $0.13\ eV/(\mathring{A})^3$, the plate with the four vacancies was split into two parts from the central plane, with the four vacancies as the start points of the failure.

If the stress is $0.11\ eV/(\mathring{A})^3$ or less, as shown in Figure 3.15a, it does not cause a fracture, but results in an expansion of the lattice and a bending at the vacancies. In Figure 3.15b, hydrogen atoms represented by

**Table 3.7**   Quantities used for determination of the potential functions and their fitted values (Reprinted with permission from Daw, M. S. and Baskes, M. I. (1983) Semiempirical, quantum mechanical calculation of hydrogen embrittlement in metals. *Physical Review Letters*, 50, 1285. Copyright 1983 by the American Physical Society)

|                                       | Experiment | Fit   |
| ------------------------------------- | ---------- | ----- |
| $a_0(\mathring{A})$                   | 3.52       | 3.52  |
| $C_{11}$ ($10^{12}$ dynes/cm$^2$)     | 2.465      | 2.452 |
| $C_{12}$ ($10^{12}$ dynes/cm$^2$)     | 1.473      | 1.452 |
| $C_{44}$ ($10^{12}$ dynes/cm$^2$)     | 1.247      | 1.233 |
| $E_s$ (eV)                            | 4.45       | 4.45  |
| $E_{1V}$ (eV)                         | 1.4        | 1.43  |
| $(E_{bcc}-E_{fcc})$ (eV)              | 0.06       | 0.14  |
| H heat of solution (eV)               | 0.16       | 0.22  |
| H migration energy (eV)               | 0.41       | 0.41  |

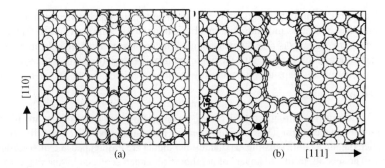

(a)                                                          (b)    [111] ⟶

**Figure 3.15**   Cross-sectional views of a nickel substrate (Reprinted with permission from Daw, M. S. and Baskes, M. I. (1984) Embedded-atom method: Derivation and application to impurities, surfaces, and other defects in materials. *Physical Review B*, 29, 6443. Copyright 1984 by the American Physical Society)

solid dots are close to the vacancies near the central plane. Calculation results show an interesting phenomenon: If hydrogen appears, a stress of $0.11\,eV/(\mathring{A})^3$ is large enough to cause the fracture.

As shown above, hydrogen atoms reduce the fracture stress of the nickel plate. One hydrogen atom in the unit reduces the fracture stress by 15%. When the density of hydrogen atoms increases, the fracture stress drops further. Therefore, hydrogen atoms reduce the material's resistance to fracture and cause hydrogen embrittlement.

## 3.11.4   Modified Embedded Atom Method (MEAM)

EAM has the physical background to consider multi-body interaction through the electronic density produced by multiple metal atoms. The assumption, however, can be considered to take only low-order terms in the gradient field of the electron cloud density. Thus, it cannot account for the effects of 3D atom distribution around the embedded atom on the potential, so a modified embedded atom method (MEAM) was proposed.[48]

### 3.11.4.1  Correction Terms of Electronic Densities

In MEAM, the EAM linear superposition of spherically averaged atomistic electron density shown in (3.11b) is adopted as the dominant contribution to a series of partial densities of background electrons. This dominant contribution was denoted by a superscript symbol 0 as

$$\rho_i^{(0)} = \sum_{j=1(j\neq i)} \rho_j^{(0)}(r_{ij}) \tag{3.12a}$$

Further, the expression is generated by defining a series of correction electron densities of $\rho_i^{(1)}, \rho_i^{(2)}, \rho_i^{(3)}$. These correction densities explicitly depend upon the relative positions of neighbors of atom i through the distance $r_{ij}$, and the direction cosine of vector $\bar{r}_{ij}$ related to the x, y, and z coordinates. Thus through these correction terms the 3D atom arrangement on the electron density at the position of atom i can be accounted for. For instance, it is shown[48] that the squares of $(\rho_i^{(1)})^2$ and $(\rho_i^{(2)})^2$ are equivalent, respectively, to a three-body cos and $\cos^2$angle dependence. For example,

$$\left(\rho_i^{(1)}\right)^2 = \sum_{j,k(\neq i)} \rho_j^{(1)}(r_{ij})\rho_k^{(1)}(r_{ik})\cos(\theta_{jik}) \tag{3.12b}$$

where j and k are atoms within the neighborhood of atom i and the angle between ij and ik is $\theta_{jik}$. From this expression, it is seen that not only the distance effects of atom j and k are accounted for but their angle $\theta_{jik}$ takes an important role. If $\theta_{jik} = 0°$, there is an additional contribution to the electron density at the location of atom i. However, if $\theta_{jik} = 90°$, the angle contribution is zero.

### 3.11.4.2 Screen Function

A "screening" function was introduced to decrease the atomic interaction of two atoms if a third atom comes in the space between them. This screening function introduced a peculiar property. When the atom is inside a lattice (i.e., in a bulk), only the first nearest neighbors are unscreened and interact directly with the host atom. The second, third, etc. layers of neighbors remained screened by the first layer of atoms, and as a result, the interaction has a very short range for a bulk atom. When some of the neighbors happen to be missing, though (for example, when the atom is at or near a free surface, or at a crack tip, or has a vacancy near by), the screening decreases and the interaction becomes of a longer range involving farthest neighbors. In this way, the atomic force becomes sensitive to the surrounding neighborhood, which allows for better fit of the material properties, particularly surface effects and defect structures.

### 3.11.4.3 Expression for the Square of Total Electron Density

Assume that the square $(\rho_i)^2$ of the total background electron density can be expressed by a weighted sum of the squares of the partial background densities, $\rho_i^{(l)}$ as:

$$(\rho_i)^2 = \sum_{i=0}^{3} t_i^{(l)} (\rho_i^{(l)})^2 \tag{3.12c}$$

This expression can be considered as a second-order Taylor expansion of the background density away from the linear superposition listed in (3.11b).

### 3.11.4.4 New Point of View to Calculate Pair Potential

To get the pair potential $V_{ij}$ in (3.11d), from a new point of view, the reference structure of atom i is emphasized. The reference structure is a crystal structure which includes detailed information about the position and behavior of atom i. It is usually the equilibrium crystal structure of type i atoms. In the analysis, the neighbor atoms j of atom i are limited only to the first neighbors. Denote the nearest neighbor distance and the nearest atom number from atom i, respectively, as R and $Z_i$. For example, for FCC $Z = 12$ and for BCC $Z_i = 8$. Atoms in this reference structure are denoted as i-type atoms.

For this type of atom, its pair potential $V_{ii}$ with any nearest neighbor is the same. In view of this situation, the atoms in the pair can all use the same subscript index "i". As a result, the summation of pair potentials for atom i in the reference structure should be a product of Z and $V_{ii}$, thus the energy $E_i^u(R)$ per atom of the reference structure as a function of nearest neighbor distance can be expressed as a special case of $V_i$ of atom i in (3.11c), reading

$$E_i^u(R) = F_i(\rho_i^0(R)/Z_i) + \frac{Z_i}{2} V_{ii}(R) \tag{3.12d}$$

where $\rho_i^0(R)$ is the background basic electron density for the reference structure of atom i and is normalized by $Z_i$. From this equation, the pair interaction for type-i atoms can be determined as follows:

$$V_{ii}(R) = \frac{2}{Z_i} \left\{ E_i^u(R) - F_i(\rho_i^0(R)/Z_i) \right\} \tag{3.12e}$$

Here, the key for determining the pair potential under the electronic background is to determine $E_i^u(R)$ for the representative atom i in the reference structure. One way to determine $E_i^u(R)$ is by the following universal energy function proposed by Rose et al.[49] as follows:

$$E_i^u(R) = -E_i^0(1 + \alpha^*)e^{-\alpha^*} \tag{3.12f}$$

$$\alpha^* = \alpha_i(R/R_i^0 - 1) \tag{3.12g}$$

where $E_i^0$ is the sublimation energy and $R_i^0$ is the equilibrium nearest-neighbor distance. In addition, $\alpha_i$ is related to the bulk modulus $B_i$, the atomic volume $\Omega_i$ of the solid elements and the diatomic force constant $K_i$ of the gaseous elements as follows:

$$\alpha_i = \sqrt{9B_i\Omega_i/E_i^0} \tag{3.12h}$$

or

$$\alpha_i = \sqrt{K_iE_i^0}R_i^0 \tag{3.12i}$$

After parameters $R_i^0, E_i^0, \alpha_i$ determined, $E_i^u(R)$ as well as pair potentials can be determined. Alternatively, there are other ways to determine $E_i^u(R)$; for instance, one could use first-principle calculations to determine the reference state equation or use high-pressure experimental data.

### 3.11.4.5 Twelve Material Parameters

When MEAM is applied to FCC or BCC, there are twelve parameters (see Table 3.8) that are used to describe embedded potential function $F_i$ and electron density $\rho_i$. Specifically, eight parameters (i.e., $t_i^{(1)}$, $\beta_i^{(1)}, 1 = 0, 1, 2, 3$) are related to electron density $\rho_i$, where $t_i^{(l)}$ is the weighting function with $t_i^{(0)} = 1$ and $\beta_i^{(1)}$ is the coefficient that makes the electron density decay exponentially with the interatomic distance $r_{ij}$.

The first three of the other four constants are linked to three universal energy functions via (3.12f) through (3.12g). They are sublimation energy $E_i^0$ (eV), the nearest neighbor distance $R_i^0$ (Å) at equilibrium, and the exponential decay factor $\alpha_i$ in the universal energy function.[49] The twelfth parameter represents the coefficient $A_i$ in the embedded energy function as:

$$F_i(\rho_i) = A_iE_i^0\rho_i \ln \rho_i \tag{3.12j}$$

Substituting the total electron density $\rho_i$ of (3.12c) into (3.12j) gives the embedded energy function $F_i(\rho_i)$ at the position of atom i. This function can be used in (3.11e) so that the whole atomistic system potential energy can be expressed and the solution can be obtained.

MEAM applications include silicon-nickel alloy and its surface analysis,[50] tensile fracture e of Al-Si interface,[51] defect energy,[52] and new EAM potential for 17 types of HCP metals.[53, 54]

## 3.11.5   Summary and Discussions

### 3.11.5.1 EAM

Interatomic potentials constructed by the EAM were first developed by Daw and Baskes in 1983.[41] In the EAM, the atoms are viewed as being embedded in the electron density clouds of their neighbors. The energy of an atom is defined as a sum of a pair energy term, and a term, $F_i(\rho_{i,tot})$ that depends on the total electron density, $\rho_{i,tot} = \sum_{j-atom\ neighbor} \rho_j(r_{ij})$. The conventional EAM potential assumes that the electron density function, $\rho_j(r_{ij})$ is centrosymmetric (i.e., not angle dependent), and does not depend on the state of the atom (e.g., if the atom forms a bond or not). Even in this simplest form, the EAM potential

**Table 3.8** Twelve parameters in MEAM

| | $E_i^0$ | $R_i^0$ | $\alpha i$ | $A_i$ | $\beta_i^{(0)}$ | $\beta_i^{(1)}$ | $\beta_i^{(2)}$ | $\beta_i^{(3)}$ | $t_i^{(0)}$ | $t_i^{(1)}$ | $t_i^{(2)}$ | $t_i^{(3)}$ |
|------|-------|------|------|------|------|------|------|------|---|-------|-------|-------|
| Cu | 3.540 | 2.56 | 5.11 | 1.07 | 3.63 | 2.2 | 6.0 | 2.2 | 1 | 3.14  | 2.49  | 2.95  |
| Ag | 2.850 | 2.88 | 5.89 | 1.06 | 4.46 | 2.2 | 6.0 | 2.2 | 1 | 5.54  | 2.45  | 1.29  |
| Au | 3.930 | 2.88 | 6.34 | 1.04 | 5.45 | 2.2 | 6.0 | 2.2 | 1 | 1.59  | 1.51  | 2.61  |
| Ni | 4.450 | 2.49 | 4.99 | 1.10 | 2.45 | 2.2 | 6.0 | 2.2 | 1 | 3.57  | 1.60  | 3.70  |
| Pd | 3.910 | 2.75 | 6.43 | 1.01 | 4.98 | 2.2 | 6.0 | 2.2 | 1 | 2.34  | 1.38  | 4.48  |
| Pt | 5.770 | 2.77 | 6.44 | 1.04 | 4.67 | 2.2 | 6.0 | 2.2 | 1 | 2.73  | −1.38 | 3.29  |
| Al | 3.580 | 2.86 | 4.61 | 1.07 | 2.21 | 2.2 | 6.0 | 2.2 | 1 | −1.78 | −2.21 | 8.01  |
| Pb | 2.040 | 3.50 | 6.06 | 1.01 | 5.31 | 2.2 | 6.0 | 2.2 | 1 | 2.74  | 3.06  | 1.20  |
| Rh | 5.750 | 2.69 | 6.00 | 1.05 | 1.13 | 1.0 | 2.0 | 1.0 | 1 | 2.99  | 4.61  | 4.80  |
| Ir | 6.930 | 2.72 | 6.52 | 1.05 | 1.13 | 1.0 | 2.0 | 1.0 | 1 | 1.50  | 8.10  | 4.80  |
| Li | 1.650 | 3.04 | 2.97 | 0.87 | 1.43 | 1.0 | 1.0 | 1.0 | 1 | 0.26  | 0.44  | −0.20 |
| Na | 1.130 | 2.73 | 3.64 | 0.90 | 2.31 | 1.0 | 1.0 | 1.0 | 1 | 3.55  | 0.69  | −0.20 |
| K  | 0.941 | 4.63 | 3.90 | 0.92 | 2.69 | 1.0 | 1.0 | 1.0 | 1 | 5.10  | 0.69  | −0.20 |
| V  | 5.300 | 2.63 | 4.83 | 1.00 | 4.11 | 1.0 | 1.0 | 1.0 | 1 | 4.20  | 4.10  | −1.00 |
| Nb | 7.470 | 2.86 | 4.79 | 1.00 | 4.37 | 1.0 | 1.0 | 1.0 | 1 | 3.76  | 3.83  | −1.00 |
| Ta | 8.089 | 2.86 | 4.90 | 0.99 | 3.71 | 1.0 | 1.0 | 1.0 | 1 | 4.69  | 3.35  | −1.50 |
| Cr | 4.100 | 2.50 | 5.12 | 0.94 | 3.22 | 1.0 | 1.0 | 1.0 | 1 | −0.21 | 12.26 | −1.90 |
| Mo | 6.810 | 2.73 | 5.85 | 0.99 | 4.48 | 1.0 | 1.0 | 1.0 | 1 | 3.48  | 9.49  | −2.90 |
| W  | 8.660 | 2.74 | 5.63 | 0.98 | 3.98 | 1.0 | 1.0 | 1.0 | 1 | 3.16  | 8.25  | −2.70 |
| Fe | 4.290 | 2.48 | 5.07 | 0.89 | 2.94 | 1.0 | 1.0 | 1.0 | 1 | 3.94  | 4.12  | −1.50 |
| C  | 7.370 | 1.54 | 4.31 | 1.80 | 5.50 | 4.3 | 3.1 | 6.0 | 1 | 5.57  | 1.94  | −0.77 |
| Si | 4.630 | 2.35 | 4.87 | 1.00 | 4.40 | 5.5 | 5.5 | 5.5 | 1 | 3.13  | 4.47  | −1.80 |
| Ge | 3.850 | 2.45 | 4.98 | 1.00 | 4.55 | 5.5 | 5.5 | 5.5 | 1 | 4.02  | 5.23  | −1.60 |
| H  | 2.225 | 0.74 | 2.96 | 2.50 | 2.96 | 3.0 | 3.0 |     | 1 | 0.20  | −0.10 | 0.00  |
| N  | 4.880 | 1.10 | 5.96 | 1.50 | 4.00 | 4.0 |     |     | 1 | 0.05  | 0.00  | 0.00  |
| O  | 2.558 | 1.21 | 6.49 | 1.50 | 6.49 | 6.5 |     |     | 1 | 0.09  | 0.10  | 0.00  |

has proved very successful in representing metallic as well as covalent bonds. Thus, it has been widely used to simulate metallic compounds and alloys.

Unfortunately, its centrosymmetric form limits its application to most FCC metals. Metals with BCC lattice (such as iron, Mo, W, etc.) or HCP lattice, especially transition metals (Co, Ti, etc.) cannot be represented by EAM in a satisfactory way. In addition, the fact that $\rho_j(r_{ij})$ is not sensitive to the surrounding neighbors (or environments) decreases the so-called transferability of the developed potentials. A potential fitted to a specific structure, such as FCC Al, may not be representative for amorphous Al or for liquid Al above the melting point. Even worse, the potential, fitted to the bulk properties, when the atom is deeply embedded into the crystal lattice, fails to reproduce correctly surface effects, such as surface energy and tension. These two issues were soon recognized by the researchers, and efforts have been made to overcome them.

### 3.11.5.2 MEAM Advantages

A solution was suggested in 1989 by Baskes et al.[55] for the particular case of Si and Ge, as a modification of the EAM, called MEAM (Modified EAM). The modification has two parts, which address both issues noted above: the centrosymmetry constraint, and the independence on the surrounding neighborhood of the electron density function $\rho_j(r_{ij})$. An angular dependent term was added to the centrosymmetric $\rho_j(r_{ij})$, so that the bonding can now depend on the angle, $\theta_{ijk}$, between two neighbors, i.e., atom $j$ and atom $k$, in addition to their distances, $r_{ij}$ and $r_{ik}$, from the host atom i.

In practice, MEAM can be fitted to any kind of compound, emulating both covalent and metallic bonds, and also to some extent ionic bonds, while the conventional EAM is mostly successful in centrosymmetric structures, like FCC metals (EAM potentials were suggested also for BCC and other structures, but with a limited accuracy). At this moment EAM or MEAM potentials exist for practically all monoatomic metals, for many non-metal elements (H, Si, etc.), and for a large number of the most common alloys and compounds (see, for example, Zhou et al., 2001[56]; Wadley et al., 2001[57] for a large database of EAM potentials).

### 3.11.5.3 Shortcomings of MEAM

A significant drawback of the MEAM potential is that it is very tedious to calculate when compared to the EAM, mostly because it requires screening to count triplets of atoms, rather than pairs of atoms. As we discussed in 3.11.4, screening is considered a major factor in the MEAM potential, and in some cases even more important than the angular terms. It is because of screening that MEAM can reproduce better defect structures and free surfaces. However, the derivative of the screening function needed for force calculations cannot be expressed in a closed analytical form and has to be done numerically.

Computationally, the energy calculation using MEAM is about 5–10 times more demanding than with EAM. This is mostly because of the screening function. For each pair of atoms one has to check all neighboring atoms to compute the screening for that pair. The force that results from that screening is an even bigger nightmare to compute, usually requiring numerical calculation of the potential gradient. Besides, computing all these squares and cos even in the most optimal way takes time, while in the EAM $\rho_j(r)$ and $F_i(\rho_i)$ are both tabulated with the distance r and no computation is needed at all. This limits the applicability of MEAM to relatively small systems (<100,000 atoms) and to molecular static simulations, where only the potential energy is need.

### 3.11.5.4 Angular Dependent Potential (ADP)

Recently, a modified expression of the angular term for the electron density function, $\rho_j(r)$ in the EAM has been suggested (Mishin et al., 2005[58]). Using a multipolar angular expansion form and eliminating the need of a screening function, makes the new potential, called ADP (Angular Dependent Potential), very simple to compute (it is just twice slower than the centrosymmetric EAM), while preserving all of the useful properties of the MEAM potential. The new ADP format seems to be very promising, but at this time it is still not widely used and so far has been fitted to a limited number of elements and compounds (such as Ni-Fe (Mishin et al., 2005[58]), Cu-Ta (Hashibon et al., 2008[59]).

## 3.12    Constructing Binary and High Order Potentials from Monoatomic Potentials

The simulation of hydrogen embrittlement of nickel described in Section 3.9.3 raises an important issue: How can binary, ternary, and higher order potentials for alloys from monoatomic potentials be constructed? As we know, pure metals are not often used, instead, alloys like steel, the airplane alloy $Al_2O_{24}$, or interfaces between two metals in soldering applications (Cu-Pb or Cu-Sn interfaces) are used. Binary and higher order potentials, Al-Cu and Al-Mn for $Al_2O_{24}$ or Fe-C potential for steel are also possible choices.

Two basic schemes will be discussed including Lorentz-Berthelot mixing rules and multi-body potential concepts such as MEAM, including the universal energy function proposed by Rose *et al.*[49] The former has a long history with pair potential and is used widely in gas, biomaterials, and some metals including the binary system of H and Ni discussed previously. The latter is especially useful when considering multi-body potential.

### 3.12.1 Determination of Parameters in LJ Pair Function for Unlike Atoms by Lorentz-Berthelet Mixing Rule

The LJ potential expressed in (2.3a) and Table 2.1 has two parameters, i.e., energy scale parameter $\varepsilon$ and collision diameter parameter $\sigma$. LJ potential was proposed over 80 years ago with a considerable amount of data about the two parameters $\varepsilon$ and $\sigma$ available for mono-atoms as shown in the literature.

To obtain the values of $\varepsilon$ and $\sigma$ for LJ potential between unlike chemical elements $A$ and $B$, the following averages are given by Lorentz-Berthelot mixing rules for parameters $\varepsilon_{A-B}$ and $\sigma_{A-B}$:[60]

$$\sigma_{A-B} = (\sigma_A + \sigma_B)/2 \tag{3.13a}$$

$$\varepsilon_{A-B} = \sqrt{\varepsilon_A \varepsilon_B} \tag{3.13b}$$

The collision diameter parameter $\sigma$ with units of length (see Figure 2.3) uses arithmetic average; while $\varepsilon$, whose units are energy, uses a geometric average process. When working with ternary and higher order potentials, it may be necessary to model several atoms as a unified site.

In practice, the rule may be extended to fit the experimental data as follows:

$$\sigma_{A-B} = \eta_1 \sigma_A + (1-\eta_1)\sigma_B \tag{3.13c}$$

$$\varepsilon_{A-B} = \eta_2 \sqrt{\varepsilon_A \varepsilon_B} \tag{3.13d}$$

where $\eta_1$ and $\eta_2$ are fitting parameters which can be determined by fitting experimental data. When $\eta_1 = 0.5$, $\eta_2 = 1.0$, formulas (3.13c) and (3.13d) return to (3.13a) and (3.13b).

### 3.12.2 Determination of Parameters in Morse and Exponential Potentials for Unlike Atoms

Morse potential given by (2.9a) and (2.9b) and the following exponential potential have three parameters, $\varepsilon$ (or D), $r_0$ and $\alpha$:

$$V_{ij}(r_{ij}) = \frac{\varepsilon}{(\alpha-6)} \left\{ 6\exp[\alpha(1-r_{ij}/r_0)] - \alpha(r_0/r_{ij})^6 \right\} \tag{3.13e}$$

The Lorentz-Berthelot mixing rules are extended to describe the cases by the following:[61]

$$\varepsilon_{AB} = k_{AB}\sqrt{\varepsilon_A \varepsilon_B} \tag{3.13f}$$

$$\alpha_{AB} = m_{AB}\sqrt{\alpha_A \alpha_B} \tag{3.13g}$$

$$r_{0AB} = l_{AB}(r_{0A} + r_{0B})/2 \tag{3.13h}$$

For the equilibrium length, $r_0$ uses arithmetic average while $\varepsilon$ and $\alpha$ with a more complicated underlying physical process use geometric average.

White[61] compares the data obtained by the QM first principle calculation and the extended mixing rules (3.13f–h) for six types of non-homogeneous mixtures including $CO/N_2$, $CO/H_2$, $CO/HF$, $N_2/HF$, $HF/H_2$, $N_2/H_2$. Interested readers may refer.

## 3.12.3   Determination of Parameters in EAM Potentials for Alloys

The determination of parameters in EAM and MEAM potential for unlike atoms i and j includes two parts. The first part is the electron density function $F(\rho_i)$ and the second part is the pair potential.

### 3.12.3.1 Embedded Energy Function for Alloys

For the first part, the key is to determine the electron density $\rho_i$. For the first rough approximation[46] it is assumed that electron density is not dependent on the type of the host atom and that the electron density at any location is taken as a linear superposition of atomistic electron densities, therefore, the electron density becomes the simple sum of the electron densities of different species of atoms (say Ni and Al) as they happen to neighbor the host atom. The obtained electron density can then be used to determine the embedding energy function, because this function is not dependent on the source of the electron density.

These electron densities are usually those of the atoms in a monoatomic EAM and MEAM for the corresponding atom. But to be comparable, both electron densities have to be scaled to match, so that the embedded function can take both electron densities as one. This is usually done by setting the equilibrium electron density (at the equilibrium distance $r_0$ in the basic crystal structure) to be unity (1.0) for both species in their pure state (FCC or BCC). Possibly, better potentials could also be obtained not by this approximate assumption but by scaling the electron density for unlike atoms. This way could possibly improve the potential, but it is still uncertain. For example, one may introduce a pairwise electron density, or a pairwise embedded function, specifically when the host and the neighbor atoms are unlike, instead of their simple sum. This is equivalent to fitting the pairwise potential function $V(r_{ij})$ for unlike atoms and is the usual practice. The question, however, is whether this helps in making better potentials. No definite conclusion can be given and there is possibly not much gain by using pairwise electron density.

After the embedded energy function is considered for unlike elements through determining the electron enery density, the remaining part is to develop the pair potentials. For the monoatomic potential in the alloy, it is assumed to be the original one of the pure atoms such as the interatomic potential of Ni-Ni. The challenge is to develop a new binary pair potential between atoms of different species. The treatment to meet this challenge is different between EAM and MEAM.

### 3.12.3.2 EAM Pair Potential for Unlike Chemical Elements

Johnson[62] suggested that all models with the EAM format be transformed to a normalized form in which the slope of the embedding function is zero at the equilibrium electron density $\rho_0$ (remark: here the zero slope of potential curve is at $\rho_0$, not at the equilibrium distance $r_0$ in Figure 2.3). In this case, the two-body potential becomes the effective two-body potentials. These potentials are negative in some range of distances such that a geometric average cannot be used. Johnson expressed (3.11b) as follows:

$$\rho_i = \sum_{j=1(\neq i)} f(r_{ij}) \tag{3.14a}$$

where $f(r_{ij})$ is the function of the distance between atoms i and j. He proved that if the form of unlike atoms $V^{ab}$ of an alloy is expressed by its monoatomic potential $V^{aa}$ and $V^{bb}$ as follows,

$$V^{ab}(r) = \frac{1}{2}\left[\frac{f^b(r)}{f^a(r)}V^{aa}(r) + \frac{f^a(r)}{f^b(r)}V^{bb}(r)\right] \tag{3.14b}$$

then alloy models are invariant to transformations in the monatomic models from which they are derived, and this invariance holds for any number of different elements in an alloy. It indicates that the pair potential for the unlike elements determined by (3.14b) can have a high transferability to be used in different conditions.

### 3.12.4   Determination of Parameters in MEAM Potentials for Alloys

To use the pair potential in MEAM for unlike atoms i and j, consider the changes of three parameters, namely $E_i^0, R_i^0, \alpha_i$ (remark: the other eight parameters related to electron density function $F(\rho_i)$ are not changed). Based on (3.12e), the pair potential can be derived for unlike atoms as follows:

$$V_{ij}(R) = \frac{1}{Z_{IJ}}\left[2E_{IJ}^u(R) - F_I(Z_{IJ}\rho_J^{a(0)}(R)/Z_I) - F_J(Z_{IJ}\rho_I^{a(0)}(R)/Z_J)\right] \tag{3.13i}$$

To get $E_{IJ}^u(R)$, the universal energy function shown in (3.12f), its parameters in (3.12g) and (3.12h) are used. The corresponding parameters, $E_{IJ}^0, \alpha_{IJ}, \Omega_{IJ}$ for unlike atoms I and J expressed in (3.12f–h) will be denoted by subscript IJ and can be given as follows:

$$E_{IJ}^0 = (E_I^0 + E_J^0)/2 - \Delta_{IJ} \tag{3.13j}$$

$$\alpha_{IJ} = (\alpha_I + \alpha_J)/2 \tag{3.13k}$$

$$\Omega_{IJ} = (\Omega_1 + \Omega_J)/2 \tag{3.13l}$$

$R_{IJ}^0$ can be calculated from the above equilibrium intermetallic atomic volume. The $\Delta_{IJ}$ can be found in Ref. [48].

## 3.13   Application Examples of Metals: MD Simulation Reveals Yield Mechanism of Metallic Nanowires

This example[63] is given to illustrate that the mechanical deformation behaviors of single-crystalline metallic nanowires are different from their bulk counterparts.

The simulations were conducted with particular interest in revealing the size effects on the mechanical properties of 1D nanowires. Initially, nearly square cross-sectioned Ni [111] nanowires with $(\bar{1}\bar{1}2)$ and $(1\bar{1}0)$ side surfaces were constructed, with atomic positions corresponding to a bulk FCC lattice. Three wires with transverse sectional sizes of 4, 8 and 16 nm were simulated. The longitudinal/transverse aspect ratio was fixed at 3 : 1. An ADP type potential of EAM method for Ni as described by Mishin *et al.*[64] was chosen in the present work because it is calibrated according to *ab initio* values of stacking fault and twin formation energies.

After the initial construction, energy minimization using the so-called conjugate gradient method was performed to obtain the equilibrium configurations with minimum potential energy. The wires were then thermally equilibrated to 2 K for 20 ps using a Nose-Hoover thermostat. The Velocity Verlet method was adopted to integrate the equation of motion with a time step of 1 fs. No periodic boundary conditions were utilized in all three dimensions. Starting from the equilibrium configuration of the nanowires, uniaxial loading was applied under the simulated quasistatic conditions.[65] To expedite the simulations, all atoms were first displaced in the first five loading steps with a prescribed uniform compressive strain increment of 1% in the length direction, and a 0.2% strain increment was applied in the remaining steps. Therefore, the strain in the first five steps was five times larger than in the remaining steps. The wires were then relaxed with their ends fixed at a constant temperature of 2 K.

To ensure that the simulation reached a quasistatic condition, the equilibrium time for each loading step was determined by monitoring the stress components of the entire system until little fluctuation remained. Typically, the equilibrium process took less than 150 ps in the 16 nm wires. (This relaxation time is for each increment step; therefore the specimen has sufficient time to relax to reach the "quasistatic" condition.) The simulation will take several hundred ps within about 10% maximum strain. The average stress over the last 5 ps of the relaxation period was taken as the stress of the nanowire. The stresses were

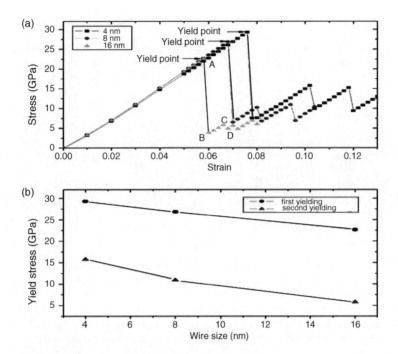

**Figure 3.16** Compressive stress-strain curves of Ni nanowires ranging in size from 4 to 16 nm (Reproduced from Cao, A., Wei, Y., and Mao, S. X. (2008) Alternating starvation of dislocations during plastic yielding in metallic nanowires. *Scripta Materialia*, 59 (2), 219, Elsevier)

calculated using the virial theorem (Section 3.9.1) which for all purposes in this example is equivalent to the Cauchy stress, a stress measured in the current configuration.

The obtained stress-strain curves for the three simulated wires are shown in Figure 3.16. The most interesting observation is the saw-like nature of the stress-strain curve, which differs significantly from the stress-strain curve of most bulk metallic materials. Note: After the elastic deformation, the initial yielding results in a precipitous stress drop, followed by an increase in the stress again; which then repeats, to result in the fluctuation of the stress-strain relations. Furthermore, from Figure 3.16b the magnitude of the compressive yield stress (the first peak) decreases as the nanowire width increases. These size effect trends relate to the experimental reports,[66] in which the diameter dependent yield stress of gold nanowires, with a size ranging from 40 to 200 nm and measured using bending experiments, was reported.

Section 3.9.1 shows a series of typical snapshots of collective dislocation dynamics processes in the 16 nm Ni wires. These snapshots clearly demonstrate that dislocations are nucleated from surface edges (see Figure caption). Once dislocations are nucleated from the surface edges, they can only travel much shorter distances before being annihilated at the free surface, thereby reducing the probability of multiplication processes, leaving a dislocation-free structure state in the interior. Continued elastic deformation is required to nucleate new dislocations, which is the underlying physics of increasing stress with further straining.

The correlation of the deformation visualization and the stress-strain curve reveals the deformation mechanism of these materials, which is difficult to determine by experiment alone. The MD simulation results show that metallic nanowires behave in a manner that is unexpected in bulk materials, offering new insight into the fundamental understanding and application of these nano-scale materials at the atomistic level.

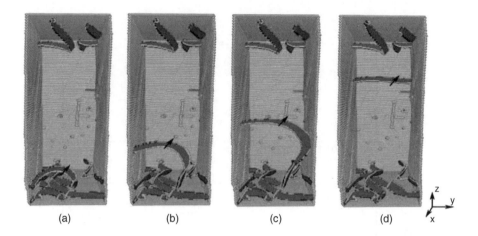

(a)                    (b)                    (c)                    (d)

**Figure 3.17** Snapshots during the yielding of the Ni 16 nm nanowire, showing collective dislocation nucleation and moving out of the wires. The colors of the atoms are assigned according to a local crystallinity classification visualized by common neighbor analysis, which permits the distinction between atoms in a local hexagonal close-packed (HCP) environment and those in an FCC environment. Perfect FCC atoms and the front surface are not shown in order to properly view the inner defects. Some dislocations are pinned by the fixed top and bottom surfaces, therefore causing them to be immobile (Reproduced from Cao, A., Wei, Y., and Mao, S. X. (2008) Alternating starvation of dislocations during plastic yielding in metallic nanowires. *Scripta Materialia*, 59 (2), 219, Elsevier)

---

**Homework**

(3.6) For a double-wall carbon nanotube, the radii of inner wall and outer wall are $R_1$ and $R_2$ respectively. Because of the van der Waals interactions between the two walls, there is pressure between the two walls. The pressure applied on the inner wall by the outer wall is $P_1$ and the pressure applied on the outer wall by the inner wall is $P_2$ assuming LJ potential is used. Which of the following relations of the pressures do you think is correct?

(a) $P_1/P_2 = R_2/R_1$;   (b) $P_1 = P_2$;   (c) none of above

Hint: (1) Note $\rho_1$ and $\rho_2$ as the number of atoms per unit surface area for the inner and outer wall respectively, and usually $\rho_1 \approx \rho_2$. (2) The distance from an atom on the outer wall to an atom on the inner wall $a = \sqrt{R_1^2 + R_2^2 + Z^2 - 2R_1R_2\cos\theta}$

---

## 3.14 Collecting Data of Atomistic Potentials from the Internet Based on a Specific Technical Requirement

The key issue in atomistic analysis is to find accurate potentials and their parameters which are problem dependent. One may start to find existing data from the internet as the first step. In the following, we take the simulation of nanoscale ceramics coating process and coating layer behavior as an example to show the rich results obtained through internet searching. To do so, it is necessary to know the technical background and requirement first, thus we briefly introduce the necessity of ceramics coating and its manufacturing process, then show what kinds of potential we need. The detailed search results are shown from Table 3A.1 to Table 3A.7 in the Appendix 3.A to this chapter.

**Figure 3.18**   Galvanic corrosion of a magnesium bracket fastened by steel bolt to car transmission box

### 3.14.1   Background About Galvanic Corrosion of Magnesium and Nano-Ceramics Coating on Steel

Corrosion is a serious problem when metals work in a corrosive environment. Among all kinds of corrosion, electric galvanic corrosion is especially harmful in engineering structures such as automobile components. Because different metals possess different chemical potentials, in a given environment, a finite electron voltage will appear to form an electric circuit and produce current if they connect to the same corrosive media. The latter includes acid, water or moist air containing chlorine ions, resulting in corrosion in anode metal. A lot of approaches have been proposed in engineering to prevent galvanic corrosion. However, it is especially difficult for steel fasteners connecting with magnesium alloy structures because magnesium is the most active metal which produces a high voltage difference with steel. Figure 3.18 illustrates the serious galvanic corrosion of magnesium parts with the steel bolts and washers.

Nano-ceramic coatings on steel fasteners is an innovative and revolutionary approach to electrically isolate steel bolts, nuts, washers from Mg parts to stop the galvanic electric current. Being insulating material, ceramic is capable of cutting off the circuit composed by iron and magnesium alloy. Therefore, it may be able to prevent electric corrosion in the steel fastener. Figure 3.19 shows the comparison of the corrosion status between a steel bolt with coating and one without coating. It is seen that the ceramic coating is effective.

### 3.14.2   Physical and Chemical Vapor Deposition to Produce Ceramics Thin Coating Layers on Steel Substrate

The formation methods of a nanoscale coating layer on steel substrate includes physical vapor deposition (PVD) and chemical vapor deposition (CVD). Traditionally, in an evaporation process of PVD, vapors of the target material will transport to the substrate to form the coating layer, they are usually generated by the target material being heated by hot wire, radiation, laser beam or electron beam. These processes correspond to the so-called thermal evaporation, electron beam evaporation and sputtering. A typical PVD process usually contains three steps.[68] Firstly, the material to be deposited is synthesized (for electric beam evaporation technique, it is the phase transformation from solid or liquid to vapor). In this step, simple reactions are allowed between the target material and the induced gas, although the process still

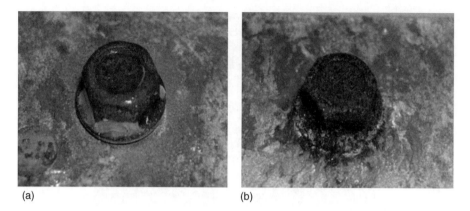

(a)                                                                    (b)

**Figure 3.19**  Comparison of corrosion status of the assembly of magnesium plates climped by steel fasteners in the spray NaCl chamber for 13 days. (a) With coating layers of $Si_3N_4$ as the first layer and UV curable $(Al_2O_3)$ plastic as the second layer. (b) No coating in the steel.

belongs to PVD. In the second step, vapors transport from the source to the substrate. The third process is the condensation of vapors, film nucleation and growth on the substrate.

On the other hand, chemical vapor deposition (CVD) is a process where the deposited materials derive from chemical reactions very close to or on the substrates. The CVD technique is excellent in producing coatings of uniform thickness and with a low porosity even on substrates of complicated shape. Unlike the PVD process, the chemical composition of the precursors is usually different from that of the resultant film. There is another coating process called PECVD, by which plasma will appear near the substrate surface to produce a good coating.

A vacuum environment (typically $10^{-5}$ to $10^{-4}$ torr or $(1.315$ to $13.15) \times 10^{-8}$ atm) is needed for the evaporation, and the evaporated atoms transport to the random locations with almost no collision with each other. When the atoms reach the substrate, scattered nucleation centers form. As the film grows, these nucleation centers become larger and coalesce with each other, forming island-like structures. As the "islands" expand, a complete film is formed on the substrate.

## 3.14.3   Technical Requirement for Potentials and Searching Results

In practice, there are several technical problems that must be solved. They include:

### Effects of coating materials
Using atomistic analysis to see which coating material has highest bonding energy with steel substrate.

### Effects of coating layer-thickness
Using atomistic analysis to see the mechanism of coating layer-thickness effects.

### Effects of processing parameters of vapor deposition
The thin-film coating layer is formed through vapor deposition process, thus the incident angle, incident rate and energy of the vapor ions are important for the quality of coatings.

To simulate these effects, the first and most important task is to find the potentials of chemical elements involved in the simulation. Suppose we choose the coating materials as iron nitrides ($Fe_4N$), silicon nitrides ($Si_3N_4$), aluminum oxide ($Al_2O_3$) and zinc oxide (ZnO), the chemical elements involved include Al, Mg, O, Si, N, Fe, Zn, H. Then we can use search machines such as Google or Google Science and key words such as "Si N Morse potential", "Mg O Buckingham potential", "magnesium oxygen potential" etc. to find related potentials. The Appendix 3.A below shows Buckingham potential (Table 3A.1), MEAM (Table 3A.2), Tersoff (Table 3A.3), LJ potential (Table 3A.4), Morse potential (Table 3A.5), EAM potential (Table 3A.6) and Stillinger-Weber potential (Table 3A.7) for these atoms as well as other atoms for materials such as $Y_3Al_5O_{12}$ garnets (YAG).

### 3.14.4 Using Obtained Data for Potential Development and Atomistic Simulation

Having found this data, one may proceed from these initial values to a determination of more accurate parameters by different methods as described previously. Specifically, (1) using GULP software to modify these parameters based on minimum potential energy (see Section 10.3); (2) using experimental data to modify these parameters (see Sections 3.4.3 and 3.14); (3) using QM or combined QM/test data (see Chapter 4).

As soon as one verifies and modifies this data, it can be used for coating processing simulation by MC and MD method, as well their combination. For the MD simulation of thin-layer coating behavior, the detailed procedure is shown in Part 10.3 of Chapter 10. Specifically, the potential parameters are given in the FIELD file when one uses Dl_Poly software (see for instance, Field file in UNIT8/2D_Non/2D_Forced directory). Furthermore, through data processing, the shearing displacement or the shearing strain can be determined (see Figure 10.12). On the other hand, we may also find the variation of shearing force (or shearing stress) during the shearing process as shown in Figure 3.20. In that figure, the top curve shows the shear force variation with shearing steps when the shearing plane is inside the iron nitride coating layer, the bottom curve shows corresponding shear force when shearing is inside the iron substrate, and the middle one is the interfacial shearing force when the shearing plane is between the iron substrate and the ceramics coating layer. As one can see, the shear resistance is much higher in the ceramics coating layer and weaker in the substrate compared with the interfacial shear strength. Furthermore, the shearing force is very large at the beginning when the shearing rate is constant, and then relaxes, specifically for shearing inside the ceramics coating layer.

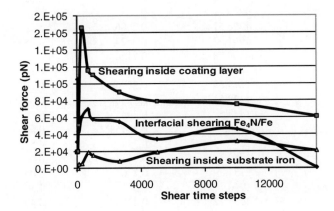

**Figure 3.20**  MD-determined shearing resistance under a constant shearing rate for a coating system in which the ceramics coating layer consists of iron nitride and the substrate is made of iron

## Appendix 3.A Potential Tables for Oxides and Thin-Film Coating Layers

**Table 3A.1**    Buckingham potential (see (3.2b))

| | A(eV) | $\rho(\text{Å})$ | C (eV Å$^6$) | Ref |
|---|---|---|---|---|
| Si-N | 7583.9473 | 0.2583 | 207.1801 | 69 |
| Si-N | 4802.28 | 0.2515 | 88.22 | 70 |
| N-N | 5937.0512 | 0.3789 | 1662.82 | 69 |
| N-N | 5916.99 | 0.3295 | 599.72 | 70 |
| Si-O | 18,003.76 | 0.2052 | 133.5381 | 69 |
| Si-O | 1672.078 | 0.241 | 0 | 70 |
| Si-O$^{2-}$ | 1283.907 | 0.32052 | 10.66158 | 71 |
| Si-O$^{1.4-}$ | 983.556 | 0.32052 | 10.66158 | 71 |
| Si$^{4+}$-O$^{2-}$ | 1283.9073 | 0.3205 | 10.6616 | 72 |
| Si$^{4+}$-O$^{1.426-}$ | 999.98 | 0.3012 | 0 | 72 |
| O-O | 1338.773 | 0.3623 | 175 | 69 |
| O-O | 30506.16 | 0.2323 | 0 | 70 |
| O-O | 22764 | 6.7114 | 27.88 | 73 |
| O-O | 22764 | 0.149 | 27.88 | 74, 75 |
| O-O | 25.41 | 0.6937 | 32.32 | 76 |
| O-O | 11782.85 | 0.234 | 30.22223 | 77 |
| O shell-O shell | 2.909182 | 0.242588 | 11.148244 | 78 |
| O$^{2-}$-O$^{2-}$ | 22764 | 0.149 | 27.88 | 79 |
| O$^{2-}$ shell-O$^{2-}$ shell | 22764 | 0.149 | 27.88 | 80 |
| O$^{2-}$-O$^{2-}$ | 9547.96 | 0.21916 | 32 | 79, 81 |
| O$^{2-}$-O$^{2-}$ | 22764.0 | 0.149 | 27.88 | 71, 72 |
| O$^{2-}$-O$^{1.4-}$ | 22764 | 0.149 | 27.88 | 71 |
| O$^{2-}$ shell-O$^{0.8-}$ shell | 22764 | 0.149 | 28.92 | 80 |
| O$^{1.4-}$-O$^{1.4-}$ | 22764 | 0.149 | 27.88 | 71 |
| O$^{2-}$-O$^{2-}$ | 676.9 | 0.3683 | 32.52 | 83 |
| O$^{2-}$-O$^{2-}$ | 9547.96 | 0.2912 | 32 | 84 |
| O$^{2-}$-O$^{2-}$ | 9547.96 | 0.2192 | 32 | 85 |
| O$^{2-}$-O$^{-}$ | 889.64 | 0.3376 | 18.81 | 83 |
| O$^{1.7-}$-O$^{-1.7}$ | 4870 | 0.267 | 77 | 79 |
| O$^{-}$-O$^{-}$ | 1396.51 | 0.2906 | 5.18 | 83 |
| O-N | 1961.5314 | 0.3623 | 174.9823 | 69 |
| N-O | 48609.78 | 0.2156 | 0 | 70 |
| Mg$^{2+}$-O$^{2-}$ | 821.6 | 0.3242 | 0 | 79 |
| Mg$^{2+}$-O$^{2-}$ | 1279.69 | 0.29969 | 0 | 79 |
| Mg$^{2+}$-O$^{2-}$ | 1428.5 | 0.2945 | 0 | 72 |
| Mg-O$^{2-}$ | 1428.5 | 0.29453 | 0 | 71 |
| Mg-O$^{1.4-}$ | 1060.5 | 0.29453 | 0 | 71 |
| Mg$^{2+}$-O$^{2-}$ | 1280 | 0.3 | 4.5 | 82 |
| Mg-O | 1280.1 | 0.3177 | 0 | 75 |
| Mg$^{2+}$-O$^{2-}$ | 2134.39 | 0.2763 | 3.05 | 83 |
| Mg$^{2+}$-O$^{1.4-}$ | 960 | 0.3 | 4.5 | 82 |
| Mg$^{2+}$-O- | 4135.77 | 0.2255 | 0.49 | 83 |
| Mg$^{1.7+}$-O$^{1.7-}$ | 929.69 | 0.29909 | 0 | 79 |
| Ti-Ti | 31120.43 | 0.154 | 5.246927 | 77 |

**Table 3A.1**   (*Continued*)

| | A(eV) | $\rho(\text{Å})$ | C (eV Å$^6$) | Ref |
|---|---|---|---|---|
| Ti-O | 16957.67 | 0.194 | 12.59259 | 77 |
| Zn-Zn | 0 | 0.1 | 0 | 73 |
| Zn-O | 700.3 | 2.9586 | 0 | 73 |
| $Zn^{2+}$-$O^{2-}$ | 529.7 | 0.3581 | 0 | 81 |
| $Zn+$-$O^{2-}$ | 470.41 | 0.3718 | 0 | 81 |
| $Zn0$-$O^{2-}$ | 180.81 | 0.4418 | 0 | 81 |
| $Zn^+$-$Cl^-$ | 9704.89 | 0.232 | 0 | 81 |
| $Al^{3+}$-$O^{2-}$ | 1460.3 | 0.29912 | 0 | 82 |
| $Al^{3+}$-$O^{2-}$ | 1283.9 | 0.32052 | 10.66 | 82 |
| $Al^{3+}$-$O^{2-}$ | 1725.2 | 0.28971 | 0 | 84 |
| $Al^{3+}$-$O^{2-}$ | 1365.79 | 0.30096 | 2.538 | 85 |
| Al-O | 741.9007 | 0.3566 | 0 | 74 |
| Al-O | 1469.3 | 0.2991 | 0 | 75 |
| Al-O | 2409.505 | 0.2649 | 0 | 76 |
| Y-O | 2036.838 | 0.3103 | 0 | 74 |
| Y-O | 1345.1 | 0.3491 | 0 | 75 |
| Y-O | 1519.279 | 0.3291 | 0 | 76 |
| $Y^{3+}$-$O^{2-}$ | 1721.23 | 0.33821 | 10.29 | 85 |
| $K^+$-$O^{2-}$ | 65269.71 | 0.213 | 0 | 72 |
| Ca-O | 1090.1 | 0.3437 | 0 | 75 |
| $Ca^{2+}$-$O^{2-}$ | 6958.3 | 0.2516 | 0 | 72 |
| $H^+$-$O^{2-}$ | 311.97 | 0.25 | 0 | 72 |
| $H$-$O^{2-}$ | 311.97 | 0.25 | 0 | 71 |
| $H$-$O^{1.4-}$ | 311.97 | 0.25 | 0 | 71 |
| $H^{0.4+}$-$O^{2-}$ | 312.0 | 0.25 | 0 | 82 |
| $H^{0.4+}$ core-$O^{0.8-}$ shell | 396.27 | 0.25 | 10 | 80 |
| $H^{0.4+}$ core-$O^{2-}$ shell | 396.27 | 0.25 | 10 | 80 |
| $P^{5+}$-$O^{2-}$ | 877.34 | 0.3594 | 0 | 82 |
| $Co^{2+}$-$O^{2-}$ | 778.02 | 0.3301 | 0 | 84 |
| $Li^+$-$O^{2-}$ | 812.57 | 0.2725 | 0.97 | 83 |
| $Li^+$-$O^{2-}$ | 828.01 | 0.2793 | 0 | 81 |
| $Li^+$-$O^-$ | 1583.43 | 0.2209 | 0.18 | 83 |
| $Fe^{2+}$-$O^{2-}$ | 1207.6 | 0.3084 | 0 | 71 |
| $Fe^{3+}$-$O^{2-}$ | 1102.4 | 0.3299 | 0 | 71 |
| $Fe^{3+}$-$O^{2-}$ | 1478.98 | 0.31306 | 6.96 | 85 |
| $Fe^{2+}$ shell-$O^{2-}$ shell | 614.97 | 0.3399 | 0 | 80 |
| Fe(III) shell-O shell | 620.985511 | 0.371059 | 0 | 78 |
| $Fe^{2+}$ shell-$O^{0.8-}$ shell | 296.23 | 0.3399 | 0 | 80 |
| $Cr^{3+}$-$O^{2-}$ | 1452.25 | 0.30918 | 4.472 | 85 |
| $Cr^{3+}$ shell-$O^{2-}$ shell | 1734.1 | 0.301 | 0 | 80 |
| $Cr^{3+}$ shell-$O^{0.8-}$ shell | 1077.86 | 0.301 | 0 | 80 |
| Sb(III) shell-O shell | 1523.76845 | 0.350672 | 5.460208 | 78 |
| Sb(V) shell-O shell | 21059.1904 | 0.225797 | 0 | 78 |

**Table 3A.2** MEAM potential (see (3.12c), (3.12f,g,j) and Table 3.8)

| | $E_c$(eV) | $r_e$(Å) | $a$ | $A$ | $\beta^{(0)}$ | $\beta^{(1)}$ | $\beta^{(2)}$ | $\beta^{(3)}$ | $t^{(0)}$ | $t^{(1)}$ | $t^{(2)}$ | $t^{(3)}$ | $C_{min}$ | $C_{max}$ | d | Ref |
|----|------|------|------|------|------|------|------|------|------|------|------|------|------|------|-----|-----|
| Ti | 4.87 | 2.92 | 1.1 | 0.66 | 2.7 | 1 | 3 | 1 | | 6.8 | $-2$ | $-12$ | 1 | 1.44 | 0 | 86 |
| N | 4.88 | 1.1 | 5.96 | 1.8 | 2.75 | 4 | 4 | 4 | | 0.05 | 1 | 0 | 2 | 2.8 | 0 | 86 |
| Fe | 4.29 | 2.48 | 5.07 | 0.89 | 2.94 | 1 | 1 | 1 | 1 | 3.94 | 4.12 | $-1.5$ | | | | 87 |
| Si | 4.63 | 2.35 | 4.87 | 1 | 4.4 | 5.5 | 5.5 | 5.5 | 1 | 4.02 | 5.23 | $-1.6$ | | | | 87 |
| O | 2.558 | 1.21 | 6.49 | 1.5 | 6.49 | 6.5 | 6.5 | | 1 | 0.09 | 0.1 | 0 | | | | 87 |
| Al | 3.58 | 2.86 | 4.61 | 1.07 | 2.21 | 2.2 | 6 | 2.2 | 1 | $-1.78$ | $-2.21$ | 8.01 | | | | 87 |
| Al | 3.36 | 2.86 | 0.79 | 1.16 | 3.2 | 2.6 | 6 | 2.6 | | 3.05 | 0.51 | 7.75 | 0.49 | 2.8 | 0.1 | 88 |
| Mg | 1.55 | 3.2 | 5.45 | 1.11 | 2.7 | 0 | 0.4 | 3 | 1 | 9.5 | 4.1 | $-2$ | | | | 89 |
| Mg | 1.55 | 3.2 | 0.37 | 0.52 | 2.3 | 1 | 3 | 1 | | 9 | $-2$ | $-9.5$ | 0.49 | 2.8 | 0 | 88 |
| N | 4.88 | 1.1 | 5.96 | 1.5 | 4 | 4 | 0 | 0 | 1 | 0.09 | 0.1 | 0 | | | | 87 |

**Table 3A.3**  Tersoff potential (see (3.9b–d))

| | Si | Si | Si | O | ZnO | N | H | C | Ge |
|---|---|---|---|---|---|---|---|---|---|
| $A$ (eV) | $1.8308 \cdot 10^3$ | $1.8308 \cdot 10^3$ | 1830.8 | $1.88255 \cdot 10^3$ | 4099.2 | $6.36814 \cdot 10^3$ | 86.712 | $1.39 \cdot 10^3$ | $1.77 \cdot 10^3$ |
| $B$ (eV) | $4.7118 \cdot 10^2$ | $4.7118 \cdot 10^2$ | 471.18 | $2.18787 \cdot 10^2$ | 209 | $5.11760 \cdot 10^2$ | 43.531 | $3.47 \cdot 10^2$ | $4.19 \cdot 10^2$ |
| $\gamma$ (Å$^{-1}$) | 2.4799 | 2.4799 | 2.4799 | 4.17108 | 3.2599 | 5.43673 | 3.7879 | 3.4879 | 2.4451 |
| $\mu$ (Å$^{-1}$) | 1.7322 | 1.7322 | 1.7322 | 2.35692 | 1.7322 | 2.7 | 1.98 | 2.2119 | 1.7047 |
| $\beta$ | $1.1000 \cdot 10^{-6}$ | $1.1000 \cdot 10^{-6}$ | $1.1000 \cdot 10^{-6}$ | $1.1632 \cdot 10^{-7}$ | $1.1000 \cdot 10^{-6}$ | $5.29380 \cdot 10^{-3}$ | 4 | $1.57 \cdot 10^{-7}$ | $9.02 \cdot 10^{-7}$ |
| $n$ | $7.8734 \cdot 10^{-1}$ | $7.8734 \cdot 10^{-1}$ | 0.78734 | 1.04968 | 0.78734 | 1.33041 | 1 | $7.28 \cdot 10^{-1}$ | $7.56 \cdot 10^{-1}$ |
| $c$ | $1.0039 \cdot 10^5$ | $1.0039 \cdot 10^5$ | 100390 | $6.46921 \cdot 10^4$ | 100390 | $2.03120 \cdot 10^4$ | 0 | $3.80 \cdot 10^4$ | $1.06 \cdot 10^5$ |
| $d$ | $1.6217 \cdot 10^1$ | $1.6217 \cdot 10^1$ | 16.22 | 4.11127 | 16.217 | $2.55103 \cdot 10^1$ | 1 | 4.348 | 15.52 |
| $h$ | $-5.9825 \cdot 10^{-1}$ | $-5.9825 \cdot 10^{-1}$ | $-5.98 \cdot 10^{-1}$ | $-8.45922 \cdot 10^{-1}$ | $-0.59825$ | $-5.62390 \cdot 10^{-1}$ | 1 | $-5.71 \cdot 10^{-1}$ | $-4.39 \cdot 10^{-1}$ |
| $R$ (Å) | 2.5 | 2.7 | 2.7 | 1.7 | 3.059 | 1.8 | 0.8 | 1.8 | 2.8 |
| $S$ (Å) | 2.8 | 3 | 3 | 2 | 0.15 | 2.1 | 1 | 2.1 | 3.1 |
| $Ref$ | 90 | 91 | 92 | 90 | 93 | 91 | 92 | 92 | 92 |

**Table 3A.4**   LJ potential (see (2.3a))

|        | $\varepsilon$(eV) | $\sigma$(Å) | ref. | Interaction | $\varepsilon$(eV) | $\sigma$(Å) | Ref. |
|--------|--------|--------|------|-------------|--------|--------|------|
| He     | 8.80E-04 | 2.56 | 94 | Cu | 0.4097 | 2.34 | 96 |
| Ne     | 3.07E-03 | 2.75 | 94 | Ni | 1.68E+00 | 2.489 | 95 |
| Ar     | 1.06E-02 | 3.4 | 94 | Pb | 9.39E-01 | 3.504 | 95 |
| Kr     | 1.44E-02 | 3.68 | 94 | Fe-C | 1861.82 | 2.6467 | 97 |
| Xe     | 1.94E-02 | 4.07 | 94 | Cr-C | 615.839 | 3.2208 | 97 |
| $N_2$  | 7.80E-03 | 3.7 | 94 | XV-C | 272.642 | 3.8221 | 97 |
| $I_2$  | 4.74E-02 | 4.98 | 94 | Ni_C | 235.111 | 3.9697 | 97 |
| Hg     | 7.33E-02 | 2.9 | 94 | Zn | 0.1574 | 2.44 | 96 |
| $CCl_4$ | 2.82E-02 | 5.88 | 94 | Zn-Mg | 0.16888 | 2.6784 | 96 |
| AL     | 1.08E+00 | 2.864 | 95 | Zn-Cu | 0.27247 | 2.6229 | 96 |
| Al     | 0.3754 | 2.61 | 96 | Zn-Li | 0.27471 | 2.2997 | 96 |
| Al-Zn  | 0.24308 | 2.5236 | 96 | Mg | 0.1812 | 2.94 | 96 |
| Al-Mg  | 0.16888 | 2.6784 | 96 | Mg-Cu | 0.27247 | 2.6229 | 96 |
| Al-Cu  | 0.27247 | 2.6229 | 96 | Mg-Li | 0.27471 | 2.2997 | 96 |
| Al-Li  | 0.27471 | 2.2997 | 96 | Li | 0.1842 | 2.26 | 96 |
| Ag     | 1.33E+00 | 2.892 | 95 | Cu-Li | 0.27471 | 2.2997 | 96 |
| Cu     | 1.37E+00 | 2.553 | 95 |  |  |  |  |

**Table 3A.5**   Morse potential (see (2.9a))

| Interaction | D(eV) | $\alpha$ (Å$^{-1}$) | $r_0$(Å) | Ref | Interaction | D(eV) | $\alpha$ (Å$^{-1}$) | $r_0$(Å) | Ref |
|-------------|-------|-------|--------|-----|-------------|-------|-------|--------|-----|
| Fe   | 0.4147 | 1.3885 | 2.845 | 98 | Ag | 0.3323 | 1.369 | 3.115 | 101 |
| Zn   | 0.17 | 1.705 | 2.793 | 99 | Cu | 0.3429 | 1.3588 | 2.866 | 101 |
| O    | 5.1655 | 2.667 | 1.21 | 100 | Ni | 0.4205 | 1.4199 | 2.78 | 101 |
| Si-O | 4.6746 | 1.865 | 1.631 | 100 | Pb | 0.2348 | 1.1836 | 3.733 | 101 |
| Si-C | 4.3503 | 4.6417 | 1.9475 | 100 | Nd | 0.6025 | 0.8306 | 2.6233 | 103 |
| Al   | 0.2703 | 1.1646 | 3.253 | 101 | Nd-H | 0.3905 | 1.4684 | 1.9803 | 103 |
| H-O$^{1.4-}$ | 7.0525 | 2.1986 | 0.9429 | 102 | H | 0.1758 | 2.5961 | 1.495 | 103 |

**Table 3A.6**   Embedded atom potential (see Ref. 94)

|                    | Ni-Ni | Al-Al |
|--------------------|-------|-------|
| D(eV)              | 2.79675 | 2.22312 |
| $r_0$(Å)           | 2.07675 | 3.21924 |
| $\alpha$(Å$^{-1}$) | 1.69648 | 3.61338 |
| Q                  | 1.83066 | 1.00881 |
| $\delta$(eV)       | −0.027537 | $0.42635 \times 10^{-1}$ |
| S                  | 0.94577 | 0.3984839 |
| $\beta$(Å$^{-1}$)  | 1.8194 | 1.45222 |
| $\varepsilon$      | −0.00066523 | $0.94920 \times 10^{-2}$ |
| a1 (eV)            | $0.32192 \times 10^2$ | $0.86105 \times 10^2$ |
| a2 (eV)            | 3.71422 | 6.27607 |
| a3 (eV)            | −2.03796 | −2.09815 |
| Ref                | 94 | 94 |

**Table 3A.7**    Stillinger-Weber potential (Reproduced from Zhou, X. W. and Wadley, H. N. G. (2007) A potential for simulating the atomic assembly of cubic AB compounds. *Computational Materials Science*, 39, 541, Elsevier)

|  | $A_{\mu\upsilon}$ (eV) | $S_{\mu\upsilon}$ | $C_{\mu\upsilon}$ (eV$^{1/2}$) | $\sigma_{\mu\upsilon}$ (Å) | $\gamma_{\mu\upsilon}$ (Å) | $r_{c,\mu\upsilon}$ (Å) | $r_{uc,\,\mu\upsilon}$ (Å) | Ref |
|---|---|---|---|---|---|---|---|---|
| Mg-Mg | 2.936977 | 1.836933 | 0 | 2.322254 | 0 | 4.511678 | 0 | 95 |
| O-O | 2.264679 | 1.030683 | 0 | 2.140179 | 0 | 3.530965 | 0 | 95 |
| Mg-O | 2.842769 | 123.5643 | 1.58378 | 0.418141 | 0.35076 | 2.979658 | 2.979658 | 95 |

# References

[1] Askeland, D. R. (1989) *The Science and Engineering of Materials*, PWS Publishing Company, Boston.

[2] Born, M. and Huang, K. (1954) *Dynamical Theory of Crystal Lattices*, Oxford University Press.

[3] Lewis, G. V. and Catlow, C. R. A. (1985) Potential models for ionic oxides. *J. Phys. C*, **18**, 1149.

[4] Dick, B. G. and Overhauser, A. W. (1958) Theory of the dielectric constants of alkali halide crystals. *Phys. Rev.*, **112**, 90.

[5] Bush, T. S., Gale, J. D., Catlow, C. R. A., and Battle, P. D. (1994) Self-consistent interatomic potentials for the simulation of binary and ternary oxides. *J. Mater. Chem.*, **4**, 831.

[6] Catlow, C. R. A. (1977) Point defect and electronic properties of uranium dioxide. *Proceedings of the Royal Society of London A*, **353**, 533.

[7] Gale, J. D. (2009) GULP Manual, Version 3, Nanochemistry Research Institute, Curtin University of Technology, Perth. Available at https://www.ivec.org/GULP/help/manuals.html.

[8] Dwivedi, A. and Cormack, A. N. (1990) A computer simulation study of the defect structure of calcia-stabilized zirconia. *Philosophia Magazine A*, **61**(1), 1.

[9] Cormack, A. N., Catlow, C. R. A., and Nowick A. S. (1989) Theoretical studies of off-centre Sc3 + impurities in CeO$_2$. *J. Phys. Chem. Solids*, **50**(2), 177.

[10] Li, P., Chen, I. W., Penner-Hahn, J. E., and Tien, T. Y. (1991) X-ray absorption studies of ceria with trivalent dopants. *J. Am. Ceram. Soc.*, **74**(5), 958.

[11] Minervini, L., Zacate, M. O., and Grimes, R. W. (1999) Defect cluster formation in M$_2$O$_3$-doped CeO$_2$. *Solid State Ionics*, **116**, 339.

[12] Ashurov, M. Kh., Voronko, Yu. K., Osiko, V. V., *et al.* (1977) Spectroscopic study of stoichiometry deviation in crystals with garnet structure. *Phys. Stat. Sol. (a)*, **42**, 101.

[13] Shackelford, J. (2000) *Introduction to Materials Science for Engineers*, 5th edn., Prentice Hall.

[14] Milanese, C., Buscaglia, V., Maglia, F., and Anselmi-Tamburini, U. (2004) Disorder and nonstoichiometry in synthetic garnets A$_3$B$_5$O$_{12}$ (A=Y, Lu-La, B=Al, Fe, Ga): A simulation study. *Chem. Mater.*, **16**, 1232.

[15] Patel, A. P., Levy, M. R., Grimes, R. W., *et al.* (2008) Mechanisms of nonstoichiometry in Y$_3$Al$_5$O$_{12}$. *Applied Physics Letters*, **93**, 191902–1.

[16] Ching, W. Y. and Xu, Y.-N. (1999) Nonscalability and nontransferability in the electronic properties of the Y-Al-O system. *Phys. Rev. B*, **59**(20), 12815.

[17] Zhilyakov, S. M., Mal'cev, V. I., and Naiden, E. P. (1980) Magnet structure Y-Fe-Sc garnets. *Sov. Solid State Phys.*, **22**(5), 1388.

[18] Asatryan, G. R., Baranov, G. P., and Zhekov, V. I. (1996) Electron paramagnetic resonance of Dy3 + ions in YAG. *Sov. Solid State Phys.*, **38**(3), 814.

[19] Kuklja, M. M. and Pandey, R. (1999) Atomistic modeling of native point defects in yttrium aluminum garnet crystal. *J. Am. Ceram. Soc.*, **82**(10), 2881.

[20] Binks, D. J. and Grimes, R. W. (1993) Incorporation of monovalent ions in ZnO and their influence on varistor degradation. *J. Am. Ceram. Soc.*, **76**(9), 2370.

[21] Levy, M. R., Grimes, R. W., and Sickafus, K. J. (2004) Disorder processes in A3 + B3 + O3 compounds: Implications for radiation tolerance. *Philos. Mag.*, **84**, 533.

[22] Devanathan, R., Weber, W. J., Singhal, S. C., and Gale, J. D. (2006) Computer simulation of defects and oxygen transport in yttria-stabilized zirconia. *Solid State Ionics*, **177**, 1251.

[23] Li, J. Y., Rogan, R. C., Ustundag, E., and Bhattacharya, K. (2005) Domain switching in polycrystalline ferroelectric ceramics. *Nature Materials*, **4**, 776.

[24] Sepliarsky, M., Phillpot, S. R., Streiffer, S. K., *et al.* (2001) Polarization reversal in a perovskite ferroelectric by molecular-dynamics simulation. *Applied Physics Letters*, **79**(26), 4417.

[25] Abell, G. C. (1985) Empirical chemical pseudopotential theory of molecular and metallic bonding. *Phys. Rev. B*, **31**(10), 6184.

[26] Tersoff, J. (1986) New empirical model for the structural properties of silicon. *Phys. Rev. Lett.*, **56**(6), 632.

[27] Tersoff, J. (1989) Modeling solid-state chemistry: Interatomic potentials for multicomponent systems. *Phys. Rev. B*, **39**(8), 5566.

[28] Brenner, D. W. (1990) Empirical potential for hydrocarbons for use in simulating the chemical vapor deposition of diamond films. *Phys. Rev. B*, **42**(15), 9458.

[29] Mowrey, R. C., Brenner, D. W., Dunlap, B. I., *et al.* (1991) Simulations of buckminsterfullerene (C60) collisions with a hydrogen-terminated diamond {111} surface. *J. Phys. Chem.*, **95**(19), 7138.

[30] Brenner, D. W., Harrison, J. A., White, C. T., and Colton, R. J. (1991) Molecular-dynamics simulations of the nanometer-scale mechanical-properties of compressed buckminsterfullerene. *Thin Solid Films*, **206**(1–2), 220.

[31] Williams, E. R., Johns, G. C. Jr., Fang, L., et al. (1992) Ion pickup of large, surface-adsorbed molecules: a demonstration of the Eley-Rideal mechanism. *J. Amer. Chem. Soc.*, **114**(9), 3207.

[32] Sinnott, S. B., Colton, R. J., White, C. T., et al. (1994) Surface patterning by atomically-controlled chemical forces: Molecular dynamics simulations. *Surface Science*, **316**(1–2), L1055.

[33] Harrison, J. A. and Brenner, D. W. (1995) Simulated tribochemistry: An atomic-scale view of the wear of diamond. *J. Amer. Chem. Soc.*, **116**, 10399.

[34] Harrison, J. A., White, C. T., Colton, R. J., and Brenner, D. W. (1992) Molecular-dynamics simulations of atomic-scale friction of diamond surfaces. *Phys. Rev. B*, **46**, 9700.

[35] Brenner, D. W. (2000) The art and science of an analytical potential. *Phys. Stat. Sol. (b)*, **217**(1), 23.

[36] Brenner, D. W. (1990) Empirical potential for hydrocarbons for use in simulating the chemical vapor deposition of diamond films. *Phys. Rev. B*, **42**, 9458.

[37] Peng, J., Wu, J., Hwang, K. C., et al. (2008) Can a single-wall carbon nanotube be modeled as a thin shell? *J. Mech. Phys. Solids*, **56**, 2213.

[38] Brenner, D. W., Shenderova, O. A., Harrison, J. A., *et al.* (2002) A second-generation reactive empirical bond order (REBO) potential energy expression for hydrocarbons. *J. Phys. Condens. Matter*, **14**, 783.

[39] Zhou, M. (2003) A new look at the atomic level virial stress: on continuum-molecular system equivalence, *Proc. R. Soc.*, London 459, 2347.

[40] LAMMPS, Users Manual http://lammps.sandia.gov.

[41] Daw, M. S. and Baskes, M. I. (1983) Semiempirical, quantum mechanical calculation of hydrogen embrittlement in metals. *Phys. Rev. Lett.*, **50**, 1285.

[42] Daw, M. S. and Baskes, M. I. (1984) Embedded-atom method: Derivation and application to impurities, surfaces, and other defects in materials. *Phys. Rev. B*, **29**, 6443.

[43] Ercolessi, F., Tosatti, E., and Parrinello, M. (1986) Au (100) surface reconstruction. *Phys. Rev. Lett.*, **57**, 719.

[44] Finnis, M. W. and Sinclair, J. E. (1984) A simple empirical N-body potential for transition metals. *Philos. Mag. A*, **50**, 45.

[45] Norskov, J. K. (1990) Chemisorption on metal surfaces. *Reports on Progress in Physics*, **53**, 1253.

[46] Norskov, J. K. and Lang, N. D. (1980) Effective-medium theory of chemical binding: Application to chemisorption. *Phys. Rev. B*, **21**, 2131.

[47] Stott, M. J. and Zaremba, E. (1980) Quasiatoms: An approach to atoms in nonuniform electronic systems. *Phys. Rev. B*, **22**, 1564.

[48] Baskes, M. I. (1992) Modified embedded-atom potentials for cubic materials and impurities. *Phys. Rev. B*, **46**, 2727.

[49] Rose, J. H., Smith, J. R., Guinea, F., and Ferrante, J. (1984) Universal features of the equation of state of metals. *Phys. Rev. B*, **29**, 2963.

[50] Baskes, M. I., Angelo, J. E., and Bisson, C. L. (1994) Atomistic calculations of composite interfaces. *Modelling Simul. Mater. Sci. Eng.*, **2**, 515.

[51] Gall, K., Horstemeyer, M. F., Van Schilfgaarde, M., and Baskes, M. I. (2000) Atomistic simulations on the tensile debonding of an aluminum–silicon interface. *J. Mech. Phys. Solids*, **48**, 2183.

[52] Huang H., Ghoniem, N. M., Wong, J. K., and Baskes, M. I. (1995) Molecular dynamics determination of defect energetics in beta-SiC using three representative empirical potentials. *Modelling Simul. Mater. Sci. Eng.*, **3**, 615.

[53] Hu, W., Deng, H., Yuan, X., and Fukumoto, M. (2003) Point-defect properties in HCP rare earth metals with analytic modified embedded atom potentials. *European Physical Journal B*, **34**, 429.

[54] Hu, W., Zhang, B., Huang, B., *et al.* (2001) Analytic modified embedded atom potentials for HCP metals. *J. Phys. Condens. Matter*, **13**, 1193.

[55] Baskes, M. I., Nelson, J. S., and Wright, A. F. (1989) Semiempirical modified embedded-atom potentials for silicon and germanium. *Phys. Rev. B*, **40**, 6085.

[56] Zhou, X. W., Wadley, H. N. G., Johnson, R. A., *et al.* (2001) Atomic scale structure of sputtered metal multilayers. *Acta Mater.*, **49**(19), 4005.

[57] Wadley, H. N. G., Zhou, X., Johnson, R. A., and Neurock, M. (2001) Mechanisms, models, and methods of vapor deposition. *Progress in Material Science*, **46**, 329 (review article).

[58] Mishin, Y., Mehl, M. J., and Papaconstantopoulos, D. A. (2005) Phase stability in the Fe–Ni system: Investigation by first-principles calculations and atomistic simulations. *Acta Mater.*, **53**(15), 4029.

[59] Hashibon, A., Lozovoi, A. Y., Mishin, Y., *et al.* (2008) Interatomic potential for the Cu-Ta system and its application to surface wetting and dewetting. *Phys. Rev. B*, **77**, 094131.

[60] Allen, M. P. and Tildesley, D. J. (1990) *Computer Simulation of Liquids*. Clarendon Press, Oxford.

[61] White, A. (2000) Intermolecular potentials of mixed systems: Testing the Lorentz-Berthelot mixing rules with *ab nitio* calculations. Australia, Defence Science and Technology Organization.

[62] Johnson, R. A. (1989) Alloy models with the embedded-atom method. *Phys. Rev. B*, **39**(17), 12554.

[63] Cao, A., Wei, Y., and Mao, S. X. (2008) Alternating starvation of dislocations during plastic yielding in metallic nanowires. *Scripta Mater.*, **59**(2), 219.

[64] Mishin, Y., Farka, D., Mehl, M. J., and Papaconstantopoulos, D. A. (1999) Interatomic potentials for monoatomic metals from experimental data and *ab initio* calculations. *Phys. Rev. B*, **59**(5), 3393.

[65] Liang, W., Zhou, M., and Ke, F. (2005) Shape memory effect in Cu nanowires. *Nano Letters*, **5**(10), 2039.

[66] Wu, B., Heidelberg, A., and Boland, J. J. (2005) Mechanical properties of ultrahigh-strength gold nanowires. *Nature Materials*, **4**(7), 525.

[67] Honeycutt, J. D. and Andersen, H. C. (1987) Molecular dynamics study of melting and freezing of small Lennard-Jones clusters. *J. Phys. Chem.*, **91**(19), 4950.

[68] Bunshah, R. F. (1994) *Handbook of Deposition Technologies for Films and Coatings: Science, Technology, and Applications*, 2nd edn., Noyes Publications, Park Ridge, NJ, pp. 6, 140, 374, 434.

[69] Yoshiya, M., Tanaka, I., and Adachi, H. (2000) Energetical role of modeled intergranular glassy film in $Si_3N_4$–$SiO_2$ ceramics. *Acta Mater.*, **48**, 4641.

[70] Liu, B., Wang, J. Y., Li, F. Z., *et al.* (2009) Atomic-scale studies of native point defect and nonstoichiometry in silicon oxynitride. *J. Phys. Chem. Solids*, **70**, 982.

[71] Walker, A. M., Wright, K., and Slater, B. (2003) A computational study of oxygen diffusion in olivine. *Phys. Chem. Minerals*, **30**, 536.

[72] Winkler, B. and Dove, M. T. (1991) Static lattice energy minimization and lattice dynamics calculations on aluminosilicate minerals. *American Mineralogist*, **76**, 313.

[73] Sun, X., Chen, Q., Wang, C., *et al.* (2004) Melting and isothermal bulk modulus of the rocksalt phase of ZnO with molecular dynamics simulation. *Physica B: Condensed Matter*, **355**, 126.

[74] Kuklja, M. M. and Pandey, R. (1999) Atomistic modeling of native point defects in yttrium aluminum garnet crystal. *J. Am. Ceram. Soc.*, **82**, 2881.

[75] Schuh, L., Metselaar, R., and Catlow, C. R. A. (1991) Computer modelling studies of defect structures and migration mechanisms in yttrium aluminum garnet. *J. Eur. Ceram. Soc.*, **7**, 67.

[76] Bush, T. S., Gale, J. D., Catlow, C. R. A., and Battle, P. D. (1994) Self-consistent interatomic potentials for the simulation of binary and ternary oxides. *J. Mater. Chem.*, **4**, 831.

[77] Alimohammadi, M. and Fichthorn, K. A. (2009) Molecular dynamics simulation of the aggregation of titanium dioxide nanocrystals: Preferential alignment. *Nano Letters*, **9**, 4198.

[78] Moore, E. A. and Widatallah, H. M. (2008) Iron(III) as a defect in diantimony tetroxide. *Materials Research Bulletin*, **43**, 2361.

[79] Henkelman, G., Uberuaga, B. P., Harris, D. J., *et al.* (2005) MgO addimer diffusion on MgO(100): A comparison of *ab initio* and empirical models. *Phys. Rev. B*, **72**, 115437.

[80] Hendy, S. C. (2004) Molecular dynamics simulations of oxide surfaces in water. *Current Applied Physics*, **4**, 144.

[81] Binks, D. J. and Grimes, R. W. (1993) Incorporation of monovalent ions in ZnO and their influence on varistor degradation. *J. Am. Ceram. Soc.*, **76**, 2370.

[82] Dorta-Urra, A. and Gulin-Gonzalez, J. (2006) A computational investigation on substitution of magnesium for aluminum in AlPO$_4$-5 microporous material. *Microporous and Mesoporous Materials*, **92**, 109.

[83] Lewis, D. W., Grimes, R. W. and Catlow, C. R. A. (1995) Defect processes at low coordinate surface sites of MgO and their role in the partial oxidation of hydrocarbons. *Journal of Molecular Catalysis A: Chemical*, **100**, 103.

[84] Su, M., Lin, C.-Y., Wang, S.-H., and Chi-Chuan, C.-C. (2004) Molecular dynamics simulations on the direct sputtering of Al$_2$O$_3$ insulating film in a magnetic tunneling junction. *Journal of Magnetism and Magnetic Materials*, **277**, 263.

[85] Levy, M. R., Grimes, R. W., and Sickafus, K. J. (2004) Disorder processes in A3 + B3 + O3 compounds: Implications for radiation tolerance. *Philos. Mag.*, **84**, 533.

[86] Yu, H. and Sun, F. (2009) *A modified embedded atom method interatomic potential for the Ti-N system. Physica B*, **404**, 1692.

[87] Baskes, M. I., (1992) Modified embedded atom potentials for cubic materials and impurities. *Phys. Rev. B*, **46**, 2727.

[88] Kim, Y.-M., Kim, N. J., and Lee, B.-J. (2009) Atomistic modeling of pure Mg and Mg–Al systems. *Calphad*, **33**, 650.

[89] Baskes, M. I. (1994) Modified embedded atom potentials for HCP metals. *Modelling Simul. Mater. Sci. Eng.*, **2**, 147.

[90] Munetoh, S., Motooka, T., Moriguchi, K., and Shintani, A. (2007) Interatomic potential for Si–O systems using Tersoff parameterization. *Computational Materials Science*, **39**, 334.

[91] de Brito Mota, F., Justo, J. F., and Fazzio, A. (1999) Hydrogen role on the properties of amorphous silicon nitride, *Journal of Applied Physics*, **86**, 1843.

[92] Tersoff, J. (1989) Modeling solid-state chemistry: Interatomic potentials for multicomponent systems. *Phys. Rev. B*, **39**, 5566.

[93] Aoumeur, F. Z., Benkabou, Kh., and Belgoumene, B. (2003) Structural and dynamical properties of ZnO in zinc-blende and rocksalt phases, *Physica B*, **337**, 292.

[94] http://www.diracdelta.co.uk/science/source/l/e/lennard-jones%20potential/source.html.

[95] Kun Qin (2009) Strain rate sensitivities of fcc metals using molecular dynamics simulation, Ph.D. Dissertation, University of Science and Technology of China.

[96] Wei Fang, Bai PuCun, Zhou Tietao, *et al.* (2004) Influence of Li on the early stage clustering behavior of Al-Zn-Mg-Cu-Li series alloy. *Journal of Aeronautical Materials*, **24**(1), 28.

[97] Liu Yuan-xia, Sun Jia-jia, Gao Yuan, *et al.* (2008) Several fitting methods of Lennard-Jones interatomic potential parameters. *Journal of Liaoning University*, **35**(3), 206.

[98] Girifalco, L.A. and Weizer, V. G. (1959) Application of the morse potential function to cubic metals. *Phys. Rev.*, **114**, 687.

[99] Nguyen Van Hung (2004) A method for calculation of morse potential for fcc, bcc, hcp crystals applied to debye-waller factor and equation of state. *Communications in Physics*, **14**, 7.

[100] Mylvaganam, K. and Zhang, L. C. (2002) Effect of oxygen penetration in silicon due to nano-indentation. *Nanotechnology*, **13**, 623.

[101] Kun Qin (2009) Strain rate sensitivities of fcc metals using molecular dynamics simulation, Ph.D. Dissertation, University of Science and Technology of China.

[102] Walker, A. M., Wright, K., and Slater, B. (2003) A computational study of oxygen diffusion in olivine. *Phys. Chem. Minerals*, **30**, 536.

[103] Yang Shiqing, Zhang Wenxu, Peng Bing, *et al.* (2000) The simulation of the hydrogen motion in the Nd crystal by molecular dynamics. *Chinese Journal of Atomic and Molecular Physics*, **17**(2), 279.

[104] Mishin, Y., Mehl, M. J., and Papaconstantopoulos, D. A. (2002) Embedded-atom potential for B2-NiAl. *Phys. Rev. B*, **65**, 224114.

[105] Zhou, X. W. and Wadley, H. N. G. (2007) A potential for simulating the atomic assembly of cubic AB compounds. *Computational Materials Science*, **39**, 541.

# 4

# Quantum Mechanics and Its Energy Linkage with Atomistic Analysis

This chapter introduces the basic concepts, features and postulates of Quantum Mechanics (QM). The single-electron solution and molecule wavefunction of the Schrödinger equation are introduced. Pauli's exclusion principle, with the Slater determinant, and the Rayleigh-Ritz variation principle via the expectation energy value of multi-electron systems are discussed. Solution schemes for multi-body systems are reviewed, including tight binding (TB), Hartree-Fock (HF) and density function theory (DFT) methods. The energy linkage of QM and atomistic analysis is emphasized and the method of using first principles to determine potential functions for atomistic simulations is exemplified at the very beginning and discussed in the last section of this chapter.

## 4.1 Determination of Uranium Dioxide Atomistic Potential and the Significance of QM

Section 3.4 discusses briefly how to determine short-distance potential by the Hartree-Fock method of QM for uranium dioxide ($UO_2$). Here further explanation is given on how to determine the potential for atomistic analysis based on QM calculations, thus to understand the energy linkage between QM and atomistic analysis and the significance of the QM study presented in this chapter.

In $UO_2$, every oxygen ion $O^{2-}$ is surrounded by four cations $U^{4+}$ located at the vertices of a tetrahedron. To better reproduce the ion $O^{2-}$-$O^{2-}$ interaction in $UO_2$, this local environment was simulated in the calculation as a tetrahedron cage of positive point charges surrounding the $O^{2-}$. The QM calculation by the Hartree-Fock method[1] was first conducted to get the energy $E_{tot}(r)$ between two cages with $O^{2-}$ inside at a given distance r. This is the total energy of the system which includes the ions' own energy $E_{iso}$ and the electrostatic energy $E_{elec}$, thus the ion $O^{2-}$-$O^{2-}$ pair potential $V(r)$ should be obtained by substracting these two kinds of energies from the total energy as follows:

$$V(r) = E_{tot} - E_{elec} - E_{iso} \qquad (4.1a)$$

More specifically, $E_{elec}$ includes energy of cage-cage and cage-$O^{2-}$ interactions. The isolated energy $E_{iso}$ of the two oxygen ions is the energy when they are apart from each other with infinite distance. This isolated energy can easily be determined by a solution using central force field (see Appendix 4.A) or found from a handbook.

The results of $E_{tot}$ obtained by Catlow (1977) using Hartree-Fock calculations and the values of $E_{iso}$ and $E_{elec}$ are shown in Table 4.1. The original results were given in atomic units (au). The conversion from the

*Multiscale Analysis of Deformation and Failure of Materials*   Jinghong Fan
© 2011 John Wiley & Sons, Ltd

**Table 4.1**   Results of Hartree-Fock calculations for ion $O^{2-}$-$O^{2-}$ interactions (Reprinted with permission from Catlow, C. R. A. (1977) Point defect and electronic properties of uranium dioxide. *Proceedings of the Royal Society of London* A, 353, 533, Royal Society of London)

| Internuclear separation (au) (Å) | System total energy $E_{tot}$ by H-F (au) | Isolated energy $E_{iso}$ (au) | Electrostatic term $E_{elec}$ (au) | $O^{2-}$-$O^{2-}$ pair potential V (au) | $O^{2-}$-$O^{2-}$ pair potential V (eV) |
|---|---|---|---|---|---|
| 2.8  1.4817 | $-149.94221627$ | $-149.5399674$ | $-0.37514030$ | $-0.027109$ | $-0.737668$ |
| 2.4  1.2700 | $-149.87535520$ | $-149.5399674$ | $-0.32141891$ | $-0.012166$ | $-0.331043$ |
| 2.3  1.2171 | $-149.83783810$ | $-149.5399674$ | $-0.30468716$ | $-0.006816$ | $-0.185484$ |
| 2.2  1.1642 | $-149.78687104$ | $-149.5399674$ | $-0.28627024$ | $-0.039367$ | $-1.07121$ |
| 2.1  1.1113 | $-149.71845755$ | $-149.5399674$ | $-0.26594991$ | $0.08746$ | $2.37989$ |
| 2.0  1.0584 | $-149.6272992$ | $-149.5399674$ | $-0.24341724$ | $0.156085$ | $4.24727$ |
| 1.9  1.0054 | $-149.5063382$ | $-149.5399674$ | $-0.21833937$ | $0.251969$ | $6.85637$ |
| 1.8  0.9525 | $-149.3461521$ | $-149.5399674$ | $-0.19029387$ | $0.384109$ | $10.4521$ |
| 1.7  0.8996 | $-149.1341203$ | $-149.5399674$ | $-0.15876197$ | $0.564609$ | $15.3637$ |

unit au to Å and eV can be found in Section 2.3.4 and Table 2.3, and the converted results are also listed in Table 4.1. Note that the energy of two isolated $O^{-2}$ is

$$E_{iso} = -149.5399674 \text{ au} \tag{4.1b}$$

Based on (4.1a) and (4.1b) the required ion $O^2$-$O^{2-}$ pair potential V is calculated and listed in the last column by unit eV. The first and last columns give the function of V(r); by curve fitting it can be fitted to the Buckingham potential discussed in Section 3.4:

$$V(r) = A\exp(-r/\rho) - Cr^{-6} \tag{4.1c}$$

with parameters

$$A = 22764 \text{ eV}, \ \rho = 0.41 \text{ Å}, \ C = 20.37 \text{ eV\AA}^6 \tag{4.1d}$$

From the above examples it is seen that:

- The QM calculation based on the Hartree-Fock method and other *ab initio* first-principle methods which will be shown in Section 4.7 is powerful and it is a useful tool to determine the potential function for atomistic analysis.
- It is better to include the environment effect of the chosen material structure for accuracy, although it will increase the amount and complexity of the calculation. Note the isolation potential energy needs to be subtracted before fitting to short-ranged interaction potential model.
- The challenging task is to obtain the energy solution for the multi-body (or multi-electrons) of the system. The basic concepts, governing equations, postulates, and solution techniques will be introduced in Sections 4.2 to 4.10. We will return to the determination of the potential function again in Section 4.12.

## 4.2   Some Basic Concepts of QM

The understanding and quantitative prediction of the electron structure plays a fundamental role in developing science and technology to reach today's level and to advance for tomorrow. Taking atomistic simulation as an example, more and more reliable potential functions for atomistic simulation are developed based on the first-principle simulation of QM. QM is a discipline which investigates the structure and motion of electrons in atoms and materials.

Within classical MD, interacting particles under interatomic forces by potentials are regarded as solid spheres. It is assumed that the internal states of atoms and molecules do not change. This assumption makes the atomistic analysis simple because it avoids accounting for the energy exchanges between particle components, i.e., electrons and nuclei. However, each atom within an atomistic system represents a complex physical entity that can evolve in time and alter its internal state by exchanging energy with its surrounding media. Most importantly, the nature of interatomic forces adopted in an atomistic system is actually determined by the evolution of electrons and nuclei as well as their states. To accurately determine the potential and the interatomic force, it is necessary to consider the evolution of the internal structure of the atom. Therefore, the basic concepts and methods of QM are fundamental for atomistic analysis.

Qualitatively, the physical nature of condensed matters such as solids can be understood easily. Within an isolated atom, motions of electrons with negative charges are bound to nuclei with positive charges by the so-called Coulomb force.

A fundamental concept of modern physics is that any matter (such as electrons) is not only a particle but also a kind of wave. Inspired by Planck's quantum physics, in 1905 Einstein proposed that light is a kind of wave and at the same time it is a particle. This concept successfully explained the photoelectric effect. He thought that light runs in space just like a particle whose energy is absorbed by an object. This particle is called a photon; and each photon has the energy $\varepsilon$ which is proportional to its wave frequency $\omega$ by the following formula:

$$\varepsilon = \hbar\omega \qquad (4.2a)$$

Where $\hbar$ is Plank's constant with the value: $\hbar = 10^{-27}$ ergs $= 10^{-34}$ J/s. In 1924, L. V. de Broglie proposed in his doctoral dissertation the general idea that a particle has a wavelength which is inversely proportional to the momentum p as follows:

$$\lambda = \hbar/p \qquad (4.2b)$$

For a general moving object, the wavelength calculated from (4.2b) is extremely small and can be neglected due to the small value of $\hbar$ and relatively large linear momentum, thus the adoption of classical theory is sufficient. However, the electron has a mass of $m_e = 9.11 \times 10^{-31}$ kg, which gives a quite small momentum p. Under this condition, the wavelength calculated from (4.2b) is at the same scale (Å) as separation distances among atoms of solids, thus the corresponding wave nature of electrons, such as wave reflection or refraction, etc. can be easily detected.

## 4.3 Postulates of QM

Because the QM theory including the Schrödinger equation, the governing equation of QM, cannot be derived from a more fundamental theory, postulates are usually given in formulating the QM framework. The descriptions of each of the postulates are from various authors.[2, 3] The following five items are examples of the postulates. Postulate 3 can be derived from a more general postulate, but so as not to involve too deep a background, here we list it as a simple postulate.

**Postulate 1. There exists for any particle system a wavefunction $\Psi$ which contains all the information that can be obtained about the system**
$\Psi$ is a function of the coordinates of all the particles in the system as well as the time. $\Psi$ by itself has no physical meaning, but $\Psi^*\Psi$ gives the probability per unit length, area, and volume of finding the particle.

Here, $\Psi^*$ is the complex conjugate function of $\Psi$. If $\Psi = \Psi_r + i\Psi_i$ then $\Psi^* = \Psi_r - i\Psi_i$ with $i = \sqrt{-1}$, so we have:

$$\Psi^*\Psi = (\Psi_r - i\Psi_i)(\Psi_r + i\Psi_i) = \Psi_r^2 + \Psi_i^2 \tag{4.3a}$$

Examples of the probability and the normalization condition are given as follows:

One particle in 1D space: In this case, $\Psi = \Psi(x, t)$, the probability of finding the particle between x and x + dx is $\Psi^*\Psi dx$ and the normalization equation is given as follows:

$$\int_{-\infty}^{\infty} \Psi^*\Psi dx = 1 \tag{4.3b}$$

One particle in 2D space: In this case, $\Psi = \Psi(x, y, t)$, $\Psi^*\Psi$ is the probability of finding the particle per unit area, thus the probability of finding the particle between x and x + dx and between y and y + dy is $\Psi^*\Psi$ dxdy and we have:

$$\int_{-\infty}^{\infty}\int_{-\infty}^{\infty} \Psi^*\Psi dx\, dy = 1 \tag{4.3c}$$

One particle in 3D space: In this case, $\Psi = \Psi(x, y, z, t)$, similarly we have the following condition:

$$\int \Psi^*\Psi d^3 r = 1 \tag{4.3d}$$

where $d^3 r$ denotes the differential volume. For Cartesian coordinates $d^3 r = dxdydz$; for spherical coordinates $d^3 r = r^2 \sin\theta dr d\theta d\phi$. The differential range will be determined by the coordinates that we choose.

Two particles in 1D space: In this case, $\Psi = \Psi(x_1, x_2, t)$ and $\Psi^*\Psi dx_1 dx_2$ is the probability of finding particle 1 in the differential domain of $(x_1, x_1 + dx_1)$ and finding particle 2 in the domain of $(x_2, x_2 + dx_2)$, and the normalization condition is:

$$\int_{-\infty}^{\infty}\int_{-\infty}^{\infty} \Psi^*\Psi dx_1 dx_2 = 1 \tag{4.3e}$$

**Postulate 2. All physical quantities in QM from classical mechanics become Hermitian operators acting upon the wavefunction $\Psi$**

An operator is a symbol which must operate on a function to have a certain meaning. Here, Hermitan operators are all applied on the wavefunction $\Psi$. The introduction of an operator will make the expressions more concise. The Hermitian operators and their QM quantities which correspond to, respectively, the position vector $\bar{r}$, linear momentum $\bar{p}$ and energy E are listed in Table 2 through their component values, x, $p_x$ and energy E itself.

In the table, the symbol $\wedge$ above a quantity denotes the Hermitian operator of that quantity. $\bar{i}, \bar{j}, \bar{k}$ are the unit vectors along the x, y, z direction of Cartesian coordinates. Hermitian operators can carry on different algebraic operations and then apply to the wavefunction. Its mathematical definition is shown at the bottom. Examples of physical quantities from CM corresponding to QM through Hermitian operators are given as follows:

1D classical Hermitian: As shown in (2.14d), Hamiltonian in classical mechanics denotes the total energy. For 1D case of a single particle we have:

$$h = T + V = E \tag{4.3f}$$

**Table 4.2** Hermitian operators connecting CM and QM physical quantities*

| Quantity of classical mechanics | QM expressions by Hermitian | Corresponding quantities of quantum mechanics |
|---|---|---|
| x, y, z and $\bar{r}$ | $\hat{x} = x,\ \hat{y} = y,\ \hat{z} = z,\ \hat{\bar{r}} = \bar{r}$ | $x\Psi,\ y\Psi,\ z\Psi,\ \bar{r}\Psi$ |
| $p_x$(x component of $\bar{p}$) | $\hat{p}_x = -i\hbar\dfrac{\partial}{\partial x}$ | $-i\hbar\dfrac{\partial\Psi}{\partial x}$ |
| $\bar{p}$ (linear momentum vector) | $\hat{\bar{p}} = -i\hbar\left(\dfrac{\partial}{\partial x}\bar{i} + \dfrac{\partial}{\partial y}\bar{j}\right.$ $\left. + \dfrac{\partial}{\partial x}\bar{k}\right) = -i\hbar\nabla$ | $-i\hbar\nabla\Psi =$ $-i\hbar\left(\dfrac{\partial\Psi}{\partial x}\bar{i} + \dfrac{\partial\Psi}{\partial y}\bar{j} + \dfrac{\partial\Psi}{\partial z}\bar{k}\right)$ |
| E (energy of particle or system) | $\hat{E} = i\hbar\dfrac{\partial}{\partial t}$ | $i\hbar\dfrac{\partial\Psi}{\partial t}$ |

* For arbitrary functions f(x) and g(x), a mathematical definition of Hermitian $\hat{\theta}$ is given as: $\int f^*(\hat{\theta}g)dx \equiv \int (\hat{\theta}f)^*g dx$

In QM this becomes:

$$\hat{h} = \hat{T} + \hat{V} = \hat{E} \tag{4.3g}$$

where:

$$T = \frac{P_x^2}{2m} \tag{4.3h}$$

Now, we would like to see what is the expression in QM corresponding to the total energy expressed by Hamiltonian in classical mechanics below:

$$h = \frac{1}{2m}(p_x)(p_x) + V(x) \tag{4.3i}$$

The way to do so is first to find the corresponding Hermitian operator of the above equation and then apply it on the wavefunction.

From Table 4.1, we can use the QM expressions of $p_x$ and V(x) in the above expressions to get the Hermitian operator for h as follows:

$$\hat{h} = \frac{1}{2m}(\hat{p}_x\hat{p}_x) + V(\hat{x}) = \frac{1}{2m}\left(-i\hbar\frac{\partial}{\partial x}\right)\left(-i\hbar\frac{\partial}{\partial x}\right) + V(x) = \frac{-\hbar^2}{2m}\frac{\partial^2}{\partial x^2} + V(x) \tag{4.3j}$$

From Table (4.1a), we also have

$$\hat{E} = i\hbar\frac{\partial}{\partial t} \tag{4.3k}$$

Substituting (4.3j) and (4.3k) into (4.3g) and then operating the obtained Hermitian operator to wavefunction $\Psi$ we get the corresponding one-dimensional time-dependent Schrödinger equation as follows:

$$\frac{-\hbar^2}{2m}\frac{\partial^2 \Psi}{\partial x^2} + V(x)\Psi = i\hbar \frac{\partial}{\partial t}\Psi(x,t) \tag{4.31}$$

For the comparison with the expressions below, this Schrödinger equation can also be written as:

$$\overset{\wedge}{h}\Psi = i\hbar \frac{\partial}{\partial t}\Psi(x,t) \tag{4.3m}$$

Find Hermitian operator for angular momentum: Vector of angular momentum in CM is defined as:

$$\overline{L} = \overline{r} \times \overline{p} = \begin{vmatrix} \overline{i} & \overline{j} & \overline{k} \\ x & y & z \\ p_x & p_y & p_z \end{vmatrix}$$

Using Table 4.2 to get the Hermitian operators for x, y, z and $p_x$, $p_y$, $p_z$, we have:

$$\overset{\wedge}{L} = \overset{\wedge}{r} \times \overset{\wedge}{p} = \begin{vmatrix} \overline{i} & \overline{j} & \overline{k} \\ \hat{x} & \hat{y} & \hat{z} \\ \hat{p}_x & \hat{p}_y & \hat{p}_z \end{vmatrix} = \begin{vmatrix} \overline{i} & \overline{j} & \overline{k} \\ x & y & z \\ -i\hbar\dfrac{\partial}{\partial x} & -i\hbar\dfrac{\partial}{\partial y} & -i\hbar\dfrac{\partial}{\partial z} \end{vmatrix} \tag{4.3n}$$

From this expression, we have:

$$\overset{\wedge}{L}_x = y\left(-i\hbar\frac{\partial}{\partial z}\right) + z\left(i\hbar\frac{\partial}{\partial y}\right) = i\hbar\left(z\frac{\partial}{\partial y} - y\frac{\partial}{\partial z}\right) \tag{4.3o}$$

This Hermitian operator acting on the wavefunction $\Psi$ produces the QM expression of the x-component of the angular momentum.

Hermitian for a helium (He) atom and its Schrödinger equation: Figure 4.1 shows the He atom. Its nucleus with charge $+2e$ has a distance from the two electrons e1 and e2, respectively, as $r_{10}$, $r_{20}$, with the distance between the two electrons being $r_{12}$. The total energy of the He atom can be written as follows:

$$E = \frac{p_0^2}{2m_0} + \frac{p_1^2}{2m_1} + \frac{p_2^2}{2m_2} + V_{10} + V_{20} + V_{12} \tag{4.3p}$$

Symbol m denotes mass, $V_{10}$, $V_{20}$ denote, respectively, the potentials between electron 1 and 2 with nucleus; $V_{12}$ denotes the potential between two electrons. Based on the Coulomb function shown in (3.1),

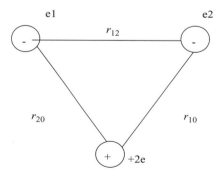

**Figure 4.1**  The relationship between protons and electrons in He atom

we have:

$$E = \frac{\bar{p}_0^2}{2m_0} + \frac{\bar{p}_1^2}{2m_1} + \frac{\bar{p}_2^2}{2m_2} + \frac{(2e)(-e)}{r_{10}} + \frac{(2e)(-e)}{r_{20}} + \frac{(-e)(-e)}{r_{12}} \tag{4.3q}$$

where the constant $\frac{1}{4\pi\varepsilon_0}$ in Coulomb potential is normalized to 1 for simplicity. Using Table 4.2 to get the Hermitian operators for E, $p_0$, $p_1$, $p_2$, and position vector through its components we have, for example, the following expressions:

$$\hat{p}_0 = -i\,\hbar\left(\frac{\partial}{\partial x}\,\bar{i} + \frac{\partial}{\partial y}\,\bar{j} + \frac{\partial}{\partial x}\,\bar{k}\right) = -i\hbar\nabla \tag{4.3r}$$

Where the symbol for gradient is: $\nabla = \dfrac{\partial}{\partial x}\,\bar{i} + \dfrac{\partial}{\partial y}\,\bar{j} + \dfrac{\partial}{\partial x}\,\bar{k}$

$$\hat{p}_0^2 = -\hbar^2\nabla_0^2 \tag{4.3s}$$

Thus (4.3l) can be expressed by Hermitian operators as follows:

$$i\hbar\frac{\partial}{\partial t} = -\frac{\hbar^2}{2m_0}\nabla_0^2 - \frac{\hbar^2}{2m_1}\nabla_1^2 - \frac{\hbar^2}{2m_2}\nabla_2^2 - e^2\left(\frac{2}{r_{10}} + \frac{2}{r_{20}} - \frac{1}{r_{12}}\right) \tag{4.3t}$$

The Schrödinger equation for He atom can then be expressed by the wavefunction $\Psi$ as follows:

$$-\left[\frac{\hbar^2}{2m_0}\nabla_0^2 + \frac{\hbar^2}{2m_1}\nabla_1^2 + \frac{\hbar^2}{2m_2}\nabla_2^2 + e^2\left(\frac{2}{r_{10}} + \frac{2}{r_{20}} - \frac{1}{r_{12}}\right)\right]\Psi = i\hbar\frac{\partial\Psi}{\partial t} \tag{4.3u}$$

Frequently, it is written by Hamiltonian operator $\hat{H}$ as follows:

$$\hat{H}\,\Psi = i\hbar\frac{\partial\Psi}{\partial t} \tag{4.3v}$$

$$\text{where } \hat{H} = -\left[\frac{\hbar^2}{2m_0}\nabla_0^2 + \frac{\hbar^2}{2m_1}\nabla_1^2 + \frac{\hbar^2}{2m_2}\nabla_2^2 - e^2\left(\frac{2}{r_{10}} + \frac{2}{r_{20}} - \frac{1}{r_{12}}\right)\right] \tag{4.3w}$$

(4.3v) is similar to (4.3m), both $\hat{h}$ and $\hat{H}$ are derived from the Hamiltonian, i.e., the total energy. They are both called Hamiltonian operator, which is one type of Hermitian operator. Each Hamiltonian operator includes terms which correspond to both the kinetic energy and potential energy. Specifically, the first three terms on the right side of (4.3w) by the product of spatial gradient such as $\nabla$(or $-\frac{\partial^2}{\partial x^2}$ in (4.3j)) with $\left(-\frac{\hbar^2}{2m}\right)$ denote the kinetic energy and the last three terms denote the potential energy.

## Discussion

From the above examples it is seen that one can obtain the Schrödinger equation by writing the classical Hamiltonian and then replacing the terms with the Hermitian operators listed in Table 4.2.

It is also seen that because classical Hamiltonian is uniquely defined as the sum of kinetic energy and potential energy (i.e., T + V), the Hermitian operator is uniquely defined. However, because every problem has its own explicit forms of kinetic and potential energy, the expanded form of the Hermitian operator and Schrödinger equation will be different; which accounts for any difference, such as different atoms and different conditions.

Because a QM quantity is just the Hermitian operator acting on wavefunction $\Psi$, frequently in QM one just uses Hermitian to describe the physical quantity, and does not explicitly write $\Psi$.

**Postulate 3. Eigenfunctions of the Hermitian operator form a basis on which any wavefunction can be written as a linear combination**

Since the operators in QM are Hermitian, one knows that the eigenfunctions $\varphi_i(x)$ (i = 1, 2, . . ., N) form a basis. Therefore, any wavefunction can be written as a linear combination of the eigenfunctions of the operator being measured. Once a measurement is performed, the wavefunction arbitrarily "collapses" into one of the eigenfunctions of the operators. In addition, there are the following important results. The eigenvalues for any Hermitian operator are real; and the eigenfunctions of any Hermitian operator are mutually orthogonal, which can be mathematically expressed as follows:

$$\int_{-\infty}^{\infty} \varphi_i^*(x)\varphi_j(x)dx = \delta_{ij} \tag{4.3x}$$

where $\delta_{ij}$ is the Kronecker delta, which is zero when $i \neq j$ and equal to 1 when $i = j$. The mathematic description that the eigenfunction of a Hermitian operator forms a basis for function f(x) can be given as follows:

$$f(x) = \sum_i a_i \varphi_i(x) \tag{4.3y}$$

This postulate is important for the QM solution to a multi-body (or multi-electron) molecular problem. It is used to create the molecular orbitals (see Section 4.6), which are expanded as a linear combination of such basis (functions) sets with the coefficients to be determined. Usually these functions are atomic orbitals in the sense that they are centered on atoms (see Appendix 4.A). Additionally, basis sets composed of sets of plane waves are often used, especially in calculations involving systems with periodic boundary conditions of crystal material; this basis set and its application will be further introduced briefly in Section 4.11.2.

**Postulate 4. All particles in nature can be classified as having integer spin or half-integer spin. The former are called bosons and the latter are called fermions. If there is a system of identical particles:**

(i) **The exchange of any two identical bosons must result in no change for wavefunction $\Psi$.**
(ii) **The exchange of any two identical fermions must result in a sign change for wavefunction $\Psi$. This shows an antisymmetric wavefunction rule for fermions.**

Electrons along with protons and neutrons are fermions, which follow the antisymmetric wavefunction rule described in (ii); this is also the requirement of Pauli's principle. The Pauli principle states that no two electrons can simultaneously exist in the same state having the same wavefunction and the same energy level. In other words, if two fermions are identical, the two wavefunctions must have opposite signs.

A very simple way of writing an antisymmetric wavefunction for identical fermions required in the postulate 4(ii) is the following Slater determinant:

$$\tilde{\Psi} = \frac{1}{\sqrt{N!}} \begin{vmatrix} \psi_1(1) & \psi_2(1) & \cdots & \psi_N(1) \\ \psi_1(2) & \psi_2(2) & \cdots & \psi_N(2) \\ \vdots & & & \\ \psi_1(N) & \psi_2(N) & \cdots & \psi_N(N) \end{vmatrix} \tag{4.4a}$$

where N is the total number of electrons in the multi-electron system, each column denoting the wavefunction for the electron i (i = 1, 2, . . ., N). $\psi_i(j)$ is the wavefunction of electron i written in terms of the variables of electron j, denoting the effect of orbital j on the wavefunction of electron i. In the special case of two electrons, (4.4a) gives the two-electron wavefunction as the second-order Slater determinant:

$$\tilde{\Psi} = \frac{1}{\sqrt{2}} \begin{vmatrix} \psi_1(1) & \psi_2(1) \\ \psi_1(2) & \psi_2(2) \end{vmatrix} = \frac{1}{\sqrt{2}} (\psi_1(1)\psi_2(2) - \psi_1(2)\psi_2(1)) \tag{4.4b}$$

The functions (4.4a) and (4.4b) comply with the Pauli principle. For example, if the two columns in the determinate are the same, which denotes the same quantum state for the two electrons, the determinate is zero, thus there is no possibility that the same electron state exists in the system. In addition, if the two columns switch their locations in the matrix, it will change the sign of the determinate following the mathematical property of a determinate. This result indicates the antisymmetric property of the electronic system. Thus the Slater determinant has a unique role in describing the electron system and will be used in the first-principle calculation.

**Postulate 5. The expectation value of any Hermitian operator $\hat{\theta}$ can be calculated by**

$$\langle \theta \rangle \equiv \frac{\int \Psi^* \theta \Psi d^3 r}{\int \Psi^* \Psi d^3 r} \tag{4.4c}$$

For simplicity, one would like to normalize the value of the wavefunction $\Psi$ so that the denominator of (4.4c) is unit value 1, thus we have the expectation value of $\hat{\theta}$ as

$$\langle \theta \rangle \equiv \int \Psi^* \hat{\theta} \Psi d^3 r \tag{4.4d}$$

As an example, the expectation value $\langle E \rangle$ of the total energy E can be expressed by its Hamiltonian operator (see (4.3f, 4.3g or 4.3v, 4.3w) as follows:

$$\langle E \rangle = \langle \hat{H} \rangle \equiv \frac{\int \Psi^* \hat{H} \Psi d^3 r}{\int \Psi^* \Psi d^3 r} \tag{4.4e}$$

Or for normalization of wavefunction $\Psi$ by

$$\langle E \rangle = \langle \hat{H} \rangle = \int \Psi^* \hat{H} \Psi d^3 r \tag{4.4f}$$

In fact, as electron motions are waves, their wavefunctions have to be mutually orthogonal solutions of the Schrödinger equation, and the electron motions must oscillate to achieve orthogonality in a multi-electron system. The postulates state these basic properties. This oscillation produces positive kinetic energy, which balances with the Coulomb potential energy to form the atomic energy state. In fact, the balance between the negative Coulomb potential energy and the kinetic energy determines the binding energy of the atom. The same principle can also be applied for multi-atom molecules and condensed matter composed of many atoms. Thus, the investigation of electron motion is a foundation of condensed physics. It is also the foundation for determining the interatomic potential and interacting forces.

## 4.4 The Steady State Schrödinger Equation of a Single Particle

If the potential function $V$ is not various with time, the wavefunction $\psi(\bar{r}, t)$ can be expressed in a form by separating variables as:

$$\psi(\bar{r}) = \varphi(\bar{r}) f(t) \tag{4.5a}$$

Substituting this formula into (4.3g), with the time function on the left side of the equation and the spatial function on the right side, we have:

$$\frac{i\hbar}{f}\frac{df}{dt} = \frac{1}{\varphi(\bar{r})}\left[-\frac{\hbar^2}{2m}\nabla^2 + V\right]\varphi(\bar{r}) \tag{4.5b}$$

where the left side is a function of time $t$ only and the right side is a function of spatial coordinate $\bar{r}$. Because $\bar{r}$ and $t$ are independent variables, the equation can be valid only when both sides are equal to a constant E, thus one obtains:

$$\frac{i\hbar}{f}\frac{df}{dt} = E \tag{4.5c}$$

By its integration we obtain:

$$f = C\exp(-iEt/\hbar) \tag{4.5d}$$

Thus, the wavefunction can be rewritten as:

$$\varphi(\bar{r}, t) = \varphi(\bar{r})\exp(-iEt/\hbar) \tag{4.5e}$$

Born's wavefunction of an unbounded particle developed in 1927 has the same exponential formula with time as (4.5e). In Born's equation the constant E is explained as the wave energy, thus the constant $E$ in (4.5e) can be immediately confirmed as the energy of the particle.

(4.5c) indicates that the right-hand formula of (4.5b) also equals $E$. Thus, we can obtain the steady state Schrödinger equation in terms of the Laplace operator $\nabla^2$ as follows:

$$\left[-\frac{\hbar^2}{2m}\nabla^2 + V\right]\varphi(\bar{r}) = E\varphi(\bar{r}) \tag{4.5f}$$

It is seen clearly that the energy E is the eigenvalue of the Schrödinger equation. Furthermore, the equation (4.5f) can be expressed by the Hamiltonian operator $\hat{h}$, which is one type of Hermitian operator, as follows:

$$\hat{h}\,\varphi(\bar{r}) = E\varphi(\bar{r}) \tag{4.5g}$$

where the Hamiltonian $\hat{h}$ can be expressed as follows:

$$\hat{h} = -\frac{\hbar^2}{2m}\nabla^2 + V \tag{4.5h}$$

In the next section, we will give a simple example of the solution to the Schrödinger equation to explore some characteristics of its solution in general and electron waves in particular.

## 4.5 Example Solution: Square Potential Well with Infinite Depth

The solution for a square-well potential field is quite simple and useful for understanding the solution of the Schrödinger equation as well as understanding characteristics of electron motion. It is a common phenomenon that particles move under the potential well. For example, when unbound electrons travel among metal crystal lattices, the attractions of the periodic positive electric fields of atom nuclei make it nearly impossible for the electrons to escape spontaneously out of the metal.

As a particular example, consider a one-dimensional potential well whose depth is infinite. This problem is equivalent to the case where a particle moves between two energy barriers with an infinite energy value. Assuming these barriers start at the position $x = 0$ and $x = a$, this case is shown in Figure 4.2 where in domain I the potential $V = 0$, and $V = \infty$ in domains II and III. This distribution can be mathematically expressed as follows:

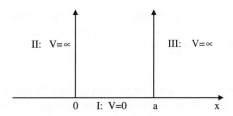

**Figure 4.2**    A schematic of square potential well with infinite depth

$$V(x) = \begin{cases} 0 & 0 \le x \le a \\ \infty & x < 0, x > a \end{cases}$$    (4.6a)

Due to $V(x) = 0$, the Schrödinger equations (4.5f) at the bottom of the well can be written as:

$$-\left(\frac{\hbar^2}{2m}\right)\frac{\partial^2 \psi}{\partial x^2} = E\psi$$    (4.6b)

It is easy to verify directly that this differential equation has the general solution:

$$\psi(x) = A\sin(k_0 x) + B\cos(k_0 x), \quad |x| \le a$$    (4.6c)

with:

$$k_0{}^2 = 2mE/\hbar^2$$    (4.6d)

Notice that outside the potential well, $\psi$ must be equal to zero because physically no particle can stay at a location where the potential energy is infinite. This greatly simplifies the solution process. To satisfy the condition that the potential $\psi = 0$ at the two ends of the well (i.e., $x = 0$ and $x = a$) the solution of (4.6c) must be $B = 0$, since $\cos(k_0 x)$ is not zero at $x = 0$. The explicit solution is then given as follows:

$$\psi(x) = A\sin(k_0 x), \ |x| \le a$$

$$k_0 = \frac{n\pi}{a}$$    (4.6e)

and:

$$\psi = A\sin\frac{n\pi}{a}x$$    (4.6f)

Substituting (4.6e) into (4.6d) we can determine the set of energy eigenvalues as follows:

$$E_n = \frac{\pi^2\hbar^2}{2ma^2}n^2 \ (n = 1, 2, 3, \ \ldots)$$    (4.6g)

## 4.5.1    Observations and Discussions

From this solution, the characteristics of particle motion between two energy barriers with infinite energy can be summarized as follows.

(i) The energy is quantized by different quantum values; and the lowest energy $E_1$ which corresponds to $n = 1$, is not zero. The state with the lowest energy is the so-called ground state. The energy value for energy level n increases rapidly with $n^2$, therefore the energy difference between adjacent energy levels becomes large rapidly. This characteristic is due to the wave nature of a particle.

(ii) The probability distribution for a particle staying within the potential barrier depends on $|\psi|^2$. (4.6f) implies that the distribution of $|\psi|^2$ is inhomogeneous, and there exist some points where the probability of a particle's occurrence is zero.

(iii) The wavefunction $\psi \sim \exp\left(-i\frac{E}{\hbar}t\right)\sin\frac{n\pi}{a}x$ represents a stationary wave. The exponential term with time $t$ represents vibration, where the amplitude equals $\sin\frac{n\pi}{a}x$. This vibration is similar to the stationary wave when a string vibrates with its two ends fixed. This indicates that a physical wave in finite space can only exist in the form of a stationary wave.

## 4.6    Schrödinger Equation of Multi-body Systems and Characteristics of its Eigenvalues and Ground State Energy

### 4.6.1    General Expression of the Schrödinger Equation and Expectation Value of Multi-body Systems

Almost all realistic systems in material science involve N interacting particles such as electrons. Fortunately, the postulates of QM are easily extended to N-body systems in which the wavefunction is a function of all the coordinates of the system, i.e.,

$$\Psi = \Psi(\bar{r}_1, \bar{r}_2, \ldots \bar{r}_N) \tag{4.7a}$$

The Schrödinger equation (4.3v) derived for the He atom including three particles is an example of a multi-body system. It is seen from (4.3v) that the change from a single-particle Schrödinger equation (i.e., (4.3m)) to (4.3v) of the multi-particle is limited only to the change of the Hamiltonian operator from $\hat{h}$ to $\hat{H}$.

Following the same procedure for a single particle as shown in (4.5a) through (4.5g), the wavefunction of a multi-body system can separate variables as follows:

$$\Psi(\bar{r}_1, \bar{r}_2 \ldots \bar{r}_n, t) = \psi(\bar{r}_1, \bar{r}_2 \ldots \bar{r}_n)\exp(-iEt/\hbar) \tag{4.7b}$$

The same type of steady state Schrödinger equation as (4.5g) can thus be written as:

$$\hat{H}\Psi = E\Psi \tag{4.7c}$$

where $\hat{H}$ is the Hamiltonian operator for the multi-body system such as the one given in (4.3w) or (4.7l) below.

Likewise, the expectation value of the Hamiltonian for the energy E can be obtained by an extension of (4.4e) and (4.4f) as follows:

$$\langle E \rangle = \langle \hat{H} \rangle = \frac{\int \ldots \int \Psi^* \hat{H}\Psi d^3 r_1 \ldots d^3 r_N}{\int \ldots \int \Psi^* \Psi d^3 r_1 \ldots d^3 r_N} \tag{4.7d}$$

If the wavefunction is normalized we have

$$\langle E \rangle = \langle \hat{H} \rangle = \int \ldots \int \Psi^* \hat{H}\Psi d^3 r_1 \ldots d^3 r_N \tag{4.7e}$$

Note that the $i^{th}$ eigenfunction and energy eigenvalue of the Hamiltonian has the following relationship (see (4.5g)):

$$\hat{H}\Psi_i = E_i\Psi_i \tag{4.7f}$$

If we know the $i^{th}$ eigenfunction (i.e., the wavefunction) then the corresponding energy $E_i$ which is the corresponding eigenvalue can be determined by the following equation:

$$E_i = \frac{\int \cdots \int \Psi_i^* \hat{H} \, \Psi_i \, d^3 r_1 \ldots d^3 r_N}{\int \cdots \int \Psi_i^* \Psi_i d^3 r_1 \ldots, d^3 r_N} \tag{4.7g}$$

This equation can be easily proved since the following equation is valid due to (4.7f).

$$\int \cdots \int \Psi_i^* \hat{H} \Psi_i d^3 r_1 \ldots \ldots d^3 r_N = \int \cdots \int \Psi_i^* E_i \Psi_i d^3 r_1 \ldots d^3 r_N \tag{4.7h}$$

Furthermore, since $E_i$ is a constant, it can be taken out of the integral on the right of the equation, thus (4.7h) can be proved by dividing both sides by $\int \cdots \int \Psi_i^* \Psi_i d^3 r_1 \ldots d^3 r_N$.

In QM, the integration over all the electron's degrees of freedom is usually expressed by a bracket $\langle \, \rangle$, thus (4.7g) can be rewritten in the concise form:

$$E_i = \frac{\langle \Psi_i | \hat{H} | \Psi_i \rangle}{\langle \Psi_i | \Psi_i \rangle} \tag{4.7i}$$

### 4.6.2   Example: Schrödinger Equation for Hydrogen Atom Systems

Figure 4.3 shows another example of a multi-body system with two hydrogen atoms.[4] They constitute one hydrogen molecule $H_2$ with two proton nuclei (a and b) and two electrons ($e_1$, $e_2$). The relative positions between two protons and two electrons are, respectively, represented by $r$ and $r_{12}$. The electron coordinates relative to the nuclei are defined by $r_{a1}, r_{a2}, r_{b1}, r_{b2}$. Because the proton is about 1800 times more massive than the electron, the nuclei are usually taken as stationary at the distance $r$, which is known as the Born-Oppenheimer approximation in QM. Under this assumption $r$ is a constant, which makes the QM solution much simpler because the system will only depend on electron distances from the nuclei and the distance $r_{12}$ between electrons. Note that not all the inter-particle distances (e.g., $r_{a1}, r_{a2}, r_{b1}, r_{b2}$) are independent variables, since many distances can be derived from others. This is the reason why a set of position vectors $(\bar{r}_1, \bar{r}_2 \ldots \bar{r}_N)$ is used, say, in (4.7a) and (4.7e) to denote the configuration of the system which gives all the inter-particle distances.

For this hydrogen molecule in Figure 4.3 the Hamiltonian operator $\hat{H}$ can be expressed in a more general way as follows:

$$\hat{H} = -\sum_i \frac{\hbar^2}{2m} \nabla_i^2 + V \tag{4.7j}$$

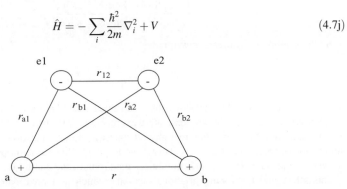

**Figure 4.3** The relationship between protons and electrons in a hydrogen molecule (Reproduced from Liu, W. K., Karpov, E. G., Zhang, S., and Park, H. S. (2004) An introduction to computational nanomechanics and materials. *Computer Methods in Applied Mechanics and Engineering*, 193, 1529, Elsevier)

where the first right term is the sum of all the kinetic energy of particles and the second term denotes the potential energy which can be expressed as follows:

$$V = -\sum_{\alpha,i} \frac{e^2}{r_{\alpha i}} + \frac{e^2}{r_{12}} \tag{4.7k}$$

Thus

$$\hat{H} = -\sum_{i} \frac{\hbar^2}{2m} \nabla_i^2 - \sum_{\alpha,i} \frac{e^2}{r_{\alpha i}} + \frac{e^2}{r_{12}} \tag{4.7l}$$

The summation covers all the Coulombic energy between the protons and the electrons, where the summation symbol $\alpha$ denotes nucleus a and b and i denotes electrons. The second term on the right of (4.7k) denotes interaction energy between two electrons.

Substituting (4.7k) into (4.7j) and then to (4.7c) and moving $E\Psi$ to the left side of the obtained equation, results in the steady state molecular multi-particle Schrödinger equation of:

$$\left[ -\sum_{i} \frac{\hbar^2}{2m} \Delta_i - \sum_{\alpha i} \frac{e^2}{r_{\alpha i}} + \frac{e^2}{r_{12}} - E \right] \Psi = 0 \tag{4.7m}$$

Note again the difference between $\hat{H}$ expressed in (4.7c) and $\hat{h}$ in (4.5g). The former describes the Hamiltonian operator of the atom or molecule with multiple electrons, while the later refers to a single electron. The same difference exists between $\Psi$ and $\psi$. $\psi$ has a readily available solution for a single electron around a single proton, while solving the multi-body wavefunction $\Psi$ is much more difficult, which will be the key issue in the next sections.

## 4.6.3 Variation Principle to Determine Approximate Ground State Energy

In principle, solving the Schrödinger equation $\hat{H}\Psi = E\Psi$ can be quite challenging. Often it is easier to guess a form of the actual wavefunction. This guessed wavefunction is called the variational wavefunction $\Psi_V$, i.e., we are assuming that $\Psi_V \approx \Psi$. As we will soon show below, if we assume that the ground state wavefunction for the Hamiltonian is $\Psi_V \approx \Psi$ and substitute $\Psi_V$ into (4.7g) we have:

$$E_V = \frac{\int \ldots \int \Psi_V^* \hat{H} \Psi_V d^3 r_1 \ldots d^3 r_N}{\int \ldots \int \Psi_V^* \Psi_V d^3 r_1 \ldots d^3 r_N} \geq E_{ground\ state} \tag{4.8a}$$

or by the symbol (4.7i) it can be written:

$$E_V = \frac{\langle \Psi_V | H | \hat{\Psi}_V \rangle}{\langle \Psi | \Psi_V | \rangle} \geq E_{ground\ state} \tag{4.8b}$$

It will shortly be proved that the above obtained value $E_V$ satisfies the above inequality relationship. This inequality indicates that any approximate wavefunction will always be larger than the energy in the ground state. In other words, the real ground state wavefunction always minimizes the system energy among all the variational functions. This is the Rayleigh-Ritz variation principle which states that the correct ground state wavefunction minimizes the ground state energy E.

This is the foundation for the variational method in which the approximate solution of wavefunction can be optimized by changing the variables to minimize the system energy.

The extremely important result of (4.8b) can be proved as follows. The $i^{th}$ eigenfunction and energy eigenvalue of the Hamiltonian (or Schrödinger equation) is given by:

$$\hat{H}\Psi_i = E_i \Psi_i \tag{4.8c}$$

Recall that the eigenfunctions of a Hermitian operator form a basis and the eigenfunctions of a Hermitian operator are orthogonal as described in Postulate 3. From (4.3y) and (4.3x) we can write respectively

$$\Psi_V = \sum_{n=0}^{\infty} a_n \Psi_n \tag{4.8d}$$

and

$$\int \Psi_j^* \Psi_i d^3 r_1 \ldots d^3 r_N = \delta_{ij} \tag{4.8e}$$

Substituting $\Psi_V = \sum_{n=0}^{\infty} a_n \Psi_n$ and (4.8c) into (4.8a) and take taking dummy index i for conjugate function $\Psi_V^*$ and dummy index j for $\Psi_V$, the expansion of the expectation value can be obtained as follows:

$$E_V = \frac{\int \ldots \int \Psi_V^* \hat{H} \Psi_V d^3 r_1 \ldots d^3 r_N}{\int \ldots \int \Psi_V^* \Psi_V d^3 r_1 \ldots d^3 r_N} = \frac{\int \ldots \int \Psi_V^* (E_V \Psi_V) d^3 r_1 \ldots d^3 r_N}{\int \ldots \int \Psi_V^* \Psi_V d^3 r_1 \ldots d^3 r_N}$$

$$= \frac{\sum_i \sum_j a_i^* a_j E_j \int \ldots \int \Psi_i^* \Psi_j d^3 r_1 \ldots d^3 r_N}{\sum_i \sum_j a_i^* a_j \int \ldots \int \Psi_i^* \Psi_j d^3 r_1 \ldots d^3 r_N} \tag{4.8f}$$

using $(4-8e)$ obtain

$$E_V = \frac{\sum_i \sum_j a_i^* a_j E_j \delta_{ij}}{\sum_i \sum_j a_i^* a_j \delta_{ij}} = \frac{\sum_i |a_i|^2 E_i}{\sum_i |a_i|^2} \geq \frac{E_o \sum_i |a_i|^2}{\sum_i |a_i|^2} = E_o \tag{4.8g}$$

where $E_0$ is the actual ground state energy of H which we do not know apriori. The last inequality is obtained because among all the eigenfunctions $E_0$, $E_1$, $E_{i,}$..., $E_N$, the ground state energy $E_0$ is the minimum as seen from the discussion about (4.6g). (4.8g) shows that the energy obtained by the guest function will also be larger than or equal to the ground energy. Therefore, following (4.7i) we have

$$E_{ground\ state} \approx \frac{\langle \Psi_V | \hat{H} | \Psi_V \rangle}{\langle \Psi_V | \Psi_V \rangle} \tag{4.8h}$$

We will make much use of this in the following sections.

## 4.7   Three Basic Solution Methods for Multi-body Problems in QM

In this section three basic solution methods in QM are introduced which can be used to carry on the coupling between MD and QM. These include the tight binding method, Hartree-Fock (HF) Method, and electron Density Function Theory (DFT) method. The basic characteristics of these methods are outlined in this section, and details of each method will be studied in the following three sections. For QM calculations, a computer package must be used. There are several well-known *ab initio* codes, for example: CASTEP, http://www.castep.org/; VASP, http://cms.mpi.univie.ac.at/vasp/; ABINIT, http://www.abinit.org/; WIEN2K, http://www.wien2k.at/. ABINIT is free. Readers are recommended to look at the manuals first and then make the decision for their needs.

## 4.7.1   First-principle or ab initio Methods

Briefly, HF and DFT methods are referred to as first-principle or *ab initio* methods. The Latin term *ab initio* means starting from the very beginning, indicating that the first-principle method is the most essential and fundamental method when starting from atomic structures. Both first-principle methods solve the multi-body problem of molecules with a single-electron approximation (see below). In addition, the multi-body effect is approximated by the electron exchange and correlation energy.

The single-electron approximation can be used first to determine the wavefunction solution for each individual electron of the molecule with many protons. By combining these solutions through a linear combination of these known functions, it is possible to obtain the multi-electron solution for the molecule. The unknown coefficients in this combination will be determined by the minimum energy requirement. The general approach will be illustrated with a two-electron system consisting of hydrogen molecules in the next section.

While the basic methodology used to determine the solution of a single-electron system is very similar to that used to determine the solution of multi-electrons of a molecul, the HF and DFT methods differ not only in dealing with the electron exchange and correlation energy, but also in their concepts and methods to obtain the single-electron wavefunction $\psi_i$ (i = 1, 2, ..., N) for all N electrons in the molecule.

The HF method directly considers the electron-electron interaction. The total energy E depends on all the degrees of freedom of the multi-electron system as shown in (4.7g). As a result, HF method is accurate but time-consuming. HF method is mainly used in small molecule computations.

DFT method is based on the electron density functional theory proposed by Hohenberg and Kohn in 1964[5] and extended by Kohn and Sham.[6] This theory can be described as follows. All electron structure aspects of an interactive electron system under an external potential and non-degenerate background can be completely determined by distribution function $\rho(\bar{r})$ of electron density. Hohenberg and Kohn also proved that the total energy of the system background state is uniquely determined by the electron density $\rho(\bar{r})$ and the energy is at a minimum when the density distribution coincides with that of real charge density under the action of an external potential.

Based on the above theory, the DFT method differs from the HF method. The latter is based on finding multi-electron wavefunctions. The former, however, is based on finding an approximate electron density function $\rho(\bar{r})$. Contrary to the case where a molecular wavefunction depends on all electron degrees of freedom, the electron density function $\rho(\bar{r})$ in any system lies only on the three coordinate components x, y, z of the position vector $\bar{r}$. This simplification indicates that the ground state of a multi-electron system is no longer determined by all electron degrees of freedom as the HF method, but only depends on spatial coordinates through electron density $\rho(\bar{r})$.

## 4.7.2   An Approximate Method

In solids, most physical properties are dependent on the valence electrons rather than on core electrons since the core electrons are very localized. Therefore, it is natural to separate the core electrons from the valence electrons. In the tight binding method, the total energy $E$ of the solid can be expressed as:

$$E = E_{bond} + E_{rep} \tag{4.8i}$$

where $E_{rep}$ is a sum of short-range repulsive pair potentials between the atomic core electrons and nuclei. This potential is a function of atomic positions only, and is independent of electron states of the atom.

$E_{bond}$ is the contribution of non-core electrons to the total energy. This energy is important and is considered in more detail by wavefunctions and Hamiltonian matrices such as the matrix related to the determinate (4.4a). The latter is introduced to express the interference between different orbitals. Each orbital belongs to an individual single-electron, thus the interaction of these orbitals shows the interaction between different electrons.

The tight binding method lies in between the first-principle and experimental methods. The Hamiltonian matrix along with its determinate elements of type (4.4a) and pair potentials are not directly calculated from first principles, but determined by experimental data and *ab initio* energies in terms of energy band structure information.

Despite a lower precision compared to first-principle methods, the tight binding method utilizes the simplest framework to investigate the atomic interaction represented by the QM bond characteristics. This method can deal with about 1000 atoms in the simulation, and is almost one order faster than the first-principle method. The limitations of tight binding rest on the approximation adopted and the necessary skill to combine Hamiltonian matrix parameters related to the $E_{bond}$ and repulsive energies. This method is successful in describing covalent bond materials, but is inadequate for describing metals such as aluminum with free electrons. Also, due to the complicated test process for choosing parameters, there is no appropriate model for alloys yet.

The following sections will introduce these three methods in more detail for interested readers.

## 4.8 Tight Binding Method

The tight binding or LCAO (linear combination of atomic orbitals) method was initially proposed by Bloch,[7] then later revised by Slater and Koster,[8] based on the periodic potential problem. Besides the points mentioned in the last section for the energy separation shown in (4.8i) and its approximation for the individual wavefunction, the essence is to construct an approximate single-electron wavefunction, under the condition that the electron is located in a non-central field of two or more protons. This type of wavefunction with multiple protons is referred to as the molecular orbital (MO). This section will concentrate on how to determine the MO.

Tight binding method is based on the assumption that the molecular orbital can be approximated by the linear combination of the relevant single-electron orbitals. This means that we can use the readily available atomic orbitals, such as the solution for a single electron under a central potential field surrounding a nucleus, to constitute the given molecular orbital.

Due to its importance in QM solutions, let us take a hydrogen molecule as an example to show how to construct an approximate molecular orbital (or wavefunction) from a single-electron solution under a central potential field. The hydrogen ion $H_2^+$, consisting of two protons and one electron, corresponds to the situation expressed in Figure 4.3, except that there is only one electron. In this case, the tight binding method gives the following approximate molecular orbital for the single electron:[9, 10]

$$\tilde{\Psi} = C_a \tilde{\psi}_a + C_b \tilde{\psi}_b \tag{4.9a}$$

where $\tilde{\psi}_a$ is the wavefunction (or orbital) of the single electron 1 surrounding proton nucleus a. This orbital is under the central potential field:

$$V(r) = -\frac{e^2}{r_{a1}} \tag{4.9b}$$

where $\tilde{\psi}_b$ is also the wavefunction (or orbital) of the same single electron (i.e., 1 in this special case). The difference is that it surrounds proton nucleus b under the central force field pointed to nucleus b under the central potential field:

$$V(r) = -\frac{e^2}{r_{b1}} \tag{4.9c}$$

where the length $r_{a1}$ and $r_{b1}$ are shown in Figure 4.3. Note that the Schrödinger equation for solving either $\tilde{\psi}_a$ or $\tilde{\psi}_b$ is given in (4.5f). Because the individual orbital is under a central potential field, the corresponding solution is spherically symmetric and central spherical coordinates can be used for the Laplace operator in the Schrödinger equation. The solution can be obtained analytically in terms of the quantum number, such as the main quantum number n, the angular number $l$ and the magnetic number m, of the electron at hand. These solutions have already been documented[11, 12] and can be effectively used. Readers can refer to the Appendix of this chapter for a brief treatment.

The physical explanation of the LCAO method shown in (4.9a) is that in the vicinities of nuclei $a$ and $b$, the molecular orbital may be approximated by a linear combination of the two overlapping, atom-centered spherical orbitals. According to QM variation theory, the energy of a given approximate wavefunction is always larger than the real or a more accurate one. Therefore, the coefficients $C_a$ and $C_b$ in (4.9a) can be calculated by minimizing $E$ in (4.7g). Substituting the $\tilde{\Psi}$ expression of (4.9a) into (4.7g) and noting the notations introduced in (4.7i) and (4.7g), we can obtain the approximate energy $\tilde{E}$ as follows:

$$\tilde{E} = \frac{\left\langle \tilde{\Psi} \middle| \hat{H} \middle| \tilde{\Psi} \right\rangle}{\left\langle \tilde{\Psi} \middle| \tilde{\Psi} \right\rangle} = \frac{\int (C_a \tilde{\psi}_a^* + C_b \tilde{\psi}_b^*) \hat{H} (C_a \tilde{\psi}_a + C_b \tilde{\psi}_b) d^3 r_1}{\int (C_a \tilde{\psi}_a^* + C_b \tilde{\psi}_b^*)(C_a \tilde{\psi}_a + C_b \tilde{\psi}_b) d^3 r_1} \tag{4.9d}$$

After the simple algebraic operation, the approximate energy for the molecular orbital can be expressed as follows:

$$\tilde{E} = \frac{C_a^2 H_{aa} + C_b^2 H_{bb} + 2 C_a C_b H_{ab}}{C_a^2 + C_b^2 + 2 C_a C_b S_{ab}} \tag{4.9e}$$

where:

$$S_{\alpha\beta} = \left\langle \tilde{\psi}_\alpha \middle| \tilde{\psi}_\beta \right\rangle, \ H_{\alpha\beta} = \left\langle \tilde{\psi}_\alpha \middle| \hat{H} \middle| \tilde{\psi}_\beta \right\rangle, \ \alpha = a, \ b; \quad \beta = a, \ b$$

$$\hat{H} = -\frac{\hbar^2}{2m} \Delta - \frac{e^2}{r_{a1}} - \frac{e^2}{r_{b1}} + \frac{e^2}{r} \tag{4.9f}$$

In deriving (4.9e), the electron orbitals are assumed to comply with the following normalization condition:

$$S_{\alpha\alpha} = \left\langle \tilde{\psi}_\alpha \middle| \tilde{\psi}_\alpha \right\rangle = 1 \tag{4.9g}$$

as well as the following symmetry conditions which are applicable to hydrogen:

$$S_{\alpha\beta} = S_{\beta\alpha}, \qquad H_{\alpha\beta} = H_{\beta\alpha}. \tag{4.9h}$$

The conditions for minimum variation are:

$$\partial \tilde{E} / \partial C_a = 0 \tag{4.9i}$$

$$\partial \tilde{E} / \partial C_b = 0 \tag{4.9j}$$

From these conditions and the symmetry nature shown in (4.9g) and (4.9h), we can obtain:

$$C_a^2 - C_b^2 = 0 \tag{4.9k}$$

By factoring the above equation, we can easily see that only two solutions exist:

$$C_a = C_b, \tag{4.9l}$$

or:

$$C_a = -C_b \tag{4.9m}$$

Thus, we can conclude that the hydrogen ion $H_2^+$ only has two molecular orbitals, with one symmetric and the other antisymmetric:

$$\tilde{\Psi}^{(+)} = N^{(+)} \left( \tilde{\psi}_a + \tilde{\psi}_b \right) \tag{4.9n}$$

$$\tilde{\Psi}^{(-)} = N^{(-)} \left( \tilde{\psi}_a - \tilde{\psi}_b \right) \tag{4.9o}$$

where $N^{(\pm)}$ is the normalization factor, determined by the orthogonality condition as (4.9f). Specifically, $\tilde{\Psi}^{(+)}$ may be considered as bonding and $\tilde{\Psi}^{(-)}$ as anti-bonding. Substituting (4.9l) into (4.9e) we have:

$$\tilde{E}^{(+)} = \frac{H_{aa} + H_{ab}}{1 + S_{ab}} \tag{4.9p}$$

This obtained energy corresponds to the molecular state with symmetric orbitals. The energy for antisymmetric orbitals can be obtained through substituting (4.9m) into (4.9e) as follows:

$$\tilde{E}^{(-)} = \frac{H_{aa} - H_{ab}}{1 - S_{ab}} \tag{4.9q}$$

In the tight binding method the approximate linear combination wavefunction $\tilde{\psi}$ of (4.9a) is used. However, it is not a wavefunction obtained through solving the molecule Schrödinger equation (4.9f), therefore, it is not a self-consistent solution. As soon as the molecular orbital (4.9a) is obtained, it can be used in the column of the Hamiltonian matrix and its determinant (4.4c) as an approximate trial function, then through an iterative process of Hartree-Fock method or other methods dealing with multi-electron systems.

## 4.9 Hartree-Fock (HF) Methods

### 4.9.1 Hartree Method for a Multi-body Problem

An intuitive and useful approximation to the multi-electron problem was proposed by Hartree in 1928.[13] It is assumed that there must be an effective, single-particle bounded-state wavefunction $\psi_a(r)$ which can describe a single electron i in state a for the multi-electron system. Correspondingly, there must be an effective, single-particle potential $V_a(r)$ whose use in the single-body Schrödinger equation shown below exactly reproduces the wavefunction $\psi_a(r)$:

$$\left[ -\frac{h^2 \nabla^2}{2m} + V_a(\bar{r}) \right] \psi_a(r) = \varepsilon_a \psi_a(r), \ (a = 1, \ \ldots, \ N) \tag{4.10a}$$

Because $V_a(\bar{r})$ must contain effects due to the other electrons which, in turn, are influenced by electron i, this potential must be determined simultaneously with the solution of wavefunctions $\psi_b(j)$ of all particles. Hartree assumed that the single-particle potential, which takes into account the effects of other electrons, can be approximately described as follows:

$$V_a(\bar{r}) \approx \int d^3 r' \frac{e^2}{|\bar{r} - \bar{r}'|} \sum_{b \neq a}^{N} |\psi_b(\bar{r}')|^2 - Z \frac{e^2}{r} \tag{4.10b}$$

where $\bar{r}$ is the coordinate for particle i, $\bar{r}'$ is the coordinate for the other single-particle wavefunctions, and the last term denotes the Coulomb potential between the nucleus and the single electron i at hand. $d^3 r$ denotes the differential volume, Z is the positive charge number in the nucleus. For Cartesian coordinates, $d^3 r = dxdydz$; for spherical coordinates $d^3 r = r^2 \sin\theta dr d\theta d\phi$.

Note that $|\psi_b(\bar{r}')|^2$ in (4.10b) denotes the probability of finding electron j in state b at $\bar{r}_j$ (see postulate 1 in Section 4.2), therefore $\frac{e}{|\bar{r}-\bar{r}'|}|\psi_b(\bar{r}')|^2$ denotes the probability of the contribution of electron j in state b on electron i at state a. Thus, (4.10b) states that the single-particle potential for electron i in state a can be expressed by the probability of all other particles and their distances, $|\bar{r}-\bar{r}'|$, from electron i. By substituting (4.10b) into the left hand side of (4.10a), the following Hartree total energy can be found, in which the first term on the right hand represents the kinetic energy:

$$\varepsilon_a\psi_a = \left[-\frac{h^2\nabla^2}{2m} + \int d^3r' \frac{e^2}{|\bar{r}-\bar{r}'|}\sum_{b\neq a}^{N}|\psi_b(\bar{r}')|^2 - Z\frac{e^2}{r}\right]\psi_a \qquad (4.10c)$$

Because all $\psi_b$ are needed to calculate $V_a$, (4.10a) and (4.10b) are coupling equations which usually cannot be solved analytically for realistic cases, they can only be solved numerically by self-consistent iterative process. It can start by taking some known wavefunctions, such as those obtained for hydrogen atoms, as the trial wavefunctions used in the first step of the iterative process. The latter includes solutions $\varphi_i$ for the hydrogen atom documented in the literature and handbook. In Hartree's approximation, the wavefunction $\Psi$ for the complete N-electron system is assumed to be a simple product of these single-particle states:

$$\Psi(1, 2, \ldots N) = \psi_a(1)\psi_b(2)\ldots\psi_{An}(N) \qquad (4.10d)$$

where the subscripts on $\psi_b$ denote the N differential states needed for N electrons which start from state a to the Nth state $A_n$.

Relations of (4.10b) are used to generate the effective potentials $V_a$ based on the initial wavefunction, for instance in the two-electron case we have:

$$\widetilde{\Psi} = \psi_1^{(0)}(1)\psi_2^{(0)}(2) \qquad (4.10e)$$

where the superscript number denotes the number of the iteration process.

The N differential equations (4.10a) are then solved with the obtained $V_a$. The improved wavefunctions are then used to determine an improved $V_a$, which can in turn determine the improved wavefunctions, and so on. This iterative process will not stop until the difference of wavefunctions and potentials at two consecutive steps is less than a pre-described tolerance. Then the approximation scheme is considered convergent and successful, and the wavefunctions and potentials obtained become self-consistent.

## 4.9.2 Hartree-Fock (HF) Method for the Multi-body Problem

While the solution technique of the Hartree method for the multi-electron Schrödinger equation is instructive, it does not follow the Pauli principle; therefore the obtained wavefunctions cannot be guaranteed to be orthogonal. Again, the Pauli principle states that two electrons cannot simultaneously exist in the same one-electron state. In the HF method, the iterative solution technique is the same as the Hartree method; however, the total multi-body function $\widetilde{\Psi}$ is composed of wavefunction $\psi_i(i = 1\ldots N)$ as given in the Slater determinate (4.4a). It is intuitively obvious that use of the N-order determinate wavefunction should improve the Hartree method, because it involves more information on the antisymmetric wavefunction of the multi-electron system.

Specifically, the Hartree-Fock equations can be written as follows:[14]

$$-\frac{h^2}{2m}\nabla^2\psi_i(\bar{r}) - Z\frac{e^2}{r}\psi_i(\bar{r}) + U^{el}(\bar{r})\psi_i(\bar{r}) - \sum_j \int d^3r \frac{e^2}{|\bar{r}-\bar{r}'|}\psi_j^*(\bar{r}')\psi_i(\bar{r}')\psi_j(\bar{r})\delta_{ij} = \varepsilon_i\psi_i(\bar{r}) \qquad (4.10f)$$

where the interaction energy of electrons is expressed as follows:

$$U^{el}(\bar{r}) = -e \int d^3r \rho(\bar{r}') \frac{1}{|\bar{r}-\bar{r}'|} \qquad (4.10g)$$

with

$$\rho(\bar{r}) = -e \sum_i |\psi_i(\bar{r})|^2 \qquad (4.10h)$$

The justification of the Hartree-Fock theory follows from the Rayleigh-Ritz variation principle discussed in Section 4.6.3. We have already used this principle in (4.9h) and (4.9i) to determine the coefficients $C_a$ and $C_b$, which was a linear combination of two single-electron wavefunctions. In general, the ground state energy is calculated as the expectation value of the multi-body Hamiltonian value for the multi-body ground state function. Following this, the variation vanishes to make the system energy minimum.

The Hartree-Fock approximation has been successfully applied to solve quantum physics chemistry problems such as $H_2O$, $CH_4$, $NH_3$. For these molecules, the calculated equilibrium bond angles and interatomic distances are within a few percent of the experimental data. For vibration frequencies, the difference is found to be within 10 percent. The limit for using HF method for geometric optimizations is in the range of about 100 carbon-like atoms.

## 4.10 Electronic Density Functional Theory (DFT)

Although the HF calculation can provide reliable results, it is not suitable for a large simulation system because it is computationally intensive and time-consuming. DFT provides an alternative method which can reduce time by allowing the application to simulate molecular systems containing hundreds of atoms.

The idea behind DFT differs from the HF method. The HF method starts with an exact Hamiltonian such as (4.7h), but utilizes an approximate trial wavefunction for a multi-body steady state Schrödinger equation (4.7i). This trial wavefunction is formed by assembling the readily available single-electron orbitals, then improving accuracy by minimizing total energy which will optimize the single-electron solutions. However, in DFT the referred ground state energy is calculated without introducing the multi-electron wavefunction, and is based on the Hohenberg-Kohn theory that the ground energy depends only on the distribution of electron density.

A mathematical description of the Hohenberg-Kohn theorem, which is the theoretical foundation of the DFT method, can be stated as follows:[15] There is a universal functional $E(\rho(\bar{r}))$ of the electronic charge density distribution $\rho(\bar{r})$ that defines the total energy of the electronic system by the expression:

$$E = \int v(\bar{r})\rho(\bar{r}) \, d^3r + F\rho(\bar{r}) \qquad (4.11a)$$

where $d^3r$ denotes the 3D differential volume for the integration. This expression is universal in the sense that no matter what the external potential $v(\bar{r})$ is, this energy expression is exact if the ground state energy has no degeneracy. For most cases of interest, the external potential is simply the Coulomb field due to the nuclei.

Unfortunately, the Hohenberg-Kohn theorem does not provide guidelines how to form $E(\rho(\bar{r}))$, and therefore the DFT applications depend on the discovery of a sufficiently accurate approximation for E $(\rho(\bar{r}))$. In order to do this, $E(\rho(\bar{r}))$ is rewritten as the Hartree total energy shown in (4.10c); with another presumably smaller, unknown function, called the exchange-correlation (xc) functional, $E_{xc}$ $(\rho(\bar{r}))$.

Specifically, the ground state energy E for a multi-electron system has the following types of dependence on the electron density $\rho(\bar{r})$:

$$E(\rho) = T_s(\rho) + V_N(\rho) + V_H(\rho) + V_{ii}(\rho) + E_{xc}(\rho) \tag{4.11b}$$

where $T_s(\rho)$ denotes the electronic kinetic energy. $V_N(\rho)$ is the Coulomb interaction energy between the electrons and the nuclei and $V_{ii}(\rho)$ arises from the interaction of the nuclei with each other. $V_H(\rho)$ is Hartree component of the electron-electron interaction energy and can be rewritten as:

$$E_H(\rho) = \frac{e^2}{2} \int d^3\bar{r}d^3\bar{r}' \frac{\rho(\bar{r})\rho(\bar{r}')}{|\bar{r}-\bar{r}'|} \tag{4.11c}$$

This expression of Coulomb electronic interaction is obvious because $e\rho(\bar{r})d^3\bar{r}$ and $e\rho(\bar{r}')d^3\bar{r}'$ denote the electronic charges in infinitesimal regions, respectively, at $\bar{r}$ and $\bar{r}'$. The value $|\bar{r}-\bar{r}'|$ denotes the distance between these two regions.

For the exchange-correlation functional $E_{xc}(\rho)$, there are different assumptions. The corresponding energy for the simplest edition of DFT, known as local density approximation (LDA), is determined by the integral of a certain function of the total electron density as follows:

$$E_{xc} = \int d^3\bar{r}\rho(\bar{r})\,\varepsilon_{xc} \tag{4.11d}$$

where $\varepsilon_{xc}(\rho)$ is the exchange-correlation energy per electron. $\varepsilon_{xc}(\rho)$ is approximated by a local function of the density, which usually reproduces the known energy of the uniform electron gas.

For an $N$-electron system, electron density function $\rho(\bar{r})$ is expressed by a sum of the modulus square of the single-electron Kohn-Sham orbital $\psi_i$, i.e.,

$$\rho(\bar{r}) = \sum_{i=1}^{N} |\psi_i(\bar{r})|^2 = \sum_{i=1}^{N} \psi^*_i(\bar{r})\psi_i(\bar{r}) \tag{4.11e}$$

The original value of Kohn-Sham $\psi_i$ is derived from a set of accurate basis functions. These basis functions are solutions to the Schrödinger equation of plane waves for an unbound electron, while the coefficients are determined by energy optimization similar to the Hartree-Fock variation method. Using the variation principle for energy minimization is rational because the electronic energy reaches a minimum for a realistic electronic density distribution of a given potential. The variation principle of DFT leads to a system of one-electron equations that can be self-consistently solved to obtain the ground state energy. If we neglect the interaction energy $V_{ii}(\rho)$ between different nuclei of (4.11b), the orbitals can be obtained by solving the following Kohn-Sham one-electron equations.

$$\left[ -\frac{\hbar^2}{2m}\hbar\Delta + V_H(r) + V_N(r_{xi}) + V_{xc}(\rho) - E_i(\rho) \right] \psi_i^{(1)}(\bar{r}) = 0, \ i = 1, 2, \cdots N \tag{4.11f}$$

where $\psi_i^{(1)}(\bar{r})$ are the one-electron Kohn-Sham orbitals and $E_i$ are the eigenvalues (or energy) of these Kohn-Sham orbitals. Compared to the (4.10f) HF governing equation, all functions in (4.11f) are expressed by coordinates $\bar{r}$ or through density function $\rho(\bar{r})$. In addition, DFT governing equation (4.11f) increases the exchange-correlation potential $V_{xc}$. It can be expressed as a functional derivative of the exchange-correlation functional $E_{xc}$ of (4.11d) with respect to the electron density $\rho$:

$$V_{xc}[\rho] = \frac{\partial E_{xc}[\rho]}{\partial \rho} \tag{4.11g}$$

Under the LDA, the explicit form of $V_{xc}(\rho)$ has been derived.[16,17]

The electron states and the electron density distribution are solved self-consistently through an iterative process similar to the Hartree or HF process described above. This iterative process does not stop until the exchange-correlation energy and electron density function converge under a certain tolerance. When solved self-consistently, the total energy can be calculated using the electron density obtained by (4.11b) or other simplified approaches developed by Harris[16] and Foulkes and Haydock.[17]

Because the electron density depends on only three coordinates as opposed to the 3N coordinates of N electrons, the computational time required to solve DFT equations is much less than that of the HF method. Generally speaking, the DFT calculation level in (4.11f) is in the order of $O(N^3)$, where $N$ is the number of electrons; as opposed to the HF and tight binding methods, that are in the order of $O(N^4)$. Thus, the computational time of DFT is one order less than the other two methods. Recently, the Car-Parrinello method[18] and conjugate gradient method[19] have made important improvements to the DFT method. The Car-Parrinello method has lowered the computation level to $O(N^2)$, while the conjugate gradient method is even more efficient.

In DFT the opening guess of the functional form is not known and has to be constructed by heuristic approximations. In the early development of DFT theory, accuracy was lacking. Through the developments of the past four decades, functions capable of sufficient accuracy and deep applications across the periodic table are available. Recently standard computer packages have been developed for first-principle (*ab initio*) calculations based on the DFT methods.[20]

Using (4.11f) in DFT, one can obtain the energy $E_a$ and $E_b$ of individual isolated atoms, the energy E of their combination system, the interatomic potential V(r) shown in (4.10b), the debonding energy, and the fracture distance.

## 4.11 Brief Introduction on Developing Interatomic Potentials by DFT Calculations

### 4.11.1 Energy Linkage Between QM and Atomistic Simulation

Obviously, the total energy $E$ of an atomistic system is the sum of energies of all unbounded atoms plus the binding energy between these atoms. Taking Figure 4.3 as an example, the system energy $E$ of the hydrogen molecule is the sum of two unbounded hydrogen atoms, $E_a$ and $E_b$, plus an atomic binding energy V(r) between the two, i.e.,

$$E = E_a + E_b + V(r) \tag{4.12a}$$

Classical MD assumes that the system is a conservative system, i.e., there is no energy absorption by the simulated atoms, but only energy transformations between kinetic and potential energy. Thus, the values $E, E_a$ and $E_b$ should relate to the atomic states with minimum energies. $E_a$ and $E_b$ can be solved easily (see Appendix 4.A). If one can obtain the energies $E_a$ and $E_b$ and the full energy $E(r)$ of the coupled system versus atomic distance $r$ by QM, the pairwise atomic potential function $V(r)$ can be determined from (4.12a) as follows:

$$V(r) = E(r) - E_a - E_b \tag{4.12b}$$

The obtained potential function $V(r)$ can be interpolated to draw a smooth curve or be expressed by an analytical function. From the above description of the general idea to link QM and MD we see that the potential function of atoms or molecules can be determined from QM calculations within a small atomic or molecular system and then can be used in MD simulations with a much larger system. Thus, QM calculation described in the last several sections is important to determine the potential for MD.

Recently, Ercolessi *et al.* have developed a force-matching method (FMM) for determining potentials by fitting experimental data and DFT atomic forces calculated for a large set of configurations including crystals, liquids, surfaces and isolated clusters.[21] The obtained potentials are used for Al and Mg alloys. Mishin *et al.* further use these combined methods to investigate the monoatomic metals of Al, Ni,[22] and Fe-Ni system by first principles and atomistic simulations.[23] In this section, we will introduce the FMM methods by which the DFT calculation can be used for potential development.

Empirical potentials are typically fitted to the experimental values of the equilibrium lattice parameter $a_0$, the cohesive energy $E_0$, elastic constants ($C_{11}, C_{12}, C_{44}$), and the vacancy formulation energy, etc. (e.g.,

Table 3.7). Unfortunately, reliable experimental information on material properties that can be directly linked to atomic interactions is very limited. Furthermore, experimental data obtained is in very small regions, thus the transferability of fitted potential parameters to a large region is also limited.

## 4.11.2  More Information about Basis Set and Plane-wave Pseudopotential Method for Determining Atomistic Potential

From Postulate 3 of Section 4.3 it is known that any wavefunction of a multi-body (multi-electron) molecule can be written as a linear combination of eigenfunctions $\varphi_i(x)$ (i = 1, 2, . . ., N). It is the basis for obtaining the system total energy, and thus the potential for atomistic analysis. Therefore, this section offers more information so readers can understand the basis set and its connection with the QM solution scheme.

If one can choose a good basis set of $\varphi_i(x)$ then the iteration process to find the molecular orbital is easy to converge. When molecular calculations are performed, it is common to use a basis composed of a finite number of atomic orbitals, centered at each atomic nucleus within the molecule (linear of atomic orbitals ansatz).

Initially, these atomic orbitals were typed as Slater orbitals. These orbitals correspond to a set of functions which decay exponentially with distance from the nuclei. Later, it was realized that these Slater-type orbitals can be approximated as linear combinations of Gaussian orbitals.[24] In the work of Catlow[1] for ion $O^2$-$O^{2-}$ pair potential the Gaussian programs of the ATOL2 package were used, the wavefunction being based on a function for the oxygen atom with both exponents and contraction coefficients obtained from the compilation of Gaussian atomic wavefunction.

Plane-wave basis sets are used in quantum simulations. Typically, a finite number of plane-wave functions are used. These basis sets are frequently used in calculations involving periodic boundary conditions. In practice, plane-wave basis sets are often used in combination with an "effective core potential" or pseudopotential. Thus, the plane waves are only used to describe the valence charge density at outermost shells. This is because core electrons tend to be concentrated very close to the atomic nuclei. This results in large wavefunction and density gradients near the nuclei which are not easily described by a plane-wave basis set.

As described in Postulate 3, all functions in the basis are mutually orthogonal, i.e., a plane-wave basis set does not exhibit basis set superposition error. Using Fast Fourier Translations, one can work in reciprocal space with a plane-wave basis. In that space, not only integrals such as kinetic energy but also derivatives are computationally less demanding, thus increasing efficiency. In addition, a plane-wave basis guarantees the calculation converges to the target wavefunction while there is no such guarantee for Gaussian-type basis sets.

At this time, DFT practitioners are divided into two communities; one employing pseudopotentials and relatively simple basis sets (particularly plane waves) and the other using methods with complex but efficient basis sets, such as the linearized augmented plane wave (LAPW). Recently, Mishin and others using the LAPW method obtained the interatomic potential for the Cu-Ta system.[25, 26] Readers interested in LAPW methods are also referred to Ref. [27].

## 4.11.3  Using Spline Functions to Express Potential Energy Functions

The FMM method has developed EAM potential for several alloys[26] involving a large amount of outputs such as positions and forces of atoms from first-principle calculations of various physical situations. By combining this data with traditional experimental fitting curves, it is possible to obtain parameters of potential functions. In the FMM approach, each potential function, such as the pair function $V_{ij}$, electron density function $\rho_i$, and embedding function $F(\rho_i)$ in (3.11d), is described as a set of points. A cubic spline function is therefore used for interpolation between these points. The spline function is given as follows.

**Table 4.3** Spline parameters for the E-A Al potential functions (Reproduced from Liu, X.-Y. (1997) The development of empirical potentials from first-principles and application to Al alloys, Ph.D. Dissertation, University of Illinois at Urbana-Champaign.)

| (a) For pair potential V(r) | | | | |
| --- | --- | --- | --- | --- |
| x(i) | y(i) | b(i) | c(i) | d(i) |
| 2.0210 | 1.9591 | −6.9356 | 8.5318 | −3.6923 |
| 2.2730 | 0.6941 | −3.3390 | 5.7404 | −3.9603 |
| 2.5257 | 0.1530 | −1.1966 | 2.7385 | −2.6205 |
| ... | ... | ... | ... | ... |
| 5.3050 | −0.0055 | 0.0394 | −0.0544 | −0.0619 |

| (b) For electron density $\rho(r)$ | | | | |
| --- | --- | --- | --- | --- |
| x(i) | y(i) | b(i) | c(i) | d(i) |
| 2.0210 | 0.0874 | 0.0780 | −0.2221 | 0.1254 |
| 2.2730 | 0.0949 | −0.0100 | −0.1273 | 0.1838 |
| 2.5257 | 0.0872 | −0.0392 | 0.0120 | −0.1081 |
| ... | ... | ... | ... | ... |
| 5.3050 | 0.0016 | −0.0071 | −0.0197 | 0.0890 |

| (c) For embedding energy function: y(i) represents$F(\rho_i)$, x(i) $\rho$ | | | | |
| --- | --- | --- | --- | --- |
| x(i) | y(i) | b(i) | c(i) | d(i) |
| 0.0000 | 0.0000 | −18.1647 | 81.6236 | −130.0891 |
| 0.1000 | −1.1303 | −5.7247 | 42.5968 | −171.9237 |
| 0.2000 | −1.4505 | −2.3810 | −8.9803 | 37.2491 |
| ... | ... | ... | ... | ... |
| 1.2000 | −2.6107 | 0.4709 | 1.2132 | −0.0242 |

For the x values in the domain: $x(i) < x < x(i + 1)$:

$$s(x) = y(i) + b(i)(x - x(i)) + c(i)(x - x(i))^2 + d(i)(x - x(i))^3 \qquad (4.12c)$$

where x(i) is the spline knot position, y(i) is the function value at the knot, and b(i), c(i) and d(i) are the derivative coefficients that construct the spline function. Their values are determined by preserving continuity of the functions and of their first and second derivatives across the junctions. An example of the E-A potential for Al is given in Table 4.3. Table 4.3a is for pair potential V(r) in which x(i) denotes the distance r (Å), and y(i) denotes pair potential V(r) with units of eV. Table 4.3b is for electron density $\rho_i(r)$. Table 4.3c is for embedding energy with y(i) as $F(\rho_i)$ with units of eV. The electron density $\rho_i$ can be any arbitrary unit. (Complete and accurate E-A potential for Al can be found in Ref. [28].)

## 4.11.4 A Systematic Method to Determine Potential Functions by First-principle Calculations and Experimental Data*

When one develops the potential, the main work is to determine the parameters in the potential expressions so that the potential can have best match with results calculated by first principles and experimental data. Let symbol $\langle \alpha \rangle$ indicate the entire set of the L parameters $\alpha 1, \alpha 2 \ldots \alpha_L$ for the potential. To determine the optimal set of these parameters, one must try to match the forces supplied by the first-principle calculations for a large set of different configurations and various experimental data. From the physical point of view, the real or optimal parameters should make the potential energy function a minimum value. Mathematically, this reduces to a nonlinear minimization problem in an L-dimensional phase with an objective function $Z(\langle \alpha \rangle)$, thus one needs to first construct the objective function Z.

This objective function should be a sum of the square of the difference between the value of the fitting variable and the data; then adjust these parameter values to make Z minimum. To use all of the valuable information, both the data obtained by the *ab initio* calculations and the experimental data are used and shown in the two parts of the right-hand side of the following objective function:[29]

$$Z(\langle \alpha \rangle) = Z_{exp}(\langle \alpha \rangle) + Z_{force}(\langle \alpha \rangle) \tag{4.12d}$$

The first part of the right-hand side is the experimental part, which can be expressed as:

$$Z_{exp}(\langle \alpha \rangle) = \sum_{l=1}^{Ne} W_l [A_l(\langle \alpha \rangle) - A_l^0]^2 \tag{4.12e}$$

where $A_L(\langle \alpha \rangle)$ are calculated physical quantities and $A_l^0$ are the corresponding reference quantities from experimental data or energetics from first-principle calculations. $W_1$ are the relative weights of the fitting data. The force part is presented as follows:

$$Z_{force} = \frac{\displaystyle\sum_{k=1}^{M} \sum_{i=1}^{Nk} |F_{ki}(\langle \alpha \rangle) - F_{ki}^0|^2}{3 \displaystyle\sum_{k=1}^{M} N_k} \tag{4.12f}$$

where M is the number of sets of atomic configurations available and $N_k$ is the number of atoms present in configuration k ($k = 1, \ldots, M$). The multiplier of 3 in the denominator denotes the degree of freedom for each particle. $F_{ki}(\langle \alpha \rangle)$ is the force on the i-th atom in set k calculated based on the $\langle \alpha \rangle$ parameters, and $F_{ki}^0$ is the reference force from the first-principle calculations.

It is desirable to include force data for different geometries and physical situations (cluster, surface, bulk, defects, liquid, etc.) in an attempt to achieve good transferability such that the results can be used in a large region. In practice, one can use samples from *ab initio* trajectories for various systems, thus obtaining a good representation of the regions of configurational space. After the minimization process of the objective function $Z(\langle \alpha \rangle)$ of (4.12d) by the least square method, one can develop a set of points for each of the parameters $\alpha(\alpha = 1, \ldots, L)$. A cubic spline function can then be used for interpolation between these points as described above.

## 4.12  Concluding Remarks

Chapter 3 introduced different potentials (Tables 3.9–3.15), Lorentz–Berthelot mixing rules (Section 3.12), and GULP software that can be effectively used to develop potentials for simulation of interactions between ions in ceramic coating layers, atoms in substrate layers, and particles between coating and substrate layers. The difficult problem, however, is how to determine the interface structure. The effective method is to use the first-principle calculation.[33] In the analysis, the bonding energy and electronic structures at the TiN(111) and VN(111) interface were investigated using the Cambridge Sequential Total Energy Package (CASTEP) within the framework of DFT. The pseudopotential was applied for describing the electron–ion interaction and the generalized gradient approximation was used to address the exchange-correlation functional. The single-particle Kohn-Sham wave functions were expanded using plane waves and a vacuum region of 10 Å was embedded into the surface and interface supercells to avoid unwanted interaction between the slab and its periodic images. Results show that the interfacial bonding is ionic, yet maintains a small amount of covalent character.

## Appendix 4.A Solution to Isolated Hydrogen Atom

The energies $E_a$ and $E_b$ for isolated hydrogen atoms a and b in (4.12b) can be determined by the Schrödinger equation. The condition for the solution is that there is no interaction between the two hydrogen atoms, or their separation distance is infinite. From the solution of Section 4.5 for the potential well problem it is seen that the Schrödinger equation involves an eigenfunction solution. According to the given boundary conditions, E in the stationary Schrödinger equation (4.5f) can be solved, which is the eigenvalue of the problem at hand. The corresponding wavefunction $\psi$ is the eigenfunction. (4.6g) and (4.6f) show examples for E and $\psi$.

The stationary Schrödinger equation (4.5f) for isolated atoms a and b can be expressed in a compact form as:

$$\hat{h}\psi_\alpha = E_\alpha\psi_\alpha \tag{4A.1}$$

where $\hat{h}$ is the single-electron Hamiltonian operator, which can be expressed with Planck's constant $\hbar$, electron mass $m$ and electric charge $e$ as follows:

$$\hat{h} = -\frac{\hbar^2}{2m}\Delta_\alpha - \frac{e^2}{r_\alpha} \quad \alpha = a, b \tag{4A.2}$$

and $r_a = r_{a1}$, $r_b = r_{b2}$. A hydrogen atom consists of one electron and one nucleus. The atom's inner motion can be assumed such that the electrons move surrounding a fixed central potential field $V(r) = -e^2/r$. The corresponding solution is spherically symmetric. In this case, Schrödinger equation (4.5f) has many solutions, corresponding to different particles' quantum states.

In fact, we know that the energy level occupied by each single electron for a single atom is determined by the quantum number, including the main quantum number n which shows the shell level, the angular quantum number $\ell$, and the magnetic number m. For the case of the hydrogen atom the main quantum number is the simplest with only one shell ($n = 1$). The number of energy levels in each quantum shell is given by the angular quantum number, i.e., the total quantum number between $-\ell$ and $\ell$.

In a spherically symmetric field, the Laplace operator is

$$\Delta = \nabla^2 = \frac{1}{r^2}\frac{\partial}{\partial r}r^2\frac{\partial}{\partial r} + \frac{1}{r^2}\left(\frac{1}{\sin\theta}\frac{\partial}{\partial\theta}\left(\sin\theta\frac{\partial}{\partial\theta}\right) + \frac{1}{\sin^2\theta}\frac{\partial^2}{\partial\phi^2}\right) \tag{4A.3}$$

Using center coordinates for the spherically symmetric field, the wavefunction $\psi$ can be solved by separation of variables as follows:[30]

$$\psi(r, \theta, \phi) = R_l(r)Y_l^m(\theta, \phi) \tag{4A.4}$$

$$Y_l^m(\theta, \phi) = \pm\sqrt{\frac{2l+1}{4\pi}\frac{l-|m|}{l+|m|}}P_l^m(\cos\theta)\exp(im\phi) \tag{4A.5}$$

Where $\ell$ and $m$, respectively, denote the angular and magnetic quantum number. $P_l^m(\cos\theta)$ is the Legendre multinomial of $\cos(\theta)$. Thus it can be seen that many solutions to the Schrödinger equation exist. The basic solution is a solution with the angular quantum number $\ell = 0$. Here the magnetic quantum number $m_\ell$ must equal zero, and the eigenfunction $\psi(r)$ for the minimum value can be obtained as:

$$Y_0^0(\theta, \varphi) = \left(\frac{1}{4\pi}\right)^{1/2} \tag{4A.6}$$

If $\ell \neq 0$, the derivation will be more complex. All the hydrogen atom's orbitals and the corresponding energy levels can be written as closed form solutions, which can be found in the relevant textbooks.[31, 32]

The solution to the wavefunction $\psi_\alpha$ in (4.5f) provides a complete description of the QM system at the corresponding energy state. Note that the wavefunction itself has no instant physical meaning. It serves as a mathematical tool to describe the QM system, but cannot be determined in experiments. The single-electron wavefunction $\psi_\alpha$ determined by (4A.4) is often quoted as the hydrogen atom's orbital.

# References

[1] Catlow, C. R. A. (1977) Point defect and electronic properties of uranium dioxide. *Proceedings of the Royal Society of London A*, **353**, 533.
[2] Shankar, R. (1987) *Principles of Quantum Mechanics*, Plenum Press, New York and London.
[3] Loucks, R. (2008) *Lecture notes for Introduction of Quantum Mechanics*, Alfred University.
[4] Liu, W. K., Karpov, E. G., Zhang, S., and Park, H. S. (2004) An introduction to computational nanomechanics and materials. *Comput. Methods Appl. Mech. Eng.*, **193**, 1529.
[5] Hohenberg, P. and Kohn, W. (1964) Inhomogeneous electron gas. *Phys. Rev.*, **136**, B864.
[6] Kohn, W. and Sham, J. (1965) Self-consistent equations including exchange and correlation effects. *Phys. Rev.*, **140**, A1133.
[7] Bloch, F. (1928) Quantum mechanics of electrons in crystals. *Zeitschrift für Physik*, **52**, 555.
[8] Slater, J. C. and Koster, C. F. (1954) Simplified LCAO method for the periodic potential problem. *Phys. Rev.*, **94**, 1498.
[9] La Paglia, S. R. (1971) *Introductory Quantum Chemistry*, Harper & Row, New York.
[10] Pilar, F. L. (1990) *Elementary Quantum Chemistry*, McGraw-Hill, New York.
[11] Harrison, W. A. (2000) *Applied Quantum Mechanics*, World Scientific, Singapore.
[12] O'Reilly, E. P. (2002) *Quantum Theory of Solids*, Taylor & Francis, London.
[13] Hartree, D. R. (1928) The wave mechanics of an atom with a non-Coulomb central field. I. Theory and methods. Proc. Cambridge Phil. Soc., **24**, 89.
[14] Ashcroft, N. W. and Bermin, N. D. (1976) *Solid State Physics*, Brooks/Cole.
[15] Park, D. (1992) *Introduction to the Quantum Theory*, McGraw-Hill, New York.
[16] Harris, J. (1985) Simplified method for calculating the energy of weakly interacting fragments. *Phys. Rev. B*, **31**, 1770.
[17] Foulkes, W. M. C. and Haydock, R. (1989) Tight-binding models and density-functional theory. *Phys. Rev. B*, **39**, 12520.
[18] Car, R. and Parrinello, M. (1985) Unified approach for molecular dynamics and density-functional theory. *Phys. Rev. Lett.*, **55**, 2471.
[19] Payne, M. C., Teter, M. P., Allen, D. C., *et al.* (1992) Iterative minimization techniques for *ab initio* total-energy calculations. Molecular dynamics and conjugate gradients. *Rev. Mod. Phys.*, **64**, 1045.
[20] VASP group, http://cms.mpi.univie.ac.at/vasp/.
[21] Ercolessi, F., Parrinello, M., and Tosatti, E. (1988) Simulation of gold in the glue model. *Philos. Mag. A*, **58**, 213.
[22] Mishin, Y., Farkas, D., Mehl, M. J., and Papaconstantopoulos, D. A. (1999) Interatomic potentials for monoatomic metals from experimental data and *ab initio* calculations. *Phys. Rev. B*, **59**(5), 3393.
[23] Mishin, Y., Mehl, M. J., and Papaconstantopoulos, D. A. (2005) Phase stability in the Fe-Ni system: Investigation by first-principles calculation and atomic simulations. *Acta Mater.*, **53**, 4029.
[24] http://en.wikipedia.org/wiki/Basis_set_(chemistry).
[25] Hashibon, A., Lozovoi, A. Y., Mishin, Y., *et al.* (2008) Interatomic potential for the Cu-Ta system and its application to surface wetting and dewetting. *Phys. Rev. B*, **77**, 094131.
[26] Mishin, Y., Mehl, M. J., and Papaconstantopoulos, D. A. (2005) Phase stability in the Fe-Ni system: Investigation by first-principles calculation and atomic simulations. *Acta Mater.*, **53**, 4029.
[27] Singh, D. J. (1994) *Planewaves, Pseudopotentials and the LAPW Method*, Kluwer Academic, Boston/Dordrecht/London.
[28] EAM potential database, http://enpub.fulton.asu.edu/cms/potentials/main/main.htm.
[29] Liu, X.-Y. (1997) The development of empirical potentials from first-principles and application to Al alloys, Ph.D. Dissertation, University of Illinois at Urbana-Champaign.
[30] Wolde, P. R. ten, Ruiz-Montero, M. J., and Frenkel, D. (1995) Numerical evidence for bcc ordering at the surface of a critical fcc nucleus. *Phys. Rev. Lett.*, **75**, 2714.
[31] Harrison, W. A. (2000) *Applied Quantum Mechanics*, World Scientific, Singapore.
[32] O'Reilly, E. P. (2002) *Quantum Theory of Solids*, Taylor & Francis, London.
[33] Yin, D., Peng, X., Qin, Y., and Wang, Z. (2010) Electronic property and bonding configuration at the TiN(111)/VN(111) interface. *J. App. Phys.*, **108**, 033714-1–033714-10.

# 5

# Concurrent Multiscale Analysis by Generalized Particle Dynamics Methods

In this chapter, a GP (generalized particle dynamics) method is used to carry out concurrent multiscale analysis. The main assumption of the GP method is that if the basic material structure can be kept the same as the real material at all scales, the multiscale analysis will be more accurate and more effective. Based on this assumption GP divides the simulation region into areas containing particles of different scales, n; where $n = 1$ is the atomistic scale and higher values of n correspond to the continuum scale. Natural non-local boundary conditions via material neighbor-link cells are used for a seamless transition between domains of different scales. An inverse mapping method from the deformed particle domain $\beta_n$ into the corresponding atomic domain $\alpha_n$ is proposed. All calculations of different scales can be conducted at the corresponding atomistic $\alpha_n$ domain by the same potential and same numerical algorithm as the atomistic scale of the model, thus this method is also called extended molecular dynamics method. The comparison between simulations of GP and MD for dislocation initiation and propagation is given. Based on the introduction of the GP method, the state of art of concurrent multiscale analysis is introduced. This may help readers to know the big picture before they learn more detail about these methods in Chapters 6 and 7.

## 5.1 Introduction

As discussed in Chapter 1, multiscale methods can be divided into two classes: hierarchical multiscale and concurrent multiscale methods. Hierarchical method is also called the sequential method where model parameters at a given length and/or time scale are determined from simulations conducted at a smaller scale. Concurrent multiscale method is also called parallel method where models which relate to different length and/or time scale are combined within a single simulation.

*Multiscale Analysis of Deformation and Failure of Materials*   Jinghong Fan
© 2011 John Wiley & Sons, Ltd

## 5.1.1 Existing Needs for Concurrent Multiscale Modeling

The success of atomistic simulation and the existing needs to extend its spatial and temporal limits have promoted the development of concurrent multiscale analysis across atomic-nano-submicron or micron meter scales, which is the first category of multiscale analysis classified in this book.

Concurrent multiscale method performs calculation simultaneously to consider strong coupling among scales, such as turbulence and elastoplastic crack propagation in a finite medium. In these strong coupling cases, it is difficult to separate the role of a single scale from others, thus concurrent calculations to transfer in-situ information between different scales is necessary to realize the coupling between different scales.

The other reason to motivate the development of concurrent multiscale methods is to reduce the degree of freedom of the simulation system. In fact, the computational demands of a suitable model size and the integration time for the simulation of practical significance are beyond the upper bound of current computer technology. The calculation of MD deals with millions of atoms on the basis of general computer options, which can be used for a model with materials roughly 0.1 micrometer in size. Traditionally, a useful representative unit size for investigating material behavior in metals has been micrometers and larger. Therefore, MD alone cannot satisfy the need for an extension of atomistic simulation to the submicron and micron scale.

## 5.1.2 Expanding Model Size by Concurrent Multiscale Methods

Concurrent multiscale methods are especially attractive to the problems that require modeling of relatively large material domains but where atomic simulation is needed only in small regions. Fortunately, for many practical problems, the need for atomistic analysis is limited to a very localized region such as the crack tip, defect neighborhood, interface of different phases, grain boundaries, narrow area under nanoindentor. In these cases, the emphasis of computational simulation is on modeling the initiation and evolution of defects and cracks, the behavior of interface and grain boundaries, dislocation initiation and interaction, nanoindentation and nanocoating, and in describing different physical properties of materials at small scale, to name a few.

In fact, by using concurrent multiscale coupled models, the size limitations of the atomistic simulation can be minimized by embedding an inner region of atomistic simulation within an outer domain. While the complex atom dynamics and large deformation processes are inside the embedded small region, the deformation gradients in the outside region are small so that a coarse grain/continuum representation of the material becomes appropriate. Thus, large degrees of freedom can be reduced in the modeling.

## 5.1.3 Applications to Nanotechnology and Biotechnology

Concurrent multiscale analyses across atomic-nano-micron scales are not only good for material simulations but also suitable for nanotechnology and biotechnology devices. A microelectromechanical system (MEMS) is a good example because of its importance in technology. MEMS is involved with scales that are too small for finite element methods and too large for MD to simulate accurately the performance and properties. Concurrent multiscale analysis, however, is efficient to meet the simulation accuracy and computation time requirements. This example shows that the emerging fields of nanotechnology and biotechnology are providing a new momentum in the study and development of multiscale analysis.

## 5.1.4 Plan for Study of Concurrent Multiscale Methods

Over the past decade, various multiscale methods have been developed to address simulations involving atomistic material domains by the concept and algorithm of coupled concurrent atom-continuum simulation. The introduction of these methods will start in detail from the generalized particle dynamics method (GP) in this chapter,[1] followed by the quasicontinuum (QC) method in the next chapter. In Chapter 7 more concurrent multiscale models with different types of methodologies will be discussed and compared with the introduction of a benchmark test involving fourteen models. The origin and main difficulties in seamless transition across boundaries of different scales of these models will also be illustrated through 1D analysis and numerical simulation.

For researchers and scientists, it may be essential to know the state of art of this field at the beginning, thus an extensive review of concurrent multiscale analysis is given in the last section of this chapter after the introduction of the GP method. Through this review they may look at the GP method and other methods to be introduced in Chapters 6 and 7 from a more general and unified point of view. For beginners, however, it may be essential for them to learn several methods in detail at the very beginning, thus it is suggested that they learn the specific methods and terminology introduced in detail from Chapter 5 to 7 first, and read comprehensively the general review in Section 5.10 with a highlight view later. This way from concrete to generality may make more sense for them to get an understanding and compare different methods deeply.

## 5.2   The Geometric Model of the GP Method

The main assumption of the GP method is that if the basic material structure can be kept the same as the real material at all scales, the multiscale analysis will be more accurate and more effective, thus the notoriously difficult problems in multiscale analysis such as seamlessly bridging different scales can be solved easily.

Based on this assumption, the GP method extends the length scale largely for the simulation model. This is achieved by assuming a material model of generalized particle domains, $\beta_n$, of different scales n $(n = 1, \ldots m)$, in an ascending order with $n = 1$ being the atomistic scale.

For truly generalized particle domain $\beta_n$ $(n = 2, \ldots m)$ there is a one-to-one correspondence of $\beta_n$ domain to the material domain $\alpha_n$ $(n = 2, \ldots m)$ with atomic structure. This one-to-one correspondence of $\alpha_n$-$\beta_n$ is shown in Figure 5.1 where the atomic lattice in $\alpha_n$ is mapping to a generalized particle lattice in $\beta_n$ with the same crystallographic structure to effectively reduce the degrees of freedom.

The BCC crystal structure can be used to show how the generalized crystal lattice and the generalized particle are formed. The generalized primitive BCC unit cell is characterized by the corner points numbered from 1 to 8; and the centroid particle, $I_0$, indicated by a dark spot in Figure 5.2. The figure shows that the primary generalized crystal cell of particles consists of 8 atomic crystal cells. The generalized lattice distance, $a_{(2)}$, is double the atomistic lattice $a_0$, i.e., $a_{(2)} = 2a_0$.

The generalized particle is defined as the particle which possesses the total mass and takes the central position of the atoms (or lower scale particles) by lumping when the particle is formed. As an example, the central generalized particle $I_0$ can be formed by lumping the eight central atoms with star symbols inside the eight atomic lattices in Figure 5.2; and the corner particle can be formed by lumping eight atoms in the corners of its connecting atomic lattices. Consequently, the mass $m_{(2)}$ of the particle in the $\beta_2$ domain is eight times the atomic mass $m_0$. The definition of the generalized particle gives a one-to-one fixed connection between a generalized particle and a specific group of atoms. This connection offers the function to use a generalized particle to represent the group behavior of the atoms intrinsically connected to that particle, and also offers the function to mutually transform information between lower and upper scales (see next Section).

Lumping eight generalized particles of the second scale $(n = 2)$ to form a larger generalized particle of the third scale $(n = 3)$, the larger particle consists of 64 atoms and the lattice distance $a_4$ is four times $a_0$. If $n = 4$, the generalized particle consists of 512 atoms lumped together. Generally, the number of atoms

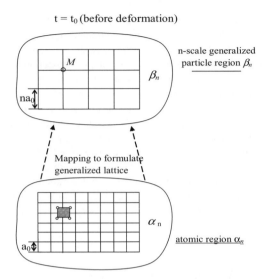

**Figure 5.1**    A schematic for mapping from the atomic lattice in the $\alpha_n$ domain to the generalized particle lattice with lattice constant $na_0$ in the $\beta_n$ domain, where a particle M consists of lumped atoms in shaded area (Fan, J. (2009) Multiscale analysis across atoms/continuum by a generalized particle dynamics method. *Multiscale Modeling and Simulation*, 8(1), 228. Copyright © 2009 Society for Industrial and Applied Mathematics. Reprinted with permission. All rights reserved)

($l_{(n)}$) that one generalized particle represents at the $n^{th}$ scale domain, $\beta_{n,}$ for BCC, FCC and other cubic crystal structures can be calculated by

$$l_{(n)} = k^{3(n-1)} \tag{5.1}$$

where k represents the ratio of the lattice constant at the $(n + 1)^{th}$ scale over that at the $n^{th}$ scale ($n = 1, \ldots$ $m - 1$) such as ratio two in Figure 5.2. Likewise, the particle mass $m_{(n)}$ and the lattice constant $a_{(n)}$ can be

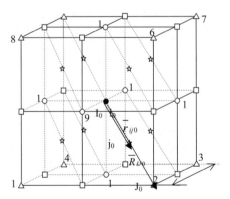

**Figure 5.2**    A generalized BCC crystal lattice and particle: 8 primitive atomistic lattices (8 atoms) form a generalized lattice (a particle) in the second-scale domain ($n = 2$) (Fan, J. (2009) Multiscale analysis across atoms/continuum by a generalized particle dynamics method. *Multiscale Modeling and Simulation*, 8(1), 228. Copyright © 2009 Society for Industrial and Applied Mathematics. Reprinted with permission. All rights reserved)

calculated as follows:

$$m_{(n)} = k^{3(n-1)}m_0 \tag{5.2}$$

$$a_{(n)} = k^{(n-1)}a_0 \tag{5.3}$$

Figure 5.3 shows a schematic of a three-scale GP model for a nano-indentation on iron, which has a BCC crystalline structure and lattice distance of $a_0$. The first scale is the atomistic scale with a radius of 90 Å, which surrounds the indenter near the load arrow. The second-scale domain is between a radius of 90 and 180 Å, in which the generalized particle lattice distance has doubled, i.e., it is $2a_0$ and the volume is as large as eight times that of the first scale. The third-scale domain is between a radius of 180 and 520 Å, in which the lattice constant is four times, and the volume is as large as 64 times that of the first scale. This indicates that 64 atoms will be lumped into one particle. The total number of atoms including the particles in the three-scale GP model in Figure 5.3 is 0.558 million. The MD simulation requires 14.014 million atoms, which is 25 times larger than the degrees of freedom in the GP model.

---

**Homework**

(5.1) Equations (5.1) and (5.3) are developed based on a BCC lattice. These rules can be proved valid for a FCC crystal. In fact, if one removes all body-center atoms in Figure 5.1 and put atoms in each face-center, then one will find, after lumping 8 atoms to one particle and lumping 8 unit cells to a generalized particle cell, several basic features including these equations are still valid for particle scales. For this exercise, prove that each unit cell of FCC crystal at the particle scale contains 4 particles as is the case in the atomistic scale.

Hint: In the atomistic scale, a unit cell of FCC crystal includes 4 atoms, 1 is from the 8 corners and 3 are from the 6 surfaces (see Figure 3.12(b)). To satisfy the requirement, one needs to prove that the 8 corner particles contribute 8 atoms (one particle) to the particle unit cell. In addition, all other particles contribute 24 atoms (3 particles) to the particle unit cell.

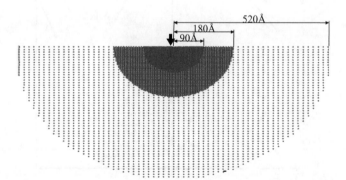

**Figure 5.3** Schematic of a three-scale nano-indentation GP model of a BCC iron crystal (Fan, J. (2009) Multiscale analysis across atoms/continuum by a generalized particle dynamics method. *Multiscale Modeling and Simulation*, 8(1), 228. Copyright © 2009 Society for Industrial and Applied Mathematics. Reprinted with permission. All rights reserved)

## 5.3 Developing Natural Boundaries Between Domains of Different Scales

Developing natural boundaries is the most important thing for seamless transition between domains of different scales. The GP method uses a unique approach to accomplish this requirement.

### 5.3.1 Two Imaginary Domains Next to the Scale Boundary

Natural boundary conditions require that the interaction forces passing through the boundary are active in the same scale of particles or atoms. Therefore atoms only see atoms and particles only see the same scale of particles at the boundary. In other words, if at the boundary, the atoms (or particles) see and directly interact with the real particles (or atoms), that boundary is not a natural boundary because atoms are inside the material and only directly interact with atoms beyond that boundary. The same is true for particles because only the interactions between atom groups represented by the same scale of particles are meaningful. The GP model satisfies this requirement by introducing two imaginary domains, W(n + 1) image and W(n) image, respectively, to the left and the right of the nth boundary.

The two imaginary domains and their overlapping with the model domains shown in Figure 5.4 are exemplified for the boundary between domain one of the atomistic scale (n = 1) and domain two of the second-scale particle domain (n = 2). The left domain of the boundary is the atomistic domain, and the right domain is the $\beta 2$ particle domain. One imaginary area with width $W_{1image}$ is located to the right of the boundary, i.e., overlapping with the second-scale particle domain. This imaginary domain consists of image atoms which connects and interacts with the real atoms on the other side of the boundary. The other

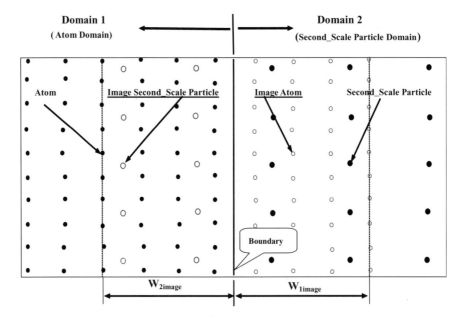

**Figure 5.4** Introducing two imaginary domains for a natural boundary between atomistic domain and the second-scale particle domain (Fan, J. (2009) Multiscale analysis across atoms/continuum by a generalized particle dynamics method. *Multiscale Modeling and Simulation*, 8(1), 228. Copyright © 2009 Society for Industrial and Applied Mathematics. Reprinted with permission. All rights reserved)

imaginary area with width $W_{2image}$ is located to the left of the boundary, i.e., overlapping with the atomistic domain. This imaginary domain consists of image particles of the second scale which connects and interacts with the real particles on the other side of the boundary.

## 5.3.2 Neighbor-link Cells (NLC) of Imaginary Particles

Each imaginary particle has a fixed cell linking to it, composed of its neighborhood real atoms in domain 1, and each imaginary atom has a fixed cell linking to it, composed of its neighborhood real particles, in domain 2. They are called neighbor-link cells (NLC) which are only used for each imaginary atom (particle), not for any real ones. These fixed cells are the key to understanding the unique mechanism for seamless transition of the GP method.

The choice of the neighborhood is controlled by a cutoff radius. Here, the value of the cutoff radius is different from the cutoff radius used in MD simulation; the latter is determined by the interaction range of potential function, but here is determined by the structure consideration. Just like the lumping process related to Figure 5.2, it usually will be eight real atoms (or lower-scale particles) in the fixed link cell for an imaginary particle and two real particles in the fixed link cell for an imaginary atom.

The locations of these imaginary particles and imaginary atoms as well as their fixed individual link cells are defined at the very beginning of the model formulation. The cell formulation is similar to the MD link-cell algorithm technique introduced in Appendix 10.D.1.2 of Chapter 10. Specifically, the group $G^I$ of these real atoms within a cutoff radius is permanently linking to a specific imaginary particle $I^{im}$ during the whole simulation process. The same is true for any imaginary atom and has a corresponding group $g^i$ of real particles to permanently link to it. As a result, in the input model file for the GP simulation, there are not only x, y, and z coordinates for each imaginary particle (or atom) but also the cell atom number and their ID numbers following that atom. The list gives the details of real atoms (particles) in the $G^I$ (or $g^i$) cell. Readers are suggested to see the input file, Model.MD in UNITS/UNIT14 of CSLI and how to produce it by design of model.in in Section 10.14.1. The following is an example for a third-scale of imaginary generalized particle in the input file of a copper nanowire:

```
174.192001342773437 7.257999897003174 1.774000167846680 3 -1
    8
 4952
 4982
 4950
 4984
 4846
81095
 4847
81126
```

The first three numbers in the first line above denote the x, y, and z coordinates of the particular particle, the fourth number 3 denotes this particle is the third-scale particle, the minus symbol in the fifth number denotes it is an imaginary particle and the number 1 denotes it is copper. The number 8 in the second line denotes that there are eight secondary real particles in its fixed link cell, and they are particles of 4952, 4982, 4950, 4984, 4846, 81095, 4847 and 81126.

## 5.3.3 Mechanisms for Seamless Transition

The two imaginary domains of scale shown in Figure 5.4 have different functions. The function of the imaginary $W_{(2)image}$ domain is to offer the interaction forces to the realistic particles. Consequently, the displacement patterns of the atoms in the atomistic scale are transferred to the upper-scale imaginary particle through this overlapping region. Specifically, the displacement of any imaginary particle $I^{im}$ in that domain is determined by the displacement (or motion) of real atoms in the $G^I$ of its NLC. It is both a spatially averaging displacement of the atoms in that group $G^I$ and a temporally averaging displacement at the past and current time step. Likewise, the displacement of any imaginary atom $i^{im}$ is determined by the displacement (or motion) of real particles in the $g^I$ of its NLC through the same averaging process.

The treatment of lumping realistic atoms on the left side of the boundary to an imaginary particle gives that particle more accurate position, and in turn, it gives more accurate inter-particle force to the realistic particle on the right of the boundary. Note that both imaginary domains do not have any free surface. They are all inside the model transition area. This is similar to the direct connecting (DC) methods introduced in Chapter 7 which do not introduce free surface at the handshaking area. Taking $W_{(2)image}$ as example, the atoms formulating the imaging particle in that domain are fully inside the atomistic domain of the model.

The same is true for the function of the $W_{(1)image}$ domain. To have natural boundary conditions for the atoms in domain 1 at the left side of the boundary, the introduced imaginary area $W_{1image}$ on the right consists of imaginary atoms. The positions of these imaginary atoms are determined by the positions of realistic particles in the overlapping area. Specifically, it is both a spatially averaging displacement of the real particles in that group $g^i$ and a temporally averaging displacement of these values at the past and the current time step. Consequently, the displacement patterns of the particles in the upper-scale domain on the right of the boundary are transferred to the imaginary atoms. The accurate position of the imaginary atom obtained through its NLC gives more accurate interatomic force to the real atoms on the left of the boundary.

## 5.3.4 Linkage of Position Vectors at Different Scales by Spatial and Temporal Averaging

As we discussed in Section 2.8.2, it is more pertinent not to use instantaneous variables such as the instant position vector to treat the atomistic data because this kind of data may not make physical sense and may include strong oscillation which is difficult to explain physically. Instead, an averaging of statistical mechanics should be used. This is important when linking the variables in the atomistic (or lower) scale to the continuum (or upper) scale. For an equilibrium or quasi-equilibrium thermodynamic system the Ergodic hypothesis shown by (2.21a) may be suitable. For a non-equilibrium system this hypothesis may not be necessarily pertinent.

In the GP method, the position vector $\bar{r}_i(t_m)$ of the imaginary atoms (or particles) in the NLC is calculated following the time averaging with previous time steps. Taking a second-scale imaginary particle I as an example, this is realized by the following procedure.

For a generic time step n, the nominal current position vector $\bar{r}_I^{nominal}(t_n)$ of the imaginary particle I at time $t_n$ is an arithmetic average of the position vectors $\bar{r}_i(t_n)$ of real atoms in its NLC, i.e.,

$$\bar{r}_I^{nominal}(t_n) = \frac{1}{N_{GI}} \sum_{i=1}^{N_{GI}} \bar{r}_i(t_n) \qquad (5.4)$$

where $N_{GI}$ is the total number of real atoms in the NLC $G^I$. The realistic position vector $\bar{r}_I^{real}(t_n)$ for imaginary particle I used in the calculation is an average of the realistic position of the last time step $\bar{r}_I^{real}(t_{n-1})$ and the current nominal one, i.e.,

$$\bar{r}_I^{real}(t_n) = [\bar{r}_I^{nominal}(t_n) + \bar{r}_I^{real}(t_{n-1})]/2 \qquad (5.5)$$

It can be easily shown that this scheme of time averaging includes the history of the particle position (or the history of the positions of real atoms in the NLC $G^I$) in the past time steps but with more weight in the steps which are close to the current time step, in other words, a fading memory type averaging of the position vector is used.

### 5.3.5 *Discussions*

There is a fully non-local QC model proposed by Knop and Ortiz (2001) which uses all atoms for nodes (see Section 6.5 for more detail). It is seamless because there is simply no transition. According to Curtin and Miller,[2] this QC model comes with a high computational cost in regions of slowly varying deformation gradient as compared to linear elasticity or even the local QC method. There is no obvious advantage when compared to the proposed GP method with fully non-local features and material neighboring link cells to transform information.

In the embedded statistical coupling method (ESCM) to be introduced in Section 7.10, the interface region consists of many interface volume cells (IVC) in which the coupling between MD and FEM is realized. Specifically, the spatial averaging displacement of atoms is taken over each IVC along with temporal averaging. The IVC averaging is with fixed spatial domain (i.e., Euler type formulation) but in the GP the spatial averaging is for fixed atoms or particles with a certain material volume (i.e., Lagrange type formulation). The averaging over the material neighboring link cells GI (or gi) are physically sound because it consistently represents the deformation of the related neighboring material elements during the whole deformation process.

For the temporal averaging, the ESCM uses the Ergodic hypothesis which uses the data in each time step equally, as is the case for an equilibrium or quasi-equilibrium system. However, in the GP method a simple time averaging of current and last step data is used for the position vector of imaginary atoms (particles), which shows a larger weighting in the steps which are close to the current time step.

## 5.4 Verification of Seamless Transition via 1D Model

To see if the transformation of deformation and force across the interface between different scales is seamless, 1D calculations similar to the 1D benchmark test of Curtin and Miller (2003) for checking different multiscale models are used. Figure 5.5 shows the 1D model with two scales. Solid spheres denote, respectively, the realistic atoms and the particles. The empty spheres denote the corresponding imaginary ones. There are eight atoms and five particles in the system. The left end is fixed and the right hand has a given displacement of 1000 units. The unstressed reference crystal corresponds to a chain of atoms with lattice constant a, and 2a for generalized particle distance since two atoms are lumped into one particle. The atom with vertical downward arrow contributes fully, the two atoms with inclined arrows each contribute half. Vice versa, one particle can decompose into two atoms as shown on the right with the

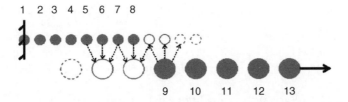

**Figure 5.5** A schematic for a two-scale 1D model (Fan, J. (2009) Multiscale analysis across atoms/continuum by a generalized particle dynamics method. *Multiscale Modeling and Simulation*, 8(1), 228. Copyright © 2009 Society for Industrial and Applied Mathematics. Reprinted with permission. All rights reserved)

three up arrows. Following Curtin and Miller (2003), a harmonic potential of interaction forces with both the first and second neighbors are considered with spring constants, respectively, as $k_1$ and $k_2$. The force equilibrium for any atom (particle) i is given as follows:

$$F_{i,i-1} + F_{i,i+1} + p_i = 0 \tag{5.6}$$

in which the first subscript denotes the atom i (or particle i) subjected to the force, the second subscript denotes which atom (or particle) applies the force. The term $p_i$ is the external applied force on atom (or particle i). All forces and displacement are positive if the direction is pointed out from the left to the right. These forces can be written in terms of the displacements $u_i$ of the atoms (particles) from their original equilibrium lattice sites as follows:

$$F_{i,i+1} = k_1(u_{i+1} - u_i) + k_2(u_{i+2} - u_i) \tag{5.7}$$

$$F_{i,i-1} = k_1(u_{i-1} - u_i) + k_2(u_{i-2} - u_i) \tag{5.8}$$

Substituting (5.7) and (5.8) into (5.6) one obtains:

$$k_1[(u_i - u_{i-1}) + (u_i - u_{i+1})] + k_2[(u_i - u_{i-2}) + (u_i - u_{i+2})] = p_i \tag{5.9}$$

Following (5.6) and the position calculation based on simple geometric relations shown in Figure 5.5, the set of governing equilibrium equations, without external applied force but with displacement boundary condition $u_{13} = 1000$, can be easily set up for the example as follows:

$$\begin{aligned} u_1 &= 0 \\ @u_2 &: k_1[(u_2 - u_1) + (u_2 - u_3)] + k_2[(u_2 - u_4)] = 0 \end{aligned} \tag{5.10}$$

$$\begin{aligned} u_i &= 3, 4, 5, 6 \\ @u_i &: k_1[(u_i - u_{i-1}) + (u_i - u_{i+1})] + k_2[(u_i - u_{i-2}) + (u_i - u_{i+2})] = 0 \end{aligned} \tag{5.11}$$

$$@u_7 : k_1[(u_7 - u_6) + (u_7 - u_8)] + k_2\left[(u_7 - u_5) + \left(u_7 - \left(\frac{u_8}{2} + \frac{u_9}{2}\right)\right)\right] = 0 \tag{5.12a}$$

$$@u_8 : k_1\left[(u_8 - u_7) + \left(u_8 - \left(\frac{u_9}{2} + \frac{u_8}{2}\right)\right)\right] + k_2[(u_8 - u_6) + (u_8 - u_9)] = 0 \tag{5.12b}$$

$$@u_9 : k_1\left[\left(u_9 - \left(\frac{u_7}{4} + \frac{u_8}{2} + \frac{1}{2}\left(\frac{u_8}{4} + \frac{u_9}{4}\right)\right)\right) + (u_9 - u_{10})\right] + k_2\left[\left(u_9 - \left(\frac{u_5}{4} + \frac{u_6}{2} + \frac{u_7}{4}\right)\right) + (u_9 - u_{11})\right] = 0 \tag{5.12c}$$

$$@u_{10} : k_1[(u_{10} - u_9) + (u_{10} - u_{11})] + k_2\left[\left(u_{10} - \left(\frac{u_7}{4} + \frac{u_8}{2} + \frac{1}{2}\left(\frac{u_8}{4} + \frac{u_9}{4}\right)\right)\right) + (u_{10} - u_{12})\right] = 0 \tag{5.12d}$$

$$\begin{aligned} u_i &= 11 \\ @u_i &: k_1[(u_i - u_{i-1}) + (u_i - u_{i+1})] + k_2[(u_i - u_{i-2}) + (u_i - u_{i+2})] = 0 \end{aligned} \tag{5.13}$$

$$\begin{aligned} @u_{12} &: k_1[(u_{12} - u_{11}) + (u_{12} - u_{13})] + k_2[(u_{12} - u_{10})] = 0 \\ u_{13} &= 1000 \end{aligned} \tag{5.14}$$

A discussion on the terms belong to atoms/particles near the interface is instructive. As is seen in Figure 5.5, from left to right i = 8 is the last atom and i = 9 is the first particle. To express the boundary condition for particle i = 9, we must use two imaginary particles with empty solid circles on its left (see Figure 5.5). Let us denote the first one in the left $P_{-1}$ and the second one $P_{-2}$. Their displacements (or positions) are, respectively, determined by real atoms as follows. For $P_{-1}$:

$$u_{P\_1} = \frac{1}{4}u_7 + \frac{1}{2}u_8 + \frac{1}{4}\left(\frac{u_8}{2} + \frac{u_9}{2}\right) \qquad (5.15a)$$

For $P_{-2}$:

$$u_{P\_2} = \frac{u_5}{4} + \frac{u_6}{2} + \frac{u_7}{4} \qquad (5.15b)$$

As is seen in the equation for $i = 9$, they are used in the equilibrium equation of that particle for the first and second neighboring reaction forces. From the construction of these terms, the following key points can be seen:

− These imaginary particles are important to determine the interaction force to the real particles, by offering them the same scale boundary condition.
− The displacements of these imaginary particles are mainly determined by the real atoms. These real atoms form the neighboring link cell: for $P_{-1}$ the cell includes real atoms 7, 8; for $P_{-2}$ the cell includes real atoms 5, 6 and 7.
− The contribution of the cell atoms to the imaginary particle has different weight with the highest weight belonging to the closest atom. In the present case, the middle one contributes a half weight, and the two side ones contribute a quarter.
− The cell constantly links to the imaginary particle, thus through this link, the deformation pattern of the lower (atom) scale transfers naturally to the upper scale through the same scale particle interaction. On the other hand, the locations of these imaginary particles are determined by the real atoms such that their location accuracy is high, thus a highly accurate interactive force on real particles can be expected.

Figure 5.6 shows the strain distribution of the system with only the first neighbor effects (i.e., $k_2 = 0$). Figure 5.7 shows the strain distribution with the first and second neighbor effects. These figures show that all the transition from the atom domain to the particle domain is seamless. There are 17 intervals for the 1000 displacement units on the right end which gives a 58.82/a homogeneous strain as shown in Figure 5.6. Slightly less strain is shown in Figure 5.7 due to the right end effects of the second neighbor. While the concrete strain values depend on $k_1$ and $k_2$, in each case, the GP result is identical to the MD values. Furthermore, the 1-D GP model is verified in the following for the same model length and force conditions of the three cases given by Curtin and Miller.[2] They are the cases of homogeneous deformation, inhomogeneous deformation, and point-force.

The problem is an extension of the simulation of the 8-atom, 5-particle system with the nearest and second nearest neighbor forces solved above (see the solution of homework 5.3 for detail). In fact, the same fixed boundary equation and the same 'transitional' equations between the atomistic and the GP regimes may be used. The forces can simply be applied to the desired atoms or particles and the resulting displacements can be converted to strains. Specifically, the model was established with 51 atoms, numbers 0 through 50, and 25 generalized particles labeled 51 through 76. Each generalized particle represents two atoms. Atom 50 represents the interface. For all loading conditions, an atomistic model is included as comparison. This atomistic model contains 101 atoms and is solved using the same framework described above. Since the values of $k_1$ and $k_2$ are not given in the Curtin and Miller paper,[2] the quantitative comparisons are shown by the direct comparison between the MD and GP calculations. A value of $a = 1$ has been assumed for the following strain calculations.

For the homogeneous case, a force of magnitude 1 was applied in tension to the free right end of the 1D chain. The inhomogeneous case was achieved by retaining the force of 1 at the free end and adding a force of 0.1 at the interface, atom 50, also adding a force of 0.1 at the two atoms adjacent to the interface (atoms 48 and 49) and the one particle adjacent to the interface (particle 51). A point force was achieved by

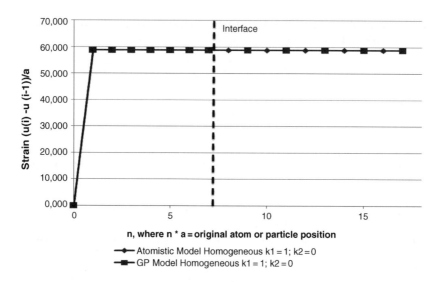

**Figure 5.6**    Seamless strain transition at interface for the model shown in Figure 5.5 when $k_2 = 0$ (Fan, J. (2009) Multiscale analysis across atoms/continuum by a generalized particle dynamics method. *Multiscale Modeling and Simulation*, 8(1), 228. Copyright © 2009 Society for Industrial and Applied Mathematics. Reprinted with permission. All rights reserved)

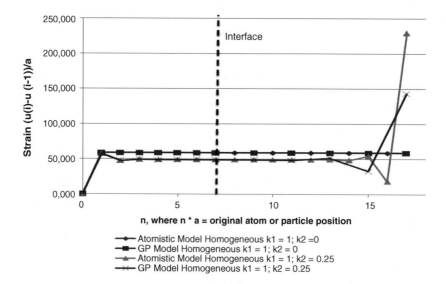

**Figure 5.7**    Seamless strain transition at interface for the model shown in Figure 5.5 with the second neighbor effects (Fan, J. (2009) Multiscale analysis across atoms/continuum by a generalized particle dynamics method. *Multiscale Modeling and Simulation*, 8(1), 228. Copyright © 2009 Society for Industrial and Applied Mathematics. Reprinted with permission. All rights reserved)

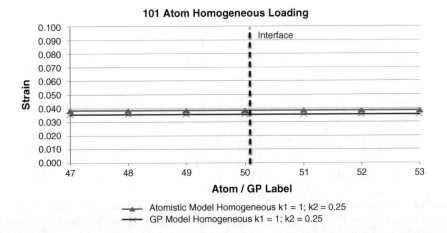

**Figure 5.8**    Seamless strain transition at interface for the 101a long model under homogeneous loading force 1 at the right atom (a = atomic lattice constant) (Fan, J. (2009) Multiscale analysis across atoms/continuum by a generalized particle dynamics method. *Multiscale Modeling and Simulation*, 8(1), 228. Copyright © 2009 Society for Industrial and Applied Mathematics. Reprinted with permission. All rights reserved)

applying a single force of 1 to the interface atom 50, while retaining the force of 1 at the free end. Applying force $p_i$ on the atom and/or particle i can be easily realized as it has been explicitly expressed on the right side of (5.6) for the equilibration of atom (particle) i.

Comparisons between the 1D GP and MD models for the three cases are given, respectively, in Figures 5.8 to 5.10. The figures show that in the first two cases the force transitions between the two scales are smooth. In the third case, a slope change is observed due to the point force applied at the interface. In all

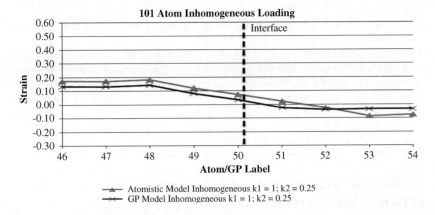

**Figure 5.9**    Seamless strain transition at interface for the 101a long model under inhomogeneous loading force near interfaces while retaining the force of 1 at the free end (Fan, J. (2009) Multiscale analysis across atoms/continuum by a generalized particle dynamics method. *Multiscale Modeling and Simulation*, 8(1), 228. Copyright © 2009 Society for Industrial and Applied Mathematics. Reprinted with permission. All rights reserved)

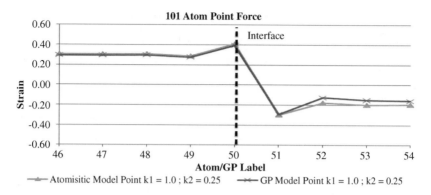

**Figure 5.10**   Seamless strain transition at interface for the 101a long model under point loading at the interface while retaining the force of 1 at the free end (Fan, J. (2009) Multiscale analysis across atoms/continuum by a generalized particle dynamics method. *Multiscale Modeling and Simulation*, 8(1), 228. Copyright © 2009 Society for Industrial and Applied Mathematics. Reprinted with permission. All rights reserved)

cases, the trends of GP simulation is consistent with the present MD simulation as well as the MD simulation by Curtin and Miller.[2] The result is quite encouraging since most models evaluated by Curtin and Miller have difficulty passing this examination.

---

**Homework**

(5.2) Determine the deformation patterns of a 1D GP multiscale model with 8 atoms and 5 particles (Figure 5.5) under the following boundary conditions: the left end is fixed and the right side is subjected to a unit force. Only first neighbor effect is considered. The atom spacing is its lattice distance "a" and the particle spacing is "2a". The strain is defined by the displacement difference between two connecting atoms/particles divided by the spacing "a" (atomic region) or "2a" (particle region).

   Hint: The deformation pattern can be expressed by displacement of each atom/particle and by strain distribution along the length of the model.

(5.3) The seamless transition results of Figures 5.8 to 5.10 are developed for a model of 51 atoms and 25 second-scale particles with a total length of 100a and interface at atom 50. Develop the basic governing equations for each atom (or particle) of the model and develop a solution scheme using Microsoft Excel to get the results for the case under inhomogeneous loading that at atoms 48, 49 and 50 and particle 51 at or near the interface there are positive forces of 0.1, and at the right free end (particle 75) there is a unit force. Finally, compare your obtained strain distribution with the results shown in Figure 5.9.

---

## 5.5   An Inverse Mapping Method for Dynamics Analysis of Generalized Particles

In the model formulation of Section 5.2 and Figure 5.1, it states that the particles are in the $\beta_n$ domain and their corresponding atoms are in the $\alpha_n$ domain (n = 1...m). The $\alpha_n$ and the $\beta_n$ domain have the same volume and the same coordinates, and they are in one-to-one correspondence. The generic particle in the $\beta_n$ domain is denoted as $I_0$, while the coincident atom is denoted as $i_0$. Figure 5.2 shows the pair $(I_0, i_0)$, its neighbors $(J_0, j_0)$ and position vectors $(\bar{R}_{IJ0}, \bar{r}_{ij0})$ when the atomic $\alpha_n$ domain is overlapped on the particle

$\beta_n$ domain in the same figure. Lower-case and upper-case letters denote, respectively, quantities in the atomic $\alpha_n$ and particle $\beta_n$ domain. $J_0$ and $j_0$ are, respectively, a generic particle and generic atom in the neighborhood of $I_0$ and $i_0$. The relative position vector between atoms $i_0$ and $j_0$ is $\bar{r}_{ij0}$, which consists of unit vector, $\bar{n}_{ij0}$, and magnitude $r_{ij0}$. The corresponding vector between particles $I_0$ and $J_0$ is $\bar{R}_{IJ0}$ consisting of unit vector, $\bar{N}_{IJ0}$, and magnitude $R_{IJ0}$.

It is important to clarify the physical meaning of lumping a $l_{(n)}$ number of atoms to form a generic particle. From a mechanics point of view, it simply means that the variables such as position, velocity, and acceleration of the $l_{(n)}$ atoms are averagely represented by the corresponding variable of that particle so that the degrees of freedom can be reduced. For example, in Figure 5.2 the acceleration of the particle $I_0$ and its coincident atom $i_0$ are used to represent the average of accelerations of the eight atoms with the star symbol.

The acceleration of the atom $i_0$ is the best representative for the average of accelerations of the $l_{(n)}$ atoms because it is located at the center of these atoms. Therefore, by determining the acceleration of atom $i_0$, the acceleration of the particle $I_0$ can be determined. The former can be determined if the current interatomic distance $r_{ij(n)}$ between the atom i and its generic neighboring atom j after a loading step is known. This inverse mapping will allow the application of the information in the current particle configuration in the $\beta_n$ domain to calculate the interatomic spacing of the corresponding realistic atomic structure around the representative atom $i_0$ in the $\alpha_n$ domain; so that the acceleration of the representative atom can be calculated, therefore the acceleration of the corresponding particle $I_0$ can be determined.

Mapping, as shown in Figure 5.1, can be defined as a transformation process from points and lines at the atomic crystal lattice in the $\alpha_n$ domain to their corresponding quantities of a generalized crystal lattice in the $\beta_n$ domain. We define inverse mapping shown in Figure 5.11 as an inverse transformation process from

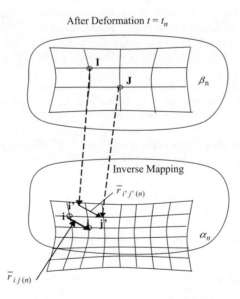

**Figure 5.11** A schematic showing points $i'$, $j'$ obtained by inverse mapping of particle I and J and their position vectors $\bar{r}_{i'j'(n)}$ versus realistic position vector $\bar{r}_{ij(n)}$ between atoms i and j (Fan, J. (2009) Multiscale analysis across atoms/continuum by a generalized particle dynamics method. *Multiscale Modeling and Simulation*, 8(1), 228. Copyright © 2009 Society for Industrial and Applied Mathematics. Reprinted with permission. All rights reserved)

points and lines in the $\beta_n$ domain to their corresponding quantities in the $\alpha_n$ domain. Note that the quantities in both the $\alpha_n$ and $\beta_n$ domain must be measured at the same time to have a correspondence. Actually, mapping or inverse mapping is one-to-one correspondence. This can easily be seen, due to the geometric similarity between primitive unit cells in the $\alpha_n$ and $\beta_n$ domains before deformation ($t = t_0$). The corresponding lines are parallel to each other and (5.3) is valid between the corresponding lengths. Specifically, the following relationship exists for the mapping process:

$$\bar{R}_{IJ0} = k^{(n-1)}\bar{r}_{ij0} \tag{5.16a}$$

$$\bar{N}_{IJ0} = \bar{n}_{ij0} \tag{5.16b}$$

The inverse mapping rule is proposed as

$$\bar{r}_{ij(n)} = k^{-(n-1)}\bar{R}_{IJ(n)} \tag{5.17}$$

where $\bar{R}_{IJ(n)}$ is the new relative position vector from particle I to J at time $t = t_0 + \Delta t$ due to the motion of the particles from $I_0$ to I and from $J_0$ to J; and $\bar{r}_{ij(n)}$ is the realistic position vector from atom i to j due to the motion of the atoms from $i_0$ to i and from $j_0$ to j. At this moment, whether (5.17) is true has not been rigorously proven. This is not as obvious as the initial case discussed below because lattice deformation changes the initial perfect lattice structure.

To understand this issue, review the relationship between (5.17) and the neighborhood around $I_0$ for the initial configuration ($t = t_0$) shown in Figure 5.2. It is seen that $\bar{r}_{ij0}$ and $\bar{R}_{ij0}$ are along the same straight line; the point $j_0$ equally divides the straight line $I_0J_0$ into two intervals so that the distance $r_{ij0}$ is just half of $R_{IJ0}$ as exactly predicted by (5.17) due to $n = 2$ and $k = 2$ in Figure 5.2. This intuitive validation for determining the true interatomic distance $\bar{r}_{ij(n)}$ from the current inter-particle distance $\bar{R}_{ij(n)}$ may not necessarily be true after deformation. For example, if the lattice is not homogeneous, the atom j may not equally divide the straight line IJ, and therefore may no longer follow rule (5.17) for proportionally reducing the magnitude of $R_{IJ(n)}$. Furthermore, if the lattice is distorted, $\bar{r}_{ij(n)}$ may not follow rule (5.16b) in paralleling to $\bar{R}_{IJ(n)}$, because $\bar{R}_{IJ(n)}$ is many times larger than $\bar{r}_{ij(n)}$ and their orientations can be different. In other words, rule (5.17) must follow certain conditions to be true.

In the following, it will be proved that the inverse mapping rule (5.17) is only valid if the local deformation in the neighborhood around the coincident pair ($I_0$, $i_0$) is homogeneous. The proof can be done geometrically or analytically. Here, we first prove (5.17) for a simple and visual case that I, J particles and i, j atoms are arranged along the x-axis when $n = 2$. Suppose I and J are lumped, respectively, with eight atoms. For simplicity only four are shown in Figure 5.12, with the positions of particles being

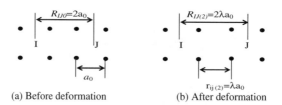

(a) Before deformation                    (b) After deformation

**Figure 5.12**    An example showing the validity of the inverse mapping distance $r_{ij(n)}$ which represents the realistic distance between atoms under the uniform deformation (dark ball: atoms; letters I and J denote the locations of generalized particles) (Fan, J. (2009) Multiscale analysis across atoms/continuum by a generalized particle dynamics method. *Multiscale Modeling and Simulation*, 8(1), 228. Copyright © 2009 Society for Industrial and Applied Mathematics. Reprinted with permission. All rights reserved)

denoted by the locations of their letters I and J, but not by dark spots as other atoms. The distance $R_{ij0}$ between the particles $I_0$ and $J_0$ before deformation can be determined by (5.16a) as follows:

$$R_{IJ0} = 2^{(2-1)}a_0 = 2a_0 \tag{5.18a}$$

Suppose local region material around $(I_0, i_0)$ subjects a uniform deformation along the x-direction so that the length $R_{IJ(2)}$ between particles I and J in the $\beta_2$ domain becomes:

$$R_{IJ(2)} = 2\lambda a_0 \tag{5.18b}$$

where $\lambda$ is the stretch ratio. Now the key issue is whether the value of $r_{ij(2)}$ in the $\alpha_2$ domain obtained by (5.17) can represent the interatomic distance between atoms i and j after the stretching deformation. If this is the case, the mapping rule (5.17) can be used to calculate the interatomic force by a potential function for atom i. By substituting (5.18b) into to (5.17) with n = 2, the following is obtained:

$$r_{ij(2)} = 2^{-(2-1)}R_{IJ(2)} = 2^{-(2-1)}(2\lambda a_0) = \lambda a_0 \tag{5.18c}$$

This distance, $\lambda a_0$, which is obtained by the inverse mapping method, is the physical distance between atom i, with its neighboring atom j, after uniformly stretching the atomic lattice by ratio $\lambda$. This situation is clearly shown in Figure 5.12. It is also seen that the acceleration obtained by this distance for atom i also represents the acceleration of its corresponding particle I. A 2D proof will be given in Homework (5.4) and a more in-depth 3D proof will be described in Chapter 6 after introducing the Cauchy-Born rule.

It is important to note that the role of inverse mapping is to find the realistic configuration of atoms in the $\alpha_n$ domain, which corresponds to the realistic configuration of particles in the $\beta_n$ domain. If the atomistic configuration underlying the current particle configuration can be found, then this information can be used to carry on the MD simulation for the generic atom i by multi-body potential, pair potential, etc. to calculate the interatomic force, which can then be used to obtain the acceleration of atom i and particle I. Two related issues are mentioned as follows:

– The inverse mapping rule does not contain an interpolation rule as other multiscale methods have. It maps back from the generalized particles of the $\beta_n$ domain into the corresponding $\alpha_n$ atomistic domain. This is consistent with the mapping rule by which the generalized particle of the $\beta_n$ domain is formed from the atoms at the $\alpha_n$ domain.
– The mathematical proof of the inverse mapping method for the 3D case in Chapter 6 relies on the Cauchy-Born rule, therefore the accuracy of the inverse mapping method also depends on the accuracy of the Cauchy-Born rule. Curtin and Miller[2] carried out an analysis for the Cauchy-Born rule, which concluded that if the deformation gradient F is sufficiently smooth over the non-local region of the interatomic interactions, the calculation using the Cauchy-Born rule is sufficiently accurate. Here, the non-local region denotes the cutoff radius $r_{cut}$. To avoid error, the GP model is designed to simulate a region with a large deformation gradient (e.g., crack tip, interface) with atoms but not with large-scale particles. This approach is the same as those used in several other multiscale models (e.g., MADD, QC and CADD discussed in Chapters 6 and 7).

---

**Homework**

(5.4) A crystallographic line ab of length $l_0$ in the xoy plane forms an angle $\theta$ with the crystalline axis x. The corresponding atomic structure maps to a third-scale domain $\beta_3$ with lattice scale ratio k = 2 (see (5.3)). After the mapping, line ab's corresponding line in the $\beta_3$ particle domain is denoted as AB. Assuming that this 2D three-scale model is subjected to uniform stretching $\lambda_1$ along the x-direction and $\lambda_2$ along the y-direction, prove mathematically that the length and orientation of the line a'b', obtained by the inverse mapping of the deformed line AB based on its rule (5.17), is the same as the line ab after the uniform stretching.

## 5.6   Applications of GP Method

In the following calculations, Morse pair potential and FCC copper and BCC iron crystalline structure are used to show how the inverse mapping method can be used to determine particle accelerations and to validate the GP method. The Morse potential, $\phi_{ij}$, between the pair atoms i and j is expressed as follows (see (2.9b) and (2.10a,b); Komanduri et al.[3]; Girifalco and Weizer[4]):

$$\phi_{ij} = D\{\exp(-2\alpha_0\eta_{ij})-2\exp(-\alpha_0\eta_{ij})\} \tag{5.19}$$

$$\eta_{ij} = \frac{r_{ij}-r_0}{r_0}, \quad \alpha_0 = \alpha r_0 \tag{5.20}$$

where $\eta_{ij}$ and $\alpha_0$ are normalized. $r_{ij}$ is the distance between the pair atoms i and j, $r_0$ is the equilibrium radius between atoms, and D and $\alpha$ are material constants. The force, $f_{ij}$, applied on atom i by atom j along the direction from atom i to j can be obtained by

$$f_{ij} = -\frac{\partial\phi_{ij}}{\partial r_{ij}} \tag{5.21}$$

For dynamic calculation of a generalized particle I in the $\beta_n$ particle domain (n = 1, 2, ..., m), the inverse mapping method requires the calculation of the interatomic position vector $\bar{r}_{ij(n)}$ from the inter-particle position vector $\bar{R}_{IJ(n)}$ by the inverse mapping rule (5.17). The obtained interatomic distance $r_{ij(n)}$ is used for the Morse potential to determine the interatomic force $f_{ij(n)}$ applied on i by atom j. Using (5.19) to (5.21), the following can be obtained:

$$\phi_{ij(n)} = D\{\exp(-2\alpha_0\eta_{ij(n)})-2\exp(-\alpha_0\eta_{ij(n)})\} \tag{5.22}$$

$$\eta_{ij(n)} = \frac{r_{ij(n)}-r_0}{r_0}, \quad \alpha_0 = \alpha r_0 \tag{5.23}$$

$$f_{ij(n)} = -\frac{d\phi_{ij(n)}}{dr_{ij(n)}} \tag{5.24}$$

Substituting (5.22) into (5.24), the following is obtained:

$$f_{ij(n)} = \frac{2D}{r_0}\alpha_0\left\{\exp(-2\alpha_0\eta_{ij(n)})-\exp(-\alpha_0\eta_{ij(n)})\right\} \tag{5.25}$$

Using Newton's second law, the acceleration of the equivalent atom i due to the interaction force $f_{ij(n)}$ can be obtained by:

$$a_{ij(n)} = \frac{d^2r_{ij(n)}}{dt^2} = \frac{2D}{r_0 m_0}\alpha_0\left\{\exp(-2\alpha_0\eta_{ij(n)})-\exp(-\alpha_0\eta_{ij(n)})\right\} \tag{5.26}$$

As long as the acceleration of the general particle I (through atom i) is obtained by the inverse mapping method, their velocity and position vector after each step can be obtained by an integration scheme, based on the previous velocity and position of particles I and J. The primary difference between different particles in different domains, $\beta_n$, comes from $\bar{R}_{IJ(n)}$ which determines a different interatomic vector $\bar{r}_{ij(n)}$. The physical nature, such as the pair potential (5.22), force equation (5.25), and acceleration equations (5.26), as well as their parameters including D, $\alpha$, $\alpha_0$, $r_0$, and mass $m_0$, are all the same.

## 5.7 Validation by Comparison of Dislocation Initiation and Evolution Predicted by MD and GP

In Section 5.3, the seamless transition of force and deformation across the interface between the atomistic and particle domains was validated for the 1D case. In this section, 3D models will be developed. To make the validation more conclusive, the loading is limited to monotonic and kept at a temperature of $T = 0\,K$. In addition, it is not necessary to use a large model to show the great saving of degrees of freedom during the validation.

The basic idea to verify the GP method in a more rigorous condition is to simulate and monitor the defect (dislocation) initiation and evolution. This validation is rigorous because dislocation is a local phenomenon; its prediction requires a high accuracy of simulation. This validation is important because plasticity and failure of ductile material including crack propagation are intrinsically related to dislocation initiation and propagation. The GP method is effective in this key validation if it has the same critical strain $\varepsilon_{ini}$ for dislocation initiation as that obtained by a simulation model with full atoms conducted by MD. In addition, the success of the GP method will depend on the consistency between dislocation patterns, respectively, predicted by the GP and MD method at incremental strain levels beyond $\varepsilon_{ini}$. The latter can demonstrate GP's effectiveness in simulation of dislocation evolution.

The first important step is to find a commonly accepted method for monitoring defect (dislocation) initiation and evolution. There are several methods in atomistic analysis to distinguish the dislocation from regular atoms. They are the CNA (common neighbor analysis) method,[5] the Ackland method and its combination with CNA[6] and the CN (coordination number) method.[7, 8] All methods commonly monitor the local structure change around the atoms. For example, in the CN method each atom inside a perfect FCC crystal material has 12 neighbor atoms, therefore its CN is 12. This number will reduce to 8 and 5 if the atom is, respectively, on the free surface and on the intersecting line between two free surfaces of a perfect FCC crystal. However, if a dislocation core occurs, the CN around the dislocation core will increase from 12 to 13 or even higher. This increase will vary the CN of atoms around that core.

The following sections will deal primarily with the CN method to monitor dislocation initiation and evolution. This is because free software such as AtomEye[9, 10] allows users to monitor any changes in the CN. As the CN changes, the atoms change color. For instance, yellow represents $CN = 12$ for a regular FCC atom, grey represents $CN = 8$ for a regular surface atom, blue represents $CN = 13$ or higher for the defect or dislocation core, and red for $CN = 5$. If there is no defect (dislocation) in the relevant atomistic region, all CN numbers in that region should be equal to or less than 12. The initiation strain of dislocation can be determined by the appearance of the first group of blue atoms with an arbitrary but fixed number, say 10 or 12 blue atoms with $CN = 13$, appearing in the first cluster of irregular atoms.

The evolution of dislocation numbers and patterns can be determined by watching the increase in number of the blue atoms at each incremental loading step. Also, AtomEye allows users to hide (make invisible) any regular atoms that have normal CN values, while keeping the irregular atoms, thus one can easily determine the position and distribution of the defects (dislocations).

To carry on the validation, a copper nanowire with the dimension of $72.9 \times 2.9 \times 11.6\,nm^3$ or $201 \times 8 \times 32\ a^3$ are designed for both the MD and a three-scale GP model. Here, $a = 3.629\,\mathring{A}$ is the atomic lattice distance. In the middle of the wire there are two tiny edge defects (notches) with a length of 1 lattice length unit, a, passing through the y-direction. This design will make the monitoring simple, since the dislocation is easy to initiate at the location near the cut. Figure 5.13 shows the GP model. The middle part of the bar contains the atomistic scale where the top edge notch can be seen. The far right portion with dilute particles is the domain of the third scale. Following equations (5.1) and (5.3) each particle consists of 64 atoms and a generalized lattice distance is 4a (14.5 Å). Between the middle atomistic domain and the domain of the third scale is a short domain in which there are particles of the second scale. Each particle consists of 8 atoms and a generalized lattice distance is 2a. The full atomistic model contains 205,296 atoms, while the three-scale GP model contains 81,392 atoms/particles. The latter includes 2154

**Figure 5.13**  A three-scale GP model for a copper tensile nanowire (Fan, J. (2009) Multiscale analysis across atoms/continuum by a generalized particle dynamics method. *Multiscale Modeling and Simulation*, 8(1), 228. Copyright © 2009 Society for Industrial and Applied Mathematics. Reprinted with permission. All rights reserved)

imaginary particles and atoms. The width of the overlapping region between different scales is given as (see Figure 5.4):

$$W_{1image} = 1a = 3.629 \text{ Å}, \ W_{2image} = 2a = 7.26 \text{ Å}, \ W_{3image} = 4a = 14.5 \text{ Å} \tag{5.27}$$

In other words, the width of the overlapping area is equal to the correspondent lattice distance of the corresponding scale.

Besides the model development, all other conditions, procedures and control variables reported below are exactly the same between MD and GP simulations, because GP is an extension of the MD method. Each end of the nanowire has a length of 2a to form a clamping area. During the equilibration period of 100 ps or 50,000 relaxation steps (time step = 0.002 ps), both ends are fixed along the y- and z-direction, but free along the x-direction. The maximum strain is controlled within 20%. There are 20 incremental loading steps with each loading step increasing in strain by 1%. Here the strain is the average strain defined as the relative elongation of the nanowire right end relative to the left end divided by the length of the wire. The left end is fixed, while the right end is moving along the [001] or x-direction. To reduce the inhomogeneous strain distribution due to the fixed end effects, each loading step contains a uniform displacement field given at the beginning along the wire length with the required displacement, say 0.714 nm, at the right end and a zero value at the left end.

The specimen is then relaxed so that every atom and particles can reach their equilibrium position. The relaxation process takes 10,000 steps, with a time step of 0.001 ps, such that the relaxation time for each 1% strain is 10 ps. Therefore the time required is 200 ps for a strain of 0.2, the average strain rate is $10^9$/S, and total simulation time is 300 ps including 100 ps for equilibration. Also, the simulation temperature is held at absolute zero (0 K) and periodic boundaries are not used along the three directions. The Morse potential is used for the interatomic force, therefore the parameters taken from Komanduri *et al.* (2001) are: $D = 0.3429$ eV, $\alpha = 1.359$/Å, and $\alpha_0 = 2.866$ Å. The stress values reported in this work were calculated using the virial theorem, which is equivalent to the Cauchy stress in the average sense (see Section 2.9).

Comparison of the stress-strain curve obtained by MD and GP is shown in Figure 5.14; where the points with the symbol × are the GP results, and the points with the symbol + are the fully atomistic simulation results. From this comparison, it seems the agreement is satisfactory.

Figure 5.15 shows the comparison of the configuration of the first cluster of blue (or royal blue) atoms with CN = 13 obtained respectively by MD (the top left window) and GP (the top right window), at the strain of 10%. These clusters denote dislocation cores inside the bulk material predicted by MD and GP respectively. To compare the position of the two clusters and their relative position to the corner point, A and A′, of the top notch, the center coordinate $\bar{r}_i$ for MD and $\bar{r}'_i$ for GP are defined, respectively, as the average coordinate values of the top left blue atom and the bottom right blue atom of each cluster. Denoting the unit vector along the x-, y- and z-directions as $\bar{i}, \bar{j}, \bar{k}$, the obtained position vectors are as follows:

For MD: The center coordinates are: $\bar{r}_i = 361.426\bar{i} + 12.3049\bar{j} + 85.3204\bar{k}$
The reference corner point of the notch: $\bar{r}_A = 360.098\bar{i} + 7.72948\bar{j} + 86.4573\bar{k}$

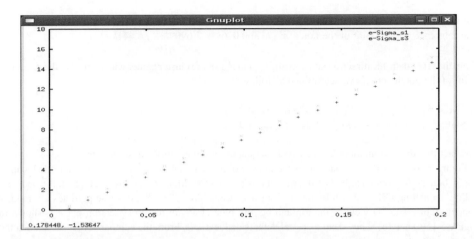

**Figure 5.14** Comparison between stress-strain data of the GP method and the fully atomistic MD simulation (Fan, J. (2009) Multiscale analysis across atoms/continuum by a generalized particle dynamics method. *Multiscale Modeling and Simulation*, 8(1), 228. Copyright © 2009 Society for Industrial and Applied Mathematics. Reprinted with permission. All rights reserved)

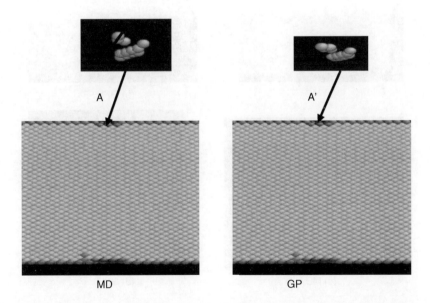

**Figure 5.15** A comparison of the dislocation core nucleation obtained by MD and GP simulation near the top small notch at a strain of 10%. The inset figures in the top denote dislocation atoms with CN = 13, the left one is for MD and the right one is for GP (Fan, J. (2009) Multiscale analysis across atoms/ continuum by a generalized particle dynamics method. *Multiscale Modeling and Simulation*, 8(1), 228. Copyright © 2009 Society for Industrial and Applied Mathematics. Reprinted with permission. All rights reserved)

For GP: The center coordinates are: $\bar{r}'_i = 361.426\bar{i} + 12.3025\bar{j} + 85.3199\bar{k}$
The reference corner point of the notch: $\bar{r}'_A = 360.076\bar{i} + 7.6909\bar{j} + 86.441\bar{k}$

From the set of data, the direction and norm of $\overline{r_{iA}}$ and $\overline{r'_{iA}}$ of the cluster center with respect to the reference point of the notch corner are determined as follows:

For MD: $\overline{r_{iA}} = \bar{r}_i - \overline{r_A} = 1.328\bar{i} + 4.5754\bar{j} - 1.1369\bar{k};\ r_{iA} = 4.8$ Å
For GP: $\overline{r_{i'A'}} = 1.35\bar{i} + 4.6064\bar{j} - 1.211\bar{k};\ r_{iA} = 4.93$ Å

The fact that the two methods give such close quantitative predictions is surprising.

Figure 5.16 shows the comparison of the inner atomic configuration and the evolution of dislocation cores (blue atoms) obtained by GP and MD at a strain of 15%. Many of the inside atoms in the body are yellow indicating CN = 12, while the grey surface atoms have CN = 8. All these particles are regular copper atoms. The 3D view of the top and bottom blue atoms in the cross-section plane is shown in Figure 5.16(c) and (d). They denote crystal defects/dislocations. From the comparison, it is seen that the distribution pattern of the defects is very close.

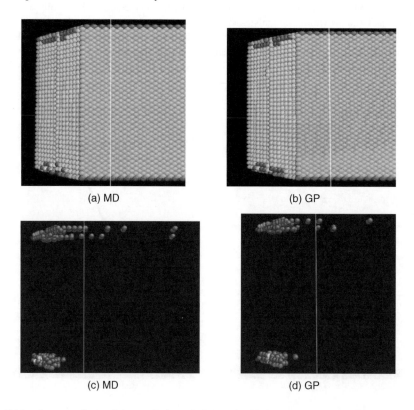

(a) MD                                              (b) GP

(c) MD                                              (d) GP

**Figure 5.16** A comparison of the evolution of dislocation/defect cores obtained by the MD and GP simulation at a strain of 15%. (a) and (b) are the inside view of the body, (c) and (d) are the inside top and bottom dislocation/core clusters for MD and GP, respectively (Fan, J. (2009) Multiscale analysis across atoms/continuum by a generalized particle dynamics method. *Multiscale Modeling and Simulation*, 8(1), 228. Copyright © 2009 Society for Industrial and Applied Mathematics. Reprinted with permission. All rights reserved)

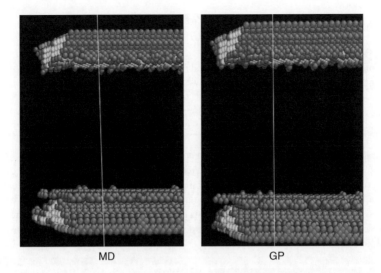

MD                                    GP

**Figure 5.17**    A comparison of extensive crystal defects shown by blue atoms (CN = 13) and grey atoms (CN = 14) obtained by the MD and GP methods at a strain of 20% (Fan, J. (2009) Multiscale analysis across atoms/continuum by a generalized particle dynamics method. *Multiscale Modeling and Simulation*, 8(1), 228. Copyright © 2009 Society for Industrial and Applied Mathematics. Reprinted with permission. All rights reserved)

Figure 5.17 shows the wide scope of defect growth. The atoms with the so-called misty rose color near the surface, neighboring the blue atoms, have CN = 14. The appearance of these high CN atoms denotes that the perfect crystal structure is seriously defective. In fact, after a small strain increase, the nanowire will sustain a first stress peak drop, a soft behavior due to defect (or damage) of the material.

In addition to the above validation for a copper nanowire, an iron plate with a short notch is also used. In this example, a notched 2D plate is subjected to a shear displacement at the top edge, but fixed at the bottom, as shown in Figure 5.18. Again the Morse potential is used. For BCC iron, the parameters are: $D = 0.4172$ eV, $\alpha = 1.389/\text{Å}$, and $r_0 = 2.845$ Å. The loading takes 60 loading steps, each of which are 10 ps long and require 10,000 relaxation steps. The nominal shear strain is 12.5% for Figures 5.18b and 5.18c. The former is the result of MD, and the latter is GP. It is seen that the dislocation patterns calculated by the GP method are close to those of the MD method even at the tip of the notches.

## 5.8    Validation by Comparison of Slip Patterns Predicted by MD and GP

Figure 5.19 shows a three-scale GP geometric model of BCC iron with $a_0 = 2.87$ Å. This model is designed to make a comparison between a full atomistic model by MD calculation and the three-scale model by the GP method under the same uniaxial loading along the horizontal axis of crystal orientation [100]. The size of the specimen is $200 \times 200 \times 200$ $a_0^3$ ($57.4 \times 57.4 \times 57.4$ nm$^3$) with a through-thickness hole along the [010] direction whose diameter is 2.87 nm ($10a_0$). To simulate the specimen, MD needs $3.36 \times 10^6$ atoms and a three-scale GP model needs $3.4814 \times 10^5$ atoms and particles. The atomistic scale (n = 1) is located at the center of the model, a $42 \times 42 \times 42$ $a_0^3$ area, which surrounds the hole. Both MD and GP use Morse potential with a cutoff radius of 4 Å.

Figure 5.20 shows the comparison between local deformation patterns around the hole of Figure 5.19 for two strain magnitudes: strain $\varepsilon = 5\%$ (a) for MD, (b) for GP; $\varepsilon = 20\%$ (c) for MD and (d) for GP. It is seen from these figures that the deformation is highly concentrated along four bands with the band angles

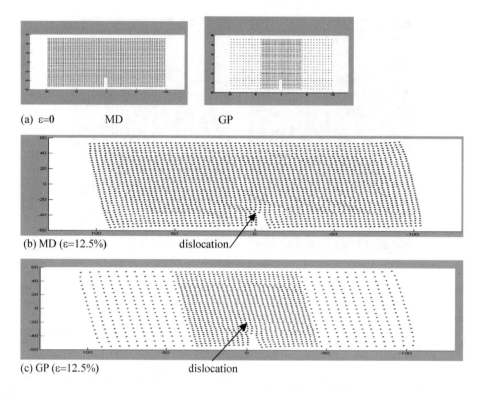

(a) ε=0              MD              GP

(b) MD (ε=12.5%)            dislocation

(c) GP (ε=12.5%)            dislocation

**Figure 5.18**  Comparisons of deformation patterns between GP and MD methods under shear strains of (a) 0, (b) 12.5% for MD, and (c) 12.5% for GP (Fan, J. (2009) Multiscale analysis across atoms/continuum by a generalized particle dynamics method. *Multiscale Modeling and Simulation*, 8(1), 228. Copyright © 2009 Society for Industrial and Applied Mathematics. Reprinted with permission. All rights reserved)

of about 45 degrees to the loading axis. The detailed deformation patterns within the four bands are the same between the results of GP and MD. At strain $\varepsilon = 5\%$, the hole sustains an elongation along the [100] direction but the shape is no longer a circle.

However, at $\varepsilon = 20\%$ the atoms on the surface of the hole diffuse to the inside of the hole. The diffusion is believed to be driven by interatomic forces related to surface tension and intensive slip near the hole. These forces around the hole with a radius of only about 1.4 nm have more effects than those due to the remote boundary elongation.

Thus, the hole elongation did not continue and the hole gradually disappeared, which is an indication of possible self-repair of flaws at the nanoscale. It is notable that the nanoscale diffusion patterns are the same for GP as those of MD. From these comparisons it is seen that the GP method has equivalent accuracy to the MD.

## 5.9   Summary and Discussions

A GP multiscale method is proposed to investigate the connections and transitions between atomistic and continuum scales. The GP model keeps the entire model with the same crystal structure, develops natural non-local boundary conditions, and uses inverse mapping methods to conduct calculations of all scales

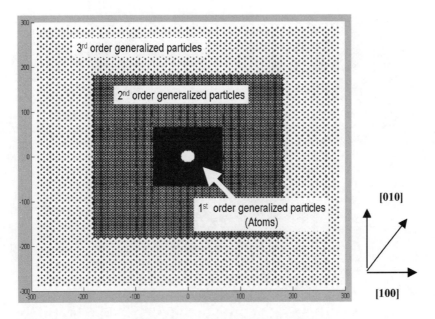

**Figure 5.19** A three-scale GP model

with the same physical laws and numerical algorithms as the MD model. The basic assumptions and features of this method include:

(1) If the basic material structure can be kept the same as the real material at all scales, the multiscale analysis will be more accurate and more effective; therefore the notoriously difficult problem in seamlessly bridging different scales can be solved easily.

(2) To keep the same structure at all scales, GP divides the simulation region into areas containing particles of different scales, n; where n = 1 is the atomistic scale and higher values of n correspond to the continuum scale. It has been proven for crystalline structures of a cubic unit cell, that a generalized particle in the $n^{th}$ scale is lumped by $k^{3(n-1)}$ atoms. The reduction in the number of degrees of freedom in a simulation system, therefore, is effective.

(3) From a physics point of view, the generalized particle at the upper scale is lumped by atoms (or lower-scale particles), therefore it represents their motion and behavior in the average sense. This intrinsic connection is especially important for the particles and atoms in the overlapping region as summarized in (4) below. This is because all the information transition between the lower and upper scale is through this intrinsic material connection. Therefore, this method is different from the DC and ESCM methods in principle in bridging the atom-continuum scales.

(4) GP uses natural boundary conditions between different scales. This can be shown by introducing two imaginary domains along both sides of the boundary. The seamless transition between different scales is not only because all scales have non-local constitutive behavior, but because the neighbor-link cell is designed for each imaginary atom and imaginary particle. This one-to-one cell-particle link begins with the atom-lumping process formulating the generalized particle. Using this link, the positions (motions) of real atoms in the cell can be averaged to get the position (motion) of the linked imaginary particle; therefore the atomistic information can be transformed through the imaginary particle to the upper scale to give the accurate interaction between the imaginary particles and real particles. It is also true that the position (motion) of any imaginary atom can be determined by averaging the positions (motions) of its

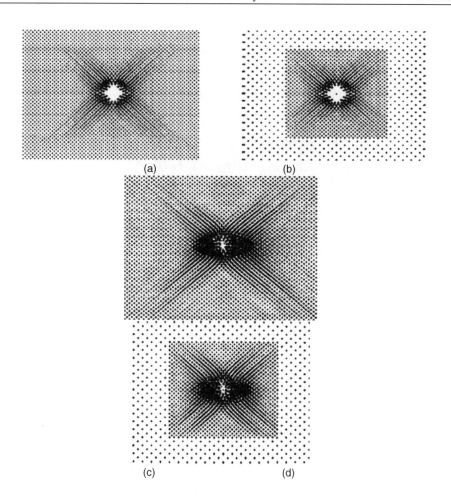

**Figure 5.20** Comparisons between deformation patterns near a nanoscale hole by MD and GP for a through-hole specimen under horizontal tensile loading along crystalline direction [010]: (a) MD results with $\varepsilon = 5\%$, (b) GP results with $\varepsilon = 5\%$, (c) MD results with $\varepsilon = 20\%$, and (d) GP results with $\varepsilon = 20\%$

linked real particles. Therefore, the upper-scale information can be transformed into the lower scale to allow the imaginary atoms near the boundary to interact with the real atoms.

(5) The following GP features are related to the natural boundary conditions:

- The connections between different scales are through the displacement of particles (atoms) and their average displacements of the linked cells. Therefore deformation patterns of either atoms or particles can be seamlessly transferred between each other, which eliminates the source of ghost force as proved from Figures 5.6 to Figure 5.10.
- The overlapping region is embedded in the model; therefore no free surface corrections of the ESCM type are needed.
- The atoms (or particles) in the imaginary region move based on the motion of real particles (or real atoms) in its neighbor-link cell. This is different from the PAD introduced by CADD where the PAD atoms are fixed to the FE mesh and their positions are determined mainly by the continuum.

(6) GP keeps potential functions, material parameters, cutoff radius, and numerical algorithms at all scales the same as those at the atomistic scale. Therefore GP is an extension of molecular dynamics used to reach high accuracy. This dominantly important feature of the GP method can be shown using the inverse mapping method. It has been proved that the position vector $\bar{r}_{ij(n)}$, obtained by the inverse mapping rule (5.17), represents atomistic configurations around atom i corresponding to the generalized particle I. It can therefore be used to determine the interatomic force and acceleration of the atom i. Due to the intrinsic geometric connections between the generic particle I with its corresponding atom i, the obtained acceleration can be used to determine the velocity and trajectories of particles in the domain $\beta_n$.

(7) By inverse mapping method, the obtained interatomic spacing $\bar{r}_{ij(n)}$ is used in the potential function to determine the force, acceleration, velocity and displacement. Therefore, this method involves all the physics of the problem, keeps the same crystalline structure at all scales, and its calculations are nonlinear in nature. Therefore, it is quite different from any interpolation process of QC and other methods using interpolation function.

(8) Validation of the GP method above is effective and encouraging, but requires additional work to verify its limitations. The stress-strain curve obtained by GP agrees with MD for the first stress drop, but not for the second stress drop for the nanowire discussed in Section 5.6. Specifically, the first drop occurs at a strain of 0.275, and the second drop at about 0.37. It is found that the agreement between MD and GP has a certain limit. In the present case, the limit of the strain value is 35%, beyond which is a large discrepancy. This may be due to an inaccurate Morse potential value, therefore an EAM potential may be required. The other limitation is the temperature. Accurate results are obtained at a temperature of 0 K, not room temperature. While the GP method is developed for different temperatures, some thermal stats must be implemented to account for the temperature effects, which will be the work of the next section.

## 5.10  States of Art of Concurrent Multiscale Analysis

From the introduction to the GP method, it is seen that a framework for a concurrent method includes the following parts: model development, boundary treatment, numerical algorithm, validations and applications. The key feature, however, is the scheme for quantitative transition of variables between different scales at their boundaries. In this section, the states of art of concurrent multiscale analysis are reviewed. It is believed that the review of different methods will help researchers to understand why the GP method is developed in the way described in this chapter. It will help readers to catch the basic feature of different methods at the early stage, so that they can look at them from a comprehensive point of view during the next two chapters.

### 5.10.1  MAAD Concurrent Multiscale Method

Concurrent multiscale analysis on large-scale ranges was done by Abraham et al.[11] in a pioneering work referred to as the "beginning of computational atomistic engineering."[12] Their method is called MAAD (macroscopic, atomistic, ab initio dynamics), also called CLS (coupling length scale) method. This method will be introduced in detail in Section 7.3.

The MAAD method distinguishes itself from all the developed multiscale methods in that it not only couples the atom region with the continuous medium region (i.e., finite element (FE) region) by interatomic potential energy, but also introduces the tight-binding (TB) region (see Sections 4.7.2 and 4.8) into the crack-tip region. This introduces coupling of the electronic motion characterized by quantum mechanics with atomic motion. The three regions are separated by two handshake regions, one between TB and MD, the other between MD and FE. Handshake regions are different from the imaginary regions introduced in Figure 5.4. They are real parts of the model. Specifically, their energy is a part of the system energy which is used to couple different regions. For instance, the MD model was coupled with the continuum model in a "handshake" domain in which two Hamiltonians (or total energy, see 2.14d) of MD and a continuum were averaged. Through the energy relationship with force such as (2.1a), the coupling between MD and continuum domains is realized.

## 5.10.2    Incompatibility Problems at Scale Boundary Illustrated with the MAAD Method

The treatment of the interface between different scales such as atomistic and continuum is the key issue for multiscale modeling. Many multiscale methods introduce artificial forces and spurious phenomena at the interface region. These are caused by the difference of material constitutive laws (stress-strain relationships) and stiffness between the materials on both sides of the interface. Taking the interface area between atomistic and continuum region as an example (see Figure 7.1), the continuum region possesses a localized response, which means that the stress depends on the local strain only. This is quite different from the atomistic region where non-localized effects exist, due to the indirect interaction forces between atoms through the interatomic potential function. The contradictory nature of material response (property) between material in the two sides of the interface produces the intrinsic incompatibility between the MD and FE region.

These incompatibilities would result in a statically non-equilibrium "ghost force", a phrase first introduced in Ref. [13]. This force is unphysical force. Ideally at the beginning of the model formulation, atoms are on lattice sites of equilibrium crystal and the finite elements of continuum are unstressed and undeformed. During the relaxation process to the model equilibrium position with minimum potential energy, forces on atoms or finite element nodes can occur due to different stress-strain relationships in the MD and FE regions. This incompatibility-produced force is unphysical and will lead to spurious distortions of the body.

For dynamics, waves are reflected at the interface between the MD and FE regions. The reflection is because the wavelength transmitted by the MD region is smaller than that which can be absorbed by the continuum region, so it is impossible for this wave to go into the continuum region. However, the energy is conservational in the entire system. Therefore, the wave has no other option but to return to the MD region. This reflection results in an artificial thermal increase, which leads to simulation failure. Generation and retention of heat in the MD region causes negative effects to the system, especially to the plastic deformation area. In some extreme circumstances, the retained heat can lead to the melting away of the entire MD region.[14]

In addition, since large finite element meshes are incapable of absorbing waves from the MD region, the FE mesh at the MD/FE interface is reduced to a size comparable to the atomic lattice. However, this reduction not only increases FE elements, but also brings numerical problems. Specifically, the temporal increment in the finite element is determined by the smallest cell, so reducing the cell size down to the atomic lattice would result in wasted time. It is therefore not reasonable and not effective to use an atomic lattice to mesh the continuum region.

To reduce spurious reflections of waves into the MD domain, damping was used in the handshake region, although the damping was not based on any rigorous theory. Because of the importance of incompatibility at the scale boundary for multiscale analysis, the nature of incompatibility due to stiffness and material physical property will be further demonstrated by a 1D model in Sections 7.6 and 7.8.

MAAD is one of the earliest attempts in multiscale analysis. Although subsequent multiscale attempts have adopted some methods and ideas of MAAD, few of them span the large range of scales like MAAD. In other words, those subsequent attempts did not follow MAAD, in order to avoid substantial difficulties and extremely intensive calculations. This is also the reason why multiscale analysis in this book is divided into two intrinsically connected categories (see Chapter 1).

## 5.10.3    Quasicontinuum (QC) Method

The method of the quasi-continuum (QC) proposed by Tadmor et al.[15] and Knap and Ortiz[16] furnishes a computational scheme for bridging the atomistic and continuum realms. This method will be introduced in detail in Part 2 of Chapter 6. In the QC, the continuum scale is not represented by generalized particles but by finite elements whose energy density (or total potential energy) is obtained by atomic potential through the Cauchy-Born rule (see Section 6.4.4). The chief objective of the method is to coarsen an atomistic description systematically by the judicious introduction of kinematic constraints. These kinematic constraints are selected and designed to preserve full atomistic resolution where required, and to treat collectively large numbers of atoms in regions where the deformation field varies slightly on the scale of the lattice.

QC has some common features with GP. For instance, both use energy-based formulation in which interatomic potential takes an essential role. In addition, both do not use a handshake region to connect atomistic and continuum scale. The main difference between QC and GP is that QC does not keep the material structures in the coarse region as GP does, but uses finite elements to interpolate the node displacements of a continuum for any spatial point in the coarse region.

Specifically, the QC method starts from an underlying conventional atomistic model that delivers the energy of the crystal as a function of the atomistic positions. The configuration space of the crystal is reduced by identifying a subset of "representative atoms" which then become the sole independent degrees of freedom of the crystal. The positions of the remaining atoms are obtained by piecewise linear interpolation of the representative atom coordinates through finite element method. Then, the effective equations are determined by minimizing the potential energy of the crystal over the reduced configuration space.

While QC uses the unified interatomic potential to determine the energy density function (or total potential energy) in both MD and finite element region, due to the local and non-local property of materials in, respectively, FE and MD region, QC produces ghost force at the boundary region between the two domains. To overcome this shortcoming, Knap and Ortiz proposed a new QC method framework.[16] This new model is non-localized throughout the simulation region, implying that limits arising from localization in the finite element calculation have been eliminated from the previous QC method. However, this is done at a high cost in the computational time used in regions of slowly varying deformation gradient as compared to a linear elastic constitutive law or even the Cauchy-Born approximation in the local QC method (see Section 6.5).

## 5.10.4    Coupling Atomistic Analysis with Discrete Dislocation (CADD) Method

Shilkrot et al.[17] introduced a CADD method, an acronym for the method that couples atomistic analysis with the discrete dislocation (DD) method. In this method, the solution for a continuum domain with dislocations under a given external loading is cast as a superposition of the following two solutions. The first one is Giessen and Needleman's solution for an infinite domain with given dislocation patterns under the condition of zero force and displacement at the infinite boundary.[18] The second is the solution of a subdomain without dislocations. The superposition will produce correct external loading and displacement boundary conditions at the finite continuum domain including dislocations. Based on the need to create the force boundary by the superposition process, this method is force-based multiscale analysis.

The focus on dislocations of the CADD method necessitates the use of linear elasticity, which is opposed to the Cauchy-Born rule used in the QC method. CADD also does not use handshake region but uses a pad consisting of atoms. These pad atoms offer interactive forces to the atoms in the atomistic region.

While CADD will be introduced in detail in Section 7.5, it may be as well to mention briefly the major differences between CADD and the GP method in transforming the deformation patterns. While both methods create an atomic boundary using either padding atoms or imaginary atoms to interact with the realistic atomistic domain, the transition mechanism between variables at different scale is different. In the GP method, the deformation pattern transformation between different scales is through the spatially and temporally averaging process of motions of real atoms which neighbor the imaginary particle at the up scale. In other words, the transformed deformation patterns are all obtained from the integration of an atom's motion equation (Newton's second law) along with statistical averaging. All pad atoms of CADD, however, are not truly connected to the motions of realistic atoms in the model and are not related any kind of statistical averaging.

## 5.10.5    Existing Efforts to Eliminate Artificial Phenomena at the Boundary

There are intensive efforts within the computational mechanics community to increase calculation accuracy at the interface area between the atomistic and the continuum scale (Rudd and Broughton 1998[19] and 2005;[20] Cai et al. 2000;[21] Huang and Huang 2001;[22] Chung and Namburu 2003;[23] Curtarolo and Cedar 2002;[24] Xiao and Belytschko 2004;[25] Wagner and Liu 2003[26]).

There are several recent developments (Saether et al. 2009;[27] Yamakov et al. 2008[28]; To and Li 2005;[29] Li et al. 2006;[30] Liu and Li 2007;[31] Li et al. 2008[32]; Chaboche 2008;[33] Medej et al. 2008;[34] Chen and Fish 2006[35]), including a method proposed by To and Li called perfectly matched multiscale simulation. This method can efficiently eliminate the spurious reflections/diffractions from the artificial atomistic/continuum interface by matching the impedance at the interface of the MD region and the perfectly matched layer. Li et al. 2008 proposed a nonequilibrium multiscale computational model. The multiscale framework couples thermomechanical equations at the coarse scale with nonequilibrium MD at the fine scale. It enables subsets of fine-scale atoms to be attracted to different coarse-scale nodes that act as thermostats.

## 5.10.6    Embedded Statistical Coupling Method (ESCM) with Comments on Direct Coupling (DC) Methods

The significant work in conceptual development of the transition between atomistic and continuum scale is Saether et al.,[27] where approaches that relate atoms and finite element nodes in a one-to-one manner or through a form of interpolation are referred to as direct coupling (DC) approaches. According to this definition, most existing multiscale methods reviewed above including MAAD (CLS), QC and CADD belong to DC methods.

### 5.10.6.1 Shortcomings of DC Multiscale Methods

While DC methods are straightforward, there is a fundamental problem in their development. Curtin and Miller[2] reviewed most of the DC multiscale models for static analysis. They concluded that almost all models have difficulty producing truly seamless transition along the boundaries between the atomistic and continuum scales. There is an intrinsic incompatibility which causes non-physical phenomena, including ghost forces. The only exception is the fully non-local QC model proposed by Knap and Ortiz[16] and the FEAt method. The former is introduced in Section 5.10.3 and 6.5, and the latter was set out by Kohlhoff et al. in 1991.[36]

Another shortcoming of DC methods is that in most cases the finite element size has to be near the interatomic distance at the atomic/continuum interface to perform well. An issue is whether it is physically sound using element size near the atomic lattice constant for FEM. This feature also causes serious difficulty when the model sizes are increased. As mentioned in Section 1.2.3, a large atomistic region embedded in a continuum of microns dimensions is needed for materials modeling. The trend in computer development with 4-core processors now in use and 100-core processors probably available within five years means multimillion atom simulations connected with continuum may be easily done on one computer in the near future. Due to the nature of the DC methods connecting the FEM domain with extremely small sizes, multiscale methods of the DC type will have difficulty meeting this new challenge.

In a 2009 paper by Miller and Tadmor,[37] a unified framework and benchmark of fourteen multiscale atomistic/continuum coupling methods of the DC type are developed. They classify these DC methods as two catalogues: one is energy-based formulation such as the QC method, the other is force-based formulation such as CADD.

### 5.10.6.2 ESCM Methods

The embedded statistical coupling method (ESCM) proposed by Saether *et al.*[27] is based on solving a coupled boundary value problem (BVP) at the MD/FE interface for a MD region embedded within a FEM domain. The method uses statistical averaging over selected time intervals and atomic subdomains at the MD/FE interface to determine nodal displacement boundary conditions for the continuum FE model. These enforced displacements then generate interface reaction forces that are applied as constant traction boundary conditions to the localized MD subdomains.

Mathematically speaking, the iterative procedure is needed for ESCM to connect a static FEA region with a dynamic atomistic region that evolves towards equilibrium. To develop the atomistic equilibrium configuration, ESCM must introduce free surface in the atomic domain that causes effects of surface tension. To eliminate this artificial effect, a surface MD region located between the interface region and the free surface must be increased. The additional region increases the stiffness of the atomistic domain. Thus, counter forces must be applied to cancel the effects caused by surface tension and the stiffness effects on the interface MD region.

Why does ESCM use the free surface but not pad atoms as CADD does? It is our understanding that the pad region greatly restricts the atomistic region for plastic deformation. Curtin's group has developed an ingenious procedure to pass dislocations through the pad region but it is a very heavy and complicated job if one changes the material. It requires an atomistic model for the dislocation core of each type of a dislocation passed through. Another requirement is to be able to embed and equilibrate the dislocation core among the atoms. One may think that replacing the pad atom region with a free surface layer gives improved possibilities for dislocation passing and for modeling other types of inelasticity deformation (diffusion, sliding, etc.).

## 5.10.7   Conclusion

Can an approach be developed which does not use the direct connection approach between atoms and FE nodes at the interface region as DC methods do, but also avoids the free surface of ESCM and the iterative process between MD and FEA used by ESCM? Our opinion is that the GP method proposed in this chapter is definitely the answer, and this is the underlying motivation for the development of the GP method.

As is seen in this chapter, the GP method assumes that materials consist of particles of different scales with the same crystallographic structure, and the higher-scale particle is lumped from atoms or lower-scale particles. It uses the natural boundary conditions between domains of different scales so that materials on both sides of the boundary have non-local responses.

In addition, it uses stationary material NLCs to mutually transfer information bottom-up and top-down to seamlessly bridge different scales. These NLCs are not only used during model formation but are kept at the overlapping region for the whole deformation process. Thus, the deformation transformation is between stationary atoms and particles of the same material elements by a statistical averaging process. It therefore does not introduce any supplemental dynamic treatment.

## 5.11 Concluding Remarks

It is difficult to predict exactly how the multiscale analysis will evolve in the coming decades. However, the historical evolution of constitutive laws of materials in the 1950–1980s may have shed light on developing multiscale analysis as discussed in Section 1.2.1. As can be seen in Section 8.4, very different and seemingly contradictory constitutive models of plasticity were finally proven to have common features. Thus, it is quite natural that various multiscale methods will be developed further and each method will serve as a brick to build a truly effective foundation of multiscale analyses.

Most concurrent multiscale models developed so far belong to the DC method that connects atomistic and continuum scales by directly linking atoms with finite element nodes. In spite of its success (see, for instance, Chapters 6 and 9, particularly Section 6.6 for the 10 findings), the introduction of concurrent multiscale methods in this book starts with the GP method, which is completely different from the DC method. It is hoped that this may help stimulate new thoughts to further advance multiscale modeling and avoid the shortcomings of the DC method as discussed in detail in Section 5.10.

GP divides the simulation region into areas containing particles of different scales, $n$, where $n = 1$ is the atomistic scale, and higher values of $n$ correspond to the continuum scale. As shown in Figure 5.4, the linkage between different scales uses imaginary particles (atoms) that overlap real atoms (particles). The GP method connects different scales by material neighbor-link cells (NLC). The NLC is composed of real particles (atoms) with scale duality. For example, the NLC of a second scale imaginary particle determines the interaction of this particle with real particles to offer natural boundary conditions. However, it consists of first scale particles (i.e., real atoms) and the position of the imaginary particle is determined by the average positions of those atoms. The NLC is formed when the geometric model is developed, which consists of real particles (real atoms) for each imaginary particle (atom). The information transfer from bottom scale-up or from top scale-down is through the constant number of real particles (or atoms) in the permanently linked cells. These particles, with the same material structure, all possess nonlocal constitutive behavior; thus, the smooth transition at the interface between different scales can be attained and validated (see Section 5.4) to avoid non-physical responses such as ghost force at the interface.

It is surprising that after mathematically proving the inverse mapping rule in Section 6.12 and the validation test for the initiation and growth of dislocations in Section 5.7, the GP method is actually an extended molecular dynamics method. This is because any generalized particle domain has a corresponding atomistic domain and the position, velocity, and acceleration for any generalized particle can be calculated in its corresponding atomistic domain with the same potential, parameters, cutoff radius, and numerical algorithm of the atomistic analysis. This may have a great impact since it makes the framework of concurrent multiscale analysis the same as the classical one of the atomistic analysis (i.e., MD, MS, and MC) that has been discussed in detail in Chapters 2–4.

There are several new aspects that should be developed and validated for the GP method, as with those of other multiscale methods during the course of their development. They include developing a suitable thermostat so that the GP method can be used in the environment of finite and varying temperatures, developing different computer codes to cover different potentials, and developing a model generation method for materials with amorphous microstructure. Readers interested in the GP method may refer to Section 10.14 and UNIT 14 of the CSLI for more information.

## References

[1] Fan, J. (2009) Multiscale analysis across atoms/continuum by a generalized particle dynamics method. *Multiscale Modeling and Simulation*, **8**(1), 228.
[2] Curtin, W. A. and Miller, R. E. (2003) Atomic/continuum coupling in computational materials science. *Modelling Simul. Mater. Sci. Eng.*, **11**, R33.

[3] Komanduri, R., Chandrasekaran, N., and Raff, L. M. (2001) Molecular dynamics (MD) simulation of uniaxial tension of some single-crystal cubic metals at nanolevel. *Int. J. Mech. Sci.*, **43**, 2237.

[4] Girifalco, L. A. and Weizer, V. G. (1959) Application of the Morse potential function to cubic metals. *Phys. Rev.*, **114**, 687.

[5] Cao, A., Wei, Y., and Mao, S. X. (2008) Alternating starvation of dislocations during plastic yielding in metallic nanowires. *Scripta Mater.*, **59**(2), 219.

[6] Ackland, G. J. and Jones, A. P. (2006) Applications of local structure measures in experiment and simulation. *Phys. Rev. B*, **73**, 054104.

[7] Vilet, K. J. V., Li, J., Zhu, T., *et al.* (2003) Quantifying the early stages of plasticity through nanoscale experiments and simulations. *Phys. Rev. B*, **67**, 104105.

[8] Li, J., Ngan, A. H. W., and Gumbsch, P. (2003) Atomistic modeling of mechanical behavior. *Acta Mater.*, **51**, 5711.

[9] Li, J. (2003) AtomEye: An efficient atomistic configuration viewer. *Modelling Simul. Mater. Sci. Eng.*, **11**, 173.

[10] Li, J. (2005) Atomistic visualization, in *Handbook of Materials Modeling* (ed. S. Yip), Springer, Berlin, 1051.

[11] Abraham, F., Broughton, J., Bernstein, N., and Kaxiras, E. (1998) Spanning the continuum to quantum length scales in a dynamic simulation of brittle fracture. *Europhysics Letter*, **44**, 783.

[12] Landau, D. P. (2005) The future of simulations in materials science, in *Handbook of Materials Modeling* (ed. S. Yip), Springer, New York, 2663.

[13] Shenoy, V. B., Miller, R., Tadmor, E. B., *et al.* (1999) An adaptive methodology for atomic scale mechanics: The quasicontinuum method. *J. Mech. Phys. Solids*, **47**, 611.

[14] Liu, W. K., Karpov, E. G., and Park, H. S. (2006) *Nano Mechanics and Materials: Theory, Multiscale Methods and Applications*, John Wiley & Sons, Ltd, New York.

[15] Tadmor, E. B., Phillips, R., and Ortiz, M. (2000) Hierarchical modeling in the mechanics of materials. *International Journal of Solids and Structures*, **37**, 379.

[16] Knap, J. and Ortiz, M. (2001) An analysis of the quasicontinuum method. *J. Mech. Phys. Solids*, **49**, 1899.

[17] Shilkrot, L. E., Miller, R. E., and Curtin, W. A. (2004) Multiscale plasticity modeling: Coupled atomistics and discrete dislocation mechanics. *J. Mech. Phys. Solids*, **52**, 755.

[18] van der Giessen, E. and Needleman, A. (1995) Discrete dislocation plasticity: A simple planar model. *Modelling Simul. Mater. Sci. Eng.*, **3**, 689.

[19] Rudd, R. E. and Broughton, J. Q. (1998) Coarse-grained molecular dynamics and the atomic limit of finite elements. *Phys. Rev. B*, **58**, R5893.

[20] Rudd, R. E. and Broughton, J. Q. (2005) Coarse-grained molecular dynamics: Nonlinear finite elements and finite temperature. *Phys. Rev. B*, **72**(14), 144104.

[21] Cai, W., DeKoning, M., Bulatov, V. V., and Yip, S. (2000) Minimizing boundary reflections in coupled-domain simulations. *Phys. Rev. Lett.*, **85**, 3213.

[22] Huang, W.E. and Z. (2001) Matching conditions in atomistic-continuum modeling of materials. *Phys. Rev. Lett.*, **87**, 135501.

[23] Chung, P. W. and Namburu, P. R. (2003) On a formulation for a multiscale atomistic-continuum homogenization method. *Int. J. Solids Struct.*, **40**, 2563.

[24] Curtarolo, G. and Cedar, S. (2002) Dynamics of an inhomogeneously coarse grained multiscale system. *Phys. Rev. Lett.*, **88**(25), 255504.

[25] Xiao, S. P. and Belytschko, T. (2004) A bridging domain method for coupling continua with molecular dynamics. *Computer Methods in Applied Mechanics and Engineering*, **93**, 1645.

[26] Wagner, G. J. and Liu, W. K. (2003) Coupling of atomic and continuum simulations using a bridging scale decomposition. *J. Comput. Phys.*, **190**, 249.

[27] Saether, E., Yamakov, V., and Glaessgen, E. H. (2009) An embedded statistical method for coupling molecular dynamics and finite elements analysis. *Int. J. Numer. Meth. Eng.*, **78**, 1292.

[28] Yamakov, V., Saether, E., and Glaessgen, E. H. (2008) Multiscale modeling of intergranular fracture in aluminum: Constitutive relation for interface debonding. *Journal of Materials Science*, **43**, 7488.

[29] To, A. C. and Li, S. (2005) Perfectly matched multiscale simulations. *Phys. Rev. B*, **72**, 035414.

[30] Li, S., Liu, X., Agrawal, A., and To, A. C. (2006) Perfectly matched simulations for discrete lattice systems: Extension to multiple dimensions. *Phys. Rev. B*, **74**, 045418.

[31] Liu, X. and Li, S. (2007) Nonequilibrium multiscale computational model. *J. Chem. Phys.*, **126**(12), 124105.

[32] Li, S., Sheng, N., and Liu, X. (2008) A non-equilibrium multiscale simulation paradigm. *Chemical Physical Letters*, **451**, 293.

[33]  Chaboche, J. L. (2008) Multi-scale analysis of polycrystalline metals and composites, in Proceedings of ICTAM 2008, CD-ROM, Adelaide.

[34] Medej, L., Mrozek, A., Kus, W., *et al.* (2008) Concurrent and upscaling methods in multiscale modeling – case studies. *Computational Methods in Materials Science*, **8**(1), 1.

[35] Chen, W. and Fish, J. (2006) A generalized space-time mathematical homogenization theory for bridging atomistic and continuum scales. *Int. J. Numer. Meth. Eng.*, **67**, 253.

[36] Kohlhoff, S., Gumbsch, P., and Fischmeister, H. F. (1991) Crack propagation in b.c.c. crystals studied with a combined finite-element and atomistic model. *Philos. Mag. A*, **64**, 851.

[37] Miller, R. E. and Tadmor, E. B. (2009) A unified framework and performance benchmark of fourteen multiscale atomistic/continuum coupling methods. *Modelling Simul. Mater. Sci. Eng.*, **17**, 053001.

# 6

# Quasicontinuum Concurrent and Semi-analytical Hierarchical Multiscale Methods Across Atoms/Continuum

In this chapter, quasicontinuum (QC) and semi-analytical multiscale methods for bridging atomistic and continuum scales are introduced. Various phenomena in nanoscale and their applications in nanotechnology are introduced. They include nanoindentation, dislocation initiation in crack tip and grain boundary, aluminum microtwinning, stress-induced phase transition, ferroelectric switching, development of atomistic-based continuum models with applications in hydrogen storage by nanocells, and mechanical, electrical and thermal properties of nanotubes. This chapter consists of three parts.

Part 6.1: Using FEM as a link, basic concepts and methods in solid mechanics are introduced, such as the basic energy principle for materials and structures and the interpolation function to reduce degrees of freedom.

Part 6.2: Introduction of the QC concurrent multiscale numerical method in terms of the atomistic-based energy density function derived from the Cauchy-Born rule and its various applications.

Part 6.3: Development, evaluation and applications of atomistic-based continuum theory of carbon nanotubes through the hierarchical multiscale method.

This chapter ends by using the Cauchy-Born rule to prove mathematically the inverse mapping rule of the GP method proposed in Section 5.5.

## 6.1 Introduction

As introduced in Chapter 1, a fundamental concept of materials science is that the properties of materials follow from their atomic and microscopic structures. This concept is well suited to finding effective approaches to enhance or modify continuum phenomenological laws by material properties based on structure and behavior at atomistic and microscopic scales. This chapter will show how this can be done both numerically and semi-analytically through both concurrent and hierarchical multiscale methods.

---

The GP method described in the last chapter is one of the efforts in this direction. In this chapter, the Cauchy-Born rule is applied to the atomic/continuum transition, using also the information of atomic bonds. By replacing phenomenological laws with interatomic potential-based constitutive models, one can investigate the collective behavior of atomistic structures.

Specifically, among several multiscale analyses using the Cauchy-Born rule, the quasicontinuum (QC) concurrent multiscale method proposed in 1996[1] will be introduced first, because of its relatively long history and great impact. The primary applications of the QC method were originally limited to 2D static equilibrium problems, then extended to 3D problems and finite temperature fields. This method has already been used to study a series of basic solid crystal problems, including fracture, crystal interfacial structure and deformation, nanoindentation and 3D dislocation analysis.

More applications of the Cauchy-Born rule are used in developing constitutive models of continuum for nanotubes. This work makes it possible to develop analytical and semi-analytical methods across atoms/continuum to investigate a wide variety of properties and phenomena at nanoscale.

It was emphasized in Chapter 1 that a new methodology based on the hierarchy of structure from the atomistic scale will bring new insights and findings. Specifically, the findings related in this chapter include that a metal type of single-wall nanotube can change to a semi-conducted type and that the thermal expansion coefficient of a graphite nanotube can change at a critical temperature from positive to negative during deformation. These examples show the possibility for exploring more interesting phenomena by use of multiscale methods in practice. Listing the details of all the significant existing findings, however, is not the main goal of this chapter.

Instead, this chapter pays more emphasis on introducing the fundamental concepts and methods in solid/structural mechanics as well as the atomistic/continuum transition based on the Cauchy-Born rule, so that readers can catch up step-by-step with these basic concepts. The philosophy behind this emphasis is that when one is knowledgeable in the interdisciplinary field one will have more freedom and capability to find new phenomena in nanotechnology and biotechnology. Based on this philosophy, the essential points of widely used finite element methods are concisely introduced.

We have tried to put the mathematics mostly in the last part of this chapter. Needless to say, however, if readers have basic knowledge of matrix algebra that will be helpful for deep understanding and practice.

## Part 6.1 Basic Energy Principle and Numerical Solution Techniques in Solid Mechanics

## 6.2 Principle of Minimum Potential Energy of Solids and Structures

For a static problem, the governing equations for determining configurations of a solid and structure or a material domain, under different loading conditions, can be determined by the principle of minimum potential energy. It states that for conservative systems, among all the possible configurations of the body which satisfy the boundary conditions, the real stable configuration of the body is the one with the minimum total potential energy. This is different but consistent with the variational principle of ground state energy introduced in Section 4.6.3.

From materials mechanics, it is known that the total potential energy, $\Pi$, can be expressed by two parts as follows:

$$\Pi = \mathcal{E} + \text{WP} \tag{6.1a}$$

where the first part denoted by $\mathcal{E}$ represents the internal energy such as strain energy stored in the body; the second part denoted by WP represents the work potential stored in the simulation system by the external loading applied at the body.

## 6.2.1 Strain Energy Density $\mathcal{E}$

The internal energy is usually expressed as follows:

$$\mathcal{E} = \int_{B^C} W dV \tag{6.1b}$$

where W denotes the strain energy density, defined as the stored energy per unit volume of material. This energy density is zero for the reference state in which there is no deformation. After the body is under deformation its value, $W(\underline{\varepsilon})$ or $W(\underline{F})$, depends on the deformation intensity which is measured by either strain tensor $\underline{\varepsilon}$ or deformation gradient $\underline{F}$; the latter two, strain and deformation gradient, are intrinsically connected as shown in Section 6.8.4.

A simple example of the strain energy density is a bar with length L, cross-section area A and volume V (=AL) under uniaxial loading. If the bar is elastic with Young's modulus E and tensile strain $\varepsilon$, from Hooke's law the stress is $\sigma = E\varepsilon$, the total force subjected by the bar is $P = \sigma A = E\varepsilon A$ and the total elongation measured at the end of the bar is $u = \varepsilon L$. The total energy stored in the bar during the loading process is

$$\mathcal{E} = \frac{1}{2} Pu = \frac{1}{2}(E\varepsilon A)(\varepsilon L) = \frac{1}{2}E\varepsilon^2 V \tag{6.1c}$$

or

$$\mathcal{E} = \frac{1}{2}\frac{EA}{L}u^2 = \frac{1}{2}ku^2 \tag{6.1d}$$

where $k = EA/L$ denotes the stiffness of the bar. From (6.1c) the strain energy density W can be obtained as follows:

$$W(\varepsilon) = \frac{1}{2}E\varepsilon^2 \tag{6.1e}$$

For a non-homogeneous deformation field, strain $\underline{\varepsilon}$ or deformation gradient $\underline{F}$ will be changed from point to point, thus it is a function of position vector $\overline{X}$ (or $\overline{r}$) and (6.1b) can be rewritten more clearly as

$$\mathcal{E} = \int_{B^C} W(\underline{F}(\overline{X}))dV = \int_{B^C} W(\underline{\varepsilon}(\overline{X}))dV \tag{6.1f}$$

## 6.2.2 Work Potential

Work potential WP can be written as a negative value of the work done by the external force on the body. The negative sign is due to the fact that if the work is done by the force along its direction the system energy is reduced. For example, if a body is at the top of a mountain its work potential energy is high; however, if it falls down by a depth h, the gravitation force, mg, does positive work mgh on it and the work potential of the body is reduced by that amount (m denotes mass, g is gravitation acceleration). For a concentrate force $\overline{P}_i$ applied at point i with correspondent displacement $\overline{u}_i$, its work potential can be written as follows:

$$WP = -\overline{P}_i \cdot \overline{u}_i = -P_i u_i \cos\theta = -(P_x u_x + P_y u_y + P_z u_z) \tag{6.1g}$$

where $P_i$ and $u_i$ are the values of the force and displacement, and the symbol "·" denotes dot product between two vectors (or tensors), which can be calculated by $\cos\theta$ or by corresponding component product as shown in (6.1g). $\theta$ is the angle between the force and displacement vector; the subscripts x, y, z denote the components along the x-, y- and z-direction for these vectors.

For a surface traction force vector $\bar{t}$ which is defined as the external force per unit area, the applied force applied at an infinitesimal area dA is $\bar{t}$dA, thus work potential over the whole surface, $\partial B_t^C$, of the body $B^C$ is given as follows:

$$WP = - \int_{\partial B_t^C} \bar{t} \cdot \bar{u} dA \tag{6.1h}$$

If there are body forces, surface tractions and concentrated loading, the work potential can be written as:

$$WP = - \int_{B^C} \bar{b} \cdot \bar{u} dV - \int_{\partial B_t^C} \bar{t} \cdot \bar{u} dA - \sum_{i=1}^{m} \bar{P}_i \cdot \bar{u}_i \tag{6.1i}$$

where $m$ is the number of total concentrated loading, $\bar{b}$ - the body force vector per unit volume. Substituting (6.1f) and (6.1i) into (6.1a), the total potential energy for the case with surface and concentrated loading can be expressed as follows:

$$\Pi = \int_{B^C} W(\underline{F}(\overline{X})) dV - \int_{B^C} \bar{b} \cdot \bar{u} \, dV - \int_{\partial B_t^C} \bar{t} \cdot \bar{u} \, dA - \sum_{i=1}^{m} \bar{P}_i \cdot \bar{u}_i \tag{6.1j}$$

According to the principle of minimum potential energy the real configuration should be determined by minimizing $\Pi$ with respect to all degrees of freedom. The next section will show how this principle can be used to develop effective simultaneous algebraic equations for structural and materials analysis. The change from using integration and/or differential equations to simultaneous algebraic equations avoids the notoriously difficult problem of finding solutions of these equations to satisfy the boundary conditions.

This great change in solution schemes became possible due to fast computer development which is especially effective for solving a large number of simultaneous algebraic equations. The most important technique related to this great change is the finite element method (FEM), which deeply changed structural engineering and has had a wide scope of applications in design of airplanes, automobiles, ships, bridges and buildings since the 1950s. Its principles and method are valid for applications in materials science and engineering and a lot of work has been done recently in this field. Furthermore, most work in concurrent multiscale analysis besides the GP method so far uses FEM for analysis in the continuum domain. Thus, learning the basics of FEM is necessary and will be introduced briefly in the next section.

## 6.3    Essential Points of Finite Element Methods

In FEM, the integration shown in (6.1f) is simplified and numerically calculated through the following several approaches and steps.[3, 4] The key is to express the continuum model by limited nodes which divide the continuum domain as finite elements. Thus, the main work of FEM is to express the total potential energy $\Pi$ by node displacement matrix $\overline{U}$ or $[U]$ which is defined by displacement vector of all nodes (see (6.2l) below).

### 6.3.1    Discretization of Continuum Domain $B^C$ into Finite Elements

In FEM, any continuum domain is discretized into many small elements with different shapes by a certain number of nodes and connection lines or planes between these nodes. For instance, one can divide a 1-meter beam into 100 small beam elements by 101 nodes, where each element is between two consecutive nodes; one can also divide a plane area by triangular elements through lines connecting three nodes. The stored strain energy $\mathbb{C}^e$ in a generic element e is expressed as:

$$\mathbb{C}^e = \int_{V^e} W(F(\overline{X}^e))dV \tag{6.2a}$$

where $V^e$ is the volume of the element. Denoting the total number of elements $n_{elem}$, the total stored strain energy $\mathbb{C}$ of (6.1f) can be expressed by a summation of the energy stored in all elements as follows:

$$\mathbb{C} = \sum_{e=1}^{n_{elem}} \int_{V^e} W(\underline{F}(\overline{X}^e))dV \tag{6.2b}$$

Different from (6.1f), the above integration of the strain energy density W and its position variable $\overline{X}^e$ is only limited to each individual element e, not defined in the whole domain $B^C$. A simple arithmetic numerical method called Gaussian quadrature will be introduced in the next paragraph to calculate the value of this integration.

## 6.3.2  Using Gaussian Quadrature to Calculate Element Energy

In general, the energy density W is a function of coordinates and an effective numerical algorithm is necessary to calculate the integration value. The element energy integration of (6.2a) can be approximately expressed by Gaussian quadrature through values of Gaussian (sample) points $X_q^e$ ($q = 1 \ldots n_q$) in that element and their weight factors $\omega_q$. Using a polynomial function of order n to approximately express the function in the integration, one can derive this kind of arithmetic summation. The values of Gaussian points $X_q^e$ and $\omega_q$ depend on the required accuracy and can be found in the textbook.[4] Taking the following 1D integration as an example:

$$I = \int_{-1}^{1} f(\xi)d\xi \tag{6.2c}$$

one can express this integration by the one-point formula

$$I = 2*f(0) \tag{6.2d}$$

where $q = 1, X_1^e = 0, \omega_q = 2$. (6.2d) is an exact solution if $f(\xi)$ is a polynomial of order 1 or linear function; in the general case, it is a mean value approximation of the integration. The integration can also be expressed by two-point Gaussian quadrature as follows:

$$I = w_1 f(\xi_1) + w_2 f(\xi_2) = \sum_{i=1}^{2} w_i f(\xi_i) \tag{6.2e}$$

where $\xi_1 = 1/\sqrt{3}$, $\xi_2 = -1/\sqrt{3}$, $\omega_1 = \omega_2 = 1.0$. (Remark: for accuracy, the values of $\xi$ are usually more precise in programming, for the two-point case, $\xi_1$, $\xi_2$ are taken as $\pm 0.5773502692$.) Corresponding to this 1D two-point case, 2D integration needs four Gaussian quadrature terms at four points inside the element to express the following integration:

$$I = \int_{-1}^{1}\int_{-1}^{1} f(\xi,\eta)d\xi d\eta = \omega_1 f(\xi_1,\eta_1) + \omega_2 f(\xi_2,\eta_2) + \omega_3 f(\xi_3,\eta_3) + \omega_4 f(\xi_4,\eta_4) \tag{6.2f}$$

$$= \sum_{1}^{4} \omega_q f(\overline{X}_q)$$

where

$$\omega_1 = \omega_2 = \omega_3 = \omega_4 = 1 \tag{6.2g}$$

$$\overline{X}_1 = (-0.571, -0.571), \quad \overline{X}_2 = (0.571, -0.571), \quad \overline{X}_3 = (-0.571, 0.571), \quad \overline{X}_4 = (0.571, 0.571) \tag{6.2h}$$

In practice, the element integration in (6.2a) usually changes the integration variable x, y, z (e.g., $dV = dxdydz$) to non-dimensional (or natural) coordinates $\xi, \eta, \zeta$. Thus a standard integration type such as (6.2c) and (6.2f) is used with integration limit from $-1$ to 1.

By using Gaussian quadrature (e.g., (6.2e) and (6.2f)) the general integration of element strain energy $\mathfrak{E}^e$ can be expressed as an arithmetic sum of strain energy density W at several Gaussian points as follows:

$$\mathfrak{E}^e = \int_{V^e} W(\underline{F}(\overline{X}^e))dV = V^e \sum_{q=1}^{n_q} \omega_q W(\overline{X}^q) \tag{6.2i}$$

Substituting it into (6.2a), the total strain energy can be expressed as follows:

$$\mathfrak{E} = \sum_{e=1}^{n_{elem}} V^e \sum_{q=1}^{n_q} \omega_q W(\overline{X}_q^e)) \tag{6.2j}$$

This equation shows that the total energy U of the solid can be expressed by strain density function W at Gaussian points of all elements, which in turn is determined by the strain or deformation gradient at these Gaussian points.

## 6.3.3    Work Potential Expressed by Node Displacement Matrix

In FEM, the work potential WP shown in (6.1i) can be approximately expressed by the node displacement vector at nodes with concentrated loading, and nodes at the surface (or body) with distributed loading. This approximation can easily be realized by equating the work done by the external forces applied at each element with the work done by node equivalent forces during the node displacement. After obtaining the equivalent node force for each element, the work potential of the system can be expressed by matrix product

$$WP = -\overline{F}^T \overline{U} \tag{6.2k}$$

where $\overline{F} = \{\overline{f}_1, \overline{f}_2 \ldots \overline{f}_N\}^T$ and $\overline{U} = \{\overline{u}_1, \overline{u}_2 \ldots \overline{u}_N\}^T$ are, respectively, total equivalent force matrix and the nodal displacement matrix. The subscript N denotes the total number of nodes which divide the body as $n_{elem}$ finite elements. The superscript T denotes the transpose of the matrix. For the purpose of developing a standard FEM formulation, these matrixes are expressed by their Cartesian components with a regular order. Taking 3D as an example, $\overline{U}$ and $\overline{F}$ can be rewritten as a column matrix with 3N entrances as follows:

$$\overline{U} \equiv [U] = \{\overline{u}_1, \overline{u}_2 \ldots \overline{u}_N\} = \{u_1, u_2, u_3, u_4, u_5, u_6, \ldots u_{3N-2}, u_{3N-1}, u_{3N}\}^T \tag{6.2l}$$

$$\overline{F} \equiv [F] = \{\overline{f}_1, \overline{f}_2 \ldots \overline{f}_N\} = \{f_1, f_2, f_3, f_4, f_5, f_6, \ldots f_{3N-2}, f_{3N-1}, f_{3N}\}^T \tag{6.2m}$$

where $\{f_1, f_2, f_3\}$ and $\{u_1, u_2, u_3\}$ are associated with node 1. They denote, respectively, the components along the x-, y-, and z-direction of the force and displacement vector at node 1. Likewise,

$\{f_{3N-2}, f_{3N-1}, f_{3N}\}$ and $\{u_{3N-2}, u_{3N-1}, u_{3N}\}$ denote the components of force and node displacement at node N. Specifically, the work potential can be written in expanded form as follows:

$$WP = -\overline{F}^T\overline{U} = -(\overline{f}_1 \cdot \overline{u}_1 + \ldots + \overline{f}_N \cdot \overline{u}_N) = -(f_1 u_1 + f_2 u_2 + f_3 u_3 +$$

$$\ldots + f_i u_i \ldots + f_{3N-2} u_{3N-2} + f_{3N-1} u_{3N-1} + f_{3N} u_{3N}) \tag{6.2n}$$

### 6.3.4  Total Potential Energy $\Pi$ Expressed by Node Displacement Matrix

#### 6.3.4.1 Displacement Vector $\overline{u}$

It is seen from Figure 6.1 that displacement vector $\overline{u}$ determines how any point $\overline{X}$ in the non-deformation reference configuration moves to the new position $\overline{x}$ in the deformed configuration, such that

$$\overline{x} = \overline{X} + \overline{u} \tag{6.3a}$$

For a given initial reference configuration, the deformation gradient $\underline{F}$ defined as $\underline{F} = d\overline{x}/d\overline{X}$ (see Section 6.8) can be determined by displacement gradient through normal strain, shear strain and rotation, which will be shown in Section 6.8.3.

Any point is uniquely determined by deformation vector $\overline{u}$. Thus, the statement that the total strain energy $\mathbb{C}$ depends on strain density function W at Gaussian points of all elements can be further stated that it is determined by the displacement vector at the Gaussian points of all elements in the system. In other words (6.2j) can be rewritten as follows:

$$\mathbb{C} = \sum_{e=1}^{n_{elem}} V^e \sum_{q=1}^{n_q} \omega_q W(\overline{u}(\overline{X}_q^e)) \tag{6.3b}$$

where $\overline{u}(\overline{X}_q^e)$ is the displacement vector at the Gaussian point q of element e whose coordinate is $\overline{X}_q^e$.

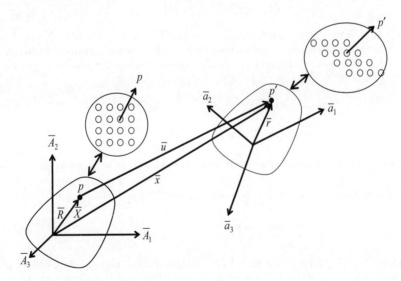

**Figure 6.1**  Schematic for displacement vector $\overline{u}$ versus position vectors $\overline{x}, \overline{X}$ and uniform deformation in the neighborhood of a generic material point

### 6.3.4.2 Expressing Displacement Field $\bar{u}(\overline{X})$ by Node Displacement Matrix Through Interpolation Function $N(\overline{X})$

(6.3b) uses displacement vector $\bar{u}_q^e = \bar{u}(\overline{X}_q^e)$ at Gaussian points of each element to express the element strain energy. For an effective numerical solution it is necessary to express the total strain energy $\mathbb{C}$ of the system by node displacement vector [U] introduced in (6.2l). To do so one needs to establish the relationship between displacement $\bar{u}(\overline{X})$ at any point $\overline{X}$ (including Gaussian points) inside the element and node displacement vectors $\overline{U}_I^e (I = 1, 2, \ldots n_{nodes})$ at element nodes 1, 2, $\ldots$ $n_{nodes}$ through the following equation:

$$\bar{u}(\overline{X}) = \sum_{I=1}^{n_{nodes}} N_I(\overline{X})\bar{u}_I^e = [N(\overline{X})][\overline{U}^e] \tag{6.3c}$$

where the interpolation matrix $[N(X)]$ and the node displacement matrix $[\overline{U}^e]$ are given as follows:

$$[N] = [N_1, N_2 \ldots N_{nodes}] \tag{6.3d}$$

$$[\overline{U}^e] = [\bar{u}_1^e, \bar{u}_2^e \ldots \bar{u}_{nodes}^e]^T \tag{6.3e}$$

Using (6.3c) for Gaussian point $\overline{X}_q^e$ one obtains

$$\bar{u}(\overline{X}_q^e) = \sum_{I=1}^{n_{nodes}} N_I(\overline{X}_q^e)\bar{u}_I^e = [N(\overline{X}_q^e)][\overline{U}^e] \tag{6.3f}$$

This may be a good place to show an example of the natural coordinate $\xi$ and the interpolation (shape) function $N(\xi)$ which are used in practice to replace the Cartesian coordinates used in (6.3f). Suppose there is a bar element with node 1 and node 2 whose Cartesian coordinates are $X_1$ and $X_2$ and displacement is $u_1$ and $u_2$. The natural (non-dimensional) coordinate $\xi$ for any point with Cartesian coordinate X inside the element can be defined as:

$$\xi = \frac{2(X - X_1)}{l_e} - 1 \tag{6.3g}$$

where $l_e = X_2 - X_1$ is the element length. It is easy to see for $X = X_1$, $\xi = -1$ and for $X = X_2$ $\xi = 1$. Thus the range of the natural coordinate of material points between the left end $(X_1)$ to the right end $(X_2)$ is from $-1$ to 1, which gives the lower and upper limit of (6.2c). The interpolation function can be given as follows:

$$N_1(\xi) = \frac{1 - \xi}{2} \tag{6.3h}$$

$$N_2(\xi) = \frac{1 + \xi}{2} \tag{6.3i}$$

Following (6.3c), the displacement at any point in the bar element can be expressed as follows:

$$u(\xi) = [N]\{U\} = [N_1, N_2]\begin{Bmatrix} u_1 \\ u_2 \end{Bmatrix} = N_1(\xi)u_1 + N_2(\xi)u_2 = \left(\frac{1 - \xi}{2}\right)u_1 + \left(\frac{1 + \xi}{2}\right)u_2$$

$$= \frac{(u_2 + u_1)}{2} + \xi\frac{(u_2 - u_1)}{2} \tag{6.3j}$$

If one knows the value of the coordinate $\xi$ of a material point in the element and the node displacement $u_1$ and $u_2$, from (6.3j) it is easy to calculate the displacement at that point. In the limit case, at the left end of the element $\xi = -1$, the value is $u(-1) = u_1$ and at $\xi = 1$ $u(1) = 1$, which is consistent with the displacement at the left and right nodes.

It is easy to prove that the displacement field u($\xi$) or u(x) produced by node displacement through linear shape functions produces constant strain $\varepsilon$ and deformation gradient $\underline{F}$ in the element. Specifically, using (6.3g) and (6.3j) gives

$$\varepsilon = \frac{du}{dx} = \frac{du}{d\xi}\frac{d\xi}{dx} = \left(\frac{u_2 - u_1}{2}\right)\left(\frac{2}{l_e}\right) = \frac{u_2 - u_1}{l_e} \tag{6.3k}$$

For uniaxial homogeneous deformation, the deformation gradient F can be expressed using (6.3a) as

$$F = \frac{dx}{dX} = \frac{dX + du}{dX} = 1 + \frac{du}{dx} = 1 + \frac{u_2 - u_1}{l_e} = 1 + \varepsilon \tag{6.3l}$$

### 6.3.4.3 Using Interpolation Function N to Express Total Potential Energy $\Pi$ by Node Displacement Matrix

Substituting (6.3f) into (6.3b), the system strain energy can be expressed by the node displacement vectors $\{\overline{U}^e\}$ of all elements (i.e., e=1, 2, ... $n_{elem}$) as follows:

$$\mathsf{E} = \sum_{e=1}^{n_{elem}} V^e \sum_{q=1}^{n_q} \omega_q W([N(\overline{X}_q^e)]\{\overline{U}^e\}) \tag{6.3m}$$

Furthermore, the node vector matrix $[\overline{U}^e]$ of (6.3e) are element variables which are a part of the global node displacement vector [U] (or $\overline{U}$) defined in (6.2l). Taking triangle element as an example, the one-to-one correspondence between $\overline{u}_1^e, .\overline{u}_2^e, \cdot \overline{u}_3^e$ in $[\overline{U}^e]$ and vectors $\overline{u}_I, .\overline{u}_J, \cdot \overline{u}_K$ in $[\overline{U}]$ are fixed through the connectivity conditions established at the beginning when dividing the model with elements by nodes. This indicates that by rearranging (6.3m) after substituting the fixed relationship between $[\overline{U}^e]$ (or $\overline{U}^e$) and $[\overline{U}]$ (or $\overline{U}$), strain energy $\mathsf{E}$ can be expressed as the function of the global node displacement vector

$$\mathsf{E} = \mathsf{E}(u_1, u_2, u_3, u_4, u_5, u_6, \ldots u_{3N-2}, u_{3N-1}, u_{3N}) \tag{6.3n}$$

Substituting (6.3n) and (6.2n) into (6.1a), the total potential energy $\Pi$ can be expressed by all node displacement components $u_i$ (i = 1, 2, ..., 3N) as follows:

$$\Pi(u_1, u_2, \ldots u_{3N}) = \mathsf{E}(u_1, u_2, \ldots u_{3N}) - \overline{F}^T \overline{U}(u_1, u_2, \ldots u_{3N}) \tag{6.3o}$$

By using the symbol of (6.2l) and (6.2m), this equation can be expressed by the global node displacement vector [U] and node force vector [F] as follows:

$$\Pi([U]) = \mathsf{E}([U]) - \overline{F}^T[U] \tag{6.3p}$$

## 6.3.5 Developing Simultaneous Algebraic Equations for Nodal Displacement Matrix

Equation (6.3p) with corresponding boundary condition is sufficient to derive simultaneous algebraic equations to determine all the node displacements. The minimum principle of potential energy requires that the infinitesimal variation $du_i$ of any node displacement $u_i$ should make the following derivative of total potential energy $\Pi$ about $u_i$ zero, thus:

$$\frac{\partial \Pi(u_1 \ldots u_i \ldots u_{3N})}{\partial u_i} = 0 \quad (1, 2, \ldots, 3N) \tag{6.3q}$$

which produces 3N simultaneous algebraic equations. The partial differential consists of two parts. The first part is the partial differential of the total strain energy $\mathcal{E}$ with respect to $u_i$; for the simple case of a bar under a load P at the end, from (6.1d) the differential is

$$d\mathcal{E}/du = \frac{d\left(\frac{1}{2}ku^2\right)}{du} = ku \qquad (6.3r)$$

The second part is the differential of work potential WP with respect to $u_i$. For the simple case of a bar under a load P at the end, $WP = -Pu$, thus

$$d(WP)/du = \frac{d(-Pu)}{du} = -P \qquad (6.3s)$$

Substituting (6.3r) and (6.3s) into (6.3q), the following equation for the bar under loading P is produced:

$$ku = P \qquad (6.3t)$$

In general, there are 3N partial differential equations, from (6.2n) it is seen

$$\partial(WP)/\partial u_i = -f_i(i = 1, 2, \dots 3N) \qquad (6.3u)$$

The partial differential of the strain energy $\mathcal{E}$ with respect to $u_i$ will produce corresponding terms related to $u_i$ and other node displacements; so, the simultaneous equations produced from (6.3q) can be written as a matrix equation:

$$[K]\{U\} = \{F\} \qquad (6.3v)$$

For a 3D problem, it can be written as:

$$[K]_{3Nx3N}[U]_{3Nx1} = [F]_{3Nx1} \qquad (6.3w)$$

where [K] is the stiffness matrix with 3N rows and 3N columns, {U} and {F} are node displacement and force matrix defined by (6.2l) and (6.2m). For linear elasticity problems with small deformation stiffness, [K] matrix is constant. For non-linear problems, [K] can be changed with loading steps.

   The solution of (6.3v) under given boundary conditions gives all the unknown node displacements in [U] of the body. This, in turn, determines the node displacement $\overline{U}_I^e(I = 1, 2 \dots n_{nodes})$ of any element through the local-global connectivity condition. By interpolation function N through (6.3c), displacement of any point inside an element in the body can be given. Further, due to the relationship between displacement components and strain tensor (see (6.8g) and (6.9c)), the normal and shear strain and corresponding stress can be determined.

---

**Homework**

(6.1) Figure 6.2 shows a system of four springs $k_1$, $k_2$, $k_3$ and $k_4$. At nodes 1, 2 and 3 there are node displacement $q_1$, $q_2$ and $q_3$ and concentrated force $F_1$, $F_2$ and $F_3$.
   (a) Determine the stored strain energy U and work potential WP in the system.
   (b) Use the principle of minimum potential energy to develop the simultaneous governing equations for the solutions of the node displacements in terms of the node forces.
   (c) Write the simultaneous governing equations in the form of a matrix.

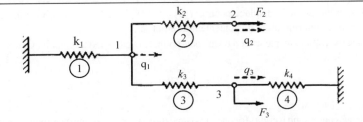

**Figure 6.2**    An equilibrium structure with four springs and two applied forces

(6.2) In the isoparametric formulation of finite elements, the displacements or coordinates inside the element can be written using the interpolation (or shape) function and the corresponding node values. This approach leads to simplicity of development of the FEM framework. Taking the triangle element of a 2D problem as an example, one can write

$$u = N_1 u_1 + N_2 u_2 + N_3 u_3$$
$$v = N_1 v_1 + N_2 v_2 + N_3 v_3$$
$$x = N_1 x_1 + N_2 x_2 + N_3 x_3$$
$$y = N_1 y_1 + N_2 y_2 + N_3 y_3$$

where the left side denotes the x, y coordinates of the point inside the element and its corresponding displacements along the x- and y-direction. In the right sides of the above equations $x_i$, $y_i$, $u_i$, $v_i$ ($i = 1$, 2, 3) denote the Cartesian coordinates and the x- and y-displacements at the three nodes. $N_i$ ($i = 1$, 2, 3) are interpolation (shape) functions whose value can be expressed as natural coordinates $\xi$, $\eta$ as follows:

$$N_1 = \xi, \ N_2 = \eta, \ N_3 = 1 - \xi - \eta$$

These functions are various in the triangle elements, For $N_1$ its value is 1 at node 1, but at the other two nodes it is 0, and the same for the other two shape functions. For instance, for $N_2$ its value is 1 at node 2, but at node 1 and node 3 it is 0, the variation from 1 to 0 inside the elements is smooth and linear.

(a)  Assuming the point P inside the element shown in Figure 6.3 is with $x = 3.96$ and $y = 4.8$, determine its natural coordinates $\xi$, $\eta$ and its corresponding interpolation (shape) function.

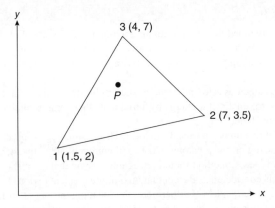

**Figure 6.3**    A point P within a triangle element

(b) Assuming the node displacements are $u_1 = 0.1$, $v_1 = 0.2$, $u_1 = 0.12$, $v_1 = 0.22$, $u_1 = 0.15$, $v_1 = 0.25$ determine the displacement at point P.

(6.3) Evaluate the following integration by the one-point, two-point and three-point Gaussian quadrature

$$J = \int_{-1}^{1} \left[ 2e^x + x^2 + \frac{1}{(x+2)} \right] dx$$

Hint: Gauss points and weights for one-, two- and three-point Gaussian quadratures are given in the table below.

| Number of points n | Location, $\xi_i$ | Weights, $w_i$ |
| --- | --- | --- |
| 1 | 0.0 | 2.0 |
| 2 | $\pm 1/\sqrt{3} = \pm 0.5773502692$ | 1.0 |
| 3 | $\pm \boxed{0.7745966692}$ | 0.5555555556 |

# Part 6.2 Quasicontinuum (QC) Concurrent Method of Multiscale Analysis

## 6.4  The Idea and Features of the QC Method

Proposed by Tadmor *et al.* in 1996,[1, 5–9] the quasicontinuum (QC) method is a multiscale method with a relatively long development history. This method has already been used to study a series of basic solid crystal problems, including fracture,[10–14] crystal interfacial structure and deformation,[1] nanoindentation[15–20] and 3D dislocation analysis.[6–8, 21] In the following eight sub-sections we will briefly introduce the essential points of the QC method.

### 6.4.1  Formulation of Representative Atoms and Total Potential Energy in the QC Method

In the QC method, it is assumed that there is an underlying atomistic model of the material which is a good description of the material's behavior. Based on this concept, in the QC literature, only "representative atoms" or simply "repatoms" are taken to be part of the set of degrees of freedom for the solution.

The goal of QC statics methods is to find the atomistic displacements that mimimize the total potential energy $\Pi(\bar{u})$ through approximating the energy by introducing these representative atoms to substantially reduce the degrees of freedom.

There are two kinds of repatoms: non-local and local. This terminology is used to distinguish between repatoms in the atomistic region and repatoms in the continuum region.[22] In other words, the term "non-local repatom region" in the QC method should be regarded as the real atomistic region, while the term "local repatom region" is considered as the continuum or finite element region. The repatom in the local region in general is the node of FEM. Furthermore, the locations of any atom in the local region are determined by location of repatoms because they are kinematically constrained by these repatoms through interpolation functions.

In atomistic analysis using semi-empirical potentials as shown in Chapter 2, there is a well-defined total energy function $E^{tot}$ determined from the relative positions of all the atoms in the system whose total number is N. Here, we use letter E to replace U in Chapter 2, etc. to avoid confusion with node displacement.

$$E^{tot} = \sum_{i=1}^{N} E_i(\bar{u}_1, \ldots \bar{u}_N) \tag{6.4a}$$

Here, N is the total number of atoms in the model.

Similar to the discussion in Section 6.2, in addition to potential energy there is work potential due to external loads applied to atoms. Thus, the total potential energy of the system (atoms plus external loads) can be written as

$$\Pi^{tot}(\bar{u}_1, \ldots \bar{u}_N) = E^{tot}(\bar{u}_1 \ldots \bar{u}_N) - \sum_{i=1}^{N} \bar{f}_{ext,i} \bar{u}_i \tag{6.4b}$$

Introducing the symbol $\tilde{u}$ to represent the set of displacement vectors of all atoms in the system

$$\tilde{u} = (\bar{u}_1, \bar{u}_2 \ldots \bar{u}_N) \tag{6.4c}$$

(6.4a) and (6.4b) can be rewritten in the following compact form where $E^{tot}$ is the summation of energy of all the atoms:

$$E^{tot} = \sum_{i=1}^{N} E_i(\tilde{u}) \tag{6.4d}$$

$$\Pi^{tot}(\tilde{u}) = E^{tot}(\tilde{u}) - \sum_{i=1}^{N} \bar{f}_{ext,i} \bar{u}_i \tag{6.4e}$$

## 6.4.2    Using Interpolation Functions to Reduce Degrees of Freedom

Let us start by reducing the number of degree of freedom in the simulation system. If the change of deformation gradient $\underline{F}$ or strain tensor $\underline{\varepsilon}$ on the atomistic scale in some part of the model is small, it is not effective to find the displacement of every atom in the region. Instead, the QC method only explicitly treats a small fraction of repatoms, with the displacements of the remaining atoms being approximately determined through interpolation. In this way, the degrees of freedom of the system are reduced to only those of repatoms.

This scheme is realized by recourse to the FEM interpolation functions described in Section 6.3.4. Here, the repatoms in the continuum region are considered the FEM nodes. Any atom not chosen as a repatom is subsequently constrained to move according to the interpolated displacements from the nodes of the element in which it resides. The scheme of the repatoms depends on the needs of the problem to hand. QC constrains the motion of most of the atoms, using a small number of repatoms in regions whose deformation gradient is small. In some critical regions such as crack tip, dislocation core, and interface all atoms are selected as repatoms.

For a generic atom i, its displacement vector $\bar{u}_i (i = 1, \ldots N)$ can be expressed by interpolation functions in the form of (6.3c). Denoting the number of total repatoms (nodes) as $N_{rep}$ the displacement can be expressed in a general form as follows:

$$\bar{u}_i = \sum_{I=1}^{N_{rep}} N_I(\bar{X}) \bar{u}_I = [N][U^{rep}] \qquad (i = 1, \ldots N) \tag{6.4f}$$

with

$$[U]^{rep} = \{\bar{u}_1, \bar{u}_2 \ldots \bar{u}_{Nrep}\}^T \tag{6.4g}$$

$$[N] = [N_I, \ldots N_{Nrep}] \tag{6.4h}$$

where $N_I$ ($I = 1 \ldots N_{rep}$) are the interpolation functions, $[U]^{rep}$ is node displacement vector, and $N_{rep}$ is the number of repatoms, $N_{rep} \ll N$.

Note that in practice the calculation of (6.4g) is much simpler due to the fact that most interpolation (or shape) functions are zero unless associated with the node of the element in which the atom is set. The QC method uses triangular constant strain elements for the continuum region. Each element has three nodes whose node displacement is $\bar{u}_1^e, \bar{u}_2^e, \bar{u}_3^e$. The displacement of any atom P inside that element can be determined by the following expression:

$$\bar{u}^h(\overline{X}_p) = N_1(\bar{x}_p)\bar{u}_1^e + N_2(\bar{x}_p)\bar{u}_2^e + N_3(\bar{x}_p)\bar{u}_3^e \tag{6.4i}$$

For the atomistic region, its associated interpolation (shape) function is 1 indicating that the atom is the repatom and the repatom displacement is the atom displacement.

Summarizing the above discussion, it is seen that displacement $\bar{u}_i$ ($i = 1 \ldots N$) of any atom can be expressed by interpolation (shape) function N through the set of displacement vector [U], thus (6.4e) can be approximated by replacing the displacement vector of any atom by (6.4g), so that (6.4e) can be rewritten as:

$$\Pi^{tot}(\tilde{u}([U]^{rep})) = E^{tot}([U]^{rep}) - \sum_{i=1}^{N} \bar{f}_{ext,i}[N][U]^{rep} \tag{6.4j}$$

$$\Pi^{tot}(\tilde{u}([U]^{rep})) = E^{tot}([U]^{rep}) - [F]^T[U]^{rep} \tag{6.4k}$$

where

$$E^{tot}(\tilde{u}([U]^{rep})) = \sum_{i=1}^{N} E_i(\tilde{u}([U]^{rep})) \tag{6.4l}$$

$$[F]^T = \sum_{i=1}^{N} \bar{f}_{ext,i}[N] \tag{6.4m}$$

From this equation it is seen that if one can find the explicit expression $E^{tot}$ in terms of $[U]^{rep}$, the displacement vector of all representative atoms can be found through the minimization principle of total potential energy, as shown in Section 6.3.5 for the solution of FEM. In the next section, we will discuss how this can be done.

## 6.4.3   Model Division

The above formulation with $[U]^{rep}$ has greatly reduced the number of degrees of freedom but does not reduce the computational burden significantly. In fact, one still needs to compute the energy over every atom in the summation calculation (6.4l). At this stage, the division of the body into several parts is necessary. In the QC model, there is a domain $B^A$ for full atomistic analysis and a region $B^C$ of continuum which uses FEM for analysis. There is no handshake domain between atomistic and continuum domain. The connection between these two domains is through a pad domain which links to the atomistic domain. Specifically, on the pad domain there are pad atoms which are neighbors to the real atoms.

While the pad domain takes the atomistic domain as its neighborhood, it is overlapped by the continuum domain. Its relative location to the atomistic and continuum domain is the same as $W_{1image}$ domain in Figure 5.4 in the GP method, thus its function is also similar to the imaginary atoms there. More specifically, pad atoms offer the atomistic boundary condition to the real atoms in the atomistic domain. The difference between pad atoms and the imaginary atoms of the GP method is that pad atoms adhere to the finite element meshes either at the FEM nodes or with the position determined through the interpolation functions, thus their locations are fully determined by the displacements of mesh deformation. On the contrary, imaginary atoms are all determined by motions of real particles which have intrinsic fixed connections with these atoms (see Section 5.3 for details).

Denoting $E^A$ and $E^B$ as the potential energy, respectively, in the atomistic and continuum region, the total potential energy $E^{tot}$ of the body can be separated as two parts as follows:

$$E^{tot}(\tilde{u}([U]^{rep})) = E^A(\tilde{u}([U]^{rep})) + E^C(\tilde{u}([U]^{rep})) \qquad (6.4n)$$

with

$$E^A([U]^{rep}) = \sum_{\alpha=1}^{N_A} E^\alpha(\tilde{u}([U]^{rep})) \qquad (6.4o)$$

$$E^C([U]^{rep}) = \sum_{\alpha=1}^{N_C} E^\alpha(\tilde{u}([U]^{rep})) \ldots \qquad (6.4p)$$

where $N_A$ and $N_C$ are the atom numbers, respectively, in the atomistic and continuum domain and $E^\alpha$ is the energy of atom $\alpha$.

The sum in (6.4o) is the atomistic energy and can be accounted for by, for instance, (2.2c) and (3.1) in which atomistic energies are expressed for atoms and ions in metals and ceramics.

It is seen that for the purpose of energy minimization one needs to find an efficient way to compute the continuum energy $E^C$ in (6.4p) on the degrees of freedom without explicitly visiting every atom. The QC method uses the Cauchy-Born rule for this. While this rule will be discussed in detail in Sections 6.8 and 6.9 along with a mathematical definition of the deformation gradient $\underline{F}$, its basic idea can be described here. In the QC method, linear interpolation functions along with triangle elements for 2D problems are used. As discussed in Section 6.3.4.2, the use of linear shape functions to interpolate the displacement field produces a uniform deformation gradient (or strain) within each element.

## 6.4.4   Using the Cauchy-Born Rule to Calculate Energy Density Function W from Interatomic Potential Energy

### 6.4.4.1  Introduction to the Cauchy-Born Rule

The Cauchy-Born rule assumes that the deformation gradient (i.e. $\underline{F} = d\bar{x}/d\bar{X}$ ) of a continuum can be used to calculate the deformation of its underlying atomistic lattice. This is achieved by using the deformation gradient to obtain the lattice constant vector $\bar{a}$ after deformation from the lattice constant $\bar{A}$ before deformation. When $\bar{a}$ is obtained the stretching length between any two atoms and the angles between any two crystallographic lines can be determined, thus the stored atomistic energy can be determined. Thus Cauchy-Born rule links the continuum and atomistic scale and enable it possible to determine the energy density of a finite element by computing the atomistic energy in a unit cell. Specifically, we have (see Figure 6.1):

$$\bar{A} = \bar{A}_1 + \bar{A}_2 + \bar{A}_3 \qquad (6.4q)$$

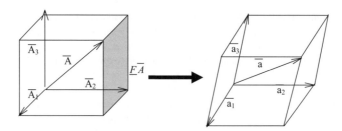

**Figure 6.4**   Cauchy-Born rule $\bar{a} = \underline{F}\bar{A}$ in linking lattice vector $\bar{a}$ after deformation with $\bar{A}$ in reference configuration

where the three vectors in the right part of the equation are the vectors starting from the origin o along the three crystallographic lines; their values equal the lattice constants along these directions. The converse is true that if one can determine the lattice vector $\bar{a}$ after deformation, the deformed crystal structure and configuration can also be determined. This is the key reason why the Cauchy-Born rule is so important for atom/continuum transition, because it offers the way to determine the new lattice vector $\bar{a}$ from the lattice vector $\bar{A}$ at the reference (undeformed) configuration by the following rule:

$$\bar{a} = \underline{F}\bar{A} \tag{6.4r}$$

with

$$\bar{a} = \bar{a}_1 + \bar{a}_2 + \bar{a}_3 \tag{6.4s}$$

(6.4r) is written as an abstract form to denote that the new lattice vector equals the product of the deformation gradient tensor with the old lattice vector. This product can be operated either by a matrix product or by tensor component. These operations will be shown in detail in Section 6.8.5 after the mathematical definition of $\underline{F}$ is introduced in (6.7j).

### 6.4.4.2 Expression of Energy Density Function in Elements Based on the Cauchy-Born Rule

Since the deformation in a local region is assumed to be uniform, the interatomic distance and relative positions of neighboring atoms can be obtained by using the lattice vector $\bar{a}$. Based on these data, interatomic potential functions such as LJ potential $V_{ij}$ and, in turn, the potential energy $E_{unit}$ of a unit cell can be determined. The numerical algorithm for this will be introduced in the next section, and will be described in detail in Section 6.8 after introducing a rigorous definition of deformation gradient $\underline{F}$.

After the atomistic energy $E_{unit}$ in a unit cell is obtained by equality of the atomic energy per unit volume with the strain energy density of the continuum, the energy density W can be obtained. Thus, it can be used in formulations such as (6.2a) or the energy $E^{\alpha}$ in (6.4p). For the latter case, assuming the generic atom is in a region of approximately uniform deformation, we have

$$E^{\alpha} \approx \Omega_0 W(\bar{u}(\bar{X}^{\alpha})) \tag{6.4t}$$

where $\Omega_0$ is the volume of the atom (in the reference configuration); here the displacement vector $\bar{u}(\bar{X}^{\alpha})$ is used to describe the strain tensor or deformation gradient as discussed in Section 6.3.4.1. Now the energy of an element is simply this energy $E^\alpha$ times $n^e$ which is the number of atoms in element e. Thus, the continuum energy $E^C$ in (6.4p) can be rewritten

$$E^C \approx \sum_{e=1}^{N_{elem}} n^e \Omega_0 W(\bar{u}(\overline{X}^\alpha)) = \sum_{e=1}^{N_{elem}} n^e \Omega_0 W([N][U]^{rep}) \tag{6.4u}$$

with $N_{elem}$ the number of elements. In the derivation of the last equation, the interpolation relationship of (6.4f) is used. Furthermore, by using (6.4n) and (6.4k), we have the expressions as follows:

$$E^{tot}([U]^{rep}) = \sum_{\alpha=1}^{N_A} E^\alpha(\bar{u}([U]^{rep})) + \sum_{e=1}^{N_{elem}} n^e \Omega_0 W([U]^{rep}) \tag{6.4v}$$

$$\Pi^{tot} = \sum_{\alpha=1}^{N_A} E^\alpha(\bar{u}([U]^{rep})) + \sum_{e=1}^{N_{elem}} n^e \Omega_0 W([U]^{rep}) - [F]^T [U]^{rep} \tag{6.4w}$$

Besides some elements at the atomistic/continuum interface, $n^e \Omega_0$ is simply taken as the total volume $\Omega_e$ of the element e, eliminating the need for explicitly counting atoms.

It is noted that the computational saving made in (6.4w) is that a sum over all atoms in the body is replaced by a sum over all elements, each one requiring an explicit energy calculation for only one atom. Since the total number N of atoms is several orders of magnitude larger than the number of elements, the computational saving is substantial.

## 6.4.5 The Solution Scheme of the QC Method

Once the framework of the energy-based QC method is formulated, by the minimum principle of total potential energy which applies to $\Pi^{tot}$ in (6.4w), simultaneous algebraic equations can be determined to find $[U]^{rep}$, which are the displacements of all the repatoms. For a 3D problem, these governing equations are derived in a similar way and with a similar form to the FEM simultaneous equations of Section 6.3.5. They can be expressed as follows:

$$[K]_{3Nx3N} [U]^{rep}_{3Nx1} = [P]_{3Nx1} \tag{6.4x}$$

The difference here from FEM is that $[U]^{rep}$ as shown in (6.4g) not only includes true FE node displacements as (6.2l), but also includes true atom displacements. In QC terminology, it includes both displacements of local repatoms (nodes) and non-local repatoms (atoms). The terminology sometimes is confusing; in the force-based QC method,[22] the authors drop the unified description by repatoms and use atoms for atomistic regions and nodes for continuum (FE) regions, as will be described in the next chapter.

The simultaneous equations in (6.4x) are nonlinear in general. Basically, the numerical solution algorithm is similar or not fundamentally different from the nonlinear FEM solution such as plasticity. To solve the nonlinear problem, an iterative process should be conducted for every node and element. The iterative process may be described as follows:

- The iteration starts from an initial displacement matrix $[U]^{0rep}$ with the initial values of all displacements of atoms/nodes. These node displacements are used to determine the displacement field inside the element by interpolation functions N and then used to determine the deformation gradient $\underline{F}$ of each element.
- The obtained $\underline{F}$ can be used in a subroutine to get the energy density W for the element. The obtained W of all elements is used to get the new total potential energy $\Pi$, and so update (6.4w) to get the new matrix $[U]^{1rep}$.

– This is then used to start the new iterative process. This iterative procedure continues until some criterion is satisfied in which the difference of some key value in the two iterations is less than the prescribed tolerance.

### 6.4.6    Subroutine to Determine Energy Density W for Each Element

Instead of phenomenological laws such as Hooke's law of elasticity used in the FEM, the QC method calculates energy density W for each element, which is the uniqueness of the QC solution. The practical way to calculate W for each element is to calculate the atomistic energy density in a "black box" which is separate from the model,[22] say by a subroutine in the solution code for (6.4x). One can give the subroutine any deformation gradient $\underline{F}$ based on the displacement vector in each element obtained from the last iterative step. The subroutine will calculate lattice constant $\bar{a}$ after deformation from its undeformed lattice constant $\bar{A}$ and $\underline{F}$ by the Cauchy-Born rule (i.e., $\bar{a} = \underline{F}\bar{A}$, see (6.4r)).

Any atomic model for interatomic potentials is in terms of atomic distances and angles, not in terms of "F" and "a", thus the task of the subroutine after calculating $\bar{a}$ is to make that translation from $\bar{a}$ to the parameters used in the potential function. The energy is then obtained by atomistic calculation for a unit cell with periodic boundary condition. Here, PBD is used to emphasize that the energy is obtained for a homogeneous deformation field and no boundary in a finite region affects its homogeneity. After the atomistic energy $E_0(\underline{F})$ in a unit cell is obtained, by dividing the unit cell volume, the energy density can be determined and then sent back to the main code. By equality of the atomic energy per unit volume with the strain energy density W of that continuum, the energy density $W(\underline{F})$ for the element at hand is obtained. This process using the subroutine will repeat for every element in the model because the deformation gradient is different from one element to another.

In a more general case where the deformation gradient is not constant in the element, the strain energy in each element may be obtained using the Gaussian quadrature method. It can be approximated by the sum of products of the energy density at Gaussian points and corresponding weight coefficients as shown in (6.3b). The strain energy density at the Gaussian point can be obtained by first finding its displacement vector at these points through formulas like (6.3f), then finding its deformation gradient $\underline{F}$ and using the subroutine to get its energy density.

### 6.4.7    Treatment of the Interface

We mentioned that the general treatment of the energy of the atomistic domain, i.e., $E^A$ in (6.4o), is the same as atomistic analysis. The only difference, however, comes from its boundary condition at the interface between atomistic domain and continuum. Since QC does not use handshaking areas, the atoms at the interface will directly connect to the element nodes. Thus, partial energy of the element has already been calculated by these atoms and the energy calculation of these elements should be reduced to avoid double counting. What QC does is to adjust the number $n^e$ in the expression (6.4v). The choice of $n^e$ is somewhat ambiguous regarding the parceling of energy near the interfaces. This is a rough treatment of the important effect which causes the so-called "ghost force" at the interface.

### 6.4.8    Ghost Force

Ghost force is the unphysical force that appears even if the system does not sustain any external loading. Ideally, if the developed model is in equilibrium, every node and atom is at its equilibrium position and no net force will appear at these nodes/atoms. Therefore, any forces on atoms or nodes that arise in the equilibrium configuration are unphysical and named as the "ghost force", a phrase first used in Ref. [23]. Under these forces, the designed model will show some spurious distortions of the body upon relaxation.

### 6.4.8.1 The Importance of Interface Treatment on the Ghost Force

The following is an analysis to the ghost force which emphasizes the importance of the treatment of interface domain. Assume the total energy $\Pi$ of a full atomistic description of a body consists of a contribution from atoms in each region, $B^A$ and $B^C$, respectively[22]

$$\Pi^{atom} = \Pi^{atom,A} + \Pi^{atom,C} \tag{6.5a}$$

Usually, in the approximate expression $\Pi^{tot}$ or $E^{tot}$ of (6.4n), $\Pi^A$ is defined as the same total potential energy of atoms in the atomistic domain A, i.e.,

$$\Pi^A = \Pi^{atom,A} \tag{6.5b}$$

but $\Pi^C$ in the continuum domain is only an approximation to $\Pi^{atom,C}$. Consider an atom $\alpha$ inside the atomistic domain $B^A$, but near the interface, such that it interacts with atoms in $B^C$. By the generic relationship between the potential energy and the force (see equation (2.1a), in the fully accurate atomistic description and the QC approximate description, respectively, we have

$$\frac{\partial \Pi^{atom}}{\partial \bar{u}^\alpha} = \frac{\partial \Pi^A}{\partial \bar{u}^\alpha} + \frac{\partial \Pi^{atom,C}}{\partial \bar{u}^\alpha} = \bar{f}^\alpha \quad \text{(accurate description)} \tag{6.5c}$$

$$\frac{\partial \Pi^{tot}}{\partial \bar{u}^\alpha} = \frac{\partial \Pi^A}{\partial \bar{u}^\alpha} + \frac{\partial \Pi^C}{\partial \bar{u}^\alpha} = \bar{f}^\alpha_g \quad \text{(approximate description)} \tag{6.5d}$$

Since a fully atomistic model is an exact model in the sense that the atoms are at equilibrium positions, by the minimization principle of total potential energy the force $\bar{f}^\alpha = 0$, i.e.,

$$\frac{\partial \Pi^A}{\partial \bar{u}^\alpha} + \frac{\partial \Pi^{atom,C}}{\partial \bar{u}^\alpha} = 0 \tag{6.5e}$$

Because $\Pi^{tot}$ is an approximate energy, it is not the true minimum potential energy, and by the same minimum energy principle, the force $\bar{f}^\alpha_g$ calculated by (6.5d) is not zero. It is the residual ghost force which applies at atom $\alpha$ and drives that atom from the designed position to another location for achieving equilibrium, which causes the model distortion.

The above analysis is suitable for any energy-based approximate formulation, but why the ghost force is strongly emphasized in QC and other related multiscale formulations is because the interface transition problem is the key issue of most multiscale models, as we introduced in the last section of Chapter 5. In fact, these ghost forces are strong enough to cause artificial phenomena at the interface to reduce the accuracy. Defining the energy error, $\Pi^{err}$ in the continuum as

$$\Pi^{Err} = \Pi^{atom,C} - \Pi^C \tag{6.5f}$$

from which $\Pi^C = \Pi^{atom,C} - \Pi^{Err}$, substituting this expression into (6.5d) we have

$$\frac{\partial \Pi^A}{\partial \bar{u}^\alpha} + \frac{\partial \Pi^{atom,C}}{\partial \bar{u}^\alpha} - \frac{\partial \Pi^{Err}}{\partial \bar{u}^\alpha} = \bar{f}^\alpha_g \tag{6.5g}$$

Due to (6.5e) the sum of the first two terms is zero. We then can express the ghost force

$$\bar{f}^\alpha_g = -\frac{\partial \Pi^{Err}}{\partial \bar{u}^\alpha} \tag{6.5h}$$

This formula provides a suitable opportunity for us to discuss why the ghost force is strong near the interface region.

First, if nodes/repatoms are not close to the interface, their contributions to the ghost force are negligible since their related interpolation functions $N_I(\overline{X}^\alpha)$ are zero (see discussion before equation (6.4i)); this implicitly indicates that the energy approximation in the area not close to the interface will have minor effects on the ghost force near the interface; on the contrary, the effects are large if close to atom $\alpha$ near the interface.

Secondly, while the QC method uses the unified formulation of repatoms to describe the modeling system, the local nature of the constitutive relationship in the continuum domain and the non-local nature in the atomistic domain still remain. Thus, even though the atom $\alpha$ in the atomistic domain has interatomic interaction with atoms in the continuum domain, the force is zero because in the continuum domain the stress is only related to the local strain, not another action at a distance from the atom. This description indicates that the source for the strong ghost force at the interface is the incompatibility of the local and non-local nature of the material property.

In the QC method, changing the $n^e$ number of the element in (6.4v) is the way to adjust the energy distribution between the element in the continuum region and the atomistic region. For instance, in the QC treatment if there are two nodes of the triangular FE element in the continuum region touching the atomistic boundary, the energy in that element will distribute 2/3 energy to the atomistic region; if only one node touches the atomistic boundary, the element energy will only distribute 1/3 to the atomistic region. Needless to say, this arbitrary treatment does not have a regular foundation, which in the sensitive area causes a strong ghost force.

### 6.4.8.2 Ways to Eliminate the Ghost Force

Several methods have been proposed to eliminate or at least mitigate ghost forces. In the original implementation of the QC method Shenoy et al. discussed the ghost force and proposed an approximate method for its correction.[23] This correction seems to involve almost negligible extra computational effort but greatly improves the accuracy. The basic idea is to explicitly compute the ghost force in some suitable configuration, and then add the negative of these forces as deadloads on the affected atoms or nodes. After applying the deadloads the total potential energy is modified from equation (6.4w) to

$$\Pi^{tot} = \sum_{\alpha=1}^{N_A} E^\alpha(\overline{u}([U]^{rep})) + \sum_{e=1}^{N_{elem}} n^e \Omega_0 W([U]^{rep}) - [F]^T[U]^{rep} - \sum_{\alpha=1}^{n_g} \overline{g}_g^\alpha \cdot \overline{u}_\alpha \qquad (6.5i)$$

In this expression, both matrix product and vector dot product are used for convenience; the matrix product covers all repatoms/nodes but the vector dot product in the last term only covers atom/nodes near the interface with ghost forces. For an undeformed model this exactly eliminates the ghost force by construction.

For non-uniformly strained loading there is an indeterminate error associated with this ghost correction method. To distinguish the ghost force from the real force, a criterion is developed based on whether the force is produced from the natural surrounding conditions of atoms/nodes. Specifically, for an atom the ghost force is defined as any force the atom would not feel if its environment was truly atomistic everywhere.[22] Using this definition, it is possible to incorporate a ghost force correction within any of the other existing methods. For instance, based on this definition ghost forces for an atom can be conceptually determined by the difference between the calculated result and the result when the atom is put in an "infinite" large domain with full atoms. The latter is truly atomistic everywhere. This concept has been used to develop an approximate way to determine the ghost force correction for the non-local QC method, see next section for the detail.

Most likely, ghost forces on the nodes defining the continuum are those forces that the nodes would not feel if their environment were truly just the continuum. When implementing these forces, it becomes clear which contributions are the ghost forces. Actually, in force-based formulations such as CADD/FEAt which will be introduced in the next chapter, only true forces are put on the atoms/nodes, thus the differences between these forces and the actual derivatives of the total potential energy $\Pi$ defines the ghost forces.

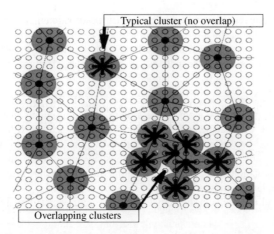

**Figure 6.5**  Scheme of QC-FNL method (Reproduced from Knap, J. and Ortiz, M. (2001) An analysis of the quasicontinuum method. *Journal of the Mechanics and Physics of Solids*, 49, 1891, Elsevier)

## 6.5  Fully Non-localized QC Method

Knap and Ortiz proposed a new QC method framework in 2001.[24] This new model is non-local formulation throughout the simulation region and there is no transition region between different scales. Thus the incompatibility between non-local in the atomistic domain and the local constitutive equation in the finite element domain has been eliminated. The method of Knap and Ortiz was originally posited as a force-based method and denoted as QC (FNL).

Recently, it was re-cast as an energy-based method[25] with the notation CQC(m)-E. Here C denotes the cluster of atoms surrounding a generic repatom and m in the parenthesis denotes the atom number in the cluster. The cluster of atoms is used in the method for calculating the node energy or force. The symbol E or F after the symbol "-" denotes whether it is an energy-based or force-based formulation. Figure 6.5 uses grey shading to show atomistic clusters surrounding the nodes. In most regular nodes in the figure there are nine atoms, thus the formula may be written CQC(9)-F if it is force-based. However, for the atomistic region, the repatom is the atom itself so m = 1. In the lower right region of Figure 6.5 nodes are crowded and atoms are connected to several clusters; by adjusting the m number these atoms can be reasonably distributed. In the following we will introduce the energy-based non-local QC. Readers interested in force-based non-local QC are referred to the paper by Knap and Ortiz.[24]

### 6.5.1  Energy-based Non-local QC Model (CQC(m)-E)

The basic idea of QC-FNL is the same as the QC (local) formulation introduced in Section 6.4. Specifically, both formulations select a handful of atoms to act as nodes to represent the whole simulation system, and these nodes are connected by a finite element mesh. In addition, both formulations keep the key rule of the QC method that all atoms between the nodes are constrained to move according to the interpolated FEM field with (6.4f) by the global node displacement vector $[U]^{rep} = \{\overline{u}_1, \overline{u}_2 \ldots \overline{u}_{Nrep}\}^T$.

The difference in the QC-FNL formulation is on how to calculate the node/repatom force or energy. In the local QC formulation, the total energy $E^{tot}(\tilde{u}([U]^{rep}))$ is separated into two parts, for the continuum part the energy was found using the Cauchy-Born rule. However, in the QC-FNL formulation, that rule is not used. The energy of any repatom/node is calculated by the average energy of the atoms in its related cluster as follows:

$$E^I \approx \frac{1}{m^I} \sum_{\alpha=1}^{m^I} E^\alpha(\tilde{u}([U]^{rep})) \tag{6.5j}$$

where $m^I$ is the number of atoms in the cluster. In (6.5i) the energy $E^I$ is expressed as the function of $[U]^{rep}$, this is clear from (6.4f) where any atom in the system, including atoms inside the cluster, can be interpolated from all node displacements $[U]^{rep}$ by interpolation functions N. In other words, while the energy $E^\alpha$ of atom $\alpha$ in the cluster depends on the positions of all atoms, it actually depends on the nodal displacements $[U]^{rep}$ as shown in (6.4f), since most atoms have constrained motions. It is noted again, however, that most interpolation (shape) functions for atoms in the cluster are zero and the node associated with the cluster has the largest weight factor.

The total energy of all the atoms in the body can be now approximated by a weighted sum of these repatom/node energies:

$$E^{tot}(\tilde{u}([U]^{rep})) = \sum_{I=1}^{N_{rep}} n^I E^I([U]^{rep}) = \sum_{I=1}^{N_{rep}} \frac{n^I}{m^I} \sum_{\alpha=1}^{m_I} E^\alpha([U]^{rep}) \tag{6.5k}$$

Here, weight function $n^I$ is introduced to every repatom (node), referring to the number of atoms whose energy is represented by repatom (node) I shown in the superscript. It is reported[22] that the details of how to choose $n^I$ and how to make the cluster average to obtain $E^I$ (equation (6.5i) do not strongly affect the speed or accuracy of the method, but the summation of $n^I$ ($I = 1, 2, \ldots, N_{rep}$) should equal the total atoms N in the entire body, i.e.,

$$\sum_{I=1}^{N_{rep}} n^I = N \tag{6.5l}$$

In regions where the distances between repatoms are near the lattice constants, the method naturally goes over to the fully atomistic limit and we have $m^I = 1$ and $n^I = 1$.

### 6.5.2   Dead Ghost Force Correction in Energy-based Non-local QC

It is recognized that increasing atom numbers in clusters will reduce the ghost force and increase the accuracy but the computational time will also increase. However, one may get good accuracy from small clusters if ghost force correction is used.

For a given cluster radius $r_{cluster}$ and a given model configuration characterized by the repatom displacement matrix $[U]^{rep}$, the force $\bar{f}$ on each repatom can be determined. Conceptually, the correct forces, denoted by $\bar{f}*$, would be the forces computed in the limit of infinite cluster radius. To compute an approximate correct force, one can recompute the forces for a given model but with a cluster radius slightly larger than the original $r_{cluster}$. The ghost force is then defined as the difference between the two sets of forces:[22, 25]

$$\bar{g} = \bar{f} - \tilde{f}* \tag{6.5m}$$

where $\tilde{f}*$ is the approximation to the correct forces. The forces $\bar{g}$ can then be added as deadloads in equation (6.5h).

## 6.6   Applications of the QC Method

Several important results in applications of the QC method, highlighted by bold letters, will be introduced in this section.

### 6.6.1　*Nanoindentation*

As shown in Figure 5.3, nanoindentation experiments are too large for a fully atomistic simulation to have realistic system sizes and boundary conditions. Multiscale analysis is a good choice which can keep the model size large enough but also keeps the atomistic analysis in some regions to trace the details of dislocation nucleation process. Tadmor *et al.*[15] performed nanoindentation simulations in 2D using a rectangular and cylindrical indenter. The focus is on plasticity incipient which involves the nucleation and motion of a few to tens of dislocations.

The significance of the QC method in studying the phenomena is that it can offer very detailed information with which one can analyze the deformation process by variables such as stress and slip distributions on the atomic scale, thus new phenomena and insights for deformation mechanisms can be obtained, which are summarized as follows:[15]

**1. Critical shear stress criterion of dislocation nucleation at nanoscale is valid for a rectangular indenter but not for a cylindrical one**

Careful analysis for the stress and strain just prior to the nucleation of the initial dislocations suggested that the criterion of critical shear stress correctly predicts the dislocation nucleation for a rectangular indenter but not for a cylindrical one. Here, the criterion of critical shear stress indicates that dislocation nucleation occurs when the resolved shear stress $\tau_{RSS}$ equals or is larger than a critical shear stress $\tau_{cr}$, i.e., when the following expression is valid:

$$\tau_{RSS} \geq \tau_{cr} \tag{6.6a}$$

where the resolved shear stress $\tau_{RSS}$ is defined as the shear stress along the slip direction on the slip plane. Let Greek letters $\lambda$ and $\phi$ denote the angles between the slip direction and the normal of slip plane, respectively, with the applied force F as shown in Figure 6.6. The resolved force $F_\lambda$ along the slip direction is $F_\lambda = F \cos \lambda$, the slip area is $A_\lambda = A/\cos \phi$, thus the resolved shear stress can be determined by

$$\tau_{Rss} = \frac{F_\lambda}{A_\lambda} = \frac{F}{A} \cos\phi \, \cos\lambda = m_s \sigma \tag{6.6b}$$

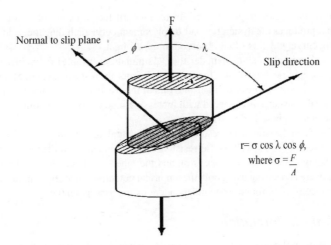

**Figure 6.6**　Resolved shear stress versus applied normal stress

with

$$m_s = \cos\phi \cos\lambda \qquad\qquad (6.6c)$$

where $\sigma$ is the normal stress and $m_s$ is the so called Schmid factor.

The same QC simulations of a rectangular indenter as given in Ref. [17] were used to test Rice's dislocation nucleation criterion based on the PN model (see Section 1.4.2.2). The latter was used by Rice for crack-tip nucleation of dislocations.[26] It was found that for a rectangular indenter the stress-based nucleation criterion is in agreement with the simulation results. In both cases, dislocations are nucleated when the resolved shear stress $\tau_{RSS}$ at the tip of the indenter reaches a critical value $\tau_{cr}$ which is independent of the width of the indenter.

**2. Aluminum microtwinning was formed during nanoindentation initiated from the sharp edge of the rectangular indenter**

The observation of deformation twinning (DT) was a surprising result since DT is uncommon in FCC metals in general and in aluminum in particular under ordinary loading condition at room temperature. In the sharp edge orientation, however, the easy glide systems for the FCC aluminum crystal are constrained, which allows the more energetic twinning mechanism to occur. A similar twinning mechanism was observed by Picu in a BCC Mo nanoindentation study using the QC method.[27]

Here, two questions may be raised. The first is why criterion (6.6a) is not valid for a cylindrical indenter. The other is related to the following issue: While the PN model does not show correct force distribution[28] as discussed in 1.4.2.2, why can it be used in criterion (6.6a) for correct predictions of dislocation nucleation of a rectangular indenter. The answers may be partially related to the ledge effects, reported by the theoretical analysis, that "in order to obtain better quantitative agreement between analysis and simulation" including ledge effects is important.[10]

**3. Stress-induced phase transformations of Si was found beneath the nanoindenter**

It was found that during the indentation process of diamond-cubic Si, a roughly hemispherical region of transformed material forms beneath the indenter, growing with increasing load.[18, 19] In the transformation region, the transformed material is composed of several different metallic phases, including BCC, bct5 and $\beta$-tin, the latter is the room temperature form of tin metal which has a body-centered tetragonal (bct) crystal structure. Interesting phenomena found in the QC simulation includes:

- During unloading, the region fragments and forms a tendril-like structure of conducting material.
- The simulation reproduces the hysteretic load-displacement curve including the sudden discontinuity in the unloading curve which is often observed in experiment.
- In conjunction with a simple analytical model, the QC simulation was able to explain the experimentally observed change in electrical resistance. In these experiments, the resistance change between electrodes is measured as an indenter is pressed either into one of the electrodes or between the two electrodes.
- A drop in electrical resistance is observed with increasing indentation, a fact often cited as proof that transition to a metallic phase is occurring.
- A comparison between the simulation results and experimental measurement data for indentation load versus electrical resistance as well as the resistance versus indentation depth is in agreement. The comparison is for two loading cases, one is on and the other is between the electrodes.
- The simulations were carried out using both the empirical potential and a TB formulation for silicon.[20] The main conclusions were found to be independent of the used potential.

## 6.6.2   Crack-tip Deformation

This is another class of applications to which multiscale analysis is particularly suited. In these problems, atomistic resolution is required in a very small region at the crack tip while the continuum boundary conditions of linear elastic fracture mechanics (LEFM) need to be applied in the far field. Miller *et al.*[11] used the QC method to study deformation at the tips of cracks in single crystal nickel. Two different orientations of cracks were tested, one is to make the crystal cleave in brittle fashion by model I loading, and the other is such that dislocations were emitted. They used the QC results to test analytical criteria for crack deformation such as the Griffith criterion for brittle fracture and Rice's criterion for dislocation emission.

### 4. Rice's ductile criterion does not predict the critical load well for dislocation nucleation

Miller *et al.* found that Rice's criterion underestimates the critical load for dislocation nucleation by 45%, and also predicted a ductile response for the brittle setting. These authors concluded that the error source is most likely tied to the PN model assumption used in Rice's work. The PN assumption indicates all nonlinearity is confined to a single slip plane. In the simulation, however, it was clear that the nonlinearity is more widely distributed. Specifically, as we discussed in Section 1.4.2, it has non-local nature and the dislocation is distributed in several locations dependent on the load direction relative to the crystal orientation.

Pillai and Miller[13] investigated crack tips near bi-material interfaces by the QC method. The goal was to systematically vary the atomistic interactions between crystals at a fully coherent, epitaxial interface and study the effect of the interactions on the fracture properties of the interface. The following conclusion was found:

### 5. A microscopic coupling of ductile-brittle can occur at the crack tip which changes the simple Griffith fracture pattern

The QC simulation[13] found that under certain conditions it was not possible to predict the interface fracture behavior from a simple Griffith picture of brittle fracture. This is because some interfaces behave in an initially ductile fashion, possibly by defects or stresses in the formation process. These preexisting stresses and defects emit dislocations and blunt the crack. This blunting created ledges at the crack tip that exposed different atomistic layers to high crack-tip stresses.

These planes, away from the true atomistic scale "interface", could act as brittle cleavage paths. Here, the "ductile-brittle coupling" is seen, i.e., brittle fracture is caused by atomistic layers produced by ductile deformation at the crack tip. As a result, brittle fracture of bi-material interfaces not only arises due to the clean cleavage of the two materials along the atomically well-defined interface. It can also include a fracture pattern such that a few atomistic layers of one material are left on the fracture surface of the other material.

### 6. A deformation twin model developed based on the QC simulation

Hai and Tadmor[14] used the QC method to study whether DT would occur at the crack tips in single-crystal aluminum. In finding 1 above, the DT observed in aluminum in nanoindentation of a rectangular indenter is introduced. In addition, DT at the tips of cracks in aluminum was observed experimentally *in situ* by transmission electron miscroscopy in two instances.[28, 29] The QC simulation results showed that the deformation mechanism at the crack tip strongly depends on the loading mode, crystallographic orientation and crack-tip morphology. For the experimental orientations where DT was observed, the QC simulation shows consistent results that the DT formation appears.

For other orientations, either DT, dislocation emission or the formation of fault were observed. The complex response behavior obtained in the QC simulation led the authors to develop a theoretical model for the DT nucleation based on the PN model which can be considered an extension of Rice's model[26] from dislocation to DT. The main idea of the extension is that DT is controlled by an energetic parameter, called the unstable twinning energy, in analogy to Rice's unstable stacking energy which plays a similar role for dislocation emission. The predictions of the analytical model are in good agreement with the QC simulation results.

## 6.6.3    Deformation and Fracture of Grain Boundaries

Just like nanoindentation, work surrounding deformation and failure at grain boundaries (GBs) is well-suited to multiscale analysis, which allows for full atomistic resolution near GBs but large enough system sizes to provide realistic boundary conditions of the model.

Nanoindentation was used as the source of dislocations which interact with a grain boundary.[8] The main result shows that after the first dislocation was nucleated, this dislocation is immediately absorbed by the GB, forming a step. The next dislocation dissociated into partials, however, stands off from the GB, forming a pile-up due to elastic interaction with the first emitted defect.

Miller et al.[11] studied the effects of an impinging crack on two different GBs in FCC aluminum, the crack is subjected to mode-I loading which is perpendicular to the crack. Interesting results include:

**7. Crack-tip induced high stress stimulates GB dislocation nucleation and motion as well as drives boundary migration and bending**
The QC simulation found that the nucleated dislocations in the interface traveled into the bulk of the two grains, and found that the grain boundary is migrated and bows to meet the crack tip.

**8. Grain boundary orientation and structure is important in fracture and slip transmission through the GB which may serve to either toughen or embrittle the polycrystal**
It was also found that under a certain crack-tip orientation in which slip on the (111) planes is constrained and the failure is brittle cleavage, the GB itself serves as a path for rapid brittle fracture.

## 6.6.4    Dislocation Interactions

3D QC has taken the study of the strength of dislocation junctions as its key application. Rodney and Phillips[30] carried out this simulation for dislocations lying between intersecting planes, and computed the critical stress required to break the dislocation junction. It is seen in the simulation that the whole process of junction stretches in response to the loading and finally breaks apart.

Miller and Rodney investigated the interactions between different dislocation groups under the nanoindentation and found non-local description is necessary.[31] The interaction between dislocations and second-phase particles was investigated in detail with the QC method.[32] Comparison of the simulation result with the continuum model shows the following conclusion.

**9. If the continuum model includes the key elements of the system, the atomistic and continuum models yield the same picture for dislocation-obstacle interaction**
Examples of the key elements to make the statement correct include line-tension effects and the presence of partial dislocations in the continuum theory.

## 6.6.5    Polarizations Switching in Ferroelectrics

The QC method is used to study the response of ferroelectric lead titanate ($PbTiO_3$) to electrical and mechanical loading.[33] This ferroelectric ceramic has a complex Bravais lattice structure and the complex lattice formulation of Tadmor et al.[21] The work constructed an effective Hamiltonian for $PbTiO_3$ with coefficients fitted to the database of DFT calculations. This Hamiltonian is a nonlinear high-order expansion with parameters of finite strain, internal degrees of freedom and electric field. It contains the nonlinearity necessary for domain switching and phase change from the ground-state tetragonal phase of $PbTiO_3$ to metastable rhombohedral and orthorhombic phases.

By incorporating it into the QC formulation, this Hamiltonian was used to study hysteresis of single-crystal $PbTiO_3$ driven by applied electric field and temperature. The related microscopic mechanisms responsible for polarization switching are analyzed. The model was also used to simulate a high-strain ferroelectric actuator.[33] The result is summarized below.

**10. Simulation of high-strain ferroelectric actuator can be used as a basis for its design**
The simulation result shows that before the external electric field the directions of polarization are randomly distributed to the left and right resulting in an overall neutral material. When the electric field reaches 120 MV/m along the vertical direction of the slab of the device, the polarization in the central region has fully re-oriented in the vertical direction; the top and bottom of the slab, however, are prevented by mechanical boundary conditions from fully re-orienting. It is seen clearly that the domain-switched slab has substantial elongated length along the vertical direction, thus it may be used as a basis for large-strain actuation.

## 6.7   Short Discussion about the QC Method

The applications of the QC method introduced in the last section are encouraging. They indicate that while multiscale methods are not perfect so far, methods such as the QC can be used to find new phenomena, new mechanisms, and develop new theories.

One of the essential contributions of the QC method is the atom/continuum energy transition via the Cauchy-Born rule. Specifically, the QC developers used this rule for the first time to calculate the strain energy density W in the finite element region by accounting the underlying atomic potential energy. This results in the essential difference of FEM used in the QC method from the traditional FEM. For the QC method the strain energy density W is not obtained from any phenomenological laws such as Hooke's law as in traditional FEM. It is obtained by the same interatomic potential used in the atomistic region with the Cauchy-Born rule. This formulation successfully imbues the continuum framework with atomistic properties such as nonlinearity and anisotropy.

This unified approach, which calculates continuum energy based on atomic potential, is very suitable for adaptive remeshing of FEM meshes. It coarsens or refines the continuum region by adjusting the interface between atoms and continuum in the simulation. This adaptive meshing method enables the allowed maximum error to be introduced into energy harmonically, rather than correlating with a typical error of a fixed interface. The QC method can enlarge the atom region based on the strain field variation rule. In this way, deformation details, such as crack evolution, can be traced; and deformation error due to deficient original mesh design can be avoided.

On the other hand, while the energy calculation is based on the same interatomic potential, the material structure and property are quite different between the atomistic and continuum region. The latter features are the basic difference from the GP method introduced in the last chapter where the material structures, properties and numerical algorithm are all kept the same between different scales, and neighbor-link cells are used to intrinsically connect material at different scales.

Specifically, in the QC method the type of force and the corresponding responses are quite different in the atomistic domain from those in the continuum domain. In the latter case the force (stress) is related to the deformation (strain) of the material at the same location so that constitutive laws are local in nature. In the atomistic region, however, the atom is subject to forces from other atoms at a distance from the atom in hand. The "non-local repatom" and "local repatom" are then used in the QC method to distinguish the non-local bonding behavior of the atomistic region from the continuum where local constitutive laws are applied.

These different structures and properties in the continuum and atomistic domains cause incompatibility at their interface. This incompatibility results in the ghost forces which cause structure distortion and other unphysical phenomena. In the next chapter, a 1D model will be used to show how the incompatibility of constitutive laws between atomistic and continuum domain will cause problems of smooth transition in the transition area. Instead of smooth transition curves of the GP method (see Figures 5.6 to 5.9), many DC multiscale methods suffer these ghost force problems and cannot show smooth transition.[22, 34] This is why the GP method is developed to link the materials on both sides of the interface by the constant neighbor-link cell connections between different scale atoms and particles to make the transition smooth.

While the fully non-local QC method is good for the smooth transition at the interface of atomistic/continuum region, its complicated computation makes it inefficient, as discussed by Curtin and Miller,[34] for the force-based non-local QC method. In fact, this kind of QC model comes with a high computational cost in regions of slowly varying deformation gradient as compared to linear elasticity or even the local QC method. Current benchmark tests for the comparison of calculation time[22] show that for the same problem and same condition, the CQC(13)-F full non-local QC method takes 23,994 seconds (s) in comparison with 45,520.11 s for full atomistic (exact) simulation and 4629.61 s for the CADD/FEAt method.

---

**Homework**

(6.4) Assuming the loading direction is shown in the [111] direction of Figure 6.6 for the fcc structure, prove or disprove that the values of Schmid factors of the following six slip systems are as given below and Schmid factors for the other six slip systems are small.

$$(\bar{1}11)[101], \quad m^s = 0.272$$

$$(\bar{1}11)[\bar{1}\bar{1}0], \quad m^s = -0.272$$

$$(1\bar{1}1)[0\bar{1}\bar{1}], \quad m^s = -0.272$$

$$(1\bar{1}1)[110], \quad m^s = 0.272$$

$$(\bar{1}\bar{1}1)[101], \quad m^s = -0.272$$

$$(\bar{1}\bar{1}1)[0\bar{1}\bar{1}], \quad m^s = 0.272$$

Hint: In a cubic crystal system, the angle between two crystal directions of two vectors of $\mathbf{D} = u\mathbf{a} + v\mathbf{b} + w\mathbf{c}$ and $\mathbf{D}' = u'\mathbf{a} + v'\mathbf{b} + w'\mathbf{c}$ can be determined by a dot product of two vectors as

$$\mathbf{D} \cdot \mathbf{D}' = |\mathbf{D}||\mathbf{D}'|\cos\delta$$

or

$$\cos\delta = \mathbf{D} \cdot \mathbf{D}'/|\mathbf{D}||\mathbf{D}'| = \frac{uu' + vv' + ww'}{\sqrt{u^2 + v^2 + w^2}\sqrt{(u')^2 + (v')^2 + (w')^2}}$$

This problem is easily solved by either Matlab or Excel through writing a simple computer code.

(6.5) From the website http://www.qc.com download the QC code and its manual. Read this manual and several QC examples and make a concise summary of what you learn from the manual. If you can download the code to your computer, try to run one of the examples.

---

# Part 6.3 Analytical and Semi-analytical Multiscale Methods Across Atomic/Continuum Scales

In this part, analytical and semi-analytical methods will be introduced. Most of these methods use the Cauchy-Born rule to make the transition between the atomistic and continuum scales in a hierarchical fashion. While that rule was given in equation (6.4r) and briefly discussed in Section 6.4.4, the key to applying that rule in practice was not introduced; partly to avoid involving readers in too much mathematics at the beginning. For practical applications and deep understanding, however, establishing quantitative relationships between the components of deformation gradient $\underline{F}$ and displacement vector $\overline{u}$ is necessary. To have the capability to calculate the lattice deformation and the energy density W($\underline{F}$) by the Cauchy-Born rule through either "black box" (subroutine) in the QC or analytical/semi-analytical method, we must be familiar with the quantitative relationships of the displacement vector $\overline{u}$ with $\underline{F}$, $\underline{\varepsilon}$ as well as the Green strain tensor $\underline{E}$. These will be introduced in detail in the next section.

## 6.8    More Discussions about Deformation Gradient and the Cauchy-Born Rule

### 6.8.1    Mathematical Definition of Deformation Gradient $\underline{F}(\overline{X})$

It is seen that the deformation gradient $\underline{F}(\overline{X})$ is important in the simulation of the continuum deformation. This function depends on the deformation and motion status of the neighborhood of the material point $p(\overline{X})$ in the reference configuration which transforms to the new position $p'(\overline{x})$ in the deformation configuration as shown in Figure 6.1. It is seen that the material point p moves from $\overline{X}$ to $\overline{x}$ through the displacement vector $\overline{u}$. To describe the deformation gradient in the neighborhood around material point p, a differential line segment $d\overline{X}$ in the neighborhood of $\overline{X}$ is taken. During the deformation process from p $(\overline{X})$ to $p'(\overline{x})$ the differential length $d\overline{X}$ is changed to another differential segment $d\overline{x}$ in the neighborhood around $p'(\overline{x})$. The value and direction of $d\overline{x}$ relative to those of $d\overline{X}$ in the reference configuration indicates the deformation intensity and orientation distribution, thus the deformation gradient $\underline{F}(\overline{X})$ can be defined by the relation as follows:

$$d\overline{x} = \underline{F}\, d\overline{X} \tag{6.7a}$$

or in a component form as

$$d\overline{x}_i = F_{iJ} d\overline{X}_j \tag{6.7b}$$

If the coordinates before and after deformation have the same Cartesian coordinates, then we have

$$dx_i = F_{iJ} dX_J \ldots \tag{6.7c}$$

where $dx_i, dX_I$ are components of $d\overline{x}$ and $d\overline{X}$ with the subscripts varying between 1, 2 and 3; convention summation is assumed for the repeat substrate index over its range of variation. As an example, since in (6.7c) the repeatable subscript is J, its value should be repeated three times from 1 to 3, thus the summation is expressed as follows:

$$dx_i = F_{i1} dX_1 + F_{i2} dX_2 + F_{i3} dX_3 \quad (i = 1, 2, 3) \tag{6.7d}$$

For the case of $i = 1$, one has

$$dx_1 = F_{11} dX_1 + F_{12} dX_2 + F_{13} dX_3$$

On the other hand, we can express $dx_i$ by differential relationship as follows:

$$dx_i = \frac{\partial x_i}{\partial X_1} dX_1 + \frac{\partial x_i}{\partial X_2} dX_2 + \frac{\partial x}{\partial X_3} dX_3$$

Comparing the last equation with (6.7d) we obtain

$$F_{ij} = \frac{\partial x_i}{\partial X_j} \quad i = 1 \ldots 3; \quad j = 1 \ldots 3 \tag{6.7e}$$

A complete form of $\underline{F}$ can be obtained by the differential relationship between the three components $dx_1$, $dx_2$, $dx_3$ of the vector $d\bar{x}$ and three components $dX_1$, $dX_2$, $dX_3$ of the original segment vector $d\bar{X}$. There are nine such relationships in all. Using (6.7e) for $d\bar{X} = [dX_1, dX_2, dX_3]^T$ arranged as a matrix form, the deformation gradient tensor $\underline{F}$ can be expressed as a matrix form as follows:

$$\underline{F} \equiv \begin{bmatrix} F_{11} & F_{12} & F_{13} \\ F_{21} & F_{22} & F_{23} \\ F_{31} & F_{32} & F_{33} \end{bmatrix} \equiv \begin{bmatrix} \dfrac{\partial x_1}{\partial X_1} & \dfrac{\partial x_1}{\partial X_2} & \dfrac{\partial x_1}{\partial X_3} \\ \dfrac{\partial x_2}{\partial X_1} & \dfrac{\partial x_2}{\partial X_2} & \dfrac{\partial x_2}{\partial X_3} \\ \dfrac{\partial x_3}{\partial X_1} & \dfrac{\partial x_3}{\partial X_2} & \dfrac{\partial x_3}{\partial X_3} \end{bmatrix} \tag{6.7f}$$

In addition, the nine entrances in the first matrix consist of nine components which are used for index calculation. These two types of expressions of $\underline{F}$ are used for both matrix and index product of the Cauchy-Born rule later. If $d\bar{X}$ denotes a column matrix, i.e., $d\bar{X} = [dX_1, dX_2, dX_3]^T$, then (6.7a) can be understood as the matrix product of $d\bar{X}$ with $\underline{F}$ defined in (6.7f) as follows:

$$d\bar{x} = \underline{F} d\bar{X} \tag{6.7g}$$

## 6.8.2 Determination of Lattice Vectors and Atom Positions by the Cauchy-Born Rule through Deformation Gradient $\underline{F}$

It has been seen from the last section that in continuum mechanics, deformation gradient $\underline{F}$ is defined by the differential relationships between infinitesimal line segment dX and dx, respectively, before and after deformation. The Cauchy-Born rule assumes that lattice vectors are infinitesimal quantities so $\underline{F}$ can be used for measuring lattice deformation.[35] Specifically, the rule states that lattice vector $\bar{a}$ after deformation can be expressed by $\bar{a}$ through $\underline{F}$ as $\bar{a} = \underline{F}\bar{A}$ shown in (6.4r). It can be conducted either by matrix product or by index operation as follows. To do it by matrix, equations (6.4q) and (6.4s) are expressed by matrix forms as

$$\bar{A} = [A_1, A_2, A_3]^T \tag{6.7h}$$

$$\bar{a} = [a_1, a_2, a_3]^T \tag{6.7i}$$

Substituting (6.7h), (6.7i) and (6.7f) into (6.4r) we have

$$\begin{bmatrix} a_1 \\ a_2 \\ a_3 \end{bmatrix} = \begin{bmatrix} F_{11} & F_{12} & F_{13} \\ F_{21} & F_{22} & F_{23} \\ F_{31} & F_{32} & F_{33} \end{bmatrix} \begin{bmatrix} A_1 \\ A_2 \\ A_3 \end{bmatrix} \tag{6.7j}$$

The matrix product for each $a_i$ can directly give the index operation as

$$a_i = F_{iJ}A_J = F_{i1}A_1 + F_{i2}A_2 + F_{i3}A_3 \quad (i = 1, 2, 3) \tag{6.7k}$$

Here, the last equation is obtained by the summation convention over the repeat index J of the index components.

In reality, the lattice vector component as shown in Figure 6.4 is with finite length, thus the Cauchy-Born rule is an approximate rule in general. It is accurate under the condition that the involved crystal lattice is subject to uniform deformation. The two inset charts with atom arrangement in Figure 6.1 illustrate this homogeneous deformation pattern. The one around point p is with original atom structure, the one around point p' is deformed but the atom deformation pattern is uniform.

As soon as the lattice vectors $\overline{A}(\overline{A}_1, \overline{A}_2, \overline{A}_3)$ and $\overline{a}(\overline{a}_1, \overline{a}_2, \overline{a}_3)$ are determined, the position vector $\overline{r}^0$ and $\overline{r}$ of any node (atom) at atomic lattice before and after deformation can be determined as follows:

$$\overline{r}^0 = l\overline{A}_1 + m\overline{A}_2 + n\overline{A}_3) \quad \text{(before deformation)} \tag{6.7l}$$

$$\overline{r} = l\overline{a}_1 + m\overline{a}_2 + n\overline{a}_3) \quad \text{(after deformation)} \tag{6.7m}$$

Using $\underline{F}$ times both sides of (6.7l) one obtains

$$\underline{F}\overline{r}^0 = l\underline{F}\overline{A}_1 + m\underline{F}\overline{A}_2 + n\underline{F}\overline{A}_3) = l\overline{a}_1 + m\overline{a}_2 + n\overline{a}_3 \tag{6.7n}$$

The comparison between (6.7n) and (6.7m) shows that

$$\overline{r} = \underline{F}\overline{r}^0 \tag{6.7o}$$

This equation shows any lattice vector before and after deformation can also be connected by the deformation gradient. In (6.7n), l, m and n are the node integer numbers which denote repeated times of the lattice vector along corresponding directions. The node is fixed in the crystal lattice, thus these numbers are fixed. During the deformation the lattice vector changes from $\overline{A}(\overline{A}_1, \overline{A}_2, \overline{A}_3)$ to $\overline{a}(\overline{a}_1, \overline{a}_2, \overline{a}_3)$ as determined by the Cauchy-Born rule. In general, these two sets of lattice vectors are not coincident, as shown in Figure 6.1.

For some complex crystals, such as the HCP lattice and some ceramic crystals, there are two or more atoms at one node of the lattice. For these atoms, additional position vectors $\overline{\xi}_k (k = 1, 2 \ldots n)$ are necessary to determine the atom positions in the undeformed (i.e., reference) configuration. Here n denotes the number of these atoms which share the same lattice node. Therefore, the real atom positions related to the undeformed frame node is $\overline{R} + \overline{\xi}_k (k = 1,2\ldots n)$. Applying the Cauchy-Born rule to this system, the atom positions after deformation can be obtained as:

$$\overline{r}_k = l\overline{a}_1 + m\overline{a}_2 + n\overline{a}_3 + \overline{\varsigma}_k \quad (k = 1 \ldots n) \tag{6.7p}$$

with

$$\overline{\varsigma}_k = \underline{F}\overline{\xi}_k \quad (k = 1, 2, \ldots, n) \tag{6.7q}$$

## 6.8.3 Physical Explanations of Components of Deformation Gradient

While the deformation gradient $\underline{F}(\overline{X})$ as a whole quantity describes the deformation and motion of the material in the neighborhood around the point $p(\overline{X})$ or simply $\overline{X}$, the nine components themselves or their combinations have meanings to denote strain or rotation in the differential area surrounding $\overline{X}$.

Let us first take uniaxial stretching $\lambda$ shown in Figure 5.12 as an example where only a horizontal axis x is used. The stretching $\lambda$ can be considered as the stretching of the continuum. Since the deformation is only along the $X_1$ direction, which are independent of $X_2$ and $X_3$. Therefore, $\frac{\partial x_1}{\partial X_2} = \frac{\partial x_1}{\partial X_3} = 0$, and the deformation gradient is uniquely determined by $dx_1/dX_1$, i.e., by the first diagonal term of the matrix of (6.7f). Suppose further the original length of a segment can be approximately taken as the lattice constant "$a_0$", i.e., $dX_1 = a_0$, and after deformation, the length changes to $dx_1 = \lambda a_0$ due to the uniformly stretching $\lambda$, then according to (6.7f) the deformation gradient can be expressed as follows:

$$F = F_{11} = dx_1/dX_1 = \frac{\lambda a_0}{a_0} = \lambda \tag{6.7r}$$

This result indicates that the first diagonal term of the deformation gradient $\underline{F}$ denotes the stretching along the x-direction. In more general cases, it can be proved that the other two terms in the diagonal of the matrix denote the stretch ratio $\lambda_2$ and $\lambda_3$ along the y- and z-direction.

For small deformation and with the same Cartesian coordinate system to express the position vector before and after deformation, it is not difficult to prove (Homework 6.1) that the summation of corresponding off-diagonal terms in (6.7f) determines engineering shear strain of material element, respectively, within the XOZ, YOZ and XOY coordinate planes as follows:

$$\gamma_{13} = \gamma_{xz} = \frac{\partial x_1}{\partial X_3} + \frac{\partial x_3}{\partial X_1} \tag{6.7s}$$

$$\gamma_{23} = \gamma_{yz} = \frac{\partial x_2}{\partial X_3} + \frac{\partial x_3}{\partial X_2} \tag{6.7t}$$

$$\gamma_{12} = \gamma_{xy} = \frac{\partial x_1}{\partial X_2} + \frac{\partial x_2}{\partial X_1} \tag{6.7u}$$

Here, both xyz and 123 index are used for the Cartesian coordinates with the correspondence of 1-x, 2-y and 3-z.

In addition, the subtraction of corresponding off-diagonal terms in (6.7f) determines the average rotation of material element, respectively, within the XOZ, YOZ and XOY coordinate planes as follows:

$$\omega_{13} = \omega_{xz} = \frac{\partial x_1}{\partial X_3} - \frac{\partial x_3}{\partial X_1} \tag{6.7v}$$

$$\omega_{23} = \omega_{yz} = \frac{\partial x_2}{\partial X_3} - \frac{\partial x_3}{\partial X_2} \tag{6.7w}$$

$$\omega_{12} = \omega_{xy} = \frac{\partial x_1}{\partial X_2} - \frac{\partial x_2}{\partial X_1} \tag{6.7x}$$

### 6.8.4    Expressions of $\underline{F}$ and $\underline{\varepsilon}$ Components in Terms of Displacement Vector

Now is a good moment to complete the study on deformation gradient $\underline{F}$ by illustrating its intrinsic relationship with displacement vector $\bar{u}$. It is seen from Figure 6.1 that the displacement vector can be expressed as follows:

$$\bar{u} = \bar{x} - \bar{X} \tag{6.8a}$$

Projecting these vectors on the same Cartesian coordinate system, say, $\bar{A}_1, \bar{A}_2, \bar{A}_3$ of the undeformed configuration one has its component relationship as follows:

$$u_i = x_i - X_i \quad (i = 1, 2, 3) \tag{6.8b}$$

or

$$dx_i = dX_i + du_i \quad (i = 1, 2, 3) \tag{6.8c}$$

Substituting this equation into (6.7h) we have

$$dx_1/dX_1 = 1 + \frac{du_1}{dX_1} = 1 + \varepsilon_{11} \tag{6.8d}$$

where $\varepsilon_{11} = \varepsilon_{xx} = du_1/dX_1$ is the normal strain component of strain tensor $\varepsilon_{ij}$ along the X-direction. For all normal strain components along the three perpendicular directions 1, 2, 3 (or X, Y, Z) we have:

$$\varepsilon_{ii} = \frac{du_i}{dX_i} \quad (i = 1, 2, 3, \text{no sum for repeat i}) \tag{6.8e}$$

The shear strain component $\varepsilon_{ij}$ of strain tensor is half the engineering strain $\gamma_{ij}$ defined in (6.7s–u). After submitting (6.8c) into these equations, the shear strain, say $\varepsilon_{13}$, can be rewritten by the displacement component as:

$$\varepsilon_{13} = \gamma_{13}/2 = \frac{1}{2}\left(\frac{\partial x_1}{\partial X_3} + \frac{\partial x_3}{\partial X_1}\right) = \frac{1}{2}\left(\frac{\partial(X_1 + u_1)}{\partial X_3} + \frac{\partial(X_3 + u_3)}{\partial X_1}\right) = \frac{1}{2}\left(\frac{\partial u_1}{\partial X_3} + \frac{\partial u_3}{\partial X_1}\right) \tag{6.8f}$$

It is easy to verify that all normal and shear components of strain tensor $\varepsilon_{ij}$ can be written in the following unified form:

$$\varepsilon_{ij} = \frac{1}{2}\left(\frac{\partial u_i}{\partial X_j} + \frac{\partial u_j}{\partial X_i}\right) \tag{6.8g}$$

Substituting equation (6.8b) into (6.7e), the components of the deformation gradient can be obtained by:

$$F_{i,J} = \partial x_i/\partial X_J = \partial X_I/\partial X_J + \partial u_i/\partial X_J \tag{6.8h}$$

The first term on the right side of the last equation is zero when the number of subscript I is not equal to the number J because $X_I$ and $X_J$ are both independent variables and there is no relationship between each other. If, however, I = J, then it equals 1. This property can be expressed by a Kronecker symbol $\delta_{IJ}$ in which

$$\delta_{IJ} = 0 \text{ if } I \neq J \tag{6.8i}$$

$$\delta_{IJ} = 1 \text{ if } I = J \tag{6.8j}$$

thus

$$\partial X_I/\partial X_J = \delta_{IJ} \tag{6.8k}$$

Further, using the subscript symbol "i,J" to express the partial derivative of coordinate $x_i$ in the deformation configuration with respect to the coordinates $X_j$ in the reference configuration, we have the following simple expressions:

$$x_{i,J} = \partial x_i/\partial X_J, \tag{6.8l}$$

$$u_{i,J} = \partial u_i/\partial X_J \tag{6.8m}$$

Substituting (6.8k–m) into (6.8h), we have

$$F_{i,J} = \partial x_i/\partial X_J = x_{i,J} = \delta_{iJ} + u_{i,J} \tag{6.8n}$$

If no deformation is present, $u_{i,J}$ is zero. Therefore, the deformation gradient matrix becomes an identity matrix with all diagonal entrances being 1 and off-diagonal entrances 0. Note that (6.8n) is only valid when the same Cartesian coordinate system is used before and after deformation. However, (6.7f) of the definition of deformation gradient and (6.4r) of the Cauchy-Born rule are valid in the general coordinate system shown in Figure 6.1.

## 6.8.5  The Relationship Between Deformation Gradient, Strain and Stress Tensors

After deformation gradient $\underline{F}(\overline{X})$ is determined, the full material deformation status around the neighborhood of point $p(\overline{X})$ is determined, thus other strain measures for deformation intensity and orientation must have a fixed relationship with $\underline{F}(\overline{X})$.

For the application below we specifically list the relationships between Green strain tensor $\underline{E}(\overline{X})$ and $\underline{F}(\overline{X})$ with various forms. In literature, Green strain tensor $\underline{E}(\overline{X})$ is frequently written in terms of a matrix form of $\underline{F}$ as follows:

$$\underline{E} = \frac{1}{2}(\underline{F}^T\underline{F} - \underline{I}) \tag{6.9a}$$

where the superscript symbol T denotes transpose of the matrix, and $\underline{I}$ is the second-order identity tensor or $3 \times 3$ unit matrix with entrance 1 in the diagonal and 0 in off-diagonal positions. Substituting (6.7f) of $\underline{F}$ into (6.9a), the following component form of the Green strain tensor can be obtained:

$$E_{IJ} = \frac{1}{2}\left(F_{KI}F_{KJ} - \delta_{IJ}\right) = \frac{1}{2}\left(\frac{\partial x_K}{\partial X_I}\frac{\partial x_K}{\partial X_J} - \delta_{IJ}\right) \tag{6.9b}$$

where the convention of summation over dummy index K by repeating it from 1 to 3 is assumed, and $F_{KI} = \frac{\partial x_K}{\partial X_I}$ is the component of $\underline{F}$ as shown in (6.7f).

Further, substituting (6.8n) into (6.9b) the component $E_{IJ}$ of the Green strain tensor can also be expressed by the component of displacement vector $\overline{u}$ as follows:

$$E_{IJ} = \frac{1}{2}\left(\frac{\partial u_I}{\partial X_J} + \frac{\partial u_J}{\partial X_I} + \frac{\partial u_K}{\partial X_I}\frac{\partial u_K}{\partial X_J}\right) \tag{6.9c}$$

If the simplified symbols of (6.8l–m) are used it can also be written as

$$E_{IJ} = \frac{1}{2}\left(u_{I,J} + u_{J,I} + u_{K,I}u_{K,J}\right) \tag{6.9d}$$

The literature also uses the so-called right Cauchy-Green strain tensor $\underline{C}$ whose definition by matrix form is

$$\underline{C} = \underline{F}^T\underline{F} \tag{6.9e}$$

Actually, it is a part of the Green strain tensor $\underline{E}$; from (6.9a) we have

$$\underline{E} = \frac{1}{2}(\underline{C} - \underline{I}) \tag{6.9f}$$

The relationship between the stress $\sigma_{ij}$ and engineering strain $\varepsilon_{ij}$ for an elastic system with small deformation is expressed through energy density W as follows:

$$\sigma_{ij} = \frac{\partial W}{\partial \varepsilon_{ij}} \tag{6.9g}$$

For finite deformation, the stress-strain relationship is written with Green tensor $\underline{E}$ as:

$$\underline{T} = \frac{\partial W}{\partial \underline{E}} \tag{6.9h}$$

or

$$T_{ij} = \frac{\partial W}{\partial E_{ij}} \tag{6.9i}$$

where $T$ is the second Piola-Kirchhoff stress tensor, which is the conjugate stress to the Green strain tensor taking undeformed configuration as the reference configuration to define the stress.

For 1D cases, equation (6.9g) is easily verifiable, substituting the expression $W(\varepsilon) = \frac{1}{2}E\varepsilon^2$ of (6.1e) into that equation and making the first derivative we obtain $\sigma = E\varepsilon$. Further, if we make the second derivative of W with respect to strain $\varepsilon$, Young's modulus is obtained, i.e.,

$$E = \frac{d^2 W}{d^2 \varepsilon}$$

For an anisotropic elastic material, Young's modulus is replaced by a tensor of elastic constants, or moduli $C_{ijkl}$ and we have

$$\sigma_{ij} = C_{ijkl}\varepsilon_{kl} \tag{6.9j}$$

$$C_{ijkl} = \partial^2 W / \partial \varepsilon_{ij}\partial \varepsilon_{kl} \tag{6.9k}$$

where stress $\sigma_{ij}$ and strain $\varepsilon_{kl}$ are second-order symmetric tensors and $C_{ijkl}$ are the fourth-order tensors. Their calculation follows the tensor index calculation including the summation convention rule of repeated index. More calculation examples and the property of the elastic constant $C_{ijkl}$ are discussed in detail in Ref. [36] for materials with different anisotropy.

## 6.9 Analytical/Semi-analytical Methods Across Atom/Continuum Scales Based on the Cauchy-Born Rule

Different from molecular dynamics simulations that trace every atom's movement, analytic/semi-analytical solutions based on the Cauchy-Born rule and other modified rules focus on the collective identity of a group of atoms, thus continuum behavior can be described. The Cauchy-Born rule indicates that all the uniformly deformed atoms change their configuration from non-deformed to deformed according to its mapping rule of (6.4r) or (6.7o). The following paragraphs will discuss how to derive the analytic solutions, such as constitutive laws and stress, by applying the Cauchy-Born rule to a centro-symmetric structure.

### 6.9.1 Application of the Cauchy-Born Rule in a Centro-symmetric Structure

In the centro-symmetric atomic structure with conjugated bonds in the positive and negative direction of every atom, the equilibrium condition is satisfied. Since the Cauchy-Born rule ensures the centro-symmetric structure after the deformation, atomic forces arise simultaneously in the positive and negative direction and mutually cancel each other, thus equilibrium of the atoms is valid. For non-centro-symmetric structures, however, the Cauchy-Born rule cannot guarantee the equilibrium and an additional condition is needed, see discussion in Section 6.10.3.1.

According to the Cauchy-Born rule, the strain energy density W is the stored energy from atomic bonds in a unit volume and can be obtained by the interatomic potential V:

$$W = \frac{1}{2} \frac{\sum_{j=1}^{n} V(r_{ij}, \theta_{ijk}; k \neq i, k \neq j)}{\Omega_0} \tag{6.9l}$$

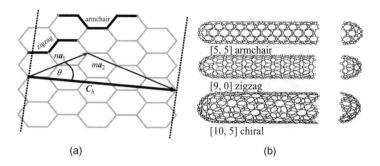

**Figure 6.7**   Chirality vector $C_h = na_1 + ma_2$ and three types of nanotubes: armchair, zigzag and chiral (Reproduced from Dresselhaus, M. S., Dresselhaus, G., and Saito, R. (1995) Physics of carbon nanotube. *Carbon*, 33, 883, Elsevier)

where the factor one half results from equipartition of bond energy by the pair atoms, and $n$ is the total number of neighbor atoms in the cut-off radius. For example, in a hexagonal lattice structure $n$ equals 3 because any atom $i$ is connected with three neighboring atoms $j$ (see Figure 6.7 below). $\Omega_0$ is the average volume of an atom, $r_{ik}$ is introduced together with $r_{ij}$ to form the angle $\theta_{ijk}$ between them, thus the above equation accounts for the effect of second-nearest-neighbor atoms via the bond angle $\theta_{ijk}$. Several potentials introduced in Chapters 2 and 3 can be used for the interatomic potential. For instance, the Brenner potential function in (3.9a) is adopted in the study of carbon atoms[37] in which the $\theta_{ijk}$ dependence is through $B_{ij}$ as shown in equations (3.9c, d).

$$V_{ij} = V_R(r_{ij}) - B_{ij}(\theta_{ijk}) V_A(r_{ij}), r_{ij} = |\bar{r}_{ij}| \tag{6.9m}$$

Once W is known, the stress can be determined by equation (6.9g) or (6.9h). Subsequently, the constitutive equations of continuous medium described by the atomic potential can be obtained to solve various problems. The keys for determining energy density W, however, are how to determine the interatomic length $r_{ij}$, $r_{ik}$ and the angle $\theta_{ijk}$ used in the interatomic potential V (see (6.9l)) after deformation. This important topic will be described in the next section.

## 6.9.2   Determination of Interatomic Length $r_{ij}$ and Angle $\theta_{ijk}$ of the Crystal after Deformation by the Cauchy-Born Rule

Let us start by determining the length a of the lattice vector $\bar{a}$ after deformation. Obviously the length or the vector norm $\|\bar{a}\|$ can be determined by the square root of the own product of vector $\bar{a}$ as

$$a = \|\bar{a}\| = \sqrt{\bar{a}^T \bar{a}} = \sqrt{(\underline{F}\bar{A})^T (\underline{F}\bar{A})} = \sqrt{\bar{A}^T \underline{F}^T \underline{F}\bar{A}} = \sqrt{\bar{A}^T \underline{C}\bar{A}} \tag{6.9n}$$

Here, the Cauchy-Born rule (6.4r) and the definition (6.9e) of $\underline{C}$ are used. During the derivation the transpose rule of two matrix product is used so that $(\underline{F}\bar{A})^T = \bar{A}^T \underline{F}^T$.

Suppose $\bar{r}_{ij}^0$ is a vector directed from atom i to atom j in the undeformed (reference) configuration, its length is $r_{ij}^0$ and unit vector is $\bar{n}_{ij}^0$, we have

$$\bar{r}_{ij}^0 = r_{ij}^0 \bar{n}_{ij}^0 \tag{6.9o}$$

After the lattice deformation, $\bar{r}_{ij}^0$ becomes $\bar{r}_{ij}$ and its length is elongated from $r_{ij}^0$ to $r_{ij}$. Now, we would like to determine the elongated value $r_{ij}$ in terms of $r_{ij}^0$, unit vector $\bar{n}_{ij}^0$ and the strain tensor $\underline{E}$. Following (6.7o) and the Cauchy-Born rule (6.4r), we have

$$r_{ij} = \|\bar{r}_{ij}\| = \sqrt{\bar{r}_{ij}^T \bar{r}_{ij}} = \sqrt{(\underline{F}\bar{r}_{ij}^0)^T (\underline{F}\bar{r}_{ij}^0)} = \sqrt{\bar{r}_{ij}^{0T} \underline{F}^T \underline{F} \bar{r}_{ij}^0} = \sqrt{\bar{r}_{ij}^{0T} \underline{C} \bar{r}_{ij}^0}$$
$$= r_{ij}^0 \sqrt{\bar{n}_{ij}^{0T} \underline{C} \bar{n}_{ij}^0} \tag{6.9p}$$

Note that here the subscript ij is not the tensor index but denotes quantities related to atoms i and j. In the derivation of the fifth and last equation above, (6.9e) and (6.9o) are, respectively, used. From (6.9f) we have

$$\underline{C} = 2\underline{E} + \underline{I} \tag{6.9q}$$

Substituting this result into the last equation of (6.9p) we have

$$r_{ij} = r_{ij}^0 \sqrt{\bar{n}_{ij}^{0T} (2\underline{E} + \underline{I}) \bar{n}_{ij}^0} \tag{6.9r}$$

In index form, this equation can be written as

$$r_{ij} = r_{ij}^0 \sqrt{(\delta_{\alpha\beta} + 2E_{\alpha\beta}) n_\alpha n_\beta} \tag{6.9s}$$

where $n_\alpha$ is the direction cosine of the unit vector of $\bar{r}_{ij}^0$ with respect to the Cartesian coordinates in the undeformed (reference) configuration. $\delta_{\alpha\beta}$ is the Kronecker symbol.

It is easy to prove

$$\delta_{\alpha\beta} n_\alpha n_\beta = 1 \quad \text{or} \quad \bar{n}_{ij}^{0T} \underline{I} \bar{n}_{ij}^0 = 1 \tag{6.9t}$$

For the first case, the convention summation over dummy index $\alpha$, $\beta$ should be repeated three times from 1 to 3 and the property of $\delta_{\alpha\beta}$ given in (6.8i, j) should be used. For the latter case, this is because the product of unit matrix $\underline{I}$ times any matrix including $\bar{n}_{ij}^0$ will equal the matrix, and by the definition of unit vector we have $\bar{n}_{ij}^{0T} \bar{n}_{ij}^0 = 1$. Substituting this result into (6.9s) we finally obtain

$$r_{ij} = \lambda r_{ij}^0 = r_{ij}^0 \sqrt{1 + 2\bar{n}_{ij}^{0T} \underline{E} \bar{n}_{ij}^0} \tag{6.9u}$$

or

$$r_{ij} = \lambda r_{ij}^0 = r_{ij}^0 \sqrt{1 + 2E_{\alpha\beta} n_\alpha n_\beta} \tag{6.9v}$$

where $\lambda$ is the stretch ratio and is expressed as

$$\lambda = \sqrt{1 + 2\bar{n}_{ij}^{0T} \underline{E} \bar{n}_{ij}^0} = \sqrt{1 + 2E_{\alpha\beta} n_\alpha n_\beta} \tag{6.9w}$$

The angle $\theta$ between the new lattice vector $\bar{a}$ and another vector $\bar{b}$ is determined by the product such as (6.1g). In matrix form, the vector dot product is expressed by the product of the transpose of the first matrix times the second matrix. The vector $\bar{b} = [b_1, b_2, b_3]$ is obtained during the deformation by the vector $\overline{B}$ defined in the undeformed crystal configuration. Its transformation still follows the Cauchy-Born rule. Thus, we have

$$\cos\theta = \frac{\bar{a}^T \bar{b}}{||a||\,||b||} = \frac{(\underline{F}\overline{A})^T (\underline{F}\overline{B})}{||a||\,||b||} = \frac{\overline{A}^T (\underline{F}^T \underline{F})\overline{B}}{||a||\,||b||} = \frac{\overline{A}^T \underline{C}\overline{B}}{||a||\,||b||} \tag{6.9x}$$

Again, here (6.4r) and (6.9e) are used.

For an angle $\theta_{ijk}$ between line $\bar{r}_{ij}$ and line $\bar{r}_{ik}$, the corresponding vectors in the reference configuration are $\overline{A} = \bar{r}_{ij}^0 = r_0 \bar{n}_{ij}$ and $\overline{B} = \bar{r}_{ik}^0 = r_0 \bar{n}_{ik}$ and $||a|| = |r_{ij}|$ $||b|| = |r_{ik}|$. Substituting these relations into (6.9x) and following the derivation of (6.9r), it is easy to obtain

$$\cos\theta_{ijk} = \frac{r_0^2 (\bar{n}_{ij}^{0T} (2\underline{E} + \underline{I})\bar{n}_{ik}^0)}{|r_{ij}||r_{ik}|} \tag{6.9y}$$

or in index form

$$\cos\theta_{ijk} = \frac{r_0^2 (\delta_{\alpha\beta} + 2E_{\alpha\beta})n_\alpha^{(1)} n_\beta^{(2)}}{|r_{ij}||r_{ik}|} \tag{6.9z}$$

where the superscripts (1) and (2) denote the two bonds $\bar{r}_{ij}$ and $\bar{r}_{ik}$ in the bond angle $\theta_{ijk}$; $n_\alpha$ $(\alpha = 1 - 3)$ and $n_\beta$ $(\beta = 1 - 3)$ are, respectively, the direction cosine of the unit vector along these two bonds.

### 6.9.3   A Short Discussion on the Precision of the Cauchy-Born Rule

The Cauchy-Born rule is used to determine the Bravais frame node position of any atom in the crystal lattice by determining the lattice vector $\bar{a}$ through the deformation gradient. This requires that the deformation field is homogeneous. In reality, the deformation gradient $\underline{F}$ may be inhomogeneous which affects the accuracy of the calculation of the atom positions and, in turn, affects the accuracy of calculation of atomistic energy and energy density W.

The offset from the calculated atom position to the real atom position is determined by $\text{Grad}(\underline{F})$ which is the gradient of the function $\underline{F}(\bar{r})$. If the gradient is smooth enough at the non-local atomic interaction region defined, say, by the cutoff radius of the potential, i.e., if $\text{Grad}(\underline{F}) \ll 1$ in that region, the obtained strain energy density W has sufficient accuracy.[34] For the far field, the displacement derived from deformation gradient $\underline{F}$ deviates from the real one at an increasing rate as it moves further away from the target point $\bar{r}$. However, this error induces very little error to the energy estimation because the short-distance potential function is usually employed. But, if a long-term potential such as the Coulomb potential is adopted, the results would be different.

---

**Homework**

(6.6) Prove for small deformation and using the same Cartesian coordinate system to investigate the deformation process, that shear strain $\gamma_{xy}$ can be expressed by off-diagonal terms $\frac{\partial x_1}{\partial X_2}$, $\frac{\partial x_2}{\partial X_1}$ of deformation gradient **F** in (6.1f) as shown in (6.7u).

Hint: Engineering shear strain $\gamma_{xy}$ is defined by the angle change measured by radian for a 90° angle in the xoy plane before deformation. It can be shown that $\frac{\partial x_1}{\partial X_2}$ denotes the distortion angle (radian) by shearing due to displacement $u_1$ alone.

## 6.10 Atomistic-based Continuum Model of Hydrogen Storage with Carbon Nanotubes

### 6.10.1 Introduction of Technical Background and Three Types of Nanotubes

Hydrogen is the cleanest energy. Vehicles propelled by electric motors supplied by oxygen/hydrogen cells such as the one described in Section 3.7.1, may generate negligible environmental pollution. One challenge is the storage and release of hydrogen in vehicles.[38]

#### 6.10.1.1 Superior Property of Carbon Nanotubes

Carbon nanotubes (CNTs) display superior mechanical, electrical and thermal properties and have many potential applications. One of them is hydrogen storage because carbon is a good adsorbent for gases; and CNTs are microporous structures with high specific surface, and have the potential to adsorb hydrogen in their nanostructures.[39–43] The adsorption capability is measured by a unit of quantity of gas with respect to a unit of quantity of adsorbent.

The goals of hydrogen storage by the US Department of Energy are 6.5 wt% and 62 kgH$_2$/m$^2$.[39] However, experiments have shown large scattering in the achievable hydrogen storage, ranging from 0.25 to 11 wt%. This scattering is caused by many factors such as defects in CNTs, and types of CNTs (e.g., open or closed end of the nanotube).

After the introduction of three types of nanotubes, in the following six sub-sections we will show how a continuum model for CNT can be developed based on the Cauchy-Born rule to transfer the atomistic energy to energy density W and how the atomic-based continuum model can be used for analysis of deformation and failure of different types of CNTs in view of the goals of hydrogen storage by the US Department of Energy.[44] Moreover, in Section 6.11.1 we will use six items to summarize this kind of method and further use them to investigate the mechanical, electrical and thermal conductivity of CNTs.

#### 6.10.1.2 Chirality and Three Types of Nanotubes

There are three types of nanotubes including armchair, zigzag and chiral as shown in Figure 6.7, defined by their chirality. The chirality vector $C_h$ is defined as

$$C_h = na_1 + ma_2 \tag{6.10a}$$

where $a_1$ and $a_2$ are grapheme crystal unit vectors and $n$ and $m$ are integral numbers. A carbon nanotube can be thought to be formed by rolling the grapheme layer along the $C_h$ direction and making the two dashed lines in Figure 6.7a coincident. Through this imaginary forming process, it is seen that the parameters $n, m$ of the nanotube chirality can describe the structural characteristics of a single-wall CNT. In other words, if the chirality vector $C_h$ is determined, the structure and its parameters are all determined. Specifically, when $n = m$ the chirality angle between chirality vector $C_h$ and crystal unit vector $a_1$ is $\theta = 30°$; the carbon distribution shows armchair shape and the CNT is the armchair type. If $m = 0$, the angle $\theta = 0°$, the carbon atom distribution shows zigzag shape and the CNT is the zigzag type. When $0° < \theta < 30°$, the mesh of the carbon atom distribution shows chiral and the CNT is the chiral type. In the following section, the CNT armchair type will be discussed as an example. The basic structural parameters of single-wall CNT are given in Table 6.1.

### 6.10.2 Interatomic Potentials Used for Atom/Continuum Transition

The second-generation interatomic potential of Tersoff type proposed by Brenner et al. is used for covalent bonds between carbon atoms on the CNT.[46] This potential was discussed in detail in Section

**Table 6.1** Structural parameters of single-wall carbon nanotube based on the Chinese dissertation of D. L. Shi, Tsinghua University, 2005.

| Symbol | Name | Formulas | Explanation |
|--------|------|----------|-------------|
| $a_1$, | Crystal unit vector | $x = \dfrac{\sqrt{3}}{2}a, \quad y = \dfrac{1}{2}a$ | $a = \sqrt{3}a_{C\text{-}C} = 2.46$ Å $a_{C\text{-}C}$-spacing between carbon atoms |
| $a_2$ | Crystal unit vector | $x = \dfrac{\sqrt{3}}{2}a, \quad y = -\dfrac{1}{2}a$ | |
| $C_h$ | Chirality vector | $C_h = na_1 + ma_2$ | $0 \le |m| \le n$ |
| $d_t$ | CNT diameter | $d_t = \dfrac{|C_h|}{\pi} = \dfrac{a\sqrt{m^2 + n^2 + mn}}{\pi}$ | |
| $\theta$ | CNT angle | $\theta = \arccos\dfrac{2n+m}{2\sqrt{m^2+n^2+mn}}$ | $0° \le |\theta| \le 30°$ |

3.8.2 and can be recalled as[44]

$$V_{ij}(r_{ij}, \theta_{ijk}; k \ne i, j) = V_R(r_{ij}) - B_{ij}(\theta_{ijk})V_A(r_{ij}) \qquad (6.11a)$$

where $V_R(r) = (1 + Q/r)Ae^{-\alpha r}f_C(r)$ and $V_A(r) = \sum_{n=1}^{3} B_n e^{-\beta_n r}f_C(r)$ are the repulsive and attractive pair terms which depend only on the interatomic distance $r_{ij}$ between atom i and j. A, Q, $\alpha$, $\beta_n$ and $B_n$ are constants, $f_c$ is the cut-off function. $B_{ij}(\theta_{ijk})$ is the multi-body coupling parameter which depends on the bond angle $\theta_{ijk}$ between i-j and j-k via the function G as follows:

$$B_{ij}(\theta_{ijk}) = \left[1 + \sum_{k(\ne i,j)} G(\cos\theta_{ijk})f_C(r_{ik})\right]^{-1/2} \qquad (6.11b)$$

The LJ 6-12 potential $V_{LJ}$ of (2.3a) is used to characterize the van der Waals interactions between a carbon atom and a hydrogen molecule inside the CNT as well as to characterize the interactions between two hydrogen molecules.

$$V(r) = 4\varepsilon\left[\left(\frac{\sigma}{r_{ij}}\right)^{12} - \left(\frac{\sigma}{r_{ij}}\right)^{6}\right] \qquad (2.3a)$$

where $r_{ij}$ is the distance between atoms (and molecules). For hydrogen molecules, the parameters $\varepsilon_{H2\text{-}H2} = 3.11 \times 10^{-3}$ eV and $\sigma_{H2\text{-}H2} = 0.296$ nm.[44] For a carbon atom and a hydrogen molecule, the Lorentz-Berthelot mixing rule (see Section 3.11.1) gives $\varepsilon_{C-H2} = \sqrt{\varepsilon_{C-C}\varepsilon_{H2-H2}} = 2.73 \times 10^{-3}$ eV and $\sigma_{C-H2} = \frac{1}{2}(\sigma_{C-C} + \sigma_{H2-H2}) = 0.319$ nm where $\varepsilon_{C\text{-}C} = 2.39 \times 10^{-3}$ eV and $\sigma_{C\text{-}C} = 0.342$ nm.

## 6.10.3    The Atomistic-based Continuum Theory of Hydrogen Storage

The CNT is modeled as a nonlinear continuum thin shell and the hydrogen inside the tube is represented by the internal pressure p. The nonlinearity results from the large stretch of C-C bonds due to hydrogen storage as characterized by the interatomic potential (6.11a). Zhang et al.[47] and Wu et al.[48] established a continuum theory for CNTs directly from the interatomic potential which accounts for the atomic structure and multi-body atomistic interactions. The atomistic-based continuum theory is briefly introduced in the following.

### 6.10.3.1 The Treatment of Non-centro-symmetric Structure of Nanotubes

In Section 6.9.1 we discussed the centro-symmetric structure where the equilibrium condition is automatically satisfied by mapping the original configuration to the deformed configuration. A schematic of a single-wall nanotube is given in Figure 6.7. A CNT plane plot shown in Figure 6.8 is a non-centro-symmetric structure, after mapping by the Cauchy-Born rule the structure cannot keep its equilibrium. Thus, the first problem for developing the atomistic-based continuum theory is how to determine an equilibrium configuration after mapping by the Cauchy-Born rule. The idea of multi-lattice atoms described in Section 6.8.2 is used to solve the problem. Specifically, the CNT structure with a bottom end tube consisting of hexagonal hollow units shown in the right part of Figure 1.1 is decomposed into two simple Bravais structures, the first consisting of solid circles and the second of open circles, respectively, in the solid and dashed hexagonal frame of Figure 6.8. Each sub-lattice follows the mapping by deformation gradient $\underline{F}$, but the two sub-lattices may have a shift. Suppose we take the solid hexagonal frame as the base whose mapping is $\bar{r}_{ij} = \underline{F}\bar{r}_{ij}^{(0)}$ by the Cauchy-Born rule, then the mapping for the dashed hexagonal one is expressed as

$$\bar{r}_{ij} = \underline{F}\bar{r}_{ij}^{(0)} + \bar{\varsigma} \tag{6.11c}$$

Where $\bar{\varsigma}$ is the shift vector from A$'$ to A as shown in Figure 6.8. The above equation may be written in a more compact form by introducing a variable $\bar{\eta}$ in reference configuration and using (6.7o) as

$$\bar{r}_{ij} = \underline{F}(r_{ij}^0(\bar{n}_{ij}^0 + \bar{\eta})) = r_{ij}^0 \underline{F}(\bar{n}_{ij}^0 + \bar{\eta}) \tag{6.11d}$$

From (6.11c) and (6.11d) and using (6.9o) it results in $\bar{\varsigma} = r_{ij}^0 \underline{F}\bar{\eta}$, or by its inverse we have

$$\bar{\eta} = \frac{(\underline{F})^{-1}\bar{\varsigma}}{r_{ij}^0} \tag{6.11e}$$

Correspondingly in the reference configuration, the two configurations have a corresponding shift. This shift offers a degree of freedom which can be determined by the principle of minimum potential energy.

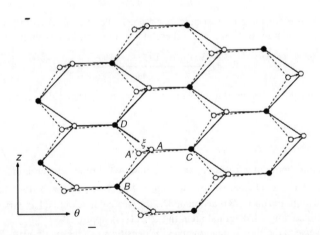

**Figure 6.8** Using vector $\bar{\varsigma}$ from A$'$ to A to link solid and dashed structures (Reproduced from Chen, Y. L., Liu, B., Wu, J., *et al.* (2008) Mechanics of hydrogen storage in carbon nanotubes. *Journal of the Mechanics and Physics of Solids*, 56, 3224, Elsevier)

### 6.10.3.2 Calculation of Atomistic-based Strain Energy Density W

Taking atom A as an example, it is seen from Figure 6.8, the bond vector $\bar{r}_{ij}$, the interatomic length $r_{ij}$ and angle $\theta_{ijk}$ between atom A and its neighboring atoms B and C after deformation are from two different sub-lattices. This fact requires modifications of (6.9v) and (6.9z). Based on the change from the regular Cauchy-Born rule to (6.11d), one just needs to change the $\bar{n}_{ij}^0$ term to $(\bar{n}_{ij}^0 + \bar{\eta})$ in these equations, i.e.,

$$r_{ij} = r_{ij}^0 \sqrt{(\bar{n}_{ij}^0 + \bar{\eta})^T (1 + 2\underline{E})(\bar{n}_{ij}^0 + \bar{\eta})} \qquad (6.11\text{f})$$

or in index form

$$r_{ij} = r_{ij}^0 \sqrt{(\delta_{\alpha\beta} + 2E_{\alpha\beta})(n_\alpha + \eta_\alpha)(n_\beta + \eta_\beta)} \qquad (6.11\text{g})$$

$$\cos\theta_{ijk} = \frac{r_0^2}{|r_{ij}||r_{ik}|}(\delta_{\alpha i} + 2E_{\alpha i})(n_\alpha^{(1)} + \eta_\alpha)(n_i^{(2)} + \eta_i) \qquad (6.11\text{h})$$

where $\eta_\alpha$ ($\alpha = 1,2,3$) are the components of the vector $\bar{\eta}$. It corresponds to the vector $\bar{\varsigma}$; both represent degrees of freedom from the configuration with solid circles to the one with open circle. Actually, $\bar{\eta}$ can be obtained by inverse mapping from vector $\bar{\varsigma}$ into the reference configuration as derived in the last section (see equation (6.11e)).

From (6.11f) to (6.11h) it is seen that the bonding energy $V(\bar{r}_{ij}, \bar{r}_{ik}, \theta_{ijk})$ is essentially the function of Green strain tensor $E_{\alpha,\beta}$ and shift vector $\bar{\eta}$, i.e.,

$$V(\bar{E}, \bar{\eta}) = V(\bar{r}_{ij}, \bar{r}_{ik}, \theta_{ijk}; k \neq i, k \neq j) \qquad (6.11\text{i})$$

When one obtains all the interatomic distances and angles between the two sub-lattices, the energy density W can be calculated as

$$W = \frac{1}{2} \frac{\sum\limits_{j=1}^{n} V(\bar{r}_{ij}, \bar{r}_{ik}, \theta_{ijk}; k \neq i, k \neq j)}{A_0} \qquad (6.11\text{j})$$

where the summation is for three nearest-neighbor atoms. Taking atom A as the generic atom, from Figure 6.8 it is seen that atom B, C and D are these atoms, thus one can explicitly express (6.11j) as

$$W = \frac{1}{2} \frac{V(\bar{r}_{AB}) + V(\bar{r}_{AC}) + V(\bar{r}_{AD})}{A_0} \qquad (6.11\text{k})$$

where $A_0$ is the unfolding projection area of CNT per atom. This value can be easily obtained as follows:

$$A_0 = \frac{3\sqrt{3}}{4}\ell^2 \qquad (6.11\text{l})$$

where $\ell$ is the atomic bond length such as the length of AB. In fact, the area of each hexagonal in Figure 6.8 is $\frac{3\sqrt{3}}{2}l^2 \left(= 6 \times \frac{l}{2} \times l \times \sin 60°\right)$ which consists of six atoms. Each atom, however, is shared by three hexagonal units and only 1/3 is accounted by the particular hexagonal, the six vortexes take two atoms for the obtained single hexagonal area, thus (6.11l) is approved.

Note that the inverse of the above equation gives the density, $\rho_C$, of the single-wall CNT carbons. Note that the above equation accounts for the effect of second-nearest-neighbor atoms via the bond angle $\theta_{ijk}$. In fact, when calculating $V(\bar{r}_{AB})$ in the equation (6.11i), i is atom A, j is atom B, and k is atom C. $\theta_{ijk}$ in the reference configuration is 120° between bonding AB and bonding AC.

The shift vector $\bar{\varsigma}$ (or $\bar{\eta}$) is determined numerically to make the total potential energy of the system minimum, as shown in Section 6.3.5, which requires

$$\frac{\partial W(\underline{E}, \bar{\eta})}{\partial \bar{\eta}} = 0, \quad \frac{\partial W(\underline{E}, \bar{\varsigma})}{\partial \bar{\varsigma}} = 0 \qquad (6.11\text{m})$$

This is an implicit equation between $\bar{\eta}$ (or $\bar{\varsigma}$) and $\underline{E}$ which can be solved by non-linear iteration process to obtain $\bar{\eta} = \bar{\eta}(\underline{E})$, thus

$$W = W(\underline{E}, \bar{\eta}(\underline{E})) = W(\underline{E})$$

---

**Homework**

(6.7)  The hexagonal lattice of a graphite sheet has the structure shown in figure below. This structure does not possess centro-symmetry. Explain how to decompose it to two sub-lattices of graphite structure in order to meet the requirement of the Cauchy-Born rule (mark 1 and 2 in the figure below to represent the atoms belonging to the two sub-lattices respectively, and pick an atom as central atom to explain centro-symmetry).

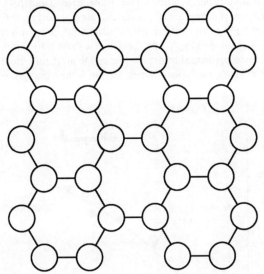

(6.8)  For a deformed graphite sheet, suppose the energy in each bond is $e$ and the bond length and thickness are $l$ and $t$ respectively. Calculate the strain energy density.

---

### 6.10.3.3 Determination of Hydrogen Pressure p(R) as a Function of Nanotube Radius

In the continuum model, where hydrogen is simulated by internal pressure p, the relationship between p and nanotube radius R is important for validation of the model and design of the nanotube for storage capability and failure.

From (6.11m) we see that to obtain the atomistic-based strain density W, we first need to obtain the Green strain tensor $\underline{E}$ for the continuum model of the nanotube. The latter is a tube and its continuum model is easy to develop. Specifically, in the case of an open-ended nanotube, there are only three important components $E_{\theta\theta}, E_{zz}, E_{z\theta}$. These components can be easily determined by the boundary conditions, such as the tensile stress $T_{zz} = 0$ along Z, the longitudinal pipe direction and the shear stress $T_{\theta z} = 0$.

As a result, they all are the function of the tube radius R, i.e., $E_{\alpha\beta} = E_{\alpha\beta}(R)$. For instance $E_{\theta\theta} \approx \frac{\Delta R}{R}$. After the strain components are determined for a given radius R, the interatomic bond length and the bond angle can be determined by (6.11f, h). Thus the strain energy density W(R) can be obtained.

The next step is to use the obtained W(R) to find the relationship p(R). From mechanics of materials we know that to make the force balance of the tube, the internal pressure p produced by the hydrogen is related to the stress $T_{\theta\theta}$ as follows:

$$p = \frac{T_{\theta\theta}t}{R(1 + E_{zz})} \tag{6.11n}$$

where t is the thickness of the nanotube, $\varepsilon_{zz} \approx E_{zz}$ when taking the reference configuration as the current configuration. This strain component causes the area increase. The circumferential stress $T_{\theta\theta}$ can be determined by (6.9i) as

$$T_{\theta\theta}(R) = \frac{dW(R)}{dE_{\theta\theta}(R)} \tag{6.11o}$$

This differential is easy to carry out. Substituting the obtained result into (6.11n), the p(R) relationship can be obtained numerically, and the results by Chen et al.[44] are given in Figure 6.9. In the figure, the result of fully atomistic simulation is given by the atomistic scale finite element methods.[49] It is seen that the atomistic-based continuum model gives results consistent with the fully atomistic analysis. The difference is that the continuum model predicted higher peak pressure than the fully atomistic model. This is because the continuum model assumes homogeneous deformation in the nanotube, but in reality, any c-c atomic

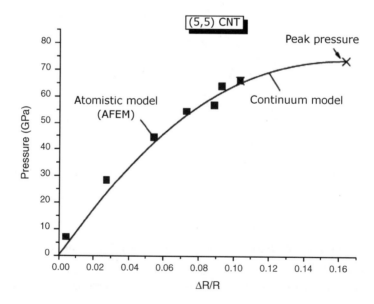

**Figure 6.9** The pressure inside the single-wall (5,5) carbon nanotube versus the percentage change of nanotube radius for the atomistic (AFEM) and continuum models (Reproduced from Chen, Y. L., Liu, B., Wu, J., *et al.* (2008) Mechanics of hydrogen storage in carbon nanotubes. *Journal of the Mechanics and Physics of Solids*, 56, 3224, Elsevier)

bond breakage will cause whole structure failure. Thus, the continuum model gives the upper bound for the nanopipe design.

### 6.10.4  Atomistic-based Continuum Modeling to Determine the Hydrogen Density and Pressure p

For hydrogen storage, it is important to get the density or average spacing $d_{H2\_H2}$ between hydrogen atoms under a given pressure p. This can be obtained by an assumption of hydrogen atom package and using the virial stress. The latter is widely used in molecular dynamics and can be written as follows (see Section 3.9.1):

$$\underline{\sigma} = \frac{1}{\Omega}\left(-m_i \bar{v}_i \bar{v}_i + \frac{1}{2}\sum_{j(\neq i)} \bar{r}_{ij}\bar{f}_{ij}\right) \tag{6.11p}$$

where $\Omega$ is average volume per atom (particle), and $m_i$ and $v_i$ are the mass and velocity of atom i, respectively. The first right term of (6.11p) is usually small and can be neglected, especially in low temperature. $\bar{f}_{ij}$ is the interatomic force which is determined by (2.1a)

$$\bar{f}_{ij} = \frac{dV_{ij}}{dr_{ij}}\frac{\bar{r}_{ij}}{r_{ij}} \tag{6.11q}$$

where the derivative on the right part gives the force value and the ratio gives the unit vector from atom i to atom j. Substituting (6.11q) into (6.11p) we have

$$\underline{\sigma} \approx \frac{1}{2\Omega}\sum_{j(\neq i)}\frac{1}{r_{ij}}\frac{dV_{ij}}{dr_{ij}}\bar{r}_{ij}\bar{r}_{ij} \tag{6.11r}$$

From this equation, any stress component can be obtained, taking the normal stress as an example

$$\sigma_{xx} = \frac{1}{2\Omega}\sum_{j(\neq i)}\frac{1}{r_{ij}}\frac{dV_{ij}}{dr_{ij}}(\bar{r}_{ij})_x(\bar{r}_{ij})_x \tag{6.11s}$$

where $(\bar{r}_{ij})_x$ denotes the x-component of the vector $\bar{r}_{ij}$; in a similar way, one can get other stress components by changing the two subscripts. For instance, if one subscript is x and the other is y the corresponding expression for shear stress is $\sigma_{xy}$. By definition of the pressure p, we have

$$p = \frac{\sigma_{xx} + \sigma_{yy} + \sigma_{zz}}{3} = \frac{1}{3\times 2\Omega}\sum_{j(\neq i)}\frac{1}{r_{ij}}\frac{dV_{ij}}{dr_{ij}}\left[(\bar{r}_{ij})_x(\bar{r}_{ij})_x + (\bar{r}_{ij})_y(\bar{r}_{ij})_y + (\bar{r}_{ij})_z(\bar{r}_{ij})_z\right]$$

$$= \frac{1}{6\Omega}\sum_{j(\neq i)}r_{ij}\frac{dV_{ij}}{dr_{ij}} \tag{6.11t}$$

In the derivation of the last equation, the value of the parenthesis is taken as $(\bar{r}_{ij})^2$. For the most packed hydrogen with the so called AB stacking fashion, there are six neighbors in the A stacking plane and three each in the top and bottom B plane, thus there are 12 molecules surrounding each molecule neighbor. If denoting the spacing between hydrogen molecules by $d_{H2\text{-}H2}$, the volume per hydrogen molecule is

$$\Omega = \frac{\sqrt{2}}{2}d_{H2-H2}^3 \tag{6.11u}$$

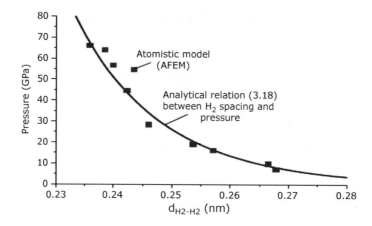

**Figure 6.10**   Pressure inside the CNT single wall versus the equilibrium spacing between hydrogen molecules inside the nanotube. Here the solid line is the continuum model by (6.11v) (Reproduced from Chen, Y. L., Liu, B., Wu, J., *et al.* (2008) Mechanics of hydrogen storage in carbon nanotubes. *Journal of the Mechanics and Physics of Solids*, 56, 3224, Elsevier)

Substituting (2.3a) and (6.11u) into (6.11t) gives the following analytical relation to determine the spacing between molecules $d_{H2\text{-}H2}$ in terms of the internal pressure p in the case of LJ interatomic potential:

$$p = \frac{48\sqrt{2}\varepsilon_{H2-H2}}{\sigma^3_{H2-H2}}\left[2\left(\frac{\sigma_{H2-H2}}{d_{H2-H2}}\right)^{15} - \left(\frac{\sigma_{H2-H2}}{d_{H2-H2}}\right)^{9}\right] \tag{6.11v}$$

Using (6.11v) the plot of pressure p versus the equilibrium spacing between hydrogen molecules inside the nanotube is given in Figure 6.10. In the figure, the solid line is calculated by (6.11v). The square data point is the fully atomistic solution. The comparison between the fully atomistic model and the atomistic-based continuum model shows agreement.

Equation (6.11v) cannot be solved analytically to give the spacing $d_{H2\text{-}H2}$ in terms of the pressure. Since the repulsive term $\left(\frac{\sigma_{H2-H2}}{d_{H2-H2}}\right)^{15}$ dominates as the internal pressure p increases, one may neglect the attractive term $\left(\frac{\sigma_{H2-H2}}{d_{H2-H2}}\right)^{9}$ to get an analytical but approximate solution as follows:

$$d_{H2-H2} \approx \left(\frac{96\sqrt{2}\varepsilon_{H2-H2}\sigma^{12}_{H2-H2}}{p}\right)^{1/15} \tag{6.11w}$$

## 6.10.5   Continuum Model of Interactions Between the CNT and Hydrogen Molecules and Concentration of Hydrogen

Concentration of hydrogen is important to calculate the capability of hydrogen storage by CNT. There is a spacing $d_{C\text{-}H2}$ between a CNT wall with hydrogen molecules. This spacing should be subtracted for the calculation of hydrogen storage. Then the number of hydrogen molecules per unit volume of the single-wall CNT is

$$N_H = \frac{\pi(R + \Delta R - d_{C-H2})^2}{A\Omega} \tag{6.12a}$$

where the total volume (area) A of the deformed nanotube is given as

$$A = \pi(R + \Delta R)^2 \tag{6.12b}$$

Substituting A of this equation and $\Omega$, the volume (area) per hydrogen molecule of (6.11u), into (6.12a), we obtain

$$N_H = \frac{\pi(R + \Delta R - d_{C-H2})^2}{\pi(R + \Delta R)^2 \frac{\sqrt{2}}{2} d_{H2-H2}^3} = \left(\frac{R + \Delta R - d_{C-H2}}{R + \Delta R}\right)^2 \frac{\sqrt{2}}{d_{H2-H2}^3} \tag{6.12c}$$

The rest of this section will describe how to calculate the spacing $d_{C-H2}$ based on the van der Waals interactions through the LJ potential (2.3a). The basic method follows the cohesive law between a CNT and polymer developed by Jiang *et al.* for polymer[50] which investigates the van der Waals interactions between a carbon differential element and an annular differential area within an infinite domain of polymer. Here, curvature effects are neglected so that the geometric model is simplified as an infinite graphene and an infinite block of hydrogen. They are placed in two parallel horizontal planes with the separation distance, h, which denotes the equilibrium distance of the hydrogen block below the graphene plane. The carbon differential element has area dA whose atomistic number can be determined by

$$n_{Carbon} = \rho_c dA \tag{6.12d}$$

where $\rho_C = 4/(3\sqrt{3}\ell_0^2)$ is the density of the carbon in single-wall CNT (see (6.11l)).

Now, we would like to integrate the stored energy by the interaction potential of the whole hydrogen atoms with this differential carbon element $\rho_c dA$. For simplicity of the integration, the coordinates of the carbon differential element are taken at the origin (i.e., 0, 0), see Figure 6.11. Any generic atom in hydrogen inside the tube is taken as (x, z) where $x \leq h$ and $z \geq 0$. Here, the positive direction of the x-axis is downwards and z is the cylindrical radius coordinate with the positive z-axis from the origin towards outside as a radius vector. Further, we take a differential annular body with volume of $(2\pi z)dzdx$ where dzdx denotes the revolving differential area and $(2\pi z)$ denotes the length of its circumferential length.

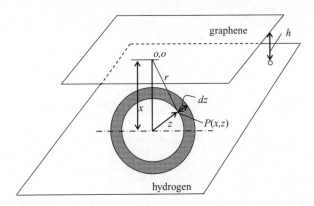

**Figure 6.11** The van der Waals interaction between a shell and the graphite units

Denoting the density of hydrogen as $\rho_{H2}$, the atomistic number of hydrogen (polymer) in this differential volume is

$$\rho_{H2}dV_{hydrogen} = \rho_{H2}2\pi z dz dx = n_{hydrogen} \tag{6.12e}$$

The differential energy can be written by volume integration as follows:

$$dU = \rho_C dA \int_{V\,hydrogen} V(r)\rho_{H2}dV_{hydrogen} \tag{6.12f}$$

with

$$r = \sqrt{x^2 + z^2} \tag{6.12g}$$

where r denotes the average distance between atoms ($\rho_c dA$) in the differential carbon group and the $n_{hydrogen}$ hydrogen atoms in the differential annular volume, and V(r) is the average potential between each of the pair atoms of these two groups. The cohesive energy $\Phi$ is the bonding energy per unit area, thus

$$\Phi = \frac{dU}{dA} = \rho_C \int_{V\,hydrogen} V(r)\rho_{H2}dV_{hydrogen} \tag{6.12h}$$

Substituting (2.3a) for V(r), (6.12e) for $\rho_{H2}dV_{hydrogen}$ and (6.12g) into (6.12h) one obtains:[50]

$$\Phi = \frac{dU}{dA} = 2\pi\rho_C\rho_{H2}\int_{-\infty}^{h}dx\int_0^\infty 4\varepsilon_{C-H2}\left[\left(\frac{\sigma_{C-H2}}{\sqrt{x^2+z^2}}\right)^{12} - \left(\frac{\sigma_{C-H2}}{\sqrt{x^2+z^2}}\right)^6\right]zdz$$

$$= \frac{2\pi}{3}\rho_C\rho_{H2}\varepsilon_{C-H2}(\sigma_{C-H2})^3\left(\frac{2\sigma_{C-H2}^9}{15h^9} - \frac{\sigma_{C-H2}^3}{h^3}\right) \tag{6.12i}$$

The equilibrium distance h between the two parallel planes of hydrogen and graphene can be determined by minimizing the cohesive energy through

$$\frac{\partial\Phi}{\partial h} = 0 \tag{6.12j}$$

This results in

$$h = \left(\frac{2}{5}\right)^{1/6}\sigma = 0.856\sigma \tag{6.12k}$$

If there is a contraction displacement v due to the compressive stress p along the negative x-axis beyond the equilibrium distance h, the cohesive energy can be obtained by simply replacing h by (h–v) as

$$\Phi = \frac{2\pi}{3}\rho_C\rho_{H2}\varepsilon_{C-H2}(\sigma_{C-H2})^3\left(\frac{2\sigma^9}{15(h-v)^9} - \frac{\sigma^3}{(h-v)^3}\right) \tag{6.12l}$$

The pressure p or the stress $\sigma_{zz}$ can be determined by the following equation:

$$p = \frac{\partial\Phi}{\partial v} = 2\pi\rho_C\rho_{H2}\varepsilon_{C-H2}(\sigma_{C-H2})^2\left[\frac{2\sigma_{C-H2}^{10}}{5(h-v)^{10}} - \frac{\sigma_{C-H2}^4}{(h-v)^4}\right] \tag{6.12m}$$

$$d_{C-H2} = h - v \tag{6.12n}$$

where $d_{C-H2}$ is the actual spacing that we want to determine. The first equation in (6.12m) is obtained from equation (2.1a) but here the energy $\Phi = \frac{dU}{dA}$ is the potential energy divided by the area, thus the obtained value is the pressure which equals the corresponding compressive force divided by the area. Replacing (h–v) by $d_{C-H2}$ we have

$$p = 2\pi \rho_C \rho_{H2} \varepsilon_{C-H2} (\sigma_{C-H2})^2 \left[ \frac{2}{5} \left( \frac{\sigma_{C=H2}}{d_{C-H2}} \right)^{10} - \left( \frac{\sigma_{C=H2}}{d_{C-H2}} \right)^4 \right] \qquad (6.12o)$$

In addition, from (6.11u) we have

$$\rho_{H2} = \frac{1}{\Omega} = \sqrt{2} d_{H2-H2}^{-3} \qquad (6.12p)$$

and from (6.11l)

$$\rho_C = \frac{4}{3\sqrt{3} r_0^2} \left( \frac{R}{R + \Delta R} \right)^2 \qquad (6.12q)$$

where the expression before the parenthesis of the right part denotes carbon density before deformation which depends on the equilibrium radius $r_0$; the parenthesis, however, denotes the density change due to the contraction where $\Delta R$ is negative. Substituting (6.12p), q) into (6.12o), the result is[44]

$$p = \frac{8\sqrt{2}\pi}{3\sqrt{3}} \left( \frac{R}{R + \Delta R} \right)^2 \frac{\varepsilon_{C-H2} (\sigma_{C-H2})^2}{r_0^2 d_{H2-H2}^3} \left[ \frac{2}{5} \left( \frac{\sigma_{C-H2}}{d_{C-H2}} \right)^{10} - \left( \frac{\sigma_{C-H2}}{d_{C-H2}} \right)^4 \right] \qquad (6.12r)$$

Now, (6.12c) can be numerically solved by using (6.11n) for $\Delta R$, (6.11v) for $d_{H2-H2}$, and (6.12r) for $d_{C-H2}$. The result is shown in the top solid line of Figure 6.12. The figure also shows the solution for the

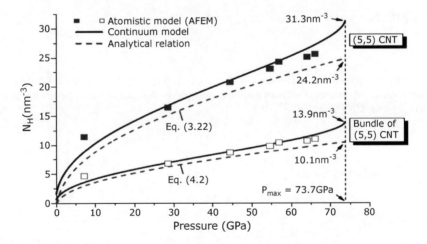

**Figure 6.12** The number of hydrogen molecules per unit volume versus the pressure inside the single-wall (5,5) CNTs or their bundles. The results are shown for the continuum model (eq. (6.12c)), atomistic model (AFEM) and (approximate) analytical solution (eq. (6.12x)) for single-wall CNTs. The solution for the bundles is given in eq. (4.2) in the original paper (Reproduced from Chen, Y. L., Liu, B., Wu, J., et al. (2008) Mechanics of hydrogen storage in carbon nanotubes. *Journal of the Mechanics and Physics of Solids*, 56, 3224, Elsevier)

bundle of single-wall CNTs given by Ref. [44] in which the distance between neighboring nanotube walls along their center lines equals $\sigma_{C-C}$, the parameter of the LJ potential.

## 6.10.6 Analytical Solution for the Concentration of Hydrogen Molecules

There is also the (approximate) analytical solution in Figure 6.9, which is obtained by dropping the attraction terms. We have already determined the approximate solution (6.11w) for $d_{H2-H2}$; the approximate solution for $d_{C-H2}$ can be also determined by dropping the attraction term in (6.12r) as follows:

$$p = \frac{8\sqrt{2}\pi}{3\sqrt{3}} \left(\frac{R}{R+\Delta R}\right)^2 \frac{\varepsilon_{C-H2}(\sigma_{C-H2})^2}{r_0^2 d_{H2-H2}^3} \left[\frac{2}{5}\left(\frac{\sigma_{C=H2}}{d_{C-H2}}\right)^{10}\right] \tag{6.12s}$$

or re-write it as

$$d_{C-H2} \approx \left[\frac{16\sqrt{2}\pi}{15\sqrt{3}}\left(\frac{R}{R+\Delta R}\right)^2 \frac{\varepsilon_{C-H2}(\sigma_{C-H2})^{12}}{r_0^2 d_{H2-H2}^3} \frac{1}{p}\right]^{1/10} \tag{6.12t}$$

For small deformation, $E_{zz}$ can be dropped from (6.11n) to get

$$T_{\theta\theta} = \frac{pR}{t} \tag{6.12u}$$

On the other hand,

$$T_{\theta\theta} = E\varepsilon_{\theta\theta} = E\left(\frac{2\pi(R+\Delta R)-2\pi(R)}{2\pi R}\right) = E\frac{\Delta R}{R} \tag{6.12v}$$

Combining (6.12u) and (6.12v) we have the approximate following equation:

$$\Delta R = \frac{pR^2}{Et} \tag{6.12w}$$

Substituting the approximate expressions of (6.11w), (6.12t) and (6.12w) into (6.12c), finally we have the (approximate) but analytical solution for the atomistic-based model of the hydrogen molecules per unit volume of the CNT as follows:[44]

$$N_H = \left(\frac{p}{24\varepsilon_{H2-H2}\sigma_{H2-H2}^{12}}\right)^{1/5} \left[1 - \frac{\left(\frac{2}{\sqrt{3}}\right)^{17/50}\left(\frac{\pi}{5r_0^2}\right)^{1/10} \frac{\left(\varepsilon_{C-H2}\sigma_{C-H2}^{12}\right)^{1/10}}{\left(\varepsilon_{H2-H2}\sigma_{H2-H2}^{12}\right)^{1/50}}}{p^{2/25}R\left(1+\frac{pR}{Et}\right)^{6/5}}\right]^2 \tag{6.12x}$$

$$= 45.0\, p^{1/5}\left[1 - \frac{0.270}{p^{2/25}R(1+0.00267\, pR)^{6/5}}\right]^2$$

The numerical solution of this equation is shown in a dashed line in Figure 6.12 (marked by Eq. (3.22) of [44]). From the comparison in that figure, it is seen that the above analytical expression agrees well with the atomistic-based continuum model and the fully atomistic models. The curve stops at the peak pressure $p_{max} = 73.7\,GPa$ obtained from Figure 6.9, which gives the maximum hydrogen storage as $24.2\,nm^{-3}$ in the (5,5) CNT. It is smaller than $31.3\,nm^{-3}$ given by the atomistic-based continuum model, but is close to $25.6\,nm^{-3}$ given by the fully atomistic model.

## 6.10.7    The Double Wall Effects on Hydrogen Storage

Suppose $R_1$ and $R_2$ are radii of the double-wall CNT for hydrogen storage. Note that while carbon atoms on the same CNT wall have covalent bonds and are characterized by the interatomic potential in (6.11a), the carbon atoms from different CNT walls have the van der Waals interactions in (2.3a), and so do carbon atoms and hydrogen molecules inside the CNT.

The continuum model to represent the inner and outer CNTs comprises thin shells of radii $R_1$ and $R_2$. The hydrogen molecules are still represented by internal pressure p on the inner shell. Using the same method as the one in Section 6.10.5 for investigating the interaction between the CNT and the hydrogen or polymer layer, one can obtain the equilibrium condition and the stored energy. Specifically, the van der Waals interaction between the two shells is characterized by the cohesive law for multi-wall CNTs,[51] which gives the force-separation relation between two CNT walls from the LJ interatomic potential.

Using the principle of minimum potential energy, it can be found that the equilibrium spacing between CNT walls is the parameter $\sigma_{C\text{-}C}$, i.e., $R_2 = R_1 + \sigma_{C\text{-}C}$. This is different from the value of $h = 0.856\sigma_{C\text{-}C}$ in (6.12k). This difference is because here the integration is related to two shell surfaces and in Section 6.10.5 the energy integration is related to one surface and one volume. The volume integration is covered by the body of either polymer or hydrogen with a certain depth.

Let $R_1 + \Delta R_1$ and $R_2 + \Delta R_2$ denote the radii of inner and outer walls after the internal pressure p is imposed. The energy per unit length of the CNT stored between two CNT walls due to van der Waals interactions is[51]

$$2\pi^2 (R_2 + \Delta R_2 + R_1 + \Delta R_1)\rho_C\rho_{C2}\varepsilon_{C\text{-}C}(\sigma_{C\text{-}C})^2 \left[ \frac{2}{5} \left( \frac{\sigma_{C\text{-}C}}{R_2 + \Delta R_2 - R_1 - \Delta R_1} \right)^{10} - \left( \frac{\sigma_{C\text{-}C}}{R_2 + \Delta R_2 - R_1 - \Delta R_1} \right)^4 \right]$$

$$(6.12y)$$

where $\rho_{C1}$ and $\rho_{C2}$ are the numbers of carbon atoms per unit area of the inner and outer CNTs, respectively, and are related to the equilibrium bond length of carbon $r_0 = 0.142\,\text{nm}$ by

$$\rho_{Ci} = \frac{4}{3\sqrt{3}r_0^2} \left( \frac{R_i}{R_i + \Delta R_i} \right)^2 \quad (i = 1, 2) \tag{6.12z}$$

The energy stored in each CNT wall is obtained in the same way as described in (6.11f) to (6.11k). The total energy is the sum of energy in each CNT and the energy between the walls in (6.12y). The minimization of total energy determines the change of inner and outer CNT radii $\Delta R_1$ and $\Delta R_2$ in terms of the internal pressure p. It can easily be seen that the double-wall design will increase the loading capability due to the van der Waals interactions between the two shells. On the other hand, this double-wall design will reduce the hydrogen storage.

In fact, for a single (5,5) armchair, the atomistic-based continuum model gives the peak pressure $p_{max} = 73.7\,\text{GPa}$ with the storage of $31.3\,\text{nm}^{-3}$. For the (5,5) (10,10) double-wall CNT, the number $N_H$ of hydrogen molecules per unit volume increases as the internal pressure p increases, but the curve stops at $p_{max} = 89.3\,\text{GPa}$ which is about a 21% strength increase. However, this double-wall CNT only gives hydrogen storage of $11.0\,\text{nm}^{-3}$. This is significantly smaller than $31.3\,\text{nm}^{-3}$ for (5,5) single-wall CNT and also smaller than $13.9\,\text{nm}^{-3}$ for the bundle of (5,5) single-wall CNTs and does not reach the goal of hydrogen storage. More information about the property and design of double-wall and multi-wall CNTs can be found in Ref. [44].

One further point needs to be mentioned. From theoretical point of view, the peak stress $p_{max}$ can be determined by (6.11n). The problem is the difficulty in finding the allowable circumferential stress $T_{\theta\theta}$. To avoid the difficulty, in Ref. [44], the peak stress determined by the atomistic-based model is to take the value of the maximum stress at which the stress-displacement curve will start to drop down. As we mentioned in the previous section this stress is larger than the failure stress determined by the fully atomistic simulation in which failure is determined by the breakage of the atomistic bonds.

---

**Homework**

(6.9) Suppose the spatial distribution of hydrogen molecules is Close-Packed Hexagonal (AB stacking), and the average spacing between two closest hydrogen molecules is 0.28 nm. If only the van der Waals interactions between the closest molecules are taken into account, calculate the average inner pressure in the hydrogen molecules.

Hint: For Close-Packed Hexagonal (ABAB stacking), each molecule has 12 closest neighboring molecules. Suppose the coordinate of the central molecule is (0,0,0) and the closest distance is $d_{H_2-H_2}$. The vectors to the closest neighboring molecules are

$$r_1 = (1,0,0)d_{H_2-H_2} \qquad r_2 = \left(\tfrac{1}{2},\tfrac{\sqrt{3}}{2},0\right)d_{H_2-H_2} \qquad r_3 = \left(-\tfrac{1}{2},\tfrac{\sqrt{3}}{2},0\right)d_{H_2-H_2}$$

$$r_4 = (-1,0,0)d_{H_2-H_2} \qquad r_5 = \left(-\tfrac{1}{2},-\tfrac{\sqrt{3}}{2},0\right)d_{H_2-H_2} \qquad r_6 = \left(\tfrac{1}{2},-\tfrac{\sqrt{3}}{2},0\right)d_{H_2-H_2}$$

$$r_7 = \left(\tfrac{1}{2},\tfrac{\sqrt{3}}{6},\tfrac{\sqrt{6}}{3}\right)d_{H_2-H_2} \qquad r_8 = \left(-\tfrac{1}{2},\tfrac{\sqrt{3}}{6},\tfrac{\sqrt{6}}{3}\right)d_{H_2-H_2} \qquad r_9 = \left(0,-\tfrac{\sqrt{3}}{3},\tfrac{\sqrt{6}}{3}\right)d_{H_2-H_2}$$

$$r_{10} = \left(\tfrac{1}{2},\tfrac{\sqrt{3}}{6},-\tfrac{\sqrt{6}}{3}\right)d_{H_2-H_2} \qquad r_{11} = \left(-\tfrac{1}{2},\tfrac{\sqrt{3}}{6},-\tfrac{\sqrt{6}}{3}\right)d_{H_2-H_2} \qquad r_{12} = \left(0,-\tfrac{\sqrt{3}}{3},-\tfrac{\sqrt{6}}{3}\right)d_{H_2-H_2}$$

And the average volume for each molecule is

$$\Omega = \frac{\sqrt{2}}{2}d^3_{H_2-H_2}$$

(6.10) For a double-wall carbon nanotube, the radii of inner wall and outer wall are $R_1$ and $R_2$ respectively. Because of the van der Waals interactions between the two walls, there is pressure between the two walls. The pressure applied on the inner wall by the outer wall is $P_1$ and the pressure applied on the outer wall by the inner wall is $P_2$. Which of the following relations of the pressures do you think is correct?

(a) $P_1/P_2 = R_2/R_1$; (b) $P_1 = P_2$; (c) none of above

Hint: Here, the cylindrical coordinate system (r, $\theta$, z) is taken; disregarding special cases, we can take the angle reference at the atom on the inner nanotube (i.e., for it $\theta = 0$), and take the angle at the atom on the outer nanotube as $\theta$. Thus, the angle between these two atoms is $\theta$. The distance from an atom on the outer wall to an atom on the inner wall is $r = \sqrt{a^2 + z^2}$ where $a = \sqrt{R_1^2 + R_2^2 - 2R_1R_2\cos\theta}$. Note $\rho_1$ and $\rho_2$ as the number of atoms per unit surface area for the inner and outer wall respectively, and usually $\rho_1 \approx \rho_2$.

---

## 6.11 Atomistic-based Model for Mechanical, Electrical and Thermal Properties of Nanotubes

In this section, we will briefly introduce how an analytical atomistic-based continuum solution can be developed to study the mechanical and thermal properties of a carbon nanotube. The results are also compared with MD simulation. Note that this method does not introduce fitting parameters, except in determining the interatomic potential function.

## 6.11.1    Highlights of the Methods

The basic approach in using the Cauchy-Born rule to investigate the properties of nanotubes is the same as introduced in Sections 6.9 and 6.10. This method will be highlighted in this section. In order to develop the interatomic potential-based constitutive laws of continuum, this theory adopts several steps to carry out the analysis of mechanical, electrical and thermal behavior of a carbon nanotube.[46, 52]

(1) With respect to a given Green strain $\underline{E}$, the elongated bond length $r$ and the bond angle $\theta_{ijk}$ are calculated by equation (6.9v, z) for centro-symmetric and (6.11g, h) for non-centro-symmetric atomic structures, respectively.
(2) The bonding energy stored in the atomistic bonds between atoms $i$ and $j$ are provided by the second generation of Brenner multi-body interatomic potential energy as given below:

$$V_{ij}(r_{ij}, \theta_{ijk}; k \neq i,j) = V_R(r_{ij}) - B_{ij}(\theta_{ijk}) V_A(r_{ij}) \tag{6.11a}$$

(3) For a hexagonal lattice structure, each atom $i$ has three neighbor atoms $j$. Therefore, the corresponding strain energy density W may be determined by equation (6.11j). It not only includes the pair energy stored in these atomic bonds but also includes multi-body effects through the bond angle $\theta_{ijk}$. Referring to the expression of (6.11i), the summation covers all of the three neighboring atoms of atom $i$, the coefficient 1/2 means that the potential energy of each bond is shared by two atoms, and $\Omega$ is the average area of each atom on the surface of the carbon nanotube.
(4) The internal variable $\xi$ or $\eta$ for intrinsic connection between sub-lattice structures should be determined by the principle of minimum potential energy through equation (6.11m). The minimum energy condition is equivalent to the condition that every atom in the non-centro-symmetric hexagonal lattice structure is in equilibrium.
(5) The second order Piola-Kirchhoff stress $T_{ij}$ is obtained by the derivative of the strain energy density $W$ with respect to Green strain $E_{ij}$ (see equation (6.9i)), while the tensor of elastic constants, or moduli $C_{ijkl}$ is the second-order derivative of $W$ with respect to Green strain $E_{ij}$ (see equation (6.9k)).
(6) The equilibrium equation and force boundary conditions are corresponding to relative conditions in classical continuum theory.

We will see in the next three sections the results of using the atomistic-based continuum theory to investigate mechanical, electrical and thermal properties of nanotubes.

## 6.11.2    Mechanical Properties

Jiang et al.[53] studied the energy stored in a carbon nanotube before deformation. They assume the energy of a flat graphite sheet is zero and want to find the energy value if a generic atom is in a nanotube with radius R. When R is given, the interatomic length and angle compared to the flat sheet can be easily determined by (6.11f, g), thus the energy density can be determined. The results are as follows.

### 6.11.2.1 Undeformed Single-wall CNT Atom Energy is Proportional to $R^{-2}$

The energy of a single atom in the carbon nanotube is inversely proportional to the square root of the tube radius R, i.e., proportional to $R^{-2}$. This result is consistent with Robertson's results, which were obtained by molecular dynamics using the Brenner potential.[52] The calculated results are compared with carbon nanotubes with radii varying from 0.2 to 1 nm. For the smallest tube, with radius 0.2 nm, there are only six atoms in the tube circumference. This indicates that the analytical solution is also applicable even in this kind of carbon nanotube composed of so few atoms.

### 6.11.2.2  Single-wall CNT Elastic Modulus Prediction is Around 0.7 TPa

The linear elastic modulus of a carbon nanotube can also be determined with this multiscale method by equation (6.9k). By employing this analytical solution, the linear elastic modulus of a single-wall nanotube was determined as 0.7 TPa,[46] which is consistent with Cornwell and Wille's calculation using the same potential function.[54] The calculation results for single-wall nanotubes with radii from 0.5 to 2 nm are also consistent with the tight-binding method.[55]

### 6.11.2.3  Prediction of Fracture Initiation Strain for Single-wall CNT

This multiscale analytical solution can also be used to study the initiation of fracture and defect. Yakobson *et al.*[56] found that the tensile fracture strain of a nanotube ranges between 25% and 55%, depending on the temperature and geometrical parameters of the tube. The strain is uniform before failure. Once the strain reaches critical failure value, one or several atomic bonds break. Because this analytical solution across atomic-continuum scales does not trace atom movements, it is incapable of studying the initiation of cracks and defects directly like molecular dynamics. Therefore the atomistic-based continuum approach must be adopted.

Zhang *et al.*[47] used the atomistic-based energy density to carry on bifurcation stability analysis, by simulating the instability at the failure region of a single-wall nanotube characterized by localized nonuniform deformation. They prove with their nonlinear bifurcation analysis on a tensile nanotube that the bifurcation strain ranges from 35% to 40%.

### 6.11.2.4  Prediction of Defect Initiation Strain for Single-wall CNT

Nardelli *et al.*[57, 58] and Yakobson[59] studied the initiation of defects in a single nanotube using molecular dynamics. They found that the carbon bond perpendicular to the tensile direction suddenly switched 90 degrees and formed a dislocation dipole when the tensile strain reached the critical value of 5%. An aberrance of lattice frame was also generated in this position.

Jiang *et al.*[60] and Song *et al.*[61] used the atomistic-based continuum theory to investigate the initiation of defects for single-wall CNT. They calculated the energy dependence of a nanotube with or without defects on axial strain. Qin *et al.*[62] found that SCT, a CNT network, can increase rupture strain more than 3 times that of single-walled CNT.

For the latter, energy is zero at zero strain. It increases gradually with increasing strain. For the nanotube with defects, the trend was found to be in inverse relation to the loading. The energy is relatively larger at zero strain due to the existence of defects. The energy then decreases with increasing strain. When plotting the two curves, it was observed that the two lines intersect at 4.95% strain. Both the bifurcation theory and minimum energy theory predict that defects occur at 4.95% strain, consistent with Yakobson's experimental results.

## 6.11.3   Electrical Property Change in Deformable Conductors

Tombler *et al.*[63] have found through various experiments that the electrical conductivity of a single-wall nanotube may change two orders in a deformed state, meaning that a metallic carbon nanotube may transfer to a semiconductor type due to deformation. Liu *et al.*[64] studied the deformation induced electrical property change in single-wall nanotubes by combining the analytical solution across atomic-continuum scales using the tight-binding method. First, they used the former method to determine atom position in a deformed nanotube. Then they used the tight-binding method to calculate the energy difference between energy bands.

If the energy difference is equal to zero or very small, the carbon nanotube shows metallic properties with finite electric conductivity. Electrons are able to move freely in this situation. On the contrary, a finite energy difference between energy bands turns the carbon nanotube into a semiconductor type because electrons must overcome the energy barrier to move in the tube. Liu *et al.* studied the relationship between energy difference and twist angle, $\theta_k$, of unit length under torsion for several kinds of single nanotubes.

They found that the energy may reach about $0.17\,\text{eV}$ when the twist angle $\theta_k$ increases to a certain value, e.g., $\pm 0.05\,\text{nm}$. They also found that the energy is zero at zero twist angle. This is consistent with the experiments which show that the electrical conductivity of a nanotube may change from metallic to semiconductor type according to its deformation.

## 6.11.4   Thermal Properties

Note that the atomistic-based energy density equation (6.11i) can only be applied where the temperature is absolute zero. This is because the interatomic potential function does not depend on temperature. Therefore, the atomistic-based continuum analytic method is a static method dealing only with equilibrium problems. It is incapable of controlling temperature by deriving the kinetic energy of all the atoms as in molecular dynamics. In order to overcome this deficiency, Jiang et al.[65] firstly calculated Helmholtz free energy at finite temperatures by referring to the localized harmonic and approximating method proposed by Shenoy et al. for the QC method in finite temperature.[9] They then used the following minimum condition of Helmholtz free energy A (r, T) to replace the condition of minimum potential energy $\Pi$ in order to determine the balance position $r^{(0)}(T)$ at finite temperatures:

$$\frac{\partial A(r,T)}{\partial r} = 0 \tag{6.13a}$$

Jiang et al. used the following approximation in a localized harmonic model to describe the entropy s:

$$s = -k_B \sum_{i=1}^{N} \sum_{k=1}^{3} \ln\left[2\sinh\left(\frac{\hbar \omega_{ik}}{4\pi K_B T}\right)\right] \tag{6.13b}$$

where $k_B$ is the Boltzmann constant, $\hbar$ is Planck constant, $\omega_{ik}$ is the vibration frequency of atom $i$, which may be determined using the following $3 \times 3$ matrix:

$$\left| \omega_{ik}^2 I_{3\times 3} - \frac{1}{m}\frac{\partial \sum V(r)}{\partial x_i 2 x_i} \right| = 0 \tag{6.13c}$$

where $I_{3\times 3}$ is a unit matrix. Here the specific heat is defined as the thermal energy needed for each unit volume to change temperature by one degree. Its value depends on the vibration frequency of atom $i$. Therefore, the specific heat is obtained as follows:[65]

$$C_V = \frac{6.022 \times 10^{23}}{16\pi^2 k_B T} \sum_{k=1}^{3} \frac{\left(\frac{1}{T} - \frac{1}{\omega_k}\frac{d\omega_k}{dT}\right)\omega_k^2}{\sinh^2\left(\frac{\omega_k}{4\pi K_B T}\right)} \tag{6.13d}$$

This expression for specific heat $C_V$(J/mole K) is consistent with experimentation on graphite in the temperature range 0 to $1600\,\text{K}$.

Jiang et al.[65] also studied the thermal expansion coefficient of a single-wall nanotube. A new bond length can be obtained by deriving the minimum Helmholtz free energy condition at a given temperature. Therefore, the thermal expansion coefficient can be determined by measuring the change of solid length corresponding to one degree temperature change. The obtained results can then be compared with experimental data of graphite along the $\alpha$ axis. In this manner, a common trend is found. The analysis provides a negative thermal expansion coefficient for the nanotube when the temperature is below $456\,\text{K}$, which means that the nanotube will shrink if the temperature increases in a range smaller than $456\,\text{K}$. A similar result has also been found in experimental observations.

### 6.11.5   Other Work in Atomistic-based Continuum Model

In addition to the above mentioned multiscale theories across atomic-continuum scales and their applications, there are also some other applications of the Cauchy-Born rule. Zhang *et al.*[66] studied the mechanical property of single-walled CNTs, including prediction of Young's modulus. The former employed a modified Cauchy-Born rule of Arroyo and Belytschko exponential type, while the latter used a higher order Cauchy-Born rule. Here, the introduction of exponential modification and a higher order Cauchy-Born rule are used to eliminate the effect of nanotube curvature. Mapping without modification would result in frame nodes that would go to the tangent plane of the shell body rather than right on the shell body. This would not only increase calculation difficulty, but also decreases precision. In general, although the analytic multiscale method is simple and clear, its application is limited, similar to other analytical solutions.

## 6.12   A Proof of 3D Inverse Mapping Rule of the GP Method

The Cauchy-Born rule and the equation (6.9 u,v) for calculating the elongation of the bond length after deformation are necessary to prove the inverse mapping rule of the GP method, thus now is a good time to consider the knowledge learned in this chapter about the rule and the equation for the required proof in Chapter 5.

The first step of a rigorous proof of the inverse mapping rule for the 3D case of the GP method introduced in the last chapter is to introduce two auxiliary points i' and j' and their relative position vector $\bar{r}_{i'j'(n)}$ in the $\alpha_n$ domain (see Figure 5.11). The latter is obtained by exactly following the inverse mapping rule (5.17), i.e.,

$$\bar{r}_{i'j'(n)} = k^{-(n-1)}\overline{R}_{IJ(n)} \tag{6.14a}$$

In general, the two points i' and j' may not necessarily be coincident with the atomistic lattice point i and j as shown in Figure 5.11.

After introducing the auxiliary points i' and j', linking their position vector $\bar{r}_{i'j'(n)}$ and comparing (5.17) with (6.14a), we see that to prove the inverse mapping rule (5.17) we only need to prove $\bar{r}_{i'j'(n)} = \bar{r}_{ij(n)}$. We will first prove that the direction of $\bar{r}_{i'j'(n)}$ is the same as that of $\bar{r}_{ij(n)}$ and then prove the equality of their magnitudes. The Cauchy-Born rule is commonly used to make a link between the deformation of a continuum and its underlying atomic crystalline lattice.[35] Because both the $\alpha_n$ and $\beta_n$ domains correspond to the same material domain they are subject to the same deformation gradient $\underline{F}$. Therefore, the Cauchy-Born rule can be applied to both the lattice vectors in the $\alpha_n$ and $\beta_n$ domains which results in

$$\bar{r}_{ij(n)} = \underline{F}\bar{r}_{ij0} = r_{ij0}\underline{F}\bar{n}_{ij0} \quad (\alpha_n \text{ domain}) \tag{6.14b}$$

$$\overline{R}_{IJ(n)} = \underline{F}\overline{R}_{IJ0} = R_{IJ0}\underline{F}\overline{N}_{IJ0} \quad (\beta_n \text{ domain}) \tag{6.14c}$$

If we substitute (5.16b) for (6.14b) and then compare it with (6.14c), it is seen that $\bar{r}_{ij(n)}$ parallels $\overline{R}_{IJ(n)}$. The latter also parallels $\bar{r}_{i'j'(n)}$ as seen by (6.14a), so that $\bar{r}_{i'j'(n)}$ has the same direction as $\bar{r}_{ij(n)}$.

Furthermore, from (6.9u), the length of $r_{ij(n)}$ can be expressed by the original length $r_{ij0}$ between the two atoms i and j as follows:

$$r_{ij(n)} = r_{ij0}\sqrt{1 + 2\bar{n}_{ij}^{0T}\underline{E}\bar{n}_{ij}^{0}} \tag{6.14d}$$

The same relationship exists between $R_{IJ(n)}$ and $R_{IJ0}$ in the $\beta_n$ domain because the Green strain tensor $\underline{E}$ is the same in the same material region (i.e., in both $\alpha_n$ and $\beta_n$ domains). It reads

$$R_{IJ(n)} = R_{IJ0}\sqrt{1 + 2\overline{N}_{IJ0}\underline{E}\overline{N}_{IJ0}} = R_{IJ0}\sqrt{1 + 2\bar{n}_{ij0}\underline{E}\bar{n}_{ij0}} \tag{6.14e}$$

The last equality is due to (5.16b). Substituting this equation into (6.14a) and then using (5.16a) to replace $R_{IJ0}$ we obtain

$$\bar{r}_{i'j'(n)} = r_{ij0}\sqrt{1 + 2\bar{n}_{ij}^{0T}\underline{\underline{E}}\bar{n}_{ij}^{0}} \tag{6.14f}$$

Comparing (6.14d) with (6.14f), it is seen that the magnitude of $\bar{r}_{i'j'(n)}$ has the same value as $\bar{r}_{ij(n)}$, thus we have the equation $\bar{r}_{i'j'(n)} = \bar{r}_{ij(n)}$. Substituting this equation to (6.14a), the inverse mapping rule (5.17) is proved, which can be used to determine the interatomic vector $\bar{r}_{ij(n)}$ of the atomistic structure corresponding to the particle I and J. Thus, it can be used for particle dynamic analysis as shown in Section 5.5.

## 6.13 Concluding Remarks

This chapter summarizes 10 findings of QC concurrent multiscale methods. Finding 2 is related to deformation twinning (DT), a surprising result since DT in Al had not been observed in previous experiments. This discrepancy was recently explained through a new model guided and validated by extensive multiscale simulations.[67] An idea to resolve the discrepancy is to consider the thermally activated nature of twinning and dislocation slip. A multiscale simulation technique that permits extensive long-time simulations of crack-tip behavior (see Chapter 9 for details) was used. The simulation demonstrates material behavior for short and longer times. It was discovered in this simulation that for aluminum there is a transition from twinning at very short times under high applied loads to dislocation slip. The latter occurs at longer times with lower applied loads. In addition, the 2D analytical Peierls model (see Section 1.4.2.2) was extended to study the competition between DT and dislocation slip. The analytical model shows the individual characteristics of the activation energy $Q(\sigma,T)$ (see Section 9.4.1 and Equation (9.1) for the connection with rate), and for full dislocation emission $Q(\sigma,T)$ becomes lower than that for twinning with decreasing load, corresponding to longer times or slower strain rates. The result indicates that twinning with high activation energy $Q(\sigma,T)$ due to high stress and low temperature corresponds to high strain rate and low observation probability. This is consistent with the observed experimental trends.

Rate dependence of material behavior has recently undergone resurgence because of the development of atomistic and multiscale modeling and simulation (see Section 9.5). The above rate criterion using activation energy $Q(\sigma,T)$ to explain the discrepancy between simulation and experiments fits within the exciting, broad trends in modern materials science. This kind of method is applicable to any thermally activated process including nucleation of stacking faults and dislocation loops, voids and microcracks, and ferroelectric and ferromagnetic domains.[68] As noted by the authors,[67] "more generally, this work demonstrates that multiscale methods and rate dependence are not just only desirable but can be essential in pursuing the proper prediction of material behavior, and are thus key components in the thrust for designing materials using computational materials science." Chapter 1 provides more details on this point.

## References

[1] and [2] Tadmor, E. B., Ortiz, M., and Phillips, R. (1996) Quasi-continuum analysis of defects in solids. *Philos. Mag. A*, **73**, 1529.

[3] Cook, R. D. (1981) *Concepts and Applications of Finite Element Analysis*, 2nd edn., John Wiley & Sons, Ltd.

[4] Chandrupatla, T. R. and Belegundu, A. D. (2002) *Introduction to Finite Elements in Engineering*, 3rd edn., Prentice Hall.

[5] Mortensen, J. J., Schiøtz, J., and Jacobsen, K. W. (2002) The quasicontinuum method revisited. *SIMU Newsletter*, **4**, 119.

[6] Shin, C. S., Fivel, M. C., Rodney, D., *et al.* (2001) Formation and strength of dislocation junctions in fcc metals: A study by dislocation dynamics and atomistic simulations. *Journal de Physique* IV, **11**, 19.

[7] Phillips, R., Rodney, D., Shenoy, V., *et al.* (1999) Hierarchical models of plasticity: Dislocation nucleation and interaction. *Modelling Simul. Mater. Sci. Eng.*, **7**, 769.

[8] Rodney, D. and Phillips, R. (1999) Structure and strength of dislocation junctions: An atomic level analysis. *Phys. Rev. Lett.*, **82**, 1704.

[9] Shenoy, V., Shenoy, V., and Phillips, R. (1999) Finite-temperature quasi-continuum methods. *Materials Research Society Symposium Proceedings*, **538**, 465.

[10] Miller, R E. and Tadmor, E. B. (2002) The quasicontinuum method: Overview, applications and current directions. *Journal of Computer-Aided Materials Design*, **9**, 203.

[11] Miller, R., Tadmor, E. B., Phillips, R., and Ortiz, M. (1998) Quasicontinuum simulation of fracture at the atomic scale. *Modelling Simul. Mater. Sci. Eng.*, **6**, 607.

[12] Miller, R., Ortiz, M., Phillips, R., *et al.* (1998) Quasicontinuum models of fracture and plasticity. *Engineering Fracture Mechanics*, **61**, 427.

[13] Pillai, A. R. and Miller, R. E. (2001) Crack behaviour at bi-crystal interfaces: A mixed atomistic and continuum approach. *Materials Research Society Symposium Proceedings*, **653**, Z.2.9.

[14] Hai, S. and Tadmor, E. B. (2003) Deformation twinning at aluminum crack tips. *Acta Mater.*, **51**, 117.

[15] Tadmor, E. B., Miller, R., Phillips, R., and Ortiz, M. (1999) Nanoindentation and incipient plasticity. *Journal of Materials Research*, **14**, 2233.

[16] Picu, P. C. (2000) Atomistic-continuum simulation of nano-indentation in molybdenum. *Journal of Computer-Aided Materials Design*, **7**, 77.

[17] Shenoy, V. B., Phillips, R., and Tadmor, E. B. (2000) Nucleation of dislocations beneath a plane strain indenter. *J. Mech. Phys. Solids*, **48**, 649.

[18] Smith, G. S., Tadmor, E. B., and Kaxiras, E. (2000) Multiscale simulation of loading and electrical resistance in silicon nanoindentation. *Phys. Rev. Lett.*, **84**, 1260.

[19] Smith, G. S., Tadmor, E. B., Bernstein, N., and Kaxiras, E. (2001) Multiscale simulations of silicon nanoindentation. *Acta Mater.*, **49**, 4089.

[20] Bernstein, N. and Kaxiras, E. (1997) Nonorthogonal tight-binding Hamiltonians for defects and interfaces in silicon. *Phys. Rev. B*, **56**, 10488.

[21] Rodney, D. and Phillips, R. (1999) Structure and strength of dislocation junctions: An atomic level analysis. *Phys. Rev. Lett.*, **82**, 1704.

[22] Miller, R. E. and Tadmor, E. B. (2009) A unified framework and performance benchmark of fourteen multiscale atomistic/continuum coupling methods. *Modelling Simul. Mater. Sci. Eng.*, **17**, 053001.

[23] Shenoy, V. B., Miller, R., Tadmor, E. B., *et al.* (1999) An adaptive finite element approach to atomic scale mechanics: The quasicontinuum method. *J. Mech. Phys. Solids*, **47**, 611.

[24] Knap, J. and Ortiz, M. (2001) An analysis of the quasicontinuum method. *J. Mech. Phys. Solids*, **49**, 1891.

[25] Eidel, B. and Stukowski, A. (2009) A variational formation of the quasicontinuum method based on energy sampling of clusters. *J. Mech. Phys. Solids*, **57**, 87.

[26] Rice, J. R. (1992) Dislocation nucleation from a crack tip: An analysis based on Peierls concept. *J. Mech. Phys. Solids*, **40**(2), 239.

[27] Picu, P. C. (2000) Atomistic-continuum simulation of nano-indentation in molybdenum. *Journal of Computer-Aided Materials Design*, **7**, 77.

[28] Cleri, F., Wolf, D., Yip, S., *et al.* (1997) Simulation of dislocation nucleation and motion from a crack tip. *Acta Mater.*, **45**(12), 4993.

[29] Pond, R. C. and Garcia-Garcia, L. M. F. (1981) Deformation twinning in aluminum. *Inst. Phys. Conf. Ser.*, **61**, 495.

[30] Rodney, D. and Phillips, R. (1999) Structure and strength of dislocation junctions: An atomic level analysis. *Phys. Rev. Lett.*, **82**, 1704.

[31] Miller, R. E. and Rodney, D. (2008) On the nonlocal nature of dislocation nucleation during nanoindentation. *J. Mech. Phys. Solids*, **56**, 1203.

[32] Hardikar, K., Shenoy, V., and Phillips, R. (2001) Reconciliation of atomic-level and continuum notions concerning the interaction of dislocations and obstacles. *J. Mech. Phys. Solids*, **49**, 1951.

[33] Tadmor, E. B., Waghmare, U. V., Smith, G. S., and Kaxiras, E. (2002) Polarization switching in PbTiO$_3$: An *ab initio* finite element simulation. *Acta Mater.*, **50**, 2989.

[34] Curtin, W. A. and Miller, R. E. (2003) Atomistic/continuum coupling in computational materials science. *Modelling Simul. Mater. Sci. Eng.*, **11**, R33.

[35] Xiao, S. P. and Belytschko, T. (2004) A bridging domain method for coupling continua with molecular dynamics. *Computer Methods in Applied Mechanics and Engineering*, **193**, 1645.

[36] Fung, Y. C. (1994) *A First Course in Continuum Mechanics*, 3rd edn., Prentice Hall.

[37] Jiang, H., Zhang, P., Liu, B., *et al.* (2003) The effect of nanotube radius on the constitutive model for carbon nanotubes. *Computational Materials Science*, **39**, 429.

[38] Darkrim, F. L., Malbrunot, P., and Tartaglia, G. P. (2002) Review of hydrogen storage by adsorption in carbon nanotubes. *Int. J. Hydrogen Energy*, **27**(2), 193.

[39] Dillon, A. C., Jones, K. M., Bekkedahl, T. A., *et al.* (1997) Storage of hydrogen in single-walled carbon nanotubes. *Nature*, **386**, 377.

[40] Liu, C., Fan, Y. Y., Liu, M., *et al.* (1999) Hydrogen storage in single-walled carbon nanotubes at room temperature. *Science*, **286**(5442), 1127.

[41] Darkrim, F. L. and Levesque, D. (1998) Monte Carlo simulations of hydrogen adsorption in single-walled carbon nanotubes. *J. Chem. Phys.*, **109**(12), 4981.

[42] Darkrim, F. L. and Malbrunot, P. (2000) High adsorptive property of opened carbon nanotubes at 77K. *J. Phys. Chem.*, **104**, 6773.

[43] Gu, C., Gao, G.-H., Yu, Y.-X., and Mao, Z.-Q. (2001) Simulation study of hydrogen storage in single-walled carbon nanotubes. *Int. J. Hydrogen Energy*, **26**(7), 691.

[44] Chen, Y. L., Liu, B., Wu, J., *et al.* (2008) Mechanics of hydrogen storage in carbon nanotubes. *J. Mech. Phys. Solids*, **56**, 3224.

[45] Dresselhaus, M. S., Dresselhaus, G., and Saito, R. (1995) Physics of carbon nanotube. *Carbon*, **33**, 883.

[46] Brenner, D. W., Shenderova, O. A., Harroson, J. A., *et al.* (2002) A second-generation reactive empirical bond order (REBO) potential energy expression for hydrocarbons. *J. Phys. Condens. Matter*, **14**, 783.

[47] Zhang, P., Jiang, H., Huang, Y., *et al.* (2004) An atomistic-based continuum theory for carbon nanotubes: Analysis of fracture nucleation. *J. Mech. Phys. Solids*, **52**, 977.

[48] Wu, J., Hwang, K. C., and Huang, Y. (2008) An atomistic-based finite-deformation shell theory for single-wall carbon nanotubes. *J. Mech. Phys. Solids*, **56**, 279.

[49] Liu, B., Jiang, H., Huang, Y., *et al.* (2005) Atomic-scale finite element method in multiscale computation with applications to carbon nanotubes. *Phys. Rev. B*, **72**(3), 035435.

[50] Jiang, L. Y., Huang, Y., Jiang, H., *et al.* (2006) A cohesive law for carbon nanotube/polymer interfaces based on the van der Waals force. *J. Mech. Phys. Solids*, **54**, 2436.

[51] Lu, W. B., Wu, J., Jiang, L. Y., and Huang, Y. (2007) A cohesive law for multi-wall carbon nanotubes. *Philos. Mag.*, **87**, 2221.

[52] Robertson, D. H., Brenner, D. W., and Mintmire, J. W. (1996) Energy of nanoscale graphite tubules. *Phys. Rev. B*, **59**, 235.

[53] Jiang, H., Zhang, P., Liu, B., *et al.* (2003) The effect of nanotube radius on the constitutive model for carbon nanotubes. *Computational Materials Science*, **39**, 429.

[54] Cornwell, C. F. and Wille, L. T. (1997) Elastic properties of single walled carbon nanotubes in compression. *Solid State Communications*, **101**, 555.

[55] Goze, C., Vaccarini, L., Henrard, L., *et al.* (1999) Elastic and mechanical properties of carbon nanotubes. *Synthetic Metals*, **103**, 2500.

[56] Yakobson, B. I., Campbell, M. P., Brabec, C. J., and Bernholc, J. (1997) High strain rate fracture and "intramolecular plasticity" in carbon nanotubes. *Computational Materials Science*, **8**, 341.

[57] Nardelli, M. B., Yakobsen, B. I., and Bernholc, J. (1998) Brittle and ductile behavior in carbon nanotubes. *Phys. Rev. Lett.*, **81**, 4656.

[58] Nardelli, M. B., Yakobson, B. I., and Bernholc, J. (1998) Mechanism of strain release in carbon nanotubes. *Phys. Rev. B*, **57**, R4277.

[59] Yakobson, B. I. (1998) Mechanical relaxation and "intramolecular plasticity" in carbon nanotubes. *Phys. Rev. Lett.*, **72**, 918.

[60] Jiang, H., Feng, X. Q., Huang, Y., *et al.* (2004) Defect nucleation in carbon nanotubes under tension and torsion: Stone–Wales transformation. *Computer Methods in Applied Mechanics and Engineering*, **193**, 3419.

[61] Song, J., Jiang, H., Shi, D. L., *et al.* (2006) Stone-Wales transformation: Precursor of fracture in carbon nanotubes. *Int. J. Mech. Sci.*, **48**, 1464.

[62] Qin, Z., Feng, X-Q, Zou, J., *et al.* (2007) Superior flexibility of super carbon nanotubes: Molecular dynamics simulations, *Appl. Phys. Lett.*, **91**, 043108-1.

[63] Tombler, T. W., Zhou, C., Alexseyev, L., *et al.* (2000) Reversible electromechanical characteristics of carbon nanotubes under local-probe manipulation. *Nature,* **405,** 769.

[64] Liu, B., Jiang, H., Johnson, H. T., and Huang, Y. (2004) The influence of mechanical deformation on the electrical properties of single wall carbon nanotubes. *J. Mech. Phys. Solids,* **52,** 1.

[65] Jiang, H., Huang, Y., and Hwang, K. C. (2005) A finite-temperature continuum theory based on interatomic potentials. *Journal of Engineering Materials and Technology,* **127,** 408.

[66] Zhang, H. W., Wang, J. B., and Guo, X. (2005) Predicting the elastic properties of single-walled carbon nanotubes. *J. Mech. Phys. Solids,* **53,** 1929.

[67] Warner, D. H., Curtin, W. A., and Qu, S. (2007) Rate dependence of crack-tip processes predicts twinning trends in f.c.c. metals. *Nat. Mater.,* **6,** 876.

[68] Shen, S., Li, J., and Wang, Y. (2008) Finding critical nucleus in solid-state transformations, *Metall. Mater. Trans. A,* **39,** 976.

# 7

# Further Introduction to Concurrent Multiscale Methods

The main focus of this chapter is an introduction to the various versions of concurrent multiscale methods as well as the difficulties in developing multiscale simulation schemes. Specifically, three direct coupling (DC) methods including MAAD, CADD, and bridge domain methods are further introduced. They belong to multiscale methods of, respectively, energy-based, force-based, and energy-based but with blended region. In addition, the embedded statistical coupling method (ESCM) is also introduced. The difficulties in a seamless transition between the atomistic and continuum scales caused by the incompatibility of stiffness and constitutive laws between materials at two sides of the interface are introduced through 1D analysis. Benchmark tests are introduced to assess the accuracy of different DC methods.

## 7.1 General Feature in Geometry of Concurrent Multiscale Modeling

Now that we have completed the study of GP and QC multiscale methods in the last two chapters, it may be a good time to highlight some common points through looking at general features of the geometry and physics of concurrent multiscale modeling. This will be helpful in distinguishing various concurrent models developed recently for multiscale analysis.

Concurrent multiscale analysis simultaneously carries out the calculations for materials with different scales. The key issue of this kind of multiscale analysis is how to handle the interface region at the boundary between different scales. Figure 5.4 has shown an example of how to use two imaginary interfacial areas for developing natural interface conditions to link variables at upper and lower scales. That treatment is used for the GP method. At this moment, most concurrent models in a coupled atomistic/continuum problem use the so-called direct coupling or direct connection (DC) method in which the basic approaches that relate atoms and finite element nodes are set out in a one-to-one manner, or through a form of interpolation.[1]

### 7.1.1 Interface Design of the DC Multiscale Models

In multiscale analysis across the atomic and continuum scale, one material region is typically described using atoms, while the other is described using continuous medium. The key to this coupling method is how to characterize the transition region at the interface between atomistic and continuum regions. This

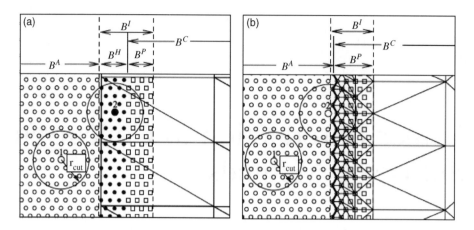

**Figure 7.1**   A generic interface schematic in a coupled atomistic/continuum problem in the DC method. The interface region $B^I$ of the model on the left includes a handshake region, $B^H$, besides the padding region $B^P$, while the model on the right does not. Padding atoms are shown as open squares, handshake atoms as black circles and regular atoms as white circles (Reproduced from Miller, R. E. and Tadmor, E. B. (2009) A unified framework and performance benchmark of fourteen multiscale atomistic/continuum coupling methods. *Modelling and Simulation in Materials Science and Engineering*, **17**, 053001. Copyright © IOP)

transition region exists in almost all concurrent multiscale models. However, different methods characterize the transition region with different details, endowing the region with different approximate features.

Figure 7.1 shows a generic interface schematic of these methods. As is seen, triangular finite elements are placed in the continuum region of $B^C$ and there are two kinds of interface region $B^I$ between the continuum region $B^C$ and the atomistic region $B^A$.

Figure 7.1(a) is a more general case which includes both handshake region $B^H$ and padding atom region $B^P$. Figure 7.1(b) only includes a padding atom region and is the case used most.

The mechanism of how padding atoms work is shown in more detail in Figure 7.1(b) in which finite element meshes and padding atoms coexist over the $B^P$ region. As we discussed in detail in Section 5.3.1, to produce natural boundary conditions for the regular atoms it is necessary to introduce some special atoms. Padding atoms and the imaginary atoms introduced in Figure 5.4 are these kinds of atoms. There are two circles around atoms 1 and 2 in Figure 7.1 whose radius is the cutoff radius $r_{cut}$. For atom 1 its circle does not intersect with the continuum, its fully atomistic interactive force can be accurately calculated. For atom 2, however, its circle interacts with the continuum, thus the overlapping part must be replaced by padding atoms to create the fully atomistic environment to calculate fully the interactomic force. This emphasizes the significance of the introduction of the padding region.

The handshake region of some models is introduced to provide a gradual transition from the atomistic to the continuum. Many models such as the QC and GP models do not have the handshake region. Other models do, but the ways to link the continuum and atomistic region are quite different. We will introduce the bridging domain method in Section 7.7 to show this type of interface.

### 7.1.2   Connection and Compatibility Between Atom/Continuum at the Interface

On the other hand, the finite element (continuum) region overlapping with the padding area plays an important role for connection and compatibility between the atomistic and continuum regions. Most DC models require a direct atom-node correspondence along the edge of the finite element region (the heavy jagged line in Figure 7.1b). This provides a way to impose displacement boundary conditions on the finite elements by the coincident atoms. These atoms are formally part of $B^A$ and their energy or force is computed for the atomistic region.[2] However, their positions define the displacements of a set of element nodes at the boundary, thus they impose a displacement boundary condition on the finite elements of $B^C$.

A fully compatible boundary condition also requires atoms' motion to follow the element displacement. This is realized by determining the displacements of padding atoms from the nodal displacements in $B^C$. The way is to connect the padding atoms directly to FE nodes or using interpolation function to determine the position of padding atoms by the node positions. In other words, the padding atoms are "adhering" to the FE nodes in the overlapping area of $B^P$. Thus the displacements of FE nodes will also affect the position of atoms at the boundary to meet the compatibility conditions in each direction.

The shortcoming of this kind of interface design is a restriction on the finite element mesh, which must be redefined to the atomic scale on the continuum side of the interface. In other words, the FE mesh must be consistent with the underlying atomistic configuration, which is normally a crystalline lattice and requires additional lattice-based algorithms.[3] There are other designs of the interface compatibility to avoid these additional requirements, where displacement boundary conditions are enforced only in some average sense, or with a lower accuracy method. The latter is called weak compatibility and the former is called strong compatibility.[2] These methods make mesh generation much easier.

Note that both the strong and weak compatibility interface design discussed here belong to the DC method, which does not consider the statistical property of the data to link the atomistic scale and the continuum scale as the ESCM does. From a statistical point of view, the data obtained at one instant of the relaxation process is not sufficient to characterize the system property including compatibility behavior and some kinds of statistical average over time and space may be needed (see Section 2.8.2.).

## 7.2 Physical Features of Concurrent Multiscale Models

Two aspects of physical features are used to distinguish concurrent multiscale models, one is the formulation principles and the other is the constitutive laws used for the continuum domain.

### 7.2.1 Energy-based and Force-based Formulation

There are two basic methods in mechanics to formulate the concurrent multiscale methods. The first is the energy-based formulation which is based on the energy principle such as the minimum potential energy for equilibrium in statics problems. This principle and its application is introduced in detail in Sections 6.2 and 6.3.5. The second is the force-based formulation which is to develop a physically motivated force system on each degree of freedom; equilibrium is reached when the resultant force reaches zero for static problem.

In the idealized case, these two methods are equivalent. Taking homework (6.1) of the four-spring structure under the action of two applied forces as an example (see Figure 6.2), the energy-based method requires the minimization of the following potential energy of the system

$$\Pi = -\frac{1}{2}k_1q_1^2 + \frac{1}{2}k_2(q_2-q_1)^2 + \frac{1}{2}k_3(q_3-q_1)^2 + \frac{1}{2}k_4q_3^2 - F_2q_2 - F_3q_3 \tag{7.1}$$

by

$$\frac{\partial \Pi}{\partial q_i} = 0 \quad (i = 1, 2, 3) \tag{7.2}$$

which produces the following three equations:

$$\frac{\partial \Pi}{\partial q_1} = -k_2(q_2-q_1) + k_1q_1 - k_3(q_3-q_1) = 0 \tag{7.3}$$

$$\frac{\partial \Pi}{\partial q_2} = k_2(q_2 - q_1) - F_2 = 0 \qquad (7.4)$$

$$\frac{\partial \Pi}{\partial q_3} = k_3(q_3 - q_1) + k_4 q_3 - F_3 = 0 \qquad (7.5)$$

However, these three equations in determining the three degree of freedom, $q_1$, $q_2$, $q_3$ can be easily obtained by the force-based method. Specifically, by writing the force equilibrium equation between the force of spring 1 and the applied force $F_1$ for node 1, equation (7.3) is obtained, and by writing the force equilibrium condition for nodes 2 and 3 equations (7.4) and (7.5) are obtained.

In reality, these two basic methods may not be equivalent due to the error caused by either or both the approximate estimation or errors in calculating the system energy or the forces at nodes/atoms. Talking about the energy-based method, in Section 6.4.8.1 we have derived (6.5g)

$$\bar{f}_g^\alpha = -\frac{\partial \Pi^{Err}}{\partial \bar{u}^\alpha} \qquad (6.5g)$$

which clearly indicates that the artificial ghost force $\bar{f}_g^\alpha$ is closely related to the error $\Pi^{Err}$ of the estimated potential energy $\Pi$. As we have seen in Section 6.4.8.2 the disadvantage of the energy-based DC method is that it is difficult and complicated to accurately eliminate the non-physical effects such as the ghost force and improper wave propagation (see next section).

On the other hand, in the force-based method, the force used in the formulation may not have corresponding well-defined system energy, thus the force introduced for the formulation may involve some unsound judgment, say, using these forces to establish the equilibrium condition but actually these forces are for unequilibrated atoms or nodes. Due to these considerations, force-based formulations can be slow to equilibrate and can converge to unstable equilibrium states such as the so-called saddle points or maxima in the energy surface. This system can be non-conservative and numerically unstable. However, in general the force-based formulation may have higher accuracy than the energy-based method.

## 7.2.2    Constitutive Laws in the Formulation

Most multiscale methods use different constitutive laws for the atomistic region and the continuum region. In the continuum region, local constitutive laws based on strain energy density W by equations (6.9g, h) is adopted. This type of constitutive law has local response (or property) in the sense that the force (or stress) is only related to the deformation (or strain) at the same location. For example, in an elastic material the local constitutive relation of the continuous medium means that the strain at any material point A is only a localized response. It is determined by local stress or the force per unit area applied by another material element at that location, which directly contacts this point.

On the other hand, the constitutive laws in the atomistic region are based on the interatomic potential. This type of constitutive law has non-local response (or property) in the sense that the interatomic forces are related to all atoms that can have a certain distance from the atom at hand. In fact, motion of an atom or deformation in its adjacent region is affected not only by the atoms which directly neighbor on this point, but also by the atoms further away. The influence is determined by the attraction force or repulsion force between atoms, governed by the potential function. Specifically, the atom $\alpha$ is subject to forces from all the atoms that are in the effect field with distance less than the cutoff radius $r_{cutoff}$. Therefore, the stress expressed by (3.10j) must be calculated by taking all of the atoms in that adjacent region into account.

Since the constitutive relation is different for the two material models (other than the interface), the force equilibrium cannot be obtained if compatible deformation is assured, resulting in a "ghost force" which has already been explained in detail in Section 6.4.8. Conversely, if equal force is assured, deformation will not be compatible and distortion is most likely to occur.

As discussed in Section 6.7, this inconsistency of the constitutive laws of local and non-local in the atomistic and continuum region produces incompatibility at the interface. This is one of the reasons why the GP method consisting of generalized particles and the fully non-localized QC method were developed so that the responses in the whole model are non-local.

Regarding the treatment of the strain energy density W for developing constitutive laws of the finite elements, different methods have different approaches. In some cases for small deformation problems one uses a linear elasticity model with elastic constants fitted to the properties of the atomistic model used in the atomistic region. In other models, a finite strain nonlinear formulation is adopted with the Cauchy-Born rule as discussed at length in Part 2 of Chapter 6 for the QC method. The latter can can be used to describe the constitutive laws in the nonlinear region.

# 7.3 MAAD Method for Analysis Across *ab initio*, Atomic and Macroscopic Scales

Abraham *et al.* developed the macroscopic, atomistic, *ab initio* dynamics method (MAAD) which is also called the coupling length scale (CLS) method.[4–6] The MAAD method spans the quantum, atomic and macroscopic scales. Although this appears to be a difficult research plan, its development has made important contributions to promoting and accelerating subsequent multiscale investigations.[7] The MAAD method focused initially on the problem of fracture in silicon, later it was applied to the vibrational response of nanoscale resonators.[8]

## 7.3.1 Partitioning and Coupling of Model Region

The basic idea of this approach is to unite tight binding (TB), molecular dynamics (MD) and finite element (FE) methods so that calculations can be performed simultaneously at the quantum, atomic and macroscopic scale. While performing the three kinds of calculations simultaneously, information such as calculation results continuously exchanges between any two of the calculations. For example, when investigating crack propagation in solids, the object will be partitioned into five regions, including the macroscopic continuum region described by FE, the vicinity of crack calculated with MD, the crack-tip region analyzed by TB, one handshaking region between the FE and MD (FE/MD), and the other handshaking region between MD and TB (MD/TB). Figure 7.2 illustrates the partition in a simulation of crack propagation in a silicon plate.[9] This partition method is not only used in crack propagation analysis in solids, but also in the analysis of problems for MEMS. In the following several paragraphs, the function of each region will be addressed.

### 7.3.1.1 Tight Binding (TB) Region

Among the millions of atoms in the simulation model, only several hundreds of atoms are in the TB region, characterized by the angstrom Å length scale. This region is characterized by the breaking of bonds and reconstruction of broken bonds. Dynamics in the TB region includes motion of particles at the crack tip as well as motion and evolution of particles under friction which are located at the contacting region of the MEMS gears. It is very important to estimate accurately the energy involved. However, because the bonds are twisted in these regions, the energy cannot be determined by empirical atomic potential. Therefore, quantum mechanics is adopted instead. Sections 4.7 and 4.8 have shown that the TB method has the fastest calculation speed in comparison to other quantum mechanics models such as HF and DFT methods. Also, the TB method includes basic physics of atomic bonding and can efficiently derive stress, therefore, it is adopted to analyze the crack-tip region.

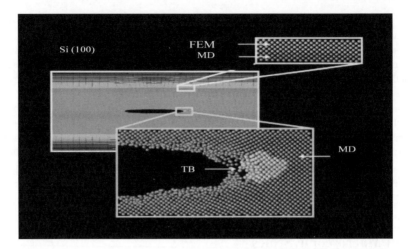

**Figure 7.2**    Partition of tight binding (TB), molecular dynamics (MD) and finite element (FE) regions in the study of crack propagation in a silicon plate using the MAAD approach (Reproduced from Rudd, R. E. and Broughton, J. Q. (2000) Concurrent coupling of length scales in solid state systems. *Physica Status Solidi* (b), 217, 251. © Wiley-VCH Verlag GmbH & Co. KGaA)

### 7.3.1.2  Molecular Dynamics (MD) Region

In addition to providing appropriate dynamic and thermal environment to the TB region, the MD region itself takes important roles which describe the physics of the system at the nanoscale. This is so because much of the coupling at the quantum scale (angstrom) can be neglected. In this case, MD becomes the smallest length scale of characterization. For example, a MEMS resonator is such a system with its important features appearing at the nanoscale and with the FE region, handshake region and MD region playing important roles together in this system.

Figure 7.3 shows that MD simulations are performed in the mid-part of the resonator specimen where strain oscillation is strong. In the region from the mid-part to clamping end the FE method is adopted. The transition region between the FE region and the MD region is considered the handshaking region. This simulation of a micron resonator reveals the important function of MD. In addition to this, MD also reveals some important physical mechanisms, which can be seen in the study of crack propagation.

### 7.3.1.3  Continuum (FEM) Region

Global material properties are studied at the macroscopic scale so that long-range fields, such as the strain field, can be studied. This kind of simulation includes the calculation of long-range force in nanoindentation, delamination of ceramic interfaces, and dynamics of MEMS. The FE method is employed here because MD is not capable of describing the mechanisms at this scale. FE analysis mainly addresses the high energy region of the system, such as the vicinity of the crack surface and the dislocation emission region in which the FE mesh should be very small and use high accuracy elements.

In addition, at the far field region near the peripheral boundary of a crack, many atoms vibrate only around their balance position. This kind of region acts to transfer the force and wave form boundaries, and therefore it cannot be neglected, but it can be studied with larger grid sizes due to its relative homogeneous deformation field.

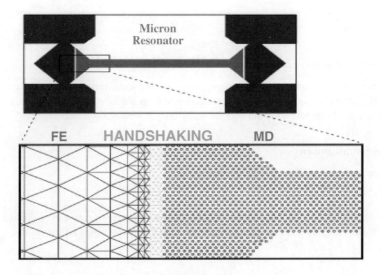

**Figure 7.3** Multiscale analysis of micron resonator. MD is employed in the region where strain oscillation is strong. The FE method is employed in the peripheral region where strain variation is small. Compatible boundary conditions are established between these two regions by the handshaking region (Reproduced from Rudd, R. E. and Broughton, J. Q. (2000) Concurrent coupling of length scales in solid state systems. *Physica Status Solidi* (b), 217, 251. © Wiley-VCH Verlag GmbH & Co. KGaA)

## 7.3.2 System Energy and Hamiltonian in Different Regions

MAAD is an energy-based multiscale method like the QC. A basic approach of combining various regions for simultaneous analysis is to express the system energy following (2.14d) to construct a Hamiltonian $H_{tot}$ for the entire system. Subsequently, velocity of atoms and particles can be obtained by the derivation of Hamiltonian $H_{tot}$ with respect to momentum $\bar{p}_i$ (see eq. (2.14e)), then their position vector and trajectory can be obtained through numerical integration.

### 7.3.2.1 System Hamiltonian

The system's Hamiltonian denotes its total energy and can be written as a sum of the individual Hamiltonians of the FE, MD, TB and two handshake regions as follows:

$$H_{tot} = H_{FE}(\boldsymbol{u}_a, \dot{\boldsymbol{u}}_a) + H_{FE/MD}(\boldsymbol{r}_j, \dot{\boldsymbol{r}}_j, \boldsymbol{u}_a, \dot{\boldsymbol{u}}_a) + H_{MD}(\boldsymbol{r}_j, \dot{\boldsymbol{r}}_j) \\ + H_{MD/TB}(\boldsymbol{r}_j, \dot{\boldsymbol{r}}_j) + H_{TB}(\boldsymbol{r}_j, \dot{\boldsymbol{r}}_j) \tag{7.6}$$

where $u_a$ and $\dot{\boldsymbol{u}}_a$ are the displacement and velocity vectors of the FE nodes, respectively, and $\boldsymbol{r}_j$ and $\dot{\boldsymbol{r}}_j$ are the atom position and velocity vectors used in MD and TB, respectively.

### 7.3.2.2 MD Hamiltonian

In the MAAD method, the Hamilton vector of MD is written as:

$$H_{MD} = \frac{1}{2}\sum_i \frac{p_i^2}{m_i} + U(\boldsymbol{r}_1, \boldsymbol{r}_2, \cdots \boldsymbol{r}_n) \tag{7.7}$$

or

$$H_{MD} = \frac{1}{2}\sum_i \frac{p_i^2}{m_i} + \sum_{i,j(\neq i)} V(r_{ij}) \tag{7.8}$$

where the first term on the right side is the kinetic energy and the second term is the potential energy of the MD region. Equation (7.8) shows only pair potential, while the potential energy U in equation (7.7) may be written as the sum of the pair potential and the multi-body potential.

### 7.3.2.3 FE Hamiltonian

The Hamilton vector of a FE region can be written as:

$$H_{FE} = U_{FE} + K_{FE} \tag{7.9}$$

where the potential energy for linear elastic solid and system kinetic energy can be written as:

$$U_{FE} = \frac{1}{2}\int_\Omega \varepsilon_{ij}(r)C_{ijkl}\varepsilon_{kl}(r)d\Omega \tag{7.10}$$

$$K_{FE} = \frac{1}{2}\int_\Omega \rho(r)\,(\dot{u})^2\,d\Omega \tag{7.11}$$

where $\varepsilon_{ij}$ is a second-order strain tensor, $C_{ijkl}$ is a fourth-order elastic stiffness tensor. As shown in (6.9j), the tensor product of $\varepsilon_{ij}$ and $C_{ijkl}$ gives the stress tensor $\sigma_{ij}$. $\rho$ is density, $\Omega$ is the volume of material region, $\rho d\Omega$ gives the differential mass of the volume $d\Omega$. $\varepsilon_{ij}$ and $\rho$ are functions of the atom position vector $r$, and $\dot{u}$ is the velocity of particles, thus $(\rho d\Omega \dot{u}^2)/2$ gives the kinetic energy of the differential volume $d\Omega$.

### 7.3.2.4 TB Hamiltonian

The expression for the kinetic energy in the TB region is the same as (7.11), the particles involved are electrons since nuclei are relatively stationary as discussed in Chapter 4. TB potential energy is expressed by (4.8h). In the MAAD method, it is expressed as:

$$U_{TB} = \sum_{n=1}^{N_{occ}} \varepsilon_n + \sum_{i,j(j>i)} V^{rep}(r_{ij}) \tag{7.12}$$

In the above equation, $\varepsilon_n$ represents the interaction energy among electrons. The first term on the right side corresponds to $E_{bond}$ in (4.8h) which is the contribution of non-core electrons to the total energy, and $V^{rep}$ describes the mutually repulsive energy. The latter is similar to the term $E_{rep}$ in (4.8h), which is a sum of short-range repulsive pair potentials between the atomic core electrons and nuclei. The summation over $N_{occ}$ terms includes all of the electrons that reach the Fermi energy level, because mutual attraction takes place only when the atom reaches the Fermi energy level.[10] Summation of $V^{rep}$ in (7.12) is performed using $i$ and $j$ with $j > i$ to avoid double counting.

## 7.3.3   Handshake Region Design

In the handshake region of the MAAD method, every finite element node coincides with an atom at the interface similar to the heavy jagged line in Figure 7.1b. In order to avoid double counting of the energy, MAAD adopts the weight method to calculate the Hamiltonian in the handshake region as follows:

$$H_{FE/MD} = \omega_{FE}H_{FE} + \omega_{MD}H_{MD} \qquad (7.13)$$

$$H_{MD/TB} = \omega_{MD}H_{MD} + \omega_{TB}H_{TB} \qquad (7.14)$$

where $\omega$ is weight factor whose choice is not with a standard rule. In most cases, these weight factors are equal to 1/2, which represents the arithmetic average. With respect to the handshake region between the FE and MD, the potential energy is obtained by averaging the $U_{FE}$ in (7.8) and the $U$ in (7.10).

There are small differences between the MAAD and the QC method in the treatment at the interface, although both are ambiguous regarding the parceling of energy near the interface. The difference is weighting of the atomistic and continuum contributions. As we discussed in Section 6.4.7, the weighting in the QC is controlled by the volume $n^e\Omega_0$ in (6.4t) which is analogous to (7.13) and (7.14). These differences are relatively minor in statics problems and lead to only slight changes in the error and the rate of convergence of the models.[2]

### 7.3.4  Short Discussion on the MAAD Method

The MAAD method distinguishes itself from all other multiscale methods in that it not only couples the atomistic region with the continuum region, but also introduces the TB region into the atomistic region. As described in Section 4.11.1, this introduction couples the electronic motion characterized by quantum mechanics with atomic motion.

The handshake region between the MD and TB is different from the handshake region between the MD and FE. The former is sandwiched between two particle-type materials that possess non-local response, while for the latter, material in the continuum region possesses a local response which is quite different from the MD region which possesses non-local response.

MAAD can only be applied where the temperature is absolute zero, causing the kinetic energy of the atoms to be zero. The system is a static system. Broughton, one of the founders of MAAD, collaborating with Rudd, proposed a coarse-grained method[11] which is capable of taking finite temperature into consideration. This method emphasizes the coupling between MD and the continuum, while ignoring the TB.

## 7.4  Force-based Formulation of Concurrent Multiscale Modeling

In Parts 1 and 2 of Chapter 6 we discussed in detail the energy-based method and then took the QC methods as examples to show how to formulate the energy-based multiscale modeling schemes. As we mentioned, the shortcoming of the energy-based DC method is the ghost force or unphysical phenomena at the interface because of the difficulty of finding an exact energy function. Now, we would like to discuss how to develop the force-based formulation of multiscale models. This alternative method starts from forces directly without recourse to the construction of the total energy, which can indeed eliminate the ghost forces. To reach this goal, the method is designed so that the forces are identically zero for the crystal lattice which is in its correct equilibrium state.

The basic difference between the energy-based method from force-based method by Newtonian laws is that in the former forces are obtained by differentiating an energy, say Hamiltonian, whereas in the latter case forces are used directly. It is important to know for a non-conservative system, Hamiltonian and Newtonian formulations are not equivalent and additional terms to (2.14d) needs to be added for Hamitonian expression. A force-based formulation cannot be guaranteed to be conservative because a total energy cannot be accurately defined. Instead, the forces are constructed with no standard, rigorous rule. Thus, while the motivation is to make the model physically sound, there is a possibility that the resulting model is ill-behaved or unstable if the actually corresponding energy is at the maximum or at saddle point on the energy surface for all the possible configurations.

Miller and Tadmor[2] used Figure 7.2(b) to discuss how to construct a force-based model. They assume that two independent potential energy functionals exist: One is $\Pi^{atom}$ that treats the entire domain $B^A$ atomistically, and the other is $\Pi^{FE}$ that treats the entire $B^C$ as a continuum using finite elements. These two are independent in the sense that the forces $f^\alpha$ for atoms in the atomistic $B^A$ domain and forces $f^I$ in the continuum $B^C$ domain can be determined individually by, respectively,

$$f^\alpha = \frac{\partial \Pi^{atom}}{\partial \bar{u}^\alpha} \tag{7.15}$$

$$f^I = \frac{\partial \Pi^{FE}}{\partial \bar{U}^I} \tag{7.16}$$

where $\bar{U}^I$ is the node displacement vector as given in (6.21). Note that these forces are not the ones obtained by minimizing the combined energy functional $\Pi^{atom} + \Pi^{FE}$ because this combination is not the total energy of the system.

The advantage of the formulation by individually developing the potential energy is that it is easy to formulate the interface compatibility condition to eliminate the ghost forces. As discussed in Section 7.1.2 what we need to do for the connection between these regions with the two individual potentials is twofold: One is to make the padding atoms in the $B^P$ region controlled by the FE nodes to transfer the information from continuum to atomistic system. The other is to make the FE nodes at the interface such as at the heavy jagged line in Figure 7.2(b) controlled by the atoms to transfer the atomistic information to the continuum.

This is to a certain extent similar to Figure 5.4 of the GP method where the domain $W_{1image}$ transfers the atomistic information to the upscale (continuum), and the domain $W_{2image}$ transfers the information of upscale (continuum) to the lower scale (atomic). As we introduced in detail in Chapter 5, the GP method is completely different from all the DC and other types of multiscale method, specifically the continuum does not use a FE mesh but uses generalized particles of different scales so all its calculations at different scales follow the MD algorithm.

## 7.5    Coupled Atom Discrete Dislocation Dynamics (CADD) Multiscale Method

### 7.5.1    Realization of Force-based Formulation for CADD/FEAt

The earliest force-based method was the FEAt method.[12] This method with no handshake region was developed to investigate crack propagation in bcc crystals with combined finite-element and atomistic model. It develops strong compatibility at the interface to make it comparable to the QC and MAAD methods. Recently, the same force-based coupling method was used in the development of the method of CADD, short for coupled atomistic analysis with discrete dislocation.[13–15] The CADD method combines atomic scale analysis with discrete dislocation analysis in a continuum to perform multiscale plastic analysis.

The first purpose in the formulation of the CADD/FEAt methods is to develop the potential $\Pi^{atom}$ and $\Pi^{FE}$ mentioned above so that forces in atoms and nodes can be, respectively, determined by (7.15) and (7.16). Following (6.4e)

$$\Pi^{atom}(\tilde{u}) = \sum_{i=1}^{N} E_i(\tilde{u}) - \sum_{i=1}^{N} \bar{f}_{ext,i}\bar{u}_i \tag{7.17}$$

Note that here the summation includes the padding atoms which is completely different from the GP and QC multiscale energy-based method. For the latter, padding atoms do not belong to the modeling system. On the other hand, for force-based methods, the contribution of the padding atoms must be included.[16]

In the literature[2] the summation domain in (7.17) is written clearly by the domain symbol $\cup$ such as $i \in (B^A \cup B^P)$ to denote the variable i covers all atoms in both the domain $B^A$ and $B^P$.

Following Eq. (6.3m) and (6.3o), the potential energy for $\Pi^{FE}$ can be written as

$$\epsilon = \sum_{e=1}^{n_{elem}} V^e \sum_{q=1}^{n_Q} \omega_q W([N(\overline{X}_q^e)]\{\overline{U}^e\}) - \overline{F}^T[U] \tag{7.18}$$

Substituting (7.17) and (7.18) respectively into (7.15) and (7.16) the forces can be obtained. These forces are then used to move the atoms and nodes either dynamically through the integration of Newton's second law (see Section 2.4.3) or towards equilibrium in statics problems by a quasi-Newtonian scheme. The calculation will be conducted step by step, in each step forces are recomputed for the new atom and node positions.

Two points may be emphasized below for a deeper understanding of CADD/FEAt and other force-based methods.

### 7.5.1.1 More Padding Atoms are Necessary

The energy-based method is based on the minimum of potential energy of the system, however, the force-based method is based on the local derivative of the potential energy which needs to involve more atoms. Thus, force-based approaches require a padding region which is about twice as thick as that for a comparable energy-based method. To make these extra atoms properly coordinated as close to natural environment as possible by additional padding in each loading step, more time is needed, so that the force-based methods will be slightly slower than energy-based methods.[2]

### 7.5.1.2 Nonlinear but not Iterative Solution Scheme

The strong compatibility in these methods requires constraint displacement boundary conditions in both FE and atomistic domains. Specifically, padding atoms are constrained by FE nodes in the padding region and the last row of atoms impose their displacements on the first row of FE nodes. It is important to note that the solution of this problem is not an iterative process. The equations of the forces through derivatives of (7.17) and (7.18) are solved simultaneously with the strong displacement boundary conditions.

## 7.5.2    Basic Model for CADD

The main advantage of the CADD method is that it can deal with dislocations. These defects in the atomistic region may enter the continuum region through the interface between the atomic and continuum scale and then move under the elastic driving force in the continuum region.

As shown in the first diagram of Figure 7.4, a body $\Omega$ is assumed to exert a predefined force $T = T_0$ at the force boundary $\partial\Omega_T$ and displacement $u = u_0$ at the displacement boundary $\partial\Omega_u$. The body $\Omega$ can be imagined as cut into two parts, $\Omega_A$ and $\Omega_c$, in order to reduce the system's degree of freedom. $\Omega_A$ is the atom region while $\Omega_c$ is the continuum region. They are combined together via interface $\partial\Omega_I$. Assuming that N dislocations are contained in the continuum region, the Burgers vector and position vector of the ith atom are recorded as $b^i$ and $d^i$, respectively. Each material particle in the continuum region, as well as each atom particle in the atomistic region has a reference position coordinate $X$ in the initial configuration without defects. After a deformation in which they experience a displacement $u$, the final position is obtained as $x = X + u$ in the deformed region as shown in Figure 6.1. If this boundary value problem is to be solved, the stress tensor $\sigma$, strain $\varepsilon$, displacement vector $u$ and the discrete dislocation position vector $d^i$ (i = 1...N) in the continuum region must be obtained. The equilibrium position of the atomic system at the given boundary conditions must be simultaneously derived.

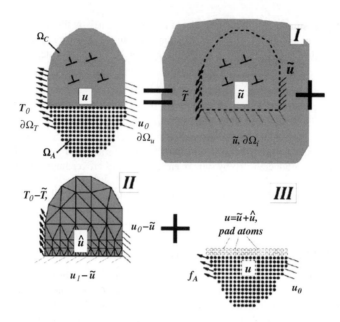

**Figure 7.4** A schematic of plastic analysis using coupled molecular dynamics (MD) and discrete dislocation (DD). The continuum couples with the atomistic region via the displacement $u_I$ of padding atoms. Atoms are coupled with continuum via the interface atoms which determine the boundary node displacement (Reproduced from Shilkrot, L. E., Miller, R.E., and Curtin, W.A. (2004) Multiscale plasticity modeling: coupled atomistics and discrete dislocation mechanics. *Journal of the Mechanics and Physics of Solids*, 52, 755, Elsevier)

## 7.5.3    Solution Scheme: A Superposition of Three Types of Boundary Value Problems

The CADD method provides a solution scheme to this multiscale boundary value problem across atom-continuum scales by decomposing it into three boundary value problems, which are shown in Figure 7.4 on the right of the equals sign.

### 7.5.3.1  Three Kinds of Boundary Value Problem

**Boundary value problem I**

This is a boundary value problem in an infinite large area. Assume that N dislocations in $\Omega_c$ occur in an infinitely large continuum region. The force and displacement is set to zero at the boundaries of the infinite region. This problem has already been solved by van der Giessen and Needleman.[17] The boundary displacement $\tilde{u}$, boundary force $\tilde{T}$ at the boundary of the continuum region $\Omega_C$ and the displacement $\tilde{u}$ at the interface $\partial\Omega_I$ can be obtained from the existing solution of this problem by van der Giessen and Needleman. $\tilde{u}$ and $\tilde{T}$ at $\Omega_C$ boundary will be used in the next boundary problem to produce a true boundary force for $\Omega_C$.

## Boundary value problem II

This is a boundary value problem in the continuum region $\Omega_C$. The rule in designing this problem is to make the superposition of this problem and boundary problem I equivalent to the real boundary conditions of region $\Omega_C$, i.e., $T_0$ at $\partial\Omega_T$ and $u_0$ at $\partial\Omega_u$. To reach this requirement, the force at the boundary condition will be set as $T_0 - \tilde{T}$, the displacement as $u_0 - \tilde{u}$ at external boundary and $u_I - \tilde{u}$ at interface $\partial\Omega_I$, thus the superposition of $(T_0 - \tilde{T})$ in Problem II and $\tilde{T}$ in Problem I produces the true boundary force in $\Omega_C$, and the same for the displacement boundary.

## Boundary value problem III

The atomic region $\Omega_A$ is subjected to an external displacement boundary condition $u_0$ and an external force boundary condition $T_0$; in addition the continuum boundary condition $u_I$ at interface $\partial\Omega_I$ also needs to be satisfied. Because the force imposed on the atomistic boundary $\partial\Omega_A$ is discrete, the force $f_A$ acting on the atoms provides an appropriate description of the force boundary of the atomistic region. In other words, the integration of $f_A$ in the unit area results in $T_0$.

### 7.5.3.2 Solution Scheme

Since the MD algorithm is discussed in detail in Chapter 2 and 3, we only need to discuss briefly the potential energy functional $\Pi^C$ of the continuum region which contains discrete dislocations and couples with the atomistic region. Problem I deals with discrete dislocations that are placed in an infinite homogeneous elastic material. CADD follows the method of van der Giessen and Needleman[17] to solve this problem. Specifically, the total stress $\tilde{\underline{\sigma}}^i$, strain tensor $\tilde{\underline{\varepsilon}}^i$ and displacement vector $\tilde{u}^i$ of this boundary problem can be expressed as:

$$\tilde{\sigma} = \sum_i^{N_d} \tilde{\underline{\sigma}}^i, \quad \tilde{\varepsilon} = \sum_i^{N_d} \tilde{\varepsilon}^i, \quad \tilde{u} = \sum_i^{N_d} \tilde{u}^i \tag{7.19}$$

where $N_d$ is the total number of dislocations in the region. These expressions indicate that the solution is the sum of all known elastic fields of discrete dislocation with position vector $d^i$ in the infinite elastic continuum medium. Because these summation results can not satisfy the boundary condition, boundary value problem II has to be employed to eliminate $\tilde{T}$ on $\partial\Omega_T$, and $\tilde{u}$ on $\partial\Omega_u$ and $\Omega_I$.

Boundary value problem II involves an elastic region without dislocation and is designed to correct the boundary error of the continuum region. The solution is free of discontinuity and singularity. Therefore, the solution can easily be obtained by FE method under the mentioned boundary loading of $(T_0 - \tilde{T})$ and $(u_0 - \tilde{u})$. Denoting the obtained displacement, stress and strain of boundary value problem II as $\hat{u}$, $\hat{\sigma}$ and $\hat{\varepsilon}$, respectively, the solution of the entire continuum region can be obtained by the summation of problems I and II, therefore:

$$u = \tilde{u} + \hat{u}, \quad \sigma = \tilde{\sigma} + \hat{\sigma}, \quad \varepsilon = \tilde{\varepsilon} + \hat{\varepsilon} \tag{7.20}$$

Note that $u = \tilde{u} + \hat{u}$ is also the boundary condition of boundary value problem III which is the sum of the displacement of problem I and II for padding atoms (Remark: $(u_I - \tilde{u})$ is for problem II).

Starting from the solution of equation (7.20), the potential energy $\Pi^C$ of the continuum region $\Omega_C$ can be written as:

$$\Pi^C = \frac{1}{2} \int_{\Omega_C} (\hat{\sigma} + \tilde{\sigma}) : (\hat{\varepsilon} + \tilde{\varepsilon}) dV - \int_{\partial\Omega} T_0(\hat{u} + \tilde{u}) dA \tag{7.21}$$

where the first item is the stored strain energy in the continuum region and the symbol: denotes the tensor product. The potential energy $\Pi^C$ is a function dependent on the displacement field $u$ in this region, the interface displacement field $\tilde{u}_I$ between this region and atom region, and the discrete dislocation distribution $d^i(i = 1, N_d)$ in this region.

The displacement field $\tilde{u}$ of boundary problem II may be obtained by minimizing the potential energy of equation (7.21) as described below:

$$\frac{\partial \Pi^C}{\partial \tilde{u}} = 0 \tag{7.22}$$

The force $p^i$ acting on the $i$th discrete dislocation can also be obtained by the derivative of the potential energy $\Pi^C$ with respect to its position vector $d^i$ as

$$\bar{p}^i = -\frac{\partial \Pi^C}{\partial \bar{d}} \tag{7.23}$$

## 7.6 1D Model for a Multiscale Dynamic Analysis

Previous sections have focused mainly on multiscale static problems. This section will discuss dynamic problems with a 1D model. The design of the interface region shown in Figure 7.1 can introduce a problem in that waves are reflected at the interface between the MD and FE regions. In fact, the wavelength of the wave transmitted from the MD region is smaller than that which can be absorbed by the continuum region, thus it is impossible for this wave to go into the FE region. However, as the energy is conservative in the entire system, it has no option but to return to the MD region. This reflection results in an artificial thermal increase, which leads to generation and retention of heat in the MD region. As a result it will increase the temperature and cause negative effects to the system, especially to the plastic deformation area. In some extreme circumstances, the retained heat can lead to the melting away of the entire MD region.[18]

Since the common size of FE meshes is incapable of absorbing waves from the MD region, the FE mesh at the MD/FE interface needs to be reduced to a size comparable to the atomic lattice constant. However, this reduction not only results in many more finite element cells, but also brings numerical problems because the temporal increment in the finite element for each step is determined by the smallest cell, so that reducing the cell size down to the atomic lattice constant would result in wasted time.

In order to better understand this problem, a 1D model proposed in Ref. [19] will be used to discuss some basic concepts and problems in dynamic analysis across the atomistic/continuum scales including the inconsistency at the FE/MD interface. It will primarily focus on the problem induced by gradual coarsening of FE mesh cells starting from the interface.

### 7.6.1 The Internal Force and Equivalent Mass of a Dynamic System

First, note that both the MD and FE models obey the Newtonian equation:

$$f = Ma \tag{7.24}$$

It turns out that force $f$ and mass $M$ of every system must be determined. The system's mass and force are dependent on each atom's mass and the subjected force. In the following, the relationships between pair atoms and individual mass (or force) will be determined. From the results, the relationship between the system's mass (or force) and single atoms can also be formulated.

#### 7.6.1.1 Pair Atom Internal Force

With respect to the MD system, force $f_{MD}$ can be derived from (2.1a)

$$\bar{F}_i = -\frac{\partial U(\bar{r}_1, \bar{r}_2 \ldots \bar{r}_N)}{\partial \bar{r}_i}, \quad i = 1, 2 \ldots N \tag{2.1a}$$

In the present 1D problem, it is assumed that the interaction between atoms obeys the following harmonic potential function:

$$U(r_{ij}) = \frac{1}{2}k(r_{ij}-r_0)^2 \tag{7.25}$$

where k is spring constant, $r_{ij}$ the distance between atoms, and $r_0$ the equilibrium length. In this situation the internal force by the interaction of a pair of atoms is given as:

$$f_i = -\frac{\partial U}{\partial r_{ij}} = -k(r_{ij}-r_0) \tag{7.26}$$

Assuming that $x_i$ and $x_j$ are the position coordinates of atoms $i$ and $j$ when they are in equilibrium, and assuming $d_i$ and $d_j$ are their displacements. Therefore $r_0 = x_j-x_i$, and their position after deformation can be determined by:

$$r_{ij} = (x_j+d_j)-(x_i+d_i) = r_0+d_j-d_i \tag{7.27}$$

Therefore:

$$r_{ij}-r_0 = d_j-d_i = \Delta x \tag{7.28}$$

By substituting it into equation (7.26):

$$f_i = -k(d_j-d_i) = -k\Delta x \tag{7.29}$$

This result indicates that the internal force actually is determined by its relative displacement, and does not depend on the original coordinates at equilibrium. It is assumed that the finite element is subject only to internal force, and any external force is absent in this specific example. The internal force is a product of the stiffness matrix $K$ and node displacement $d$ of the finite element which is similar to the left side of (6.3v), therefore:

$$f_{FE} = K_{FE}d_{FE} \tag{7.30}$$

For a 1D linear elastic system, with the matrix $\mathbf{d}_{FE} = [d_j \ d_i]^T$, by equation of node displacement similar to (7.29), the stiffness matrix can be written as:

$$K_{FE} = -k\begin{bmatrix} 1 & -1 \\ -1 & 1 \end{bmatrix} \tag{7.31}$$

### 7.6.1.2 Pair Atom Mass

In MD, the mass matrix is a diagonal matrix. As far as a two-particle system is concerned, the mass matrix $M_{MD}$, force matrix $F_{MD}$ and acceleration matrix $a_{MD}$ can be given as:

$$M_{MD} = \begin{pmatrix} m_1 & 0 \\ 0 & m_2 \end{pmatrix} \tag{7.32}$$

$$F_{MD} = \begin{pmatrix} F_1 \\ F_2 \end{pmatrix} \tag{7.33}$$

$$a_{MD} = \begin{pmatrix} a_1 \\ a_2 \end{pmatrix} \tag{7.34}$$

For a finite element system, the mass matrix by the consistent mass definition can be written as:[20]

$$M_{FE} = \int_{\Omega_0} \rho_0 N^T N d\Omega_0 \tag{7.35}$$

where $\rho_0$ is the material's original density before deformation, and $\Omega_0$ its original volume. $N$ is the shape function or interpolation function of the finite elements. Through this function, deformation at a material point in the finite element can be calculated by interpolating to the deformation of FE mesh nodes as shown in (6.3c). $N$ usually is a low-order power function. If a linear function is adopted such as the one given by (6.3h, i), the quality matrix of an element can be written as:

$$M_{FE} = \frac{\rho_0 A_0 l_0}{2} \begin{bmatrix} 1 & 0 \\ 0 & 1 \end{bmatrix}$$

(7.36)

where $A_0$ and $l_0$ are the initial area and length of a finite element. This equation shows that each finite element node supports half of the mass of the element.

## 7.6.2 Derivation of the FE/MD Coupled Motion Equation

Now, we would like to look at a MD/FEM coupling case in which a MD region composed of 101 atoms is on the left part and a FEM region with 100 finite elements is on the right part. The elements are with variable size and there is a handshake region between the MD and FEM regions similar to Figure 7.1a. This problem can be investigated by first establishing the coupling equation for a simplest case with four points shown in Figure 7.5 and then extend it to the complicated case. In the simplest case, three atoms are put at points 1, 2 and 3; two elements 23 and 34 are put between points 2 and 3 and 3 and 4 respectively, thus the space between points 2 and 3 is the handshake region; because FE nodes coincide exactly with the atoms in the atom region, this element may be regarded as the handshake finite element.

The main focus will be to define the mass matrix and force vector of the handshake element. For the former, the mass matrix may be defined similar to that in equation (7.32). For the latter, the force must be redefined in another way to ensure that the total force is equal to the sum of the forces imposed on the FE and MD at the same location. Here, the weighting method is adopted to distribute the total force to the MD and FE. Because the FE nodes coincide exactly with atoms at points 2 and 3, the total force can be uniformly distributed to MD and FE. Because the atoms and finite elements are of the same mass, the total mass matrix may be written as:

$$M = \begin{bmatrix} m & 0 & 0 & 0 \\ 0 & m & 0 & 0 \\ 0 & 0 & m & 0 \\ 0 & 0 & 0 & m \end{bmatrix}$$

(7.37)

In Figure 7.5, the atomistic spacing $h_a$ is also the spacing $l_0$ between two adjacent finite element nodes. This means that the atomic spacing is also the length designed for the finite elements that directly connect to the handshake region. Here, displacements of the atoms and nodes at the four points are assumed to be $d_1$, $d_2$, $d_3$ and $d_4$. Following (7.29) the force between the first pair of atoms can be written as

$$f = -k\Delta x = -k(d_2 - d_1)$$

(7.38)

Therefore, the force matrix may be written as:

$$\begin{bmatrix} f_1 \\ f_2 \end{bmatrix} = \begin{bmatrix} k(d_2 - d_1) \\ -k(d_2 - d_1) \end{bmatrix}$$

(7.39)

where the positive direction is assumed to be along $x$ (from left to right). For the finite element nodes 3 and 4, the force may be written as the product of the stiffness matrix and displacement, as[20]

$$\begin{bmatrix} f_3 \\ f_4 \end{bmatrix} = -\frac{k h_a}{l_0} \begin{bmatrix} 1 & -1 \\ -1 & 1 \end{bmatrix} \begin{bmatrix} d_3 \\ d_4 \end{bmatrix}$$

(7.40)

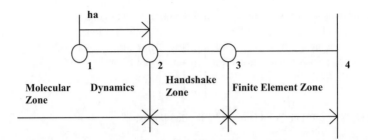

**Figure 7.5** Description of the coupling between MD and FEM in 1D problem (Reproduced from Park, H. S. and Liu, W. K. (2004) An introduction and tutorial on multiple-scale analysis in solids. *Computer Methods in Applied Mechanics and Engineering*, 193, 1733, Elsevier)

When the atomic spacing $h_a$ is equal to the FE length $l_0$, Eq. (7.40) is identical to Eq. (7.39); otherwise, the FE stiffness is less than atom's. The longer of the $l_0$ is the weak of the stiffness. First, we only consider the contribution of the inter-atomic and inter-node force, the corresponding force matrix can be written as

$$f = \begin{bmatrix} f_1 \\ f_2 \\ f_3 \\ f_4 \end{bmatrix} = \begin{bmatrix} k(d_2-d_1) \\ -k(d_2-d_1) \\ k(d_4-d_3) \\ -k(d_4-d_3) \end{bmatrix} \tag{7.41}$$

where the first two rows are contributed by two atoms at points 1 and 2, the third and fourth rows are contributed by element 3–4. Now, we are in a good position to add the contribution of the MD/FEM interaction force at handshake region 2–3 to force matrix (7.41). Because the FE nodes coincide with the MD atoms in this region, by assuming that each model undertakes half of the point force, the additional forces $f_2', f_3'$, which are contributed by coupled MD and FE and imposed on the handshake region nodes 2 and 3, are:[19]

$$\begin{bmatrix} f_2' \\ f_3' \end{bmatrix} = \frac{1}{2}\begin{bmatrix} f_{MD2} \\ f_{MD3} \end{bmatrix} + \frac{1}{2}\begin{bmatrix} f_{FE2} \\ f_{FE3} \end{bmatrix} \tag{7.42}$$

Following the form used in (7.39), the following results are given:

$$\begin{bmatrix} f_{MD2} \\ f_{MD3} \end{bmatrix} = k\begin{bmatrix} d_3-d_2 \\ -(d_3-d_2) \end{bmatrix} \tag{7.43}$$

If the atomic spacing $h_a$ is equal to the FE length $l_0$, following the form of (7.40), we have the following results:

$$\begin{bmatrix} f_{FE2} \\ f_{FE3} \end{bmatrix} = k\begin{bmatrix} d_3-d_2 \\ -(d_3-d_2) \end{bmatrix} \tag{7.44}$$

Substitute (7.43) and (7.44) into (7.42), resulting in:

$$\begin{bmatrix} f_2' \\ f_3' \end{bmatrix} = k\begin{bmatrix} d_3-d_2 \\ d_2-d_3 \end{bmatrix} \tag{7.45}$$

This force can be added to corresponding forces $f_2$ and $f_3$ in (7.41) to obtain the revised force matrix as

$$f = \begin{bmatrix} f_1 \\ f_2 \\ f_3 \\ f_4 \end{bmatrix} = \begin{bmatrix} k(d_2-d_1) \\ k(d_3-2d_2+d_1) \\ k(d_4-2d_3+d_2) \\ k(d_3-d_4) \end{bmatrix} \tag{7.46}$$

The acceleration matrix **a** can be expressed by the second derivative of $d_i$ ($i = 1,2\ldots4$) with respect to time as follows:

$$\mathbf{a} = \begin{bmatrix} a_1 \\ a_2 \\ a_3 \\ a_4 \end{bmatrix} = \begin{bmatrix} \ddot{d}_1 \\ \ddot{d}_2 \\ \ddot{d}_3 \\ \ddot{d}_4 \end{bmatrix} \tag{7.47}$$

Substituting (7.37), (7.46) and (7.47) into (7.24) one obtains:[19]

$$\begin{bmatrix} m & 0 & 0 & 0 \\ 0 & m & 0 & 0 \\ 0 & 0 & m & 0 \\ 0 & 0 & 0 & m \end{bmatrix} \begin{bmatrix} \ddot{d}_1 \\ \ddot{d}_2 \\ \ddot{d}_3 \\ \ddot{d}_4 \end{bmatrix} = \begin{bmatrix} k(d_2 - d_1) \\ k(d_3 - 2d_2 + d_1) \\ k(d_4 - 2d_3 + d_2) \\ k(d_3 - d_4) \end{bmatrix} \tag{7.48}$$

where the double dot symbol denotes the second derivatives with respect to time. The above equations are obtained based on the linear spring assumption, as well as the condition that the atomistic spacing is equal to the finite element size $l_0$. In this circumstance, MD and FEM contribute the equal node force as one can see from the comparison between (7.39) and (7.40) as well as comparison between (7.41) and (7.42).

### 7.6.3   *Numerical Example of the Coupling Between MD and FE*

In this section a numerical example by Park and Liu[19] will be given which considers the interaction between MD and FE at the handshake region. In the model, all the atoms/particles are symmetrically placed about $x = 0$. There are 101 atoms in the MD region from $x = -2$ to 2 (see Figure 7.6). In addition to the MD region, two FE regions have been symmetrically placed, each of which contains 50 elements positioned from $x=-10$ to $-1.96$ or from $x=1.96$ to 10. This structure contains two handshake regions where atoms coincide with FE nodes in the range from $x=-2$ to $-1.96$ and from $x=1.96$ to 2.

Following the method used in developing equation (7.46), governing equations may be established for this system. The same numerical integration method described in Sections 2.4.3 to 2.4.4 may be performed with respect to the incremental time step, such as the Verlert, Beeman or Gear method. The Ruth leapfrog method is employed by Park and Liu, which updates speed and displacement in two steps.

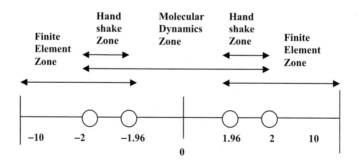

**Figure 7.6**   Schematic of the numerical example on 1D coupling between MD and FE (Reproduced from Park, H. S. and Liu, W. K. (2004) An introduction and tutorial on multiple-scale analysis in solids. *Computer Methods in Applied Mechanics and Engineering*, 193, 1733, Elsevier)

The numerical algorithm is given in (7.47) to (7.50). Subscript 0 represents the initial value, subscript 1 represents the value after the first step and subscript 2 represents the value for the second step.

$$\bar{V}_1 = \bar{V}_0 \tag{7.49}$$

$$\bar{d}_1 = \bar{d}_0 + \frac{1}{2}\bar{V}_1 \Delta t \tag{7.50}$$

$$\bar{V}_2 = \bar{V}_1 + \bar{f}(d_1)\Delta t \tag{7.51}$$

$$\bar{d}_2 = \bar{d}_1 + \frac{1}{2}\bar{V}_2 \Delta t \tag{7.52}$$

where $\bar{f}(d_1)$ denotes the force or acceleration at the new time step 1 (or after one step from the original step) where the mass is assumed to be 1, a unit mass. This method is superior to other methods in that force $\bar{f}$ is only calculated once. Because the calculation of force is the most time-consuming in MD, this method shows a remarkable advantage. In addition, due to the small number of calculations, this method requires very little computer memory.

### 7.6.4 Results and Discussion

First, a symmetric Gauss wave is introduced to the MD region at the central point $(x = 0)$ of the model. There are two kinds of calculations. In the first case, the spacing between the FE nodes is equal to the atomic lattice constant $h_a$, therefore $l_0 = h_a$ in equation (7.40). However, in the second case, the spacing $l_0$ between adjacent FE nodes gradually increases as the position moves away from the MD region. For simplicity, the mass of the atoms and the spring constant are both assumed to be 1.

The calculation result for the first case is illustrated in Figure 7.7, which shows that the wave propagates smoothly from the center to the two sides. Because the FE element size $l_0$ is the same as the atomistic

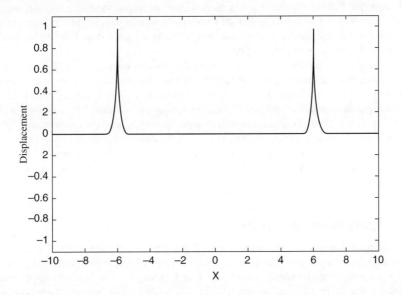

**Figure 7.7** Wave propagation in the FE region when the FE size is equal to the atomistic spacing, i.e., $l_0 = h_a$ (Reproduced from Park, H. S. and Liu, W. K. (2004) An introduction and tutorial on multiple-scale analysis in solids. *Computer Methods in Applied Mechanics and Engineering*, 193, 1733, Elsevier)

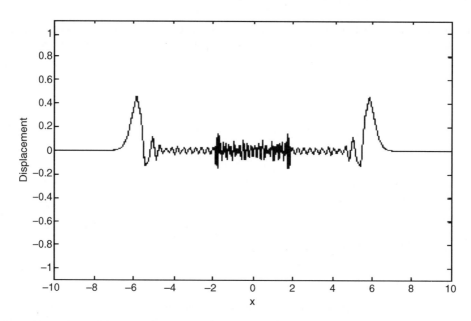

**Figure 7.8**  Wave propagation in the FE region and reflecting back to the MD region when the FE mesh size $l$ is larger than atomistic spacing $h_a$ (Reproduced from Park, H. S. and Liu, W. K. (2004) An introduction and tutorial on multiple-scale analysis in solids. *Computer Methods in Applied Mechanics and Engineering*, 193, 1733, Elsevier)

spacing $h_a$ throughout the model, the coupling motion equation is consistent with the motion equation of different regions. Whether in the MD region or FE region, the wave propagates in a uniform environment. Therefore, the simulation shows a smooth transition from the MD region to the FE region without reflected waves from the FE region to the MD region.

However, the result for the calculation of the second case shows a contrary result. Figure 7.8 shows that the wave distorted when propagating into the MD/FE interface. Simultaneously, reflected waves with very high frequency are produced. If the FE mesh size becomes larger away from the MD region, the reflected waves can be immediately observed in the MD region. In fact, the stiffness of the FE element is reduced when the FE spacing $l_0$ becomes larger than the atomistic spacing $h_a$, as shown in equation (7.40). Under this condition, the FE element is not stiff enough to sustain the wave propagation. However, the wave must propagate to keep energy conservative for the conservative system. If it cannot enter the FE element region it must be reflected back to the MD region, thus the unexpected heating phenomenon described at the beginning of this section occurs.

## 7.7   Bridging Domains Method*

One way to mitigate the unphysical phenomena, such as the artificial heating in the MD region by wave reflection and the ghost force at the interface, is to make the transition from atomistic region to continuum less abrupt by using a handshake region as shown in Figure 7.1a. The desired smooth transition is somewhat ad hoc; it can be achieved, however, by the simple blending of energy in the handshake region. Xiao and Belytschko proposed a method which is called the bridging domain (BD) multiscale method.[21] Here, the bridging domain is the handshake region. We will briefly use the BD method as an example to show the so-called Arlequin method[22] which is a general approach to coupling different models based on a blended energy functional. Actually, the BD method may be the first application of the blending energy method within an energy-based scheme.

For energy-based method with handshake region, the total energy should include the Hamiltonian in the atomistic, continuum and handshake region. If the atomistic energy in MD is denoted by superscript M and continuum energy by C, the total Hamiltonian of the entire simulation system may be written as:[21]

$$H = \int_{\Omega_0} \beta(X)(K^C + U^C)d\Omega_0 + \sum_I (1-\beta(X_I))(K_I^M + U_I^M)$$
$$+ \sum_{I \in H} \lambda(X_I) \cdot \left\| \bar{u}^C(X_I, t) - \bar{u}_I^M(t) \right\|^2 \tag{7.53}$$

where $\beta(X)$ is equal to 0 in the MD region, 1 in the continuum region, and increases from 0 to 1 smoothly in the bridging region. K is the kinetic energy; U is the potential energy; and $\bar{u}$ is the displacement. $\lambda(X_I)$ is the Lagrange multiplier. The first term on the right side of the equation includes all area with continuum (or FEM) formulation. When it is inside the pure continuum domain $B^C$, due to $\beta(X) = 1$ the Hamiltonian expression is the same as the traditional one such as (7.9) to (7.11); when it is in the handshake region, it varies, with the value at the boundary between continuum and handshake region smoothly reducing to zero when reaching the boundary of the atomistic domain; the variation is fully controlled by the function $\beta(X)$.

The same is true for the second term on the right of (7.53), when it is inside the fully atomistic domain $B^A$, due to $\beta(X)=0$ the Hamiltonian expression is the same as the traditional one such as (7.7). When it is in the handshake region, it varies with the value at the boundary between fully atomistic and handshake region smoothly reducing to zero when reaching the boundary of the continuum domain; the variation is fully controlled by the function of $1-\beta(X)$.

The last term of (7.53) is the constraint term controlled by the Lagrange multiplier $\lambda(X_I)$. The displacement constraints in the BD method are introduced to force the displacement $\bar{u}_I^M$ for every atom I in the handshake region to follow the displacement field $\bar{u}^C$ of a continuum element whose position in the reference configuration is $\bar{X}^I$. Specifically, following (6.3c) $\bar{u}^C$ can be expressed using the coordinates and shape function as follows:

$$\bar{u}^C = [N(\bar{X}^I)][U] \tag{7.54}$$

Through these constraints by the Lagrange multiplier to make the atom position coincident with the original continuum element, it is hoped that the compatibility condition can be improved.

The coupling equations of the FE and MD region are derived from the coupling Hamiltonian given by equation (7.54):

$$M_I \ddot{u}_I = F_I^{ext} - F_I^{int} - F_I^L \tag{7.55}$$

$$m_I \ddot{d}_I = f_I^{ext} - f_I^{int} - f_I^L \tag{7.56}$$

where $F_I^{ext}, F_I^{int}$ are the external and internal force imposed on the FE nodes; $f_I^{ext}, f_I^{int}$ are the force imposed on the atoms in the atomistic region; and $F_I^L, f_I^L$ are the restriction force based on the Lagrange multiplier. The advantage of this method is that different temporal increment steps may be employed in the MD region and FE region. The bridging method has been used in 2D wave and crack propagation problems. For static problems, the use of handshake region has the effect of smearing out the ghost forces and making them smaller on a given atom or node, but it may introduce ghost forces on more atoms and nodes.

For static cases, one can follow (6.4u) to write the total potential energy $\Pi$ for the BD method. If denoting the norm of (7.51) as $\|h^\alpha\|$, $\Pi$ can be written as follows:

$$\Pi^{BD} = \sum_{I \in B^H} (1-\beta(\bar{X}_I))E^I + \sum_{e \in B^H} \beta(\bar{X}_{cent}^e)W(\bar{X}_{cent}^e)V^e + \sum_{\alpha \in B^H} \left[ \beta_1 \bar{\lambda}^I \cdot \bar{h}^I + \frac{\beta_2}{2} \bar{h}^I \cdot \bar{h}^I \right] \tag{7.57}$$

where the symbol $I \in B^H$ and $e \in B^H$ on the first and second terms on the right side, respectively, denote that the atom I and the element e both belong to the handshake region. $\bar{X}^e_{cent}$ denotes the position vector at the center point of element e, thus $W(\bar{X}^e_{cent})$ denotes the strain energy density at the element's center. $\bar{\lambda} = [\lambda^\alpha_1, \lambda^\alpha_2, \lambda^\alpha_3]$ are Lagrange multipliers for the degree of freedom of atom I. $\beta_1, \beta_2$ are called penalty functions that can be chosen to optimize computational efficiency. This is a convenient tool in computational mechanics for finding a constrained minimum energy configuration, and the magnitude of the ghost forces can be determined. The use of a finite handshake region has the effect of reducing ghost forces on a given atom or node, but it may introduce ghost forces on more atoms and nodes.

## 7.8   1D Benchmark Tests of Interface Compatibility for DC Methods

In Section 5.3.5 we discussed in detail the verification of seamless transition of the GP method via a 1D model. Curtin and Miller[23] carried out a systematic verification on the interface transition behavior for various DC multiscale method via a 1D model. What they used is the energy method which is different from the force method used in Section 5.4. In the following we introduce their analysis and the benchmark testing results.

We can write the total atomistic energy as:

$$E^\alpha = \sum_i E^a_i \tag{7.58}$$

$E^\alpha_I$ is the energy of the ith atom and can be expressed as:

$$E^\alpha_I = \frac{1}{2}\left[\frac{1}{2}k_1(u_i - u_{i-1})^2 + \frac{1}{2}k_1(u_{i+1} - u_i)^2 + \frac{1}{2}k_2(u_i - u_{i-2})^2 + \frac{1}{2}k_2(u_{i+2} - u_i)^2\right] \tag{7.59}$$

where the first two terms on the right side denote the energy stored by the ith atom with its first neighbors, i.e., the (i − 1) and (i + 1) atoms, and the third and the fourth terms are the ith atom's energy stored with its second neighbors, i.e., the (i − 2) and (i + 2) atoms. The coefficient 1/2 before the parenthesis indicates the equal share of energy between the ith atom with its first and second neighbors.

For continuum part of the 1D model, the strain energy density W and strain can be written as

$$W = \frac{1}{2}C_c\varepsilon^2 \tag{7.60}$$

$$\varepsilon = \frac{(u_i - u_{i-1})}{a} \tag{7.61}$$

where a is the interatomic distance. The total continuum energy $E^C$ is given as the sum of all elements:

$$E^C = \sum_i E^C_i \tag{7.62}$$

where $E^C_i$ is the energy stored in element i with volume $a^3$. This element energy can be determined as

$$E^C_i = W \times a^3 = \frac{1}{2}C_c\frac{(u_i - u_{i-1})^2}{a^2} \cdot a^3 = \frac{1}{2}K_c(u - u_{i-1})^2 \tag{7.63}$$

$$K_c = C_c a \tag{7.64}$$

where $K_c$ is the continuum effective spring constant. In the homogeneous deformation case, the energy $E^\alpha_i$ at point i calculated by the atomistic model should equal the energy $E^C_i$ calculated by the continuum model, i.e.,

$$E^a_i = E^C_i \tag{7.65}$$

This condition was used by Curtin and Miller[23] to determine the effective material constant $K_C$. Using U to denote the homogeneous deformation field as

$$U = u_i - u_{i-1} \qquad (7.66)$$

and substituting (7.66) into (7.59) and (7.63), then substituting the results into (7.65), we obtain

$$\frac{1}{2} \left( k_1 U^2 + 4k_2 U^2 \right) = \frac{1}{2} K_C U^2 \qquad (7.67)$$

thus

$$K_C = k_1 + 4k_2 \qquad (7.68)$$

Curtin and Miller applied equations (7.59) and (7.68), developed for atomistic and continuum region, respectively, for a coupling atomistic/continuum 1D model. This model includes a chain of 101 atoms at initial positions, $r_i = ia$, $i = 0–100$ where a is the interatomic distance. This chain is used for a fully atomistic simulation to get an accurate solution. The coupling model is with the interface at $i = 50$, the region from the left end to the interface is the atomistic region and that from $i = 51$ to $i = 100$ is the continuum region. The boundary and loading conditions are the same as those described in Section 5.4. In short, the left end (atom 0) is fixed, the right end is applied a unit load for all the loading conditions. The three loading conditions include uniform deformation, nonuniform deformation, and single point force at the interface besides the unit force applied at the right end. The transition behavior at the interface is different due to different design at the interface for different models. For the QC model, the system energy is given as

$$E_{QC} = \sum_{i \leq I} E_i^a + \frac{1}{2} E_{I+1}^C + \sum_{i > I+1} E_i^C \qquad (7.69)$$

where the integer I denotes the interface (e.g., $I = 50$ in this case), or the location of the last atom. The first term in the right part is the atomistic energy, the last term denotes the energy in the continuum region from $i = I + 2$ to $i = 100$. For $i = I + 1$ the weighting function is $1/2$ because part of the energy of material at that interval region is contributed by the interface atoms. The energy $E_{QC}$ is minimized with respect to the displacements $u_i$ of atoms and FE nodes and then the solution is obtained.

For the CADD, the transition region contains the pad atoms $I + 1$ and $I + 2$ but they are not connected to the corresponding continuum nodes, they are extra degrees of freedom even if they are initially put at the location $I + 1$ and $I + 2$ to overlap with the continuum node $I + 1$ and $I + 2$, thus the energy for the continuum should start from $i = I + 1$, and the atom energy should include $I + 1$ and $I + 2$ as the following:

$$E_{CADD} = \sum_{i \leq I+2} E_i^a + \sum_{i > I} E_i^C \qquad (7.70)$$

The energy $E_{CADD}$ is minimized not only with respect to the displacements $u_i$ of real atoms and FE nodes in the system but also with respect to the two pad atoms and then the solution is obtained.

For the QC-GFC (ghost force corrected) method discussed in Section 6.4.8.2, there are two coupled problems to solve, the first one is to get the ghost force and then use it as the deadload force in (6.5g, h) in the second coupled problem. The energy expression of the atomistic region also includes the atom $I + 1$ and $I + 2$ as follows:

$$E_{QC-GFC} = \sum_{i \leq I+2} E_i^a + \sum_{i > I} E_i^C \qquad (7.71)$$

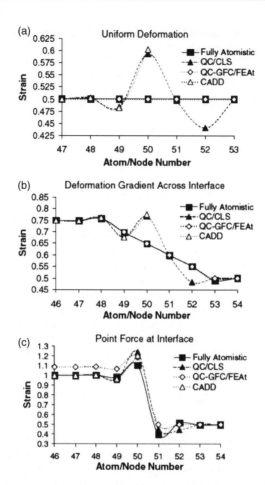

**Figure 7.9**  Results of simple 1D model to compare the transition regions of the various models. The CADD model is identical to the QC-GFC/FEAt model in the continuum region and is thus not shown in this region. (a) Uniform deformations of the purely atomistic model are reproduced only by the QC-GFC model. (b) Errors in the models in the presence of non-uniform deformation. (c) Errors in the models in the presence of a single point force applied at the interface (Reproduced from Curtin, W. A. and Miller, R. E. (2003) Atomistic/continuum coupling in computational materials science. *Modelling and Simulation in Materials Science and Engineering*, 11, R33. Copyright © IOP)

The minimization of energy $E^a$ is only over $i \leq I$ with the positions of atoms $I + 1$ and $I + 2$ fixed by the continuum nodal solutions. Similarly, $E^C$ is constrained such that the position of node $I$ is fixed by the atomistic solution. This is described in detail in Section 7.1.2.

Figure 7.9 shows the results of behavior at the interface of different models in comparison with fully atomistic simulation. It is seen that apart from the QC-GFC and FEAt methods, most DC multiscale models including the QC, MAAD (or CLS) and CADD methods do not truly meet the requirement of seamless transition. For homogeneous deformation expressed in Figure 7.9(a), they give additional tension strain in the interface region but compressive strain in the continuum region. The same trends are observed for inhomogeneous case shown in Figure 7.9(b) (Remark: for the CADD method there is no loading applied at the padding atoms 51 and 52.)

It is seen from Figure 7.9(c) that the QC-GFC and FEAt methods have a long-range error under the interface point force. These errors arise because in the QC-GFC and FEAt approach, there is no restricted requirement of the self-equilibrium of the local forces; their discontinuity in treatment of the atoms and nodes leads to unbalanced forces that are then balanced by long-range forces. On the other hand, the QC and CADD/CLS methods show relatively good behavior under this sharp gradient of deformation.

Note that the basic formula of (7.68) used by Curtin and Miller for the atomistic/continuum coupling analysis is only valid for homogeneous deformation; however, in the interface analysis, the continuum is also involved in inhomogeneous deformation, which may affect the accuracy of the quantitative comparison. This is the difference of the 1-D model here from the 1-D GP model in Section 5.4.

## 7.9    Systematic Performance Benchmark of Most DC Atomistic/Continuum Coupling Methods

Recently, Miller and Tadmor[2] presented a unified framework in which fourteen leading DC multiscale methods could be presented as special cases. Their unified framework is introduced in some parts of Sections 7.1 and 7.2 and will be further treated in this section.

### 7.9.1    The Benchmark Computation Test

#### 7.9.1.1  Description of the Geometric Model for Benchmark Test

The research uses a common framework as a platform to test the accuracy and efficiency of the fourteen methods on one test problem. This test model is a FCC block of single crystal aluminum containing a dipole of Lomer dislocations. The crystal is roughly $400 \text{ Å} \times 400 \text{ Å}$ in the $X_1 X_2$ plane and periodic in the $X_3$ direction (with a periodic length of 2.85 Å). The lattice constant is 4.032 Å and the model region contains 27,760 atoms.

The model is under simple external shear at the top and bottom surface. It is studied both fully atomistically and with the various multiscale methods. The multiscale models use meshes that are approximately the same across the methods. In all the cases, the Lomer dipole lies on the $X_2 = 0$ plane along the [110] direction, with the two cores initially at $X_1 = \pm 20$ Å. For the multiscale models there are three sizes for the atomistic regions, i.e., 10, 20 and 30 in the $X_2$ direction but with the same size, $\pm 200$ Å, along the $X_1$ direction. The model-30 (or model-20) is three (or two) times thicker in the atomistic region than the model-10 along the $X_2$ direction.

Obviously, the larger number denotes a larger atomistic region, thus the model is more fine. The multiscale models tested are briefly introduced in Table 7.1. Most of these models have been introduced in the present and previous chapters, but suitable references are given. Among these models, five have their origin in the QC method; the ghost-force corrected model of concurrent AtC coupling is not published yet but introduced in Ref. [2], specifically, in the paper a 1D model is used to explain why the force-based model still has ghost force. It is stated that the ghost-force corrected method introduced in Section 6.4.8.2 for the energy-based method is valid for correcting the ghost force appearing in the force-based method.

#### 7.9.1.2  The Definition for the Error in Displacements

The fully atomistic simulation is the benchmark against which one compares the multiscale models, the corresponding atomistic solution is considered as the exact solution in Ref. [2]. Denoting the exact displacement vector of atom $\alpha$ to be $\bar{u}^\alpha_{exact}$ and the displacement obtained from a multiscale model as $\bar{u}^\alpha$, the displacement error $e^\alpha$ at each atom is defined as the so-called $L_2$ norm, as

$$e^\alpha = ||\bar{u}^\alpha - \bar{u}^\alpha_{exact}|| \tag{7.72}$$

**Table 7.1** The 14 DC multiscale models in the benchmark test (Reproduced from Miller, R. E. and Tadmor, E. B. (2009) A unified framework and performance benchmark of fourteen multiscale atomistic/continuum coupling methods. *Modelling and Simulation in Materials Science and Engineering*, 17, 053001. Copyright © IOP)

| Method | Acronym | Continuum Model | Handshake | Formulation |
|---|---|---|---|---|
| Quasicontinuum | QC[24, 25] | Cauchy-Born | None | Energy-based |
| Coupling of length scales | CLS or MAAD[9] | Linear elasticity | None | Energy-based |
| Bridging domain | BD[21] | Cauchy-Born | Linear mixing of energy | Energy-based |
| Bridging scale method | BSM[26, 27] | Cauchy-Born | None | Energy-based with dead GFC |
| Composite grid atomistic continuum method | CACM[28] | Linear elasticity | None | |
| Cluster-energy Quasicontinuum | CQC(m)-E[29] | Averaging of atomic clusters | None | Energy-based |
| Ghost-force corrected quasicontinuum | QC-GFC[30] | Cauchy-Born | None | |
| Ghost-force corrected cluster-energy QC | CQC(m)-GFC[29] | Averaging of atomic clusters | None | Energy-based With dead load GFC |
| Finite-element/atomistic method | FEAt[12] | Nonlinear, non-local elasticity | None | Force-based |
| Coupled atomistic and discrete dislocation | CADD[13, 15] | Linear elasticity | None | Force-based |
| Hybrid simulation Method | HSM[31] | Nonlinear elasticity | atomic averaging for nodal BC | Force-based |
| Concurrent AtC coupling | AtC[32-35] | Linear elasticity | Linear mixing of stress & atomic force | Force-based |
| Ghost-force corrected concurrent AtC coupling | AtC-GFC[2] | Linear elasticity | Linear mixing of stress & atomic force | Force-based |
| Cluster-force Quasicontinuum | CQC(m)-F[36] | Averaging of atomic clusters | None | Force-based |

The global displacement error for a given model is similarly defined as the $L_2$ norm of the difference between the global displacement vectors normalized by the number of atoms (in the case of Ref. [2], recall $N_A = 27{,}760$). This can be written as

$$e = |\sqrt{\frac{\sum_{\alpha=1}^{N_A}(e^{\alpha})^2}{N_A}}$$

(7.73)

To turn the global error into a per cent error, e%, one has

$$e = \frac{e}{u_{avg}} \times 100$$

(7.74)

where $u_{avg}$ is the average of the atomistic displacement norm in the exact solution defined by

$$u_{avg} = \frac{1}{N_A}\sum_{\alpha=1}^{N_A}||\bar{u}^{\alpha}_{exact}||$$

(7.75)

For the model in Ref. [2] the value is $0.05755\,\text{Å}$.

### 7.9.1.3 The Comparison Results for Displacement Error

Table 7.2 shows a summary of the per cent displacement errors, e%, and the normalized error, $e/e_{min}$ for each method at the different levels of refinement of Model-10 (%), Model-20 (%) and Model-30 (%), with Model-30 being the finest. The normalized error is the minimum error $e_{min} = 2.262 \times 10^{-3}\,\text{Å}$ which corresponds to the minimum error of 3.93% for the Model-30 of CADD/FEAt, thus for that model the normalized error is 1.00 and is taken as the reference.

**Table 7.2**  Summary of the errors for each method at the different levels of model refinement (Reproduced from Miller, R. E. and Tadmor, E. B. (2009) A unified framework and performance benchmark of fourteen multiscale atomistic/continuum coupling methods. *Modelling and Simulation in Materials Science and Engineering*, 17, 053001. Copyright © IOP)

| Method | Model-10 (e%) | Model-20 (e%) | Model-30 (e%) | Model-10 ($e/e_{min}$) | Model-20 ($e/e_{min}$) | Model-30 ($e/e_{min}$) |
|---|---|---|---|---|---|---|
| CADD/FEAt | 11.42 | 5.13 | 3.93 | 2.91 | 1.30 | 1.00 |
| QC-GFC | 12.42 | 5.72 | 4.05 | 3.16 | 1.46 | 1.03 |
| AtC-GFC | 14.59 | 8.28 | 5.04 | 3.71 | 2.11 | 1.28 |
| BD | 14.63 | 9.09 | 6.44 | 3.72 | 2.31 | 1.64 |
| QC/CLS | 14.66 | 9.54 | 8.50 | 3.73 | 2.43 | 2.16 |
| BSM | 17.59 | 12.07 | 10.17 | 4.47 | 3.07 | 2.59 |
| HSM | 22.03 | 15.44 | 12.25 | 5.61 | 3.93 | 3.12 |
| CQC(13)-E | 22.59 | 16.62 | 16.55 | 5.75 | 4.23 | 4.21 |
| CQC(1)-GFC | 40.06 | 20.10 | 20.19 | 10.19 | 5.11 | 5.14 |
| CQC(1(-E | 86.75 | 43.61 | 38.48 | 22.07 | 11.09 | 9.79 |
| CACM | 42.59 | 40.13 | 39.92 | 10.84 | 10.21 | 10.16 |
| CQC(13)-F | 70.84 | 60.75 | 46.42 | 18.02 | 15.45 | 11.81 |
| AtC | 55.11 | 70.17 | 83.65 | 14.02 | 17.85 | 21.28 |

The models are presented in the order of decreasing accuracy based on the Model-30 column. It is seen that all models except the AtC method appear to converge with increased model refinement. The bad trend of the AtC model is successfully corrected by the correct convergence of the AtC-GFC method. The good or fair results come from CADD/FEAt, QC-GFC, BD, QC/CLS, AtC-GFC and HSM, while the accuracy of the remaining models is relatively poor.

## 7.9.2    Summary and Conclusion of the Benchmark Test

Besides the displacement error discussed in the last section, other comparisons such as energy error and calculation speed of different multiscale models are also introduced. The systematic comparison of the benchmark test is summarized in Ref. [2]. In the following, the summary and conclusion of that work is given.

### 7.9.2.1  Efficiency Measured by Computational Speed

Generally (not exclusively) energy-based methods are faster than force-based, strong compatibility is faster than weak compatibility, and the absence of a handshake region is faster than the presence of one. The low speed of the force-based method may be improved by developing a good numerical algorithm.

### 7.9.2.2  Displacement Error

Force-based methods are more accurate than energy-based, strong compatibility is more accurate than weak, and the presence of a handshake region reduces the accuracy.

### 7.9.2.3  Energy Error

It seems that none of the methods are especially inaccurate in their estimate of the energy as the absolute energy error for most of the methods was generally less than 5%. Even so, it seems that even this slight error can have profound effects on the resulting dislocation motion.

### 7.9.2.4  Error of Models with Handshake Region and Weak Compatibility

Intuitively, one may expect that using a handshake region would improve accuracy because it may provide a more gradual transition from the atomistic to the continuum scale. In fact, this is not the case, as handshake methods tended to be both slower and less accurate. This is also the case for methods using weak compatibility. The reason is that in both handshake region or the weak compatibility model the finite element mesh is easy to develop because it does not require mesh refinement down to the atomistic scale. This convenience reduces the accuracy, which indicates that the necessary condition for high accuracy in the DC method is down to the mesh size of the atomistic lattice spacing, as we discussed in several sections including Section 5.9.

### 7.9.2.5  Error of Models Using Cluster-based Methods

Cluster-based methods are generally slower and less accurate than methods which follow traditional continuum laws such as linear elasticity or the Cauchy-Born rule. This finding is consistent with a recent mathematical study comparing cluster-based and element-based models.[37] The cluster representation in the continuum region induces a large error which significantly contributes to the overall error of these methods.

## 7.10    The Embedded Statistical Coupling Method (ESCM)

ESCM, short for embedded statistical coupling method, is an alternative type of multiscale method proposed by Saether, Yamakov and Glaessgen.[1] This method replaces DC's direct linkage with a statistical averaging over selected time interval of atomistic displacements in local atomistic volumes associated with each FE node in the interface region.

### 7.10.1    Why Does ESCM Use Statistical Averaging to Replace DC's Direct Linkage?

While the DC method is straightforward, the fundamental difficulty in its development lies in the inherent difference between the atomistic and continuum constitutive relationships. This difference and its induced difficulty in naturally satisfying the interface compatibility condition are described in several places including Section 5.9.2. In short, the physical state of atomistic region is described through non-local interatomic forces between discrete atoms, while the physical state of the continuum region is described through stress-strain relationships. The latter is a local behavior in the sense that it reflects local statistical averages of atomic interactions at large length and time scales. This inconsistency between materials at two sides of the interface causes ghost force and unphysical phenomena.

In general, the physical connection between the upper scale (continuum) and lower scale (e.g., atoms) can only be achieved through an adequate averaging over scales where the lower scale (e.g., atomistic) structure can be approximated by the upper-scale material (e.g., continuum). Based on this requirement, the DC methods are forced to reduce the finite element size to be nearly the interatomic distance at the atomistic/continuum interface to perform well. This is confirmed by the benchmark test described in the last section. This feature causes substantial difficulty when the order of model sizes is further increased.

In many cases one needs a large atomistic region embedded in a continuum of microns dimensions for materials modeling. Following the trend in CPU development with quad-core processors on the market and even hundreds-core in the next five years, multimillion atom simulations should easily be done even on one computer soon. Because DC methods need to have an extremely small mesh size at the interface of the FEM domain, they require a large number of FE nodes at the interface. This will cause substantial difficulty in meeting the new challenge of multimillion atoms coupled with continuum.

### 7.10.2    The ESCM Model

The ESCM model embeds an inner atomistic MD system within a surrounding continuum FEM domain. This reduces the degrees of freedom of the system while investigating important phenomena at the atomistic domain. At this stage, mainly to show a different methodology from the DC methods, the model is kept simple. Specifically, the embedding MD domain is a circular region and a linear elastic constitutive relationship is assumed for the continuum.

As shown in Figure 7.10, the structure of the ESCM model consists of four regions: (1) an inner MD region; (2) an interface MD region wherein MD and finite elements are superimposed; (3) a surface MD region that does not interact with the FE nodes but is used to compensate for atomic free-edge effects; and (4) a FEM domain in which standard FEM is used. Together, these regions constitute the complete MD system. The partitioning of the MD system into different regions is not a physical separation; an atom assigned to a particular location freely interacts with atoms in its neighborhood that may reside in a different region. Thus, any conventional MD technique can be used for performing the overall simulation of the atomistic region. It is important to note that the ESCM performs the atomistic simulation and FEM calculation separately, not simultaneously. The result for a MD run will feed to the FEA boundary through interface region; vice versa the FEA result will feed back to the MD boundary to carry on the coupling between ME/FEA regions.

### 7.10.3    MD/FE Interface

The role of the MD/FE interface is to provide a computational link between the MD region and the FEM domain. The atoms that surround a given FE node at the interface are partitioned to form a cell, called an interface volume cell (IVC) as shown in Figure 7.10. A similar partitioning is also applied to the surface

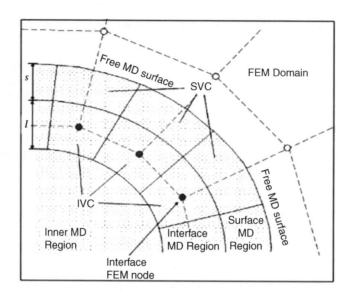

**Figure 7.10** Four regions of the MD/FE interface in the ESCM method (Reproduced from Saether, E., Yamakov, V., and Glaessgen, E. H. (2009) An embedded statistical method for coupling molecular dynamics and finite elements analysis. *International Journal for Numerical Methods in Engineering*, 78, 1292, John Wiley & Sons, Ltd.)

MD Region, forming surface volume cells (SVCs). Note that the partitioning is to some extent arbitrary, unlike the generalized particle method described in Section 5.3 which has constant intrinsic connection with realistic atoms (or lower-scale particles) through neighbor-link cells.

The IVCs compute averaged MD displacements at their mass center, then use them as displacement boundary conditions to the associated interface FE nodes. The IVCs need not coincide in size or shape with the finite element to which the FE node belongs. Typically, one finite element encompasses a region of several hundred to several thousand atoms. As described in Ref. [1], a lower bound for the number of atoms associated with each FE node is determined by the requirement of obtaining a minimally fluctuating average of atomic displacements. With this effective average at this scale, the atomic structure is homogenized enough so that the FE node at the interface represents the average behavior of the underlying atoms.

The following equation shows a spatial atomistic average within each IVC to yield the displacement $\bar{\delta}_{CM,k}^{MD}$ at the mass center of the $k^{th}$ IVC:

$$\bar{\delta}_{CM,k}(t_j) = \frac{1}{N_{kIVC}} \sum_{i=1}^{N_{iIVC}} \bar{u}_i(t_j) = \frac{1}{N_{kIVC}} \sum_{i=1}^{N_{iIVC}} (\bar{r}_i(t_j) - \bar{r}_i(0)) = \bar{r}_{CM,k}(t_j) - \bar{r}_{CM,k}(t_0) \qquad (7.76)$$

where $\bar{r}_{CM}(t_j)$ expressed by

$$\bar{r}_{CM}(t_j) = \frac{1}{N_{kIVC}} \sum_{i=1}^{N_{iIVC}} \bar{r}_i(t_j) = \frac{1}{N_{kIVC}} \sum_{i=1}^{N_{iIVC}} \bar{r}_i(t_j) \qquad (7.77)$$

is the mass center position vector of the $k^{th}$ IVC at the time $t_j$ of the $j^{th}$ step, $N_{kIVC}$ denotes the total number of atoms in the $k^{th}$ IVC region and $\bar{r}_i(t_j)$ is the position vector of atom i at time $t_j$ of the $j^{th}$ step. The mass center displacement $\bar{\delta}_{CM,k}^{MD}$ is calculated relative to the initial zero-displacement position $\bar{r}_i(0)$ of the atom i ($i = 1,2 \ldots N_{kIVC}$).

Following the Ergodic hypothesis of statistical mechanics described in Section 2.8.2 and (2.21a), the statistically meaningful displacement vector $\bar{\delta}_{I,k}$ of the interface $k^{th}$ element should be obtained by further averaging over a certain period of M time steps of the MD simulation as follows:

$$\bar{\delta}_{I,k} = \langle \bar{\delta}_{CM,k} \rangle_t = \frac{1}{M} \sum_{j=1}^{M} (\bar{r}_{CM,k}(t_j) - \bar{r}_{CM,k}(t_0)) \tag{7.78}$$

The obtained atomistic-averaging displacements over IVCs are used for the FE node displacements in the interface domain which formulate the displacement boundary condition of the FE domain. In turn, the FE nodes apply reaction forces on the corresponding atoms within the IVCs. Thus, ESCM is forced to involve an iterative process based on solving boundary value problems for both MD and FEM systems at their common interface. The FEM system is loaded by far-field loads applied to the external boundaries and along the MD/FE interface by nodal displacement conditions. The MD system, in turn, is simulated under periodic updated constant traction boundary conditions that are obtained from the FEM system as reaction forces to the MD displacements at the interface.

The iterative process is necessary because the atomistic-averaging displacements, say $\bar{\delta}_{I,k}^{n}$ at the nth iteration, should make the FEM produce suitable node forces on the atomistic domain such that the IVCs' node displacement $\bar{\delta}_{I,k}^{n+1}$ obtained by MD's (n + 1) iteration is very close to the original $\bar{\delta}_{I,k}^{n}$, such that their difference is smaller than the prescribed tolerance. The iteration process will continue until the tolerance is satisfied. This is the way ESCM calculates the coupling between MD and FEM regions. The iterative process allows the continuity at the MD/FE interface to be achieved at different length and time scales inherent to both systems.

## 7.10.4 Surface MD Region

In order for the MD domain to deform freely under the applied reaction forces from the FE nodes at the interface, it is necessary to eliminate the free surface effects, because in ESCM the MD region is separated from the continuum region and a free surface will appear. This is completely different from all the DC models shown in Figure 7.1 as well as different from the GP model shown in Figure 5.4. The free surface introduces several undesirable effects in the MD system. First, it creates surface tension forces. Second, as described in Section 5.6, the coordination number of the surface atoms is reduced so they are less strongly

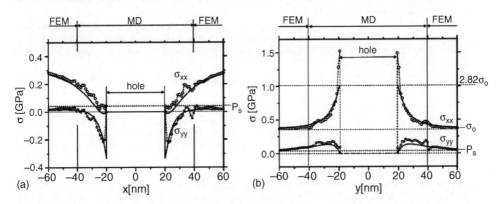

**Figure 7.11**  Stress profiles in the open hole specimen for $\sigma_{xx}$ and $\sigma_{yy}$ along (a) the x-axis and (b) the y-axis scanned through the center of the hole (Reproduced from Saether, E., Yamakov, V., and Glaessgen, E. H. (2009) An embedded statistical method for coupling molecular dynamics and finite elements analysis. *Int. J. Numer. Meth. Eng.*, 78, 1292, John Wiley & Sons, Ltd.)

bonded to the surrounding atomic field than those within the interior. To mitigate these free surface effects and to stabilize the atoms in the interface MD region, an additional volume of outlying atoms constituting a surface MD region is introduced as shown in Figure 7.10.

While the introduced surface MD region eliminates the free surface effects within the inner MD region, it also introduces an undesirable fictitious stiffness that elastically constrains the deformation of the inner MD region. The combined effects of surface tension and the fictitious stiffness are defined as a resultant force, $\bar{f}_s$.[1] This force acts at the boundary between the surface MD region and the interface MD region. Further, the authors propose to apply a compensatory force $\bar{f}_C$ at each SVC to cancel the $\bar{f}_s$ effects in the corresponding IVC.

## 7.10.5    Validation

Here, we will not explain how the compensation force $\bar{f}_C$ is determined and how the iterative process of the coupling boundary value problems of MD and FEM are solved. Readers who are interested are referred to Ref. [1]. Figure 7.11 shows a comparison between the ESCM simulation data and the FEM results. The symbols represent ESCM simulation data while the full lines represent fully continuum results. For reference, the surface tension induced pressure $p_s$ is also shown. From the comparison, it is seen that the stress at the interface ($x = \pm 40$ nm) or ($y = \pm 40$ nm) is well preserved and the stress profiles of the coupled model closely follow that of the fully continuum FEM analysis of the equivalent model.

## References

[1] Saether, E., Yamakov, V., and Glaessgen, E. H. (2009) An embedded statistical method for coupling molecular dynamics and finite elements analysis. *Int. J. Numer. Meth. Eng.*, **78**, 1292.

[2] Miller, R. E. and Tadmor, E. B. (2009) A unified framework and performance benchmark of fourteen multiscale atomistic/continuum coupling methods. *Modelling Simul. Mater. Sci. Eng.*, **17**, 053001.

[3] Arndt, M., Sorkin, V., and Tadmor, E. B. (2009) Efficient algorithms for discrete lattice calculations. *J. Comput. Phys.*, **228**, 4858.

[4] Abraham, F. F., Broughton, J. Q., Bernstein, N., and Kaxiras, E. (1998) Spanning the continuum to quantum length scales in a dynamic simulation of brittle fracture. *Europhysics Letters*, **44**, 783.

[5] Abraham, F. F., Broughton, J. Q., Bernstein, N., and Kaxiras, E. (1998) Spanning the length scales in dynamic simulation. *Computers in Physics*, **12**, 538.

[6] Abraham, F. F., Bernstein, N., Broughton, J. Q., and Hess, D. (2000) Dynamic fracture of silicon: Concurrent simulation of quantum electrons, classical atoms, and the continuum solids. *Mater. Res. Soc. Bull.*, **25**, 27.

[7] Landau, D. P. (2005) The future of simulations in materials science, in *Handbook of Materials Modeling* (ed. S. Yip), Springer, New York, 2663.

[8] Rudd, R. E. and Broughton, J. Q. (1999) Atomistic simulation of MEMS resonators through the coupling of length scales. *J. Modeling Simul. Microsyst.*, **1**, 29.

[9] Rudd, R. E. and Broughton, J. Q. (2000) Concurrent coupling of length scales in solid state systems. *Phys. Stat. Sol. (b)*, **217**, 251.

[10] Fermi, E., Pasta, J., and Ulam, S. (1955) Studies on nonlinear problems, Los Alamos preprint LA-1940.

[11] Rudd, R. E. and Broughton, J. Q. (1998) Coarse-grained molecular dynamics and the atomic limit of finite elements. *Phys. Rev. B*, **58**, R5893.

[12] Kohlhoff, S., Gumbsch, P., and Fischmeister, H. F. (1991) Crack propagation in bcc crystals studied with a combined finite-element and atomistic model. *Philos. Mag. A*, **64**, 851.

[13] Shilkrot, L. E., Miller, R. E., and Curtin, W. A. (2002) Coupled atomistic and discrete dislocation plasticity. *Phys. Rev. Lett.*, **89**, 025501.

[14] Shilkrot, L. E., Curtin, W. A., and Miller, R. E. (2002) A coupled atomistic/continuum model of defects in solids. *J. Mech. Phys. Solids*, **50**, 2085.

[15] Shilkrot, L. E., Miller, R. E., and Curtin, W. A. (2004) Multiscale plasticity modeling: coupled atomistics and discrete dislocation mechanics. *J. Mech. Phys. Solids*, **52**, 755.

[16] Fish, J., Nuggehally, M. A., Shephard, M. S.,*et al.* (2007) Concurrent AtC coupling based on a blend of the continuum stress and the atomistic force. *Computer Methods in Applied Mechanics and Engineering*, **196**, 4548.

[17] van der Giessen, E. and Needleman, A. (1995) Discrete dislocation plasticity: A simple planar model. *Modelling Simul. Mater. Sci. Eng.*, **3**, 689.

[18] Hirth, J. P. and Loche, J. (1992) *Theory of Dislocations*, Krieger Publishing, Malabar, Florida.

[19] Park, H. S. and Liu, W. K. (2004) An introduction and tutorial on multiple-scale analysis in solids. *Computer Methods in Applied Mechanics and Engineering*, **193**, 1733.

[20] Chandrupatla, T. R. and Belegundu, A. D. (2002) *Introduction to Finite Elements in Engineering*, 3rd edn., Prentice Hall.

[21] Xiao, X. P. and Belytschko, T. (2004) A bridging domain method for coupling continua with molecular dynamics. *Computer Methods in Applied Mechanics and Engineering*, **193**, 1645.

[22] Bauman, P. T., Dhia, H. B., Elkhodja, N.,*et al.* (2008) On the application of the Arlequin method to the coupling of particle and continuum models. *Comput. Mech.*, **42**, 511.

[23] Curtin, W. A. and Miller, R. E. (2003) Atomistic/continuum coupling in computational materials science. *Modelling Simul. Mater. Sci. Eng.*, **11**, R33.

[24] Shenoy, V. B., Miller, R., Tadmor, E. B.,*et al.* (1998) Quasicontinuum models of interfacial structure and deformation. *Phys. Rev. Lett.*, **80**, 742.

[25] Tadmor, E. B., Ortiz, M., and Phillips, R. (1996) Quasi-continuum analysis of defects in solids. *Philos. Mag. A*, **73**, 1529.

[26] Wagner, G. J. and Liu, W. K. (2003) Coupling of atomistic and continuum simulations using a bridging scale decomposition. *J. Comput. Phys.*, **190**, 249.

[27] Qian, D., Wagner, G. J., and Liu, W. K. (2004) A multiscale projection method for the analysis of carbon nanotubes. *Computer Methods in Applied Mechanics and Engineering*, **193**, 1603.

[28] Datta, D. K., Picu, R. C., and Shephard, M. S. (2004) Composite grid atomistic continuum method: An adaptive approach to bridge continuum with atomistic analysis. *Int. J. Multiscale Comput. Eng.*, **2**, 71.

[29] Eidel, B. and Stukowski, A. (2009) A variational formation of the quasicontinuum method based on energy sampling of clusters. *J. Mech. Phys. Solids*, **57**, 87.

[30] Shenoy, V. B., Miller, R., Tadmor, E. M.,*et al.* (2002) An adaptive finite element approach to atomic-scale mechanics: The quasicontinuum method. *J. Mech. Phys. Solids*, **47**, 611.

[31] Luan, B. Q., Hyun, S., Molinari, J. F.,*et al.* (2006) Multiscale modeling of two-dimensional contacts, *Phys. Rev. E*, **74**, 046710.

[32] Badia, S., Bochev, P., Lehoucq, R.,*et al.* (2007) A force-based blending model for atomistic-to-continuum coupling. *Int. J. Multiscale Comput. Eng.*, **5**, 387.

[33] Badia, S., Parks, M., Bochev, P.,*et al.* (2008) On atomistic-to-continuum coupling by blending. *Multiscale Modeling Simulation*, **7**, 381.

[34] Fish, J., Nuggehally, M. A., Shephard, M. S.,*et al.* (2007) Concurrent AtC coupling based on a blend of the continuum stress and the atomistic force. *Computer Methods in Applied Mechanics and Engineering*, **196**, 4548.

[35] Parks, M., Bochev, P. B., and Lehoucq, R. (2008) Connecting atomistic-to-continuum coupling and domain decomposition. *Multiscale Modeling Simulation*, **7**, 362.

[36] Knap, J. and Ortiz, M. (2001) An analysis of the quasicontinuum method. *J. Mech. Phys. Solids*, **49**, 1899.

[37] Luskin, M. and Ortner, C. (2009) An analysis of node-based cluster summation rules in the quasicontinuum method. *Journal on Numerical Analysis*, **47**, 3070.

# 8

# Hierarchical Multiscale Methods for Plasticity

This chapter introduces principles and methods of hierarchical (sequential) multiscale analysis of plasticity across micro/meso/macroscopic scales. It includes a multiscale method by developing constitutive equations of meso-cells to quantitatively link the variables between microscopic and macroscopic scales of continuum; a method which links dislocation analysis to the second class of multiscale analysis so that size effects of microstructure can be realized through bridging and transition from atomic dislocation to upscale plasticity parameters; the basic concepts of self-consistent schemes which consistently link the stress and strain in inclusions (i.e., meso-cells) with those in their aggregates (i.e., macro-cells); and the basic concepts of the plasticity theory of continuum using a mechanical model.

The explanation includes how to fabricate specimens with different microstructure sizes for multiscale analysis, how to confirm the size effects by experiments, and how to carry on two kinds of iterative numerical calculation across three-scale nonlinear multiscale analysis. Furthermore, the significant findings in the microscopic scale by the multiscale analysis are described and discussed. While the multiscale method is specifically exemplified through analysis of cyclic plasticity (ratcheting) of dual-phase rail steel, the general method is summarized in detail in the last section.

## 8.1 A Methodology of Hierarchical Multiscale Analysis Across Micro/meso/macroscopic Scales and Information Transformation Between These Scales

In Chapter 6, we discussed a hierarchical multiscale method by developing a continuum theory of nanotubes based on the bottom-up information obtained from the atomistic scale. This chapter will continue systematically to introduce the principles and methods of hierarchical (sequential) multiscale analysis of plasticity across micro/meso/macroscopic scales of a continuum.

### 8.1.1 Schematic View of Hierarchical Multiscale Analysis

Figure 8.1 shows a method of hierarchical (sequential) multiscale analysis.[1] The multiscale method takes the development of constitutive equations of mesoscopic scale as a means to connect parameters in the

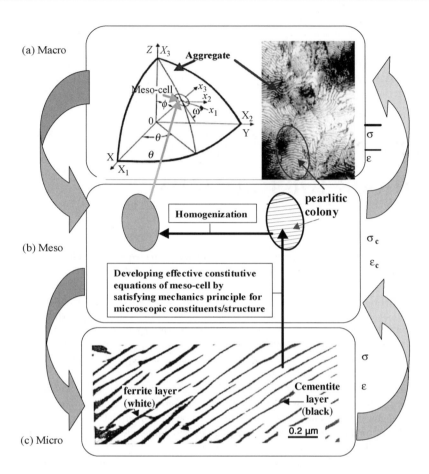

**Figure 8.1** Schematic methodology of multiscale analysis across micro/meso/macroscopic scales within the continuum domain (Reproduced from Fan, J., Gao, Z., and Zeng X. (2004) Cyclic plasticity across micro/meso/macroscopic scales. *Proceedings of Royal Society of London A*, 460 (2045), 1477, Royal Society of London)

microscopic scale to the macroscopic scale. The idea is explained through micro/meso/macroscopic multiscale analysis of cyclic plasticity for dual-phase materials which gives quantitative relationships of variables between the three scales. In the figure, the mesoscopic scale is the intermediate scale and is expressed in the middle box; the macroscopic scale is denoted by the upper box and consists of a great number of randomly spatially distributed pearlitic colonies; and the microscopic scale is shown in the lower box. The structure of the microscopic scale includes many alternately arranged fragile thin cementite (black) layers and soft thick ferrite (white) layers.

In Figure 8.1, the variables are stress and strain. The stresses and strains at each scale are indicated by different symbols respectively. For stress and strain at the micro-scale the plain letters $\sigma$ and $\varepsilon$ are used, at the meso-scale a superscript C is added, and at the macro-scale there is a bar over $\sigma$ and $\varepsilon$. For simplicity the subscript of tensor is ignored in the figure. Note that all subscript letters such as i, j and ij no longer represent atoms i and j, but denote components of vectors and tensors in the corresponding axis.

The goal of the multiscale method is to establish quantitative relationships between stress and strain at these three scales. Through this analysis, microscopic stress and strain as well as macroscopic responses under a given loading condition can be simultaneously determined. The macroscopic constitutive relationship obtained in this way is a material constitutive equation based on the micro-mesoscopic analysis and it is different from the common phenomenological relationships. The different findings obtained in the microscopic scale from this multiscale analysis are valuable since the related phenomena at the microscopic "world" were difficult to find previously by experiment. Thus these findings may constitute a basis for developing new mechanisms/theory or upgrade/design new materials.

### 8.1.2 Using Two-face Feature of Meso-cell to Link Both Microscopic and Macroscopic Scales

In Figure 8.1, the function of the meso-cell in bridging the three scales is twofold. It is seen that in the middle box its unit cell on the right is a real pearlitic colony with heterogeneous microstructure, but the microstructure of the unit cell on the left is smeared out, thus the unit cell is homogeneous. As we can see, for the success of this method, this two-face feature of the intermediate meso-cell is not only reasonable but also effective in quantitatively linking variables at microscopic and macroscopic scales. Basically, when thousands of meso-cells consist of the macroscopic structure, the important property of the meso-cell on macroscopic property is directly related to its average behavior, not the detail of its microstructure, thus a smeared homogeneous cell is sufficient. On the other hand the average behavior of a meso-cell is determined by the geometry and property of its constituents, thus the related detailed information is essential when investigating the micro-meso linkage and transitions.

The up and down curved arrows on the left and right sides of Figure 8.1 between (a) and (b) show the connections and information transformation between variables at the mesoscopic and macroscopic scale. The quantitative linkage of physical variables at the mesoscopic scale (e.g., $\sigma_{IJ}^C$ and $\varepsilon_{IJ}^C$) with those at the macroscopic scale (e.g., $\bar{\sigma}_{IJ}$ and $\bar{\varepsilon}_{IJ}$) are realized by using a self-consistent scheme (SCS) and will be introduced in the next section.

On the other hand, the up and down curved arrows on the two sides of Figure 8.1 between (b) and (c) indicate the linkage and information transformation between variables at the mesoscopic and microscopic scale. Their quantitative relationship can be determined through the development of relevant constitutive equations of representative meso-cells by satisfying mechanical and physical principles at the microscopic scale. This will be introduced in section 8.5.

## 8.2 Quantitative Meso-macro Bridging Based on Self-consistent Schemes

This section shows how to develop quantitative linkage between the stress and strain of the homogenized meso-cell and that of the macroscopic cell through a self-consistent scheme (SCS).

### 8.2.1 Basic Assumption

Now consider a representative macroscopic unit cell of a statistically homogeneous dual-phase material. It is assumed that the macroscopic representative unit cell is an aggregate of a large number of randomly oriented ellipsoidal or spherical inclusions, and that each inclusion is composed of a relatively large number of constituent elements such as two alternately arranged and perfectly bonded layered materials shown in Figure 8.1(c).

This assumption makes it reasonable to obtain the macroscopic stress and strain by volume average over all orientations. It also makes it reasonable to consider an inclusion such as a pearlitic colony to be an effectively homogeneous media in which the detailed microstructure is smeared out as shown on the left of the middle box of Figure 8.1. Therefore, the interaction of every inclusion with its aggregate is calculated individually by taking the inclusion under consideration as an effectively homogeneous ellipsoid or sphere.

Furthermore, materials at all scales are assumed to be plastically incompressible. As we know, volume strain equals the sum of normal strain components along three perpendicular directions. The volume strain can be separated into elastic and plastic components. The elastic one can be determined by Hooke's law (see (8.5b) below), the plastic one is assumed as zero for plastically incompressible materials. This plastic incompressibility assumption is based on the experimental result of a famous test conducted by P. W. Bridgman (1882–1961) at Harvard. He built a test chamber in which high hydrostatic pressure, approaching the sea-water pressure at 10,911 m deep around the Mariana trench in the Pacific Ocean, was achieved.[2] The test results indicate that the tensile-loading curve with large plastic deformation of a steel beam is virtually unaffected by the hydrostatic pressure and no plastic volume deformation was found. Thus the plasticity theory of metals usually assumes that plastic strain is independent of hydrostatic stress and takes plastic incompressibility as a basic assumption. This, in turn, makes the concepts of deviatoric stress and deviatoric stress space important for plasticity theory, and they will be introduced in Section 8.3.2.

## 8.2.2   Introduction to Self-consistent Schemes (SCS)

### 8.2.2.1  SCS Equation in Linking Variables of Inclusions and Their Aggregate

SCS is a scheme which establishes the quantitative relationship between the overall variables (e.g., $\bar{\sigma}_{IJ}$ and $\bar{\varepsilon}_{IJ}$) of the aggregate and those (e.g., $\sigma_{IJ}^C$ and $\varepsilon_{IJ}^C$) of its inclusion. Here, the term "inclusion" is the basic building unit of the aggregate, in other words the aggregation of all these inclusions forms the aggregate. In multiscale analysis an inclusion is equivalent to the homogenized meso-scale unit cell (or meso-cell for simplification); its physical quantity in the case of iron steel is the pearlitic colony. In the description below, the terms "inclusion" and "meso-cell" will be used if the description is related to the macro-meso transition.

The strains of inclusions are constrained by the strain of the aggregate; on the other hand, the latter strain is a volume averaging of the strains of inclusions, thus there must be a consistent relationship between the variables in inclusion and those in the aggregate. These self-consistent macroscopic (aggregate) stresses/strains and the mesoscopic (inclusion) stresses/strains can be expressed by different types of SCS. Here the one which can be used in the cyclic loading case with accuracy and convenience of applications is given as follows:[3]

$$d\bar{\sigma}_{IJ} - d\sigma_{IJ}^C = -A(d\bar{\varepsilon}_{KK} - d\varepsilon_{KK}^C)\delta_{IJ} - B(d\bar{\varepsilon}_{IJ} - d\varepsilon_{IJ}^C) \tag{8.1a}$$

where subscript IJ denote the tensor components with $I = 1, 2, 3$ and $J = 1, 2, 3$. Again, $\delta_{IJ}$ is the Kronecker delta, which is

$$\delta_{IJ} = 0 \text{ for } I \neq J \tag{8.1b}$$

$$\delta_{IJ} = 1 \text{ for } I = J \tag{8.1c}$$

### 8.2.2.2  Simplification Based on Plastic Incompressibility Assumption

By notation convention, the dummy index KK in (8.1a) denotes the sum of repeated terms for kk = 11, 22 and 33. It forms the sum of three normal strains along the three perpendicular axes, 11, 22 and 33 (or xx, yy and zz) as

$$d\bar{\varepsilon}_{KK} = d\bar{\varepsilon}_{11} + d\bar{\varepsilon}_{22} + d\bar{\varepsilon}_{33} \tag{8.2a}$$

$$d\varepsilon_{KK}^C = d\varepsilon_{11}^C + d\varepsilon_{22}^C + d\varepsilon_{33}^C \tag{8.2b}$$

Thus, $d\bar{\varepsilon}_{KK}, d\varepsilon_{KK}^C$ denote respectively the volume strain of the aggregate and the inclusion. The decomposition of total strain $d\bar{\varepsilon}_{IJ}$ into elastic strain $d\bar{\varepsilon}_{IJ}^e$ and plastic strain $d\bar{\varepsilon}_{IJ}^p$ gives

$$d\bar{\varepsilon}_{IJ} = d\bar{\varepsilon}_{IJ}^e + d\bar{\varepsilon}_{IJ}^p \qquad (8.3a)$$

Based on this decomposition (8.2a) can be written as

$$d\bar{\varepsilon}_{KK} = (d\bar{\varepsilon}_{11}^e + d\bar{\varepsilon}_{22}^e + d\bar{\varepsilon}_{33}^e) + (d\bar{\varepsilon}_{11}^p + d\bar{\varepsilon}_{22}^p + d\bar{\varepsilon}_{33}^p) = d\bar{\varepsilon}_{KK}^e + d\bar{\varepsilon}_{KK}^p = d\bar{\varepsilon}_{KK}^e \qquad (8.3b)$$

The last equation in (8.3b) is obtained by the plastic incompressibility assumption; since $\bar{\varepsilon}_{KK}^p$ denotes the plastic volume strain this assumption requires $d\bar{\varepsilon}_{KK}^p = 0$. The same kind of result such as $d\varepsilon_{KK}^{pC} = 0$ can be obtained for the strain in the meso-cell (inclusion), giving

$$d\varepsilon_{KK}^C = d\varepsilon_{KK}^{eC} \qquad (8.3c)$$

Substituting (8.3b, c) into (8.1a) we obtain

$$d\bar{\sigma}_{IJ} - d\sigma_{IJ}^C = -A(d\bar{\varepsilon}_{KK}^e - d\varepsilon_{KK}^{eC})\delta_{IJ} - B(d\bar{\varepsilon}_{IJ}^e + d\bar{\varepsilon}_{IJ}^p - d\varepsilon_{IJ}^C) \qquad (8.4a)$$

## 8.2.3 Weakening Constraint Effect of Aggregate on Inclusion with Increase of Plastic Deformation

The two coefficients A and B in (8.1a) are given as[3]

$$A = \frac{\mu^t(3 - 5v^t)}{4 - 5v^t} \qquad (8.4b)$$

$$B = \frac{\mu^t(7 - 5v^t)}{4 - 5v^t} \qquad (8.4c)$$

where $\mu^t$ and $v^t$ are the macroscopic tangential elastoplastic shear modulus and Poisson's ratio of the assumed isotropic aggregate, respectively. Since $v^t$ does not change much, but $\mu^t$ decreases quickly, A and B are decreasing coefficients when the macroscopic plastic strain increases. This decrease denotes a weakening constraint power of the aggregate on the inclusion with increase of plastic deformation. The simplification of Berveiller and Zaoui[4] that the aggregate is a macroscopically isotropic material is taken so that Hooke's elastic relationships between strain and stress tensors of the aggregate can be expressed through the macroscopic elastic constants $\bar{G}, \bar{v}$, and $\bar{K}$ as follows:[3, 5]

$$d\bar{\varepsilon}_{IJ}^e = \frac{1}{2\bar{G}}(d\bar{\sigma}_{IJ} - \frac{\bar{v}}{1 + \bar{v}}d\bar{\sigma}_{KK}\delta_{IJ}) \qquad (8.5a)$$

$$d\bar{\varepsilon}_{KK}^e = \frac{1}{3\bar{K}}d\bar{\sigma}_{KK} \qquad (8.5b)$$

It is clear that (8.5b) denotes that the elastic volume change is proportional to the hydraulic stress defined by $d\bar{\sigma}_{KK}/3$ and inversely proportional to the volume modulus $\bar{K}$. Substituting (8.5a) and (8.5b) into (8.4a), setting the collection of all terms which are related to the aggregate to be $d\bar{Q}_{IJ}$, and putting all the terms which are related to the meso-cell (inclusion) on the right part of the equation, (8.4a) can be rewritten as:

$$d\bar{Q}_{IJ} = d\sigma_{IJ}^c + A d\varepsilon_{KK}^{ec}\delta_{IJ} + B d\varepsilon_{IJ}^c \qquad (8.6a)$$

The explicit expression of the collection for $d\overline{Q}_{IJ}$ is given as:

$$d\overline{Q}_{IJ} = \left(1 + \frac{B}{2G}\right)d\overline{\sigma}_{IJ} + \left(\frac{A}{3\overline{K}} - \frac{B\overline{v}}{2G(1+\overline{v})}\right)d\overline{\sigma}_{KK}\delta_{IJ} + Bd\overline{\varepsilon}_{IJ}^{p} \qquad (8.6b)$$

(8.6a) and (8.6b) together formulate the constraint law which is used in the SCS proposed by Fan (1999).[3]

## 8.2.4 Quantitative Linkage of Variables Between Mesoscopic and Macroscopic Scales

The SCS equation (8.6a) offers a link between the overall deformation of the aggregate and that of the homogenized meso-cell. More specifically, after the applied loading is given, $d\overline{Q}_{IJ}$ can be obtained by (8.6b) assuming an initial value of plastic strain $d\overline{\varepsilon}_{IJ}^{p}$. This value will be used to start a global iterative process. Each value of $d\overline{Q}_{IJ}$ is taken to be an input for the calculation of stresses and strains in the meso-cell. The orientation of each meso-cell is different, thus this analysis is called a local analysis. The first step in carrying out the local analysis for each meso-cell is to transform (8.6a) from the global coordinate system $X_I$ to the individual local coordinate system $x_i$ (see both coordinates in Figure 8.1(a)) by a coordinate transformation. After this transformation, the SCS equation (8.6a) can be written in the following matrix form in the local coordinate system in which the $x_1 - x_2$ plane (or $x_3$ axis) is parallel (or perpendicular) to the individual layer plane:

$$\{dQ^c\} = \{d\sigma^c\} + Ad\varepsilon_{kk}^{ec}\{\delta\} + B\{d\varepsilon^c\} \qquad (8.7a)$$

where

$$\{dQ^c\} = \left\{dQ_{11}^c, dQ_{22}^c, dQ_{33}^c, dQ_{23}^c, dQ_{13}^c, dQ_{12}^c\right\} \qquad (8.7b)$$

$$\{d\sigma^c\} = \left\{d\sigma_{11}^c, d\sigma_{22}^c, d\sigma_{33}^c, d\sigma_{23}^c, d\sigma_{13}^c, d\sigma_{12}^c\right\} \qquad (8.7c)$$

$$\{d\varepsilon^c\} = \left\{d\varepsilon_{11}^c, d\varepsilon_{22}^c, d\varepsilon_{33}^c, d\varepsilon_{23}^c, d\varepsilon_{13}^c, d\varepsilon_{12}^c\right\} \qquad (8.7d)$$

$$\{\delta\} = \{1, 1, 1, 0, 0, 0\}^T \qquad (8.7e)$$

(8.7a) is a local matrix form of the constraint law and can be specified after the constitutive law of the generic inclusion is derived. For elastoplastic materials, this constitutive law may be conveniently described by either a tangential form with tangential stiffness matrix $[D]_{\text{tan}}^{ep}$:[6]

$$\{d\sigma^C\} = [D]_{\text{tan}}^{ep}\{d\varepsilon^C\} \qquad (8.7f)$$

or a non-tangential matrix form:[1, 7]

$$\{d\sigma^C\} = [D]^{ep}\{d\varepsilon^C\} + \{d\overline{H}\} \qquad (8.7g)$$

$$\{d\varepsilon^C\} = [M]^{ep}\{d\sigma^C\} + \{dH\} \qquad (8.7h)$$

where $[D]^{ep}$ and $[M]^{ep}$ are, respectively, the elastoplastic stiffness and compliant matrix. In this chapter, form (8.7h) is used in which $[M]^{ep}$ accounts for instantaneous responses and $\{dH\}$ a matrix that accounts for the history-dependent plastic responses.

These matrices in the meso-scale can be expressed by microscopic parameters such as elastoplastic material parameters, volume fractions $v^{(m)}$ of soft and hard phases, the accumulative plastic strain $p^{(m)}$ in these layers. Through the analysis of quantitative meso-micro linkage in Section 8.5, we can get the relationship of $\{d\varepsilon^C\}$, $\{d\sigma^C\}$ with these microscopic parameters. We can, in turn, through the meso-macro connections of (8.6a) and (8.6b), establish relationships between macroscopic and microscopic variables by taking variables $\{d\sigma^C\}$ and $\{d\varepsilon^C\}$ of the meso-cell as an intermediate means. Relationships between these variables will be introduced after elastoplastic properties of constituent phases are discussed in the next two sections and quantitative micro-meso linkage is established in section 8.5.

## 8.3 Basics of Continuum Plasticity Theory

This section will start to introduce continuum plasticity theory as a basis for connecting the second class of multiscale analysis in the continuum domain with the atomistic analysis of the first class of multiscale analysis through dislocation theory. Needless to say, the continuum plasticity theory is phenomenological in nature. Some theory such as the theory of internal state variables (see below) is introduced to indirectly consider the microstructure effects on material property, but it does not change the phenomenological nature of the theory.

### 8.3.1 Several Basic Elements of Continuum Plasticity Theory

When the maximum normal stress in a tensile bar reaches its yield stress $\sigma_y$ or the maximum shear stress in a shear specimen reaches its shearing yield stress $\tau_y$, the material starts permanent plastic deformation. These critical stresses $\sigma_y$ and $\tau_y$ as well as their evolution after the yielding can be directly determined, respectively, by tensile and thin-tube torsional experiments. If a material element subjects to complicated loading with different stress components, a purely experimental method to determine when and how plastic deformation appears is not applicable, and plasticity theory must be used instead. Plasticity theory usually includes the following basic elements.

#### 8.3.1.1 Criterion for Initiation of Plastic Strain

For a 3D stress state, the criterion is usually described by an initial yield surface in a stress space. Instead of x, y, z as Cartesian coordinate axes, stress space takes stress components as its coordinate axes, thus the number of axes is six independent stress components. If any stress point is within the yield surface, its corresponding physical status is elastic and no plastic deformation occurs. This is exactly similar to a 1D case, when stress is less than yield stress the stress-strain relationship is elastic, as described by Hooke's law. Furthermore, if a 3D case reduces to 1D, the initial yield surface will reduce to a single point which is the yield point in the stress-strain curve. Based on this one-to-one correspondence, the parameters of the yield surface such as the center and radius of a spherical yield surface can be determined by 1D tensile or shear experiment.

#### 8.3.1.2 Hardening Rule

This rule describes how the subsequent yield surface moves with the applied loading after the initial plastic deformation occurs. Usually, it will involve the so-called deformation induced hardening which is consistent with the stress increase due to deformation increase. Just like the correspondence between initial yield surface and yield point in the 1D case, each point on the tensile curve beyond the yield stress corresponds to a subsequent yield surface.

#### 8.3.1.3 Criterion for Loading and Unloading

Even if the point of stress state is located on the yield surface in the stress space, there are still three possibilities for material deformation behavior which depends on the direction of stress increment in relation to the current yield surface. If the stress increment is towards the inside of the yield surface, material stress status is under unloading and towards the elastic region. If the increment is towards the outside of the yield surface, material is under loading and plastic deformation will occur. If the stress increment keeps the stress point moving along the yield surface, the material is under neutral loading and no plastic deformation occurs.

#### 8.3.1.4 Rule of Plastic Flow

This flow rule determines the components of plastic strain increment. In classical plasticity theory, an associate flow rule with the yield surface is given. Based on the so-called Drucker's postulate,[8] the yield surface is convex and the vector of plastic strain increment is perpendicular to the yield surface in the stress space (see next section).

### 8.3.2  Description of Continuum Plasticity Theory Within Deviatoric Stress Space

This section will briefly review the mathematical description of yield surface, flow rule and hardening rule of continuum plasticity theory. Following Bridgman's test deviatoric stress must be used.

#### 8.3.2.1  Yield Surface Described by Deviatoric Stress

When material such as constituent phases of microstructure initiates plastic deformation, the stress state described by its deviatoric stress component $S_{ij}$ must be located at a spherical yield surface described mathematically as follows:

$$(S_{ij} - R_{ij})(S_{ij} - R_{ij}) = S_{y0}^2 \qquad (8.8a)$$

In this equation, $S_y$ represent the radius of a spherical yield surface in the deviatoric stress space that takes deviatoric stresses as its coordinates. $R_{ij}$ is called back stress and it is the center of the spherical yield surface (see Figure 8.2).

Here, deviatoric stress $S_{ij}$ is related to the common stress $\sigma_{ij}$ by

$$s_{ij} = \sigma_{ij} - \delta_{ij}\sigma_{kk}/3 \qquad (8.8b)$$

The introduction of the symbol of Kronecker delta $\delta_{ij}$ defined in (8.1b, c) in this equation has the following results:

- If $i \neq j$ then $s_{ij} = \sigma_{ij}$ which indicates that deviatoric stress is identical to the stress (e.g., shear stress) itself.
- If $i = j$ then $s_{ij} = \sigma_{ij} - \sigma_{kk}/3$. Here, $\sigma_{kk}/3 = (\sigma_{11} + \sigma_{22} + \sigma_{33})/3$ denotes the average hydrostatic stress. This definition indicates that the deviatoric stress equals the stress after it minus the hydrostatic stress. Why it is necessary to subtract the hydrostatic stress is because the Bridgman test data showed that plasticity behavior does not depend on the hydrostatic stress, as discussed in Section 8.2.1.

#### 8.3.2.2  Flow Rule of Plastic Strain Described in Deviatoric Stress Space

Similarly, the deviatoric strain $e_{ij}$ of strain $\varepsilon_{ij}$ and its elastic part can be defined, respectively, as follows:

$$e_{ij} = \varepsilon_{ij} - \delta_{ij}\varepsilon_{kk}/3 \qquad (8.8c)$$

$$e_{ij}^P = \varepsilon_{ij}^P - \delta_{ij}\varepsilon_{kk}^P/3 \qquad (8.8d)$$

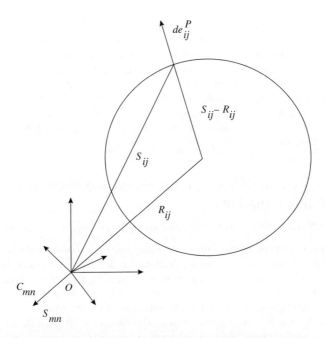

**Figure 8.2**   Yield stress surface and flow direction of plastic strain increment $de_{ij}^{p}$ within the deviatoric stress space (Reproduced by permission of [IPPT PAN], Valanis, K. C. (1980) Foundamental consequences of new intrinsic time measure plasticity as a limit of the endochronic theory. *Archives of Mechanics*, 23, 171–190, Polish Academy of Sciences)

Due to the plastic incompressibility assumption, $\varepsilon_{kk}^{P} = 0$, thus $e_{ij}^{P} = \varepsilon_{ij}^{P}$ and $de_{ij}^{P} = d\varepsilon_{ij}^{P}$. In relation to this, an important variable called accumulated plastic strain p is defined by its increment dp as follows:

$$dp = \sqrt{de_{ij}^{P} de_{ij}^{P}} \tag{8.8e}$$

which combines the contribution of all components of plastic strain increment due to the sum convention for the repeated dummy index.

(8.8a) can be schematically described in Figure 8.2. It is seen that $R_{ij}$ denotes the center of the yield surface and the increment of plastic strain $de_{ij}^{p}$ is normal to the yield surface and along the direction of $(S_{ij} - R_{ij})$, thus we have the following flow rule of plastic strain (see (8.9f) below):

$$de_{ij}^{P} = \frac{(S_{ij} - R_{ij})}{S_y} dp \tag{8.8f}$$

### 8.3.2.3  Isotropic Hardening Rule

The hardening of plasticity can be basically divided into isotropic hardening and kinematic hardening. Isotropic hardening is described by the increase of $S_y$ with loading process which makes the spherical yield surface expanded. From Figure 8.2 it is seen that the square root of $(S_{ij} - R_{ij})$, i.e., $S_y = \sqrt{(S_{ij} - R_{ij})(S_{ij} - R_{ij})}$ defines the radius of the subsequent yield surface. $S_y$ may be related to the initial yield stress $S_{y0}$ as follows:

$$S_y = S_{y0} f(p) \tag{8.8g}$$

where f(p) is a monotonic increase function of accumulative plastic strain p, when $p = 0$, $f(0) = 1$. Thus, the isotropic hardening rule makes the yield surface enlarge equally in all directions. Furthermore, these rules can also be used to describe material softening in which $S_y$ may be reduced to cause yield surface shrink due to, say, internal damage. In that case, f(p) is a decreasing function at some softening stage.

On the other hand, kinematic hardening is described by evolution of back stress $R_{ij}$, thus makes the yield surface move in the space. There is certainly a mixed hardening rule in which both $S_y$ and $R_{ij}$ are changed.

For kinematic hardening, the evolution equation of back stress $R_{ij}$ in terms of different mechanisms and different deformation stages is dominantly important, and will be discussed after a brief introduction on the internal variable theory of plasticity.

## 8.4  Internal Variable Theory, Back Stress and Elastoplastic Constitutive Equations

### 8.4.1  Internal Variable Theory Expressed by a Mechanical Model

In the development of phenomenological plasticity theory, one tries to describe the fact that the internal structure of material is changed due to plastic deformation. The way to consider this fact within the phenomenological framework is to introduce a certain number, say n, of internal state variables in addition to observable thermodynamic variables such as temperature T, pressure p and volume v. While these internal variables do not have any direct connection with atomistic or microscopic structure, their effects on material property can be investigated through approaches of irreversible thermodynamics or mechanical models with spring and blocks, or dashpots to collectively and averagely describe the effects of the change of internal structures on material hardening or softening behavior.

For simplicity, let us briefly describe the mechanical model which was used by Kelvin and Maxwell to investigate material behavior. Figure 8.3(a) shows a mechanical model consisting of a main spring in the top and n Maxwell elements arranged parallel in the bottom. The main spring has the material shear modulus $\mu$ as its spring constant, thus under the applied deviatoric stress $S_{ij}$, such as shear stress $\tau$, its deviatoric (or shear) elastic strain $e^e_{ij}$ can be expressed as

$$e^e_{ij} = \frac{S_{ij}}{2\mu} \tag{8.9a}$$

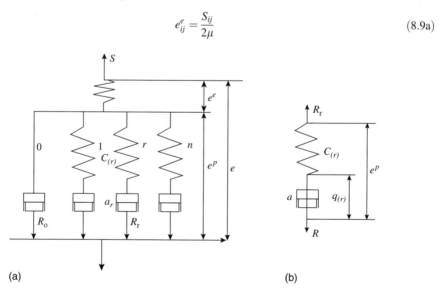

(a)                                                                      (b)

**Figure 8.3**   Mechanical model for description of plastic deformation

Each Maxwell element consists of a spring with spring constant $C_{(r)0}$ and a dashpot-like sliding block. The block r (r = 1...n) has a friction coefficient $a_{r0}$ and its displacement $q_{(r)ij}$ is the r-th internal variable which is a second-order tensor. (In Figure 8.3 the tensor index ij is omitted for simplification.) The collection set of internal variable $q_r$ (r = 1, 2, ..., n) denotes the change of material internal structure due to dislocations in plastic deformation.

While the block dissipates useful energy for the irreversible deformation, the spring $C_{(r)0}$ plays a role storing energy in a local region for driving dislocation nucleation, regeneration and motion. It is easy to see from Figure 8.3(a) that the applied stress $S_{ij}$ equal to the sum of the resistance offered by all Maxwell elements and a special block element at the left-most part, it can be expressed mathematically as

$$S_{ij} = S_{ij}^0 + R_{ij} \tag{8.9b}$$

$$R_{ij} = \sum_{r=1}^{n} R_{(r)ij} \tag{8.9c}$$

When the stress $S_{ij}$ is within the initial yield surface all strain components are elastic and $e_{ij}^p$ is zero, thus no mechanical element in the bottom of Figure 8.3(a) has any displacement. However, when the stress $S_{ij}$ reaches the initial yield surface described in (8.8a), plastic strain suddenly occurs, but only the left-most block has movement immediately, all the Maxwell elements (r = 1...n) only have elastic deformation by spring because a much larger stress is needed to move blocks.

Thus, $S_{ij}^0$ of the left-most block represents the initial resistance to the motion of the left-most sliding block. Physically, it corresponds to the critical resistance of the first group dislocations. Specifically, when dislocation driving stress reaches $S_y^0$ the dislocation-pin resistance will be overcome and dislocation starts motion which produces initial plastic strain. During this process, the required stress component $S_{ij}^0$ can be considered proportional to the ratio of plastic strain increment $de_{ij}^p$ with respect to dp, the increment of total plastic strain. In other words, the initial plastic status can be reached when the following equation is given:

$$S_{ij}^0 = S_y^0 \frac{de_{ij}^p}{dp} \tag{8.9d}$$

In fact, if both sides multiply themselves, we have

$$S_{ij}^0 S_{ij}^0 = (S_y^0)^2 \left( \frac{de_{ij}^p}{dp} \frac{de_{ij}^p}{dp} \right) = (S_y^0)^2 \tag{8.9e}$$

The last equation is due to $(dp)^2 = de_{ij}^p de_{ij}^p$ given in (8.8e). Comparing this equation with (8.8a), it is seen that this equation describes the initial yield surface of isotropic material when the center of the yield surface is at the coordinate origin. Substituting (8.9d) into (8.9b), we have

$$S_{ij} - R_{ij} = S_y^0 \frac{de_{ij}^p}{dp} \tag{8.9f}$$

Rearranging this equation, it is easy to obtain the following equation:

$$de_{ij}^p = \frac{S_{ij} - R_{ij}}{S_y^0} dp \tag{8.9g}$$

which is the flow rule expressed in (8.8f). Furthermore, if both sides of (8.9f) multiply themselves, the equation (8.8a) of initial yield surface can be obtained.

### 8.4.2   Calculation of Back Stress $R_{ij}$ in Terms of Plastic Strain

The increment of back stress $dR_{ij}$ is the sum of terms $dR_{(r)ij}$ ($r = 1,\ldots n$):

$$dR_{ij} = \sum_{r=1}^{n} dR_{(r)ij} \tag{8.9h}$$

Here $R_{(r)ij}$ is the resistance to the motion of the block $a_{(r)0}$, representing physically the material resistance to the subsequent r-th dislocation group; it can be determined by the r-th Maxwell element when the continuum plastic strain is $e_{ij}^p$. The value $R_{(r)ij}$ is different for different Maxwell elements and different at different deformation stages since the corresponding dislocation conditions may be quite different.

The resistance value $R_{(r)ij}$ depends on the value of $C_{(r)0}$ and $a_{(r)0}$ during the concert deformation process of the local spring and the block; their equations can be given as follows:

$$\text{For spring: } R_{r(ij)} = c_{(r)0}(e_{ij}^p - q_{(r)ij}) \tag{8.9i}$$

$$\text{For block: } R_{(r)ij} = a_{(r)0}\frac{dq_{(r)ij}}{dp} \tag{8.9j}$$

(8.9j) is based on the assumption that the resistance to the block movement is proportional to the generalized rate of the block displacement $q_{(r)ij}$ with respect to the accumulative plastic strain p. Changes of internal variable $dq_{(r)ij}$ indicate dissipative plastic deformation which is described by the block movement in which $a_{(r)0}$ is the frictional coefficient. This assumption is consistent with (8.9d) and is reasonable because for a given increment of cumulative plastic strain, the one with larger change of internal structure will sustain larger resistance.

Differentiating (8.9i) we obtain

$$dR_{r(ij)} = c_{(r)0}de_{ij}^p - c_{(r)0}dq_{(r)ij}$$

From (8.9j) $dq_{(r)ij} = R_{(r)ij}dp/a_{(r)0}$, and substituting this expression to replace $dq_{(r)ij}$ in the last equation, we get

$$dR_{(r)ij} = C_{(r)0}de_{ij}^p - \alpha_{(r)0}R_{(r)ij}dp \tag{8.9k}$$

where

$$\alpha_{(r)0} = \frac{c_{(r)0}}{a_{(r)0}} \tag{8.9l}$$

The parameter $\alpha_{(r)0}$ which combines the local spring and block behavior is important. It intrinsically relates to the local driving force (through spring) and dissipation (through block) for the characteristics of the evolution of the dislocation group which are expressed by the r-th Maxwell element. Thus, this parameter is the characteristic parameter which can describe the behavior of the Maxwell element and its corresponding dislocation group during the deformation process.

Substituting (8.9k) into (8.9h) we obtain

$$dR_{ij} = \sum_{r=1}^{n} dR_{(r)ij} = de_{ij}^p \sum_{r=1}^{n} C_{(r)0} - dp \sum_{r=1}^{n} \alpha_{(r)0}R_{ij(r)} \tag{8.9m}$$

Denoting

$$\rho_0 = \sum_{r=1}^{n} C_{(r)0} \tag{8.9n}$$

and

$$h_{ij} = -\sum_{r=1}^{n} \alpha_{(r)0} R_{ij(r)} \qquad (8.9o)$$

(8.9m) can be rewritten as

$$dR_{ij} = \rho_0 de_{ij}^p + h_{ij}dp \qquad (8.9p)$$

This equation shows how the back stress $R_{ij}$ moves with an increment of plastic strain.

Note that the evolution equations of the back stress given by (8.9k) and (8.9p) with the mechanical model are consistent with the constitutive equations derived from other approaches, such as Valanis,[9] Chaboche et al.,[10] Ohno and Wang,[11] McDowell,[12] and Jiang and Sehitoglu.[13] Readers interested in the similarity of these plasticity theories are referred to Ref. [3].

## 8.4.3 Expressing Elastoplastic Constitutive Equations for Each Constituent Phase

In the following, the superscript (m) denotes the number of constituent phase in material microstructure. For instance, m = 1 denotes the cementite phase, m = 2 denotes the ferrite phase. Elastic material such as cementite can be considered as a special case of elastoplastic material where plastic deformation is zero.

Differentiating (8.9f) we have

$$dS_{ij} - dR_{ij} = S_y^0 \frac{d^2 e_{ij}^p}{d^2 p} dp \qquad (8.10a)$$

Replacing $dR_{ij}$ with (8.9p) we obtain

$$dS_{ij} = \rho_0 de_{ij}^p + h_{ij}^0 dp \qquad (8.10b)$$

where

$$h_{ij}^0 = h_{ij} + S_{y0} \frac{d^2 e_{ij}^p}{d^2 p} \qquad (8.10c)$$

Combining (8.10b) with Hooke's law and using the definition of deviatoric stress and strain given in (8.8b), (8.8c), the following elastoplastic constitutive equation in the index form can be developed for each constituent phase as follows:[3]

$$d\sigma_{ij}^{(m)}/2\mu_p^{(m)} = d\varepsilon_{ij}^{(m)} + a_m d\varepsilon_{kk}^{(m)} \delta_{ij} + h_{ij}^{0(m)} dp^{(m)}/\rho_0^{(m)} \quad (i = 1, 2) \qquad (8.10d)$$

with

$$2\mu_p^{(m)} = \frac{\rho_0^{(m)}}{1 + \rho_0^{(m)}/2G^{(m)}} \qquad (8.10e)$$

$$a_m = v^{(m)}/\left(1 - 2v^{(m)}\right) + K^{(m)}/\rho_0^{(m)} \qquad (8.10f)$$

where $v^{(m)}$, $G^{(m)}$, and $K^{(m)}$ denote Poisson's ratio, shear modulus and bulk modulus of hard cementite phase (m = 1) and soft ferrite phase (m = 2). If $\rho_0^{(m)} \to \infty$ the last terms in both (8.10d) and (8.10f) vanish, $\mu_p^{(m)} = G^{(m)}$, and (8.10d) reduces to Hooke's law.

## 8.5 Quantitative Micro-meso Bridging by Developing Meso-cell Constitutive Equations Based on Microscopic Analysis

In this section, the relationships between microscopic variables of continuum medium $(\sigma_{ij}^{(m)}, \varepsilon_{ij}^{(m)})$, microstructure parameters (volume fraction $V_m$ and layer thickness t) and mesoscopic variables $(\sigma_{IJ}^c, \varepsilon_{IJ}^c)$ are developed. They are dependent on characteristics of a pearlitic colony which consists of alternately arranged thin hard cementite layers and thick soft ferrite layers.

### 8.5.1 Developing Meso-cell (Inclusion) Constitutive Equations

The layer-structured meso-cell (a pearlitic colony) may be simplified as a rectangular parallelepiped as shown in Figure 8.4. The coordinate of Figure 8.4 is the local coordinate of the pearlitic colony. This parallelepiped connects an ellipse from its inside but it is totally encompassed by the ellipse, taking the maximum volume possible with the length $2b_X$, width $2b_Y$, and the height $2b_Z$. Because these dimensions are much larger than the layer-thickness, t, this simplification is reasonable for determining stress and strain in the hard and soft layers.

The direct stress (strain) interactions between the two phases with very thin layers can be approximately considered through their equilibrium conditions of out-plane components:

$$d\sigma_{ij}^{(1)} = d\sigma_{ij}^{(2)} = d\sigma_{ij}^C \quad ij = 33, 13, 23 \tag{8.11a}$$

where index 3 indicates the axis perpendicular to the layer plane 12. The in-plane stress components can be determined by a mix law:

$$d\sigma_{ij}^C = \sum_{m=1}^{2} V_m d\sigma_{ij}^{(m)} \quad ij = 11, 22, 12 \tag{8.11b}$$

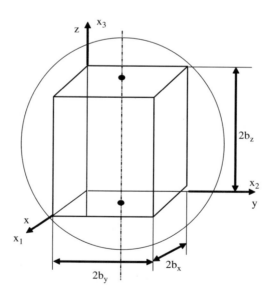

**Figure 8.4**  A rectangular parallelepiped of size $2b_x \times 2b_y \times 2b_z$ within an ellipse (Reproduced from Fan, J., Gao, Z., and Zeng X. (2004) Cyclic plasticity across micro/meso/macroscopic scales. *Proceedings of Royal Society of London A*, 460 (2045), 1477, Royal Society of London)

where $V_m$ (m = 1, 2) are volume fractions of each individual phase. In-plane strain components $d\varepsilon_{11}^C$, $d\varepsilon_{22}^C$, and $d\varepsilon_{12}^C$ are determined by the in-plane geometric constraint conditions at their interface. For very thin layers these conditions can be approximately described by the equality conditions of in-plane components:

$$d\varepsilon_{ij}^{(1)} = d\varepsilon_{ij}^{(2)} = d\varepsilon_{ij}^C \quad ij = 11, 22, 12 \tag{8.11c}$$

Their out-plane strain components can be expressed by the mix law:

$$d\varepsilon_{ij}^C = \sum_{m=1}^{2} V_m d\varepsilon_{ij}^{(m)} \quad ij = 33, 13, 23 \tag{8.11d}$$

It merits notice that the total number of stress and strain components of the two phases and the inclusion together is 36 with each one having six stress and six strain components. There are altogether 30 equations, six from (8.11a), six from (8.11c), six from (8.11b) and (8.11d), and 12 from (8.10d). Combining these equations will result in six independent equations that are the effective constitutive equations of the inclusion. They can be expressed explicitly in matrix form as follows:

$$\{d\varepsilon^C\} = [M]^{ep}\{d\sigma^C\} + \{dH\} \tag{8.12a}$$

This equation is the same as (8.7h), but the explicit matrix forms of $[M]^{ep}$, $\{dH\}$ and their related parameters are given in Appendix 8.A. From (8A.6–8A.11) we can see that $\{dH\}$ is closely related to the increment, dp, of the accumulated plastic strain defined in (8.8e).

$[M]^{ep}$ is an elastoplastic matrix. From (8.10e) it is seen when $\rho_0^{(m)} \to \infty$, $\mu_p^{(m)} \to G^{(m)}$, and $[M]^{ep}$ reduces to the elastic compliance matrix $[M]^e$. As a consequence of this reduction and dp = 0, the elastic strain matrix $[d\varepsilon^{eC}]$ of the inclusion can be expressed as:

$$[d\varepsilon^{eC}] = [M]^e\{d\sigma^C\} \tag{8.12b}$$

Subtracting (8.12b) from (8.12a), we express the plastic strain matrix of the homogenized inclusion by

$$[d\varepsilon^{pC}] = [d\varepsilon^C] - [d\varepsilon^{eC}] = ([M]^{ep} - [M]^e)\{d\sigma^C\} + \{dH\} \tag{8.12c}$$

## 8.5.2 Bridging Micro- and Macroscopic Variables via the Meso-cell Constitutive Equation

The solution technique for stress and strain in the meso-cell and constituent phases for a given macroscopic input $\{dQ\}^C$ can be described as follows. Substituting $\{d\varepsilon^C\}$ from constitutive equation (8.12a) into the SCS constraint equation (8.7a), we obtain

$$\{dQ^c\} = \{d\sigma^c\} + Ad\varepsilon_{kk}^{ec}\{\delta\} + B([M]^{ep}\{d\sigma^c\} + \{dH\}) \tag{8.12d}$$

in which the term $Ad\varepsilon_{kk}^{ec}\{\delta\}$ can also be expressed by $\{d\sigma^c\}$ as follows:

$$Ad\varepsilon_{kk}^{ec}\{\delta\} = A[\overline{\overline{M}}]^e\{d\sigma^C\} \tag{8.12e}$$

The detail of the derivation and the parameters in the matrix $[\overline{\overline{M}}]^e$ are given in Appendix 8.B. Substituting (8.12e) into (8.12d) we have

$$\{dQ^c\} = [M^*]\{d\sigma^c\} + B\{dH\} \tag{8.12f}$$

where

$$[M^*] = [[I] + A[\overline{\overline{M}}]^e + B[M]^{ep}] \tag{8.12g}$$

Since $[M]^{ep}$ relates microscopic variables in each constituent phase and $\{dQ^C\}$ relates to the macroscopic variables and the local orientation, the above equation bridges microscopic and macroscopic variables through the meso-scale constitutive equation (8.12a).

### 8.5.3  Solution Technique

Basically, the solution can be obtained through a locally iterative process. The procedure in each iteration process includes the following steps.[3]

**Step 1:** Calculate $d\overline{Q}_{IJ}$ by (8.6b) based on the given loading condition and the previous macroscopic plastic strain increment.

**Step 2:** Determine $\{dQ^C\}$ from $d\overline{Q}_{IJ}$ by a coordinate transformation from global coordinates to the local coordinates of inclusion at hand.

**Step 3:** Rearrange the equation (8.12f) by introducing matrix $\{dQ^*\}$ as:

$$\{dQ^*\} = \{dQ^c\} - B\{dH\} \tag{8.12h}$$

From (8.12f) we obtain

$$\{dQ^*\} = [M^*]\{d\sigma^C\} \tag{8.12i}$$

Thus

$$\{d\sigma^c\} = [M^*]^{-1}\{dQ^*\} \tag{8.12j}$$

where $[M^*]^{-1}$ is the inverse matrix of $[M^*]$ and

$$\{dQ^C\} = \{dQ_{11}, dQ_{22}, dQ_{33}, dQ_{23}, dQ_{13}, dQ_{12}\} \tag{8.12k}$$

$$\{d\sigma^C\} = \{d\sigma_{11}, d\sigma_{22}, d\sigma_{33}, d\sigma_{23}, d\sigma_{13}, d\sigma_{12}\} \tag{8.12l}$$

$$\{d\varepsilon^C\} = \{d\varepsilon_{11}, d\varepsilon_{22}, d\varepsilon_{33}, d\varepsilon_{23}, d\varepsilon_{13}, d\varepsilon_{12}\} \tag{8.12m}$$

The local iteration will not be ended until the relative error of dp, which is proportional to dH (see (8A.6) to (8A.11)), between two succeeding iterations is less than a tolerance. The details for the three-scale iteration process to find the solution and the comparison of simulation results with experimental data will be introduced in Sections 8.8 and 8.9.

The content of the next section is important as it quantitatively links variables between the first and second group of multiscale analysis and gives a theoretical treatment for the important size effects of microstructure.

## 8.6  Determining Size Effect on Yield Stress and Kinematic Hardening Through Dislocation Analysis

In Section 8.2 we established quantitative relationships of variables between meso- and macroscopic scales, in the last section we further established the quantitative relationships between variables of microscopic and macroscopic scales by taking constitutive equations of meso-scale as an intermediary means. Thus, the framework of the multiscale analysis of the second class in the continuum region is completed. However, the bridging of variables of this class with variables of the first class of multiscale analysis, which involves atomistic-nanoscopic scale, remains to be established.

In the following, we will emphasize this bridging by dislocation analysis which is closely related to the determination of parameters of constituent constitutive equations of microstructure. This bridging is based on both the dislocation pile-up theory proposed by Eshelby *et al.*[14] and Li and Chou[15] and the plasticity theory in Section 8.4. As a result, the important size effects on material properties, such as the one with cyclic plasticity described in Figure 1.3, can be quantitatively described. Note that the approach here is fundamentally different from other approaches such as the so-called gradient strain theory of plasticity in which a characteristic length is assumed to introduce the size effect. The other difference is that other theories concern size effects mainly for the yield stress but here we also emphasize the size effects on the material hardening behavior of plasticity which are related to deformation or further dislocation motion after the initial yielding.

## 8.6.1 Basic Idea to Introduce Size Effects in Plasticity

The idea is to use dislocation pile-up theory and the plasticity constitutive laws described in Section 8.4 for a same problem, i.e., the deformation of a layered structure under a given shear force. The idea is similar to using shearing test (or uniaxial test) for determining material constants and then applying them to a 3D analysis. More specifically, the first step uses both theories to investigate stress concentration near the interface between different constituent phases, and then transfers the size effect expressed in the dislocation model to the continuum constitutive laws.

In (8.9i) to (8.9o), we use subscript 0 under plastic parameters to denote material parameters. These material parameters are without size effects; it can be the case when the layer thickness is relatively large that the size effects can be neglected. In the remaining part of this chapter, we will use $S_y$, $C_{(r)}$ and $\alpha_{(r)}$ without subscript 0 to denote apparent material parameters with size effects of thin layers. These apparent material parameters are functions of $S_{y0}$, $C_{(r)0}$, $\alpha_{(r)0}$ and the layer-thickness t. Specifically, $S_y$ is related to size effects on yield stress; $C_{(r)}$ and $\alpha_{(r)}$ are related to size effects on dislocation back stress $R_{(r)}$ or on kinematic hardening of materials.

## 8.6.2 Expressing Size Effects on Yielding and Hardening Behavior by Dislocation Pile-up Theory

When dislocation pile-ups form at the interface of the thin soft (e.g., ferrite) lamellae with the hard layer, the shearing force at the tip of the pile-up can be magnified sufficiently. The first key in determining the layer-thickness effects on material parameters by dislocation theory is to determine the magnified stress, $\tau_{\mathrm{tip}}$, at the tip of dislocation pile-ups.

### 8.6.2.1 Size Effects on Yield Stress and Hall-Petch Rule

Let us consider a pile-up of dislocations under the action of shear stresses $\tau$ as shown in Figure 8.5. Based on the theory of Eshelby *et al.* we have the following two equations. One is for the length, l, of the dislocation pile-ups; the other is for the concentrated tip stress, $\tau_{\mathrm{tip}}$, exerted on the pinned dislocation or at the edge of the pile-ups:

$$l = 2nD/\tau \tag{8.13a}$$

$$\tau_{tip} = n\tau \tag{8.13b}$$

where n is the number of dislocations, and $D = \mu b/2\pi(1 - \upsilon)$ where b is the Burgers vector and $\mu$, $\upsilon$ are shear module and Poisson's ratio. Combining (8.13a) and (8.13b) we obtain

$$\tau_{tip} = \frac{\tau^2 l}{2D} \tag{8.13c}$$

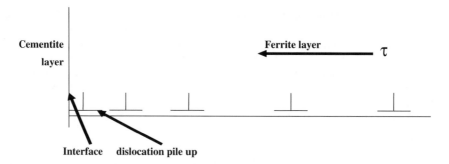

**Figure 8.5**   Dislocations pile up at the interface between ferrite and cementite layer under shear stress $\tau$ thickness (Reproduced from Fan, J., Gao, Z., and Zeng X. (2004) Cyclic plasticity across micro/meso/ macroscopic scales. *Proceedings of Royal Society of London A*, 460 (2045), 1477, Royal Society of London)

When $\tau_{tip}$ reaches a critical stress $\tau_{Cr}$, i.e.,

$$\tau_{tip} = \tau_{Cr} \tag{8.13d}$$

dislocations can overcome the pin resistance and start moving to produce plastic deformation. Substituting $\tau_{tip}$ of (8.13c) into (8.13d) and rearranging it, the corresponding applied stress, $\tau$, to produce dislocation motion or plastic deformation can be given as follows:

$$\tau = \sqrt{2D\tau_{tip}}\,l^{-1/2} = \sqrt{2D\tau_{Cr}}\,l^{-1/2} \tag{8.13e}$$

This applied shear stress is the deviatoric yield stress $S_{y0}$ (i.e., $\tau = S_{y0}$). For thin layers, the dislocation pile-up length l can approximately take the corresponding layer thickness t, so that the deviatoric yield stress can be approximately described as follows:

$$S_{y0} = \sqrt{2D\tau_{Cr}}\,t^{-n1} \tag{8.13f}$$

where $n1 = 0.5$, indicating a Hall-Petch relationship in which the yield stress of constituent phase has a relation of reciprocal square root with its grain size.[16, 17]

## 8.6.2.2 Size Effects on Kinematic Hardening Behavior

The other key in determining layer-thickness effects on apparent material parameters is to get the necessary increment $d\tau_{tip}$ of the tip stress to continue dislocation movement after the initial yielding. The increase of tip stress is necessary due to increased material resistance near the interface between hard (cementite) and soft (ferrite) phases; microscopically it is caused by dislocation enhanced resistance and macroscopically by material kinematic hardening. After differentiating (8.13c) we have

$$d\tau_{tip} = \frac{l}{D}\tau d\tau \tag{8.13g}$$

Substituting (8.13e) into the above equation results in

$$d\tau_{tip} = \sqrt{2l\tau_{Cr}/D}\,d\tau = \sqrt{2\tau_{cr}/D}\,t^{1/2}d\tau \tag{8.13h}$$

Note that in the last equation we approximately replace the dislocation length l by the layer thickness t as is done by (8.13f). This equation shows that to continue the dislocation motion after the initial yielding the necessary tip stress increment $d\tau_{tip}$ is proportional to the applied stress increment $d\tau$ and square root of the layer thickness t.

### 8.6.3 Tangential Modulus and Hardening Behavior Under Shear Force by Continuum Plasticity Theory

(8.10b) shows continuum plasticity theory for simple shearing. In this case, $d\tau = dS_{12}$, $dp = \pm\sqrt{2}de_{12}^p$, where the minus symbol is for the case $de_{12}^p < 0$ since dp is always assumed positive (see (8.8e)). Since in this case $\dfrac{de_{12}^p}{dp} = \pm\sqrt{2}$, thus $\dfrac{d^2 e_{12}^p}{d^2 p} = 0$, and $h^0 = h$. Substituting these expressions into (8.10b) we have

$$d\tau = (\rho_0 \pm \sqrt{2}h)de^p \tag{8.14a}$$

In this equation, the subscript 12 under $\tau$, h and $e^p$ is dropped for simplicity. Using (8.14a) for a material element at the tip of dislocation pile-ups, we have

$$d\tau_{tip} = (\rho_0 \pm \sqrt{2}h_{tip})de_{tip}^p \tag{8.14b}$$

### 8.6.4 Equating Dislocation-obtained Shear Stress Increment with that Obtained by Continuum Plasticity Theory

(8.13h) and (8.14b) are both expressions for $d\tau_{tip}$, the shear stress increment needed to continue plastic deformation (or dislocation motion), obtained respectively from dislocation pile-up theory and continuum plasticity theory. They should be the same, and thus the right-hand part of these two equations is equal. Equating these two right parts, we have

$$\sqrt{2\tau_{cr}/D}t^{1/2}d\tau = (\rho_0 \pm \sqrt{2}h_{tip})de_{tip}^p \tag{8.14c}$$

or

$$\frac{d\tau}{de_{tip}^p} = \frac{(\rho_0 \pm \sqrt{2}h_{tip})}{\sqrt{2\tau_{cr}/D}}t^{-1/2} \tag{8.14d}$$

It is seen that the size effect is brought into this equation through layer thickness t. To see the size effect more clearly, we may introduce concentration coefficients $\gamma$ and $\gamma_1$ to connect the plastic strain and back stress (i.e., $de_{tip}^p$, $R_{(r)tip}$) in (8.14b) at the tip of dislocation pile-ups with the average values $(de^p, R_{(r)})$ over the layer thickness t for thin layers, i.e.,

$$de_{tip}^p = \gamma_0 de^p \tag{8.14e}$$

$$R_{(r)tip} = \gamma_1 R_{(r)} \quad (r = 1, 2 \ldots n) \tag{8.14f}$$

Combining this with (8.9o) we obtain:

$$h_{tip} = \gamma_1 h \tag{8.14g}$$

Substituting (8.14e) and (8.14f, g) into (8.14d) and rearranging it, the plastic tangential modulus $G^t$ can be expressed as follows:

$$G^t = \frac{d\tau}{de^p} = \frac{(\rho_0 \pm \gamma_1\sqrt{2}h)\gamma_0}{\sqrt{2\tau_{Cr}/D}}t^{-1/2} \tag{8.14h}$$

### 8.6.5 Explicit Expressions of Size Effects on Tangential Modulus and Kinematic Hardening Behavior

Replacing $\rho_0$ and $h_{ij}$ by (8.9n) and (8.9o), $G^t$ can be rewritten as:

$$G^t = \frac{d\tau}{de^p} = \sum_{r=1}^{n} C_{(r)} \pm \sum_{r=1}^{n} - \alpha_{(r)} R_{ij(r)} \tag{8.15a}$$

with

$$C_{(r)} = C_{(r)0} \frac{t^{-1/2}\gamma_0}{d_D}, \tag{8.15b}$$

$$\alpha_{(r)} = \alpha_{(r)0}\gamma_0\gamma_1 \frac{t^{-1/2}}{d_D} \tag{8.15c}$$

where $d_D = \sqrt{2\tau_{Cr}/D}$. From (8.14e) and (8.14f) we can see that $\gamma_1$ and $\gamma_0$ have the physical meaning of concentration factors of stress and plastic strain at the tip of dislocation pile-ups. According to Langford[18] "at very small interlamellar spacing, generation of the dislocations required in the pile-ups will absorb a significant fraction of the work of deformation." This statement implicitly indicates that the thinner the layer, the higher the stress concentration factor $\gamma_1$, and the thinner the layer, the higher the plastic strain concentration factor $\gamma_0$.

The relationships between these concentration factors and layer thickness t may be expressed by reciprocal power laws as follows:

$$\gamma_0(t) = d_1 t^{-\eta_0} \tag{8.15d}$$

$$\gamma_1(t) = d_2 t^{-\eta_1} \tag{8.15e}$$

Substituting (8.15d,e) into (8.15b,c), we obtain

$$C_{(r)} = C_{(r)0} \frac{t^{-n_2}}{d_{D1}}, \tag{8.15f}$$

$$\alpha_{(r)} = \alpha_{(r)0} \frac{t^{-n_3}}{d_{D2}} \tag{8.15g}$$

where $n_2 = 0.5 + \eta 0$, $n_3 = 0.5 + \eta 0 + \eta_1$ are power indexes; $d_{D1} = d_D/d_1$, and $d_{D2} = d_D/(d_1 d_2)$ are constants.

A direct application of these results is that if we know a set of apparent parameters $S_{yt_0(r)}$, $C_{t_0(r)}$, $\alpha_{t_0(r)}$, for a material with a baseline thickness $t_0$ then by using (8.15f) and (8.15g) we can get the set of apparent parameters $S_{yt(r)}$, $C_{t(r)}$, $\alpha_{t(r)}$ for other material with layer thickness t through the following equations:

$$S_{yt(r)} = S_{yt_0(r)} \left(\frac{t}{t_0}\right)^{-n_1} \quad (n_1 = 0.5) \tag{8.15h}$$

$$C_{t_{(r)}} = C_{t_0} \left(\frac{t}{t_0}\right)^{-n_2} \tag{8.15i}$$

$$\alpha_{t(r)} = \alpha_{t_0(r)} \left(\frac{t}{t_0}\right)^{-n_3} \tag{8.15j}$$

These equations are the size laws accounting for size effects of microstructures on yield stress and on kinematic hardening parameters. These laws will be used for determining material parameters of different layer thicknesses in multiscale numerical analyses in Section 8.8.

## 8.7 Numerical Methods to Link Plastic Strains at the Mesoscopic and Macroscopic Scales

In this section, numerical methods are introduced to complete the iterative process to obtain the solution for three-scale multiscale analysis.

### 8.7.1 Bridging Plastic Variables at Different Scales from Bottom-up and Top-down to Complete the Iterative Process

We established the quantitative relationship between dislocation pile-ups with microscopic plastic strain through (8.14d) or (8.14h), then derived the governing equation (8.12j) for bridging micro/meso/macroscale variables, with relevant parameters from Appendix 8.A and 8.B. Thus via a local iteration process using (8.12j) we can obtain the mesoscopic plastic strain increment $d\varepsilon_{ij}^{pc}$ of the meso-cell under given $d\overline{Q}_{IJ}^{K}$ by (8.12c) after determining $\{d\sigma^C\}$ and $\{dH\}$.

Based on the result of local iterative process for each representative meso-cell, the macroscopic responses of material can be determined through a global iteration process. As we can see in the following description, the key for the global iteration is to establish the quantitative relationship between the obtained meso-scale plastic strain $d\varepsilon_{ij}^{pc}$ and the macroscopic plastic strain $\Delta\bar{\varepsilon}_{IJ}^{p}$. In fact, through this multiscale simulation process, plastic variables at different scales, namely from atomistic scale to microscopic scale, to mesoscopic scale, and to macroscopic scale are quantitatively related and transformed from the bottom scale to the up scales.

Suppose the current step is the k-th global iteration of the I-th incremental loading, in which the stress increment $\Delta\sigma_{IJ}$ is given and fixed. The value of macroscopic plastic strain $\Delta\bar{\varepsilon}_{IJ}^{p}$ in equation (8.6b) is unknown; it must be obtained through an iteration process from bottom-up and top-down between different scales so that plastic strains obtained at different scales can satisfy all equations.

### 8.7.2 Numerical Procedure for the Iterative Process

Specifically, the values of $\Delta\bar{\varepsilon}_{IJ}^{p}$ obtained from the last step, (k − 1), are taken to calculate $\Delta\overline{Q}_{IJ}^{k}$ of the k-th iteration step by (8.6b), then we can get the relevant $dQ_{IJ}^{C}$ of meso-cells with different orientation through coordinate transform of $\Delta\overline{Q}_{IJ}^{k}$. $dQ_{IJ}^{C}$ is used to start a local iteration with (8.12j) and (8.12c) by which $\Delta\varepsilon_{IJ}^{pc}$ of each inclusion can be obtained, the macroscopic plastic strain $\Delta\bar{\varepsilon}_{IJ}^{p}$ of the aggregate is determined by a volume average of $\Delta\varepsilon_{ij}^{pc}$ as follows:

$$\Delta\bar{\varepsilon}_{ij}^{p} = \frac{1}{V}\int_{v}(\Delta\varepsilon_{ij}^{pc})dv \tag{8.16a}$$

where V is the volume of the macro-cell. The obtained $\Delta\bar{\varepsilon}_{IJ}^{p}$ combined with $\Delta\bar{\sigma}_{ij}$ are then used to determine $\Delta\overline{Q}_{IJ}^{*k}$ at the end of the k-th iteration by using (8.6b) again. Whether the iteration should be finished is determined by the difference in value between $\Delta\overline{Q}_{IJ}^{*k}$ and $\Delta\overline{Q}_{IJ}^{k}$ at the beginning and at the end of the k-th iteration. The process is repeated until the following criterion of convergence is satisfied:

$$\frac{\left\|\Delta\overline{Q}_{IJ}^{*k} - \Delta\overline{Q}_{IJ}^{k}\right\|}{\left\|\Delta\overline{Q}_{IJ}^{*k}\right\|} < \varepsilon \tag{8.16b}$$

where $\left\|\Delta Q_{IJ}^{k}\right\| = (\Delta Q_{IJ}^{k}\Delta Q_{IJ}^{k})^{1/2}$, and $\varepsilon$ is the tolerance, say, 0.01% for controlling the error.

**Table 8.1** Euler coordinates $\theta$ and $\phi$ for the centers of 20 spatially and uniformly distributed planes $x_1$ and $x_2$

|  | A | B | C | D | E | F | G | H | I | J |
|---|---|---|---|---|---|---|---|---|---|---|
| $\theta$ (degree) | 0 | 72 | 144 | 216 | 288 | 0 | 72 | 144 | 216 | 288 |
| $\phi$ (degree) | 37.4 | 37.4 | 37.4 | 37.4 | 37.4 | 79.2 | 79.2 | 79.2 | 79.2 | 79.2 |
|  |  |  |  |  |  |  |  |  |  |  |
|  | K | L | M | N | O | P | Q | R | S | T |
| $\theta$ (degree) | 36 | 108 | 180 | 252 | 324 | 36 | 108 | 180 | 252 | 324 |
| $\phi$ (degree) | 142.6 | 142.6 | 143 | 142.6 | 142.6 | 100.8 | 101 | 100.8 | 100.8 | 100.8 |

### 8.7.3 How to Carrying on the Volume Averaging of Meso-cell Plastic Strain to Find Macroscopic Strain

In (8.16a) the volume averages are necessary to obtain the macroscopic plastic strain $\Delta\bar{\varepsilon}^p_{IJ}$. Since inclusions are assumed to be randomly oriented, these volume averages can be done by an average over all directions. To make the volume average in the $\theta$ and $\phi$ direction of the Euler angles more accurate than the Gaussian quadrature commonly used,[19] an icosahedron with 20 spatially and uniformly distributed directions is chosen, see Figure 8.5b for each plane and Table 8.1 for the corresponding angles $\theta$ and $\phi$. This will replace arbitrary inclusions with arbitrary Euler angles of $\theta$, $\phi$ and $\omega$ over all directions (Figure 8.6a) by an icosahedron with 20 spatially uniformly placed planes. These planes themselves are taken as the basic planes of $x_1$, $x_2$ in which the angle $\omega$ varies. This choice makes the average over the 20 directions an arithmetic one. It is assumed that in each of these planes there are infinite inclusions with randomly distributed orientations so that

$$\Delta\bar{\varepsilon}^p_{IJ} = \frac{1}{20}\sum_{i=1}^{20}\left[\frac{1}{2\pi}\int_{-\pi}^{\pi}\Delta\varepsilon^{pc}_{IJ}(\theta_i, \phi_i, \omega)d\omega\right] \tag{8.16c}$$

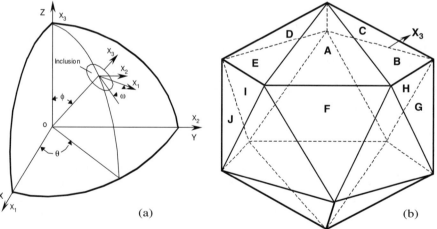

(a)                                            (b)

**Figure 8.6** Replace arbitrary inclusions over all directions by an icosahedron with 20 spatially uniformly placed planes. (a) An inclusion with arbitrary Euler angles of $\theta$, $\phi$ and $\omega$ in the aggregate. (b) An icosahedron to limit inclusions on its plane with fixed $\theta$, $\phi$ but various $\omega$ (Reproduced from Fan, J., Gao, Z., and Zeng, X. (2004) Cyclic plasticity across micro/meso/macroscopic scales. *Proceedings of Royal Society of London A*, 460 (2045), 1477, Royal Society of London)

**Table 8.2**  Specimen chemical composition of high carbon pearlitic rail steel

| C (% wt) | Si (% wt) | Mn (% wt) | P (% wt) | S (% wt) | V(% wt) |
|----------|-----------|-----------|----------|----------|---------|
| 0.78 | 0.78 | 0.91 | 0.015 | 0.016 | 0.09 |

The details of this averaging process and the comparison of the excellent accuracy compared with other several numerical algorithms can be seen, respectively, from Fan[3] and Peng and Fan.[20]

## 8.8 Experimental Study on Layer-thickness Effects on Cyclic Creep (Ratcheting)

As mentioned in Section 1.4.1, cyclic creep occurs when the cyclic loading is asymmetric and the cyclic creep will develop in the mean stress direction. In the following we will first introduce the method of specimen fabrication to produce different sizes of microstructure. Then we will explain how asymmetric stresses were applied on the specimen to produce cyclic creep.

The method used to generate different sizes of microstructures in a layered material was to change the cooling rate during its heat-treatment. High carbon rail steel with the chemical composition described in Table 8.2 was used. Cooling rates of 0.5, 1, 2 and 5 °C per second during heat-treatment of specimens were controlled by a Gleeble-1500 thermal testing system (Dynamic Systems Inc., Poestenkill, NY, USA), shown in Figure 8.7. This system can make the gauge length of specimens fall under a homogeneous temperature field.

**Figure 8.7**  Gleeble-1500 thermal testing system used to generate different sizes of microstructures (Reproduced from Fan, J., Gao, Z., and Zeng, X. (2004) Cyclic plasticity across micro/meso/macroscopic scales. *Proceedings of Royal Society of London A*, 460 (2045), 1477, Royal Society of London)

**Table 8.3** Specimen layer thickness, t, of high carbon pearlitic rail steel at different cooling rates

| No. of measurement | t at cooling rate of 0.5 °C/sec (nm) | t at cooling rate of 1 °C/sec (nm) | t at cooling rate of 2 °C/sec (nm) | t at cooling rate of 5 °C/sec (nm) |
|---|---|---|---|---|
| 1 | 373 | 313 | 294 | 199 |
| 2 | 340 | 275 | 305 | 216 |
| 3 | 292 | 381 | 261 | 235 |
| 4 | 282 | 324 | 264 | 259 |
| 5 | 340 | 250 | 251 | 237 |
| 6 | 344 | 350 | 226 | 239 |
| 7 | 308 | 265 | 317 | 193 |
| 8 | 255 | 213 | 288 | 280 |
| 9 | 366 | 249 | 264 | 272 |
| 10 | 292 | 252 | 206 | 223 |
| Average value | 319 | 287 | 267 | 235 |

Specimens were cooled from 850 °C to room temperature and characterized by an XL-30 scanning electron microscope. Figure 1.2 shows the microstructure of a specimen with a cooling rate of 2 °C/sec. For a cooling rate of 0.5 °C/sec, the average layer thickness has a value of 319 nm that will be taken as the baseline thickness for the comparison of size effects. The layer thickness, t, is defined as the sum of the thickness of one ferrite layer and one cementite layer and was obtained by measuring the layer spacing between adjacent cementite layers over ten different places and then taking their average value. The average layer thickness reduces to 287, 267 and 235 nm when the cooling rates are increased to 1, 2 and 5 °C/sec respectively (Table 8.3). The higher the cooling rate, the smaller the layer thickness, due to shorter times available for reorganization movement.

After microstructure characterization, the specimens were cyclically tested to about several hundred cycles by an MTS 809 testing machine under asymmetric loading of $\sigma_{max} = 720$ MPa and $\sigma_{min} = -490$ MPa (or $115 \pm 605$ MPa). Figure 1.3 shows, respectively, the cyclic stress-strain curves of the first seven cycles for specimens with average layer thicknesses of 267, 287 and 319 nm. It can be seen that the specimens strained along the longitudinal direction, indicating extensive cyclic creep. It can also be seen that the thicker the layer, the higher the ratcheting, as described in Section 1.4.1.2.

## 8.9 Numerical Results and Comparison Between Experiments and Multiscale Simulation

In this section, numerical analyses were conducted to simulate size effects of microstructures on cyclic creep behavior through a three-scale multiscale analysis described in the previous sections. We take high carbon pearlitic rail steel as a typical dual-phase material for numerical simulation so that the simulation results can compare with experimental results under the same conditions.

### 8.9.1 General Features of the Numerical Simulation

The multiscale analysis was through equations (8.6a, b) and (8.12j) by a combination of global and local iterative processes. The tolerance in both global and local iterations was taken to be 0.01%. This choice

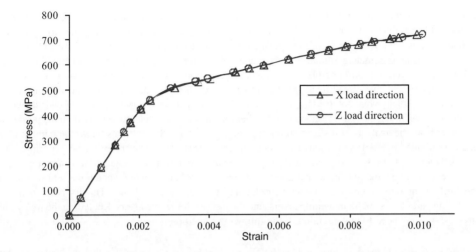

**Figure 8.8** Comparison of tensile stress-strain curves obtained with applied loading along the X-direction and Z-direction under the same loading of $\sigma_{max} = 720$ MPa (Reproduced from Fan, J., Gao, Z., and Zeng, X. (2004) Cyclic plasticity across micro/meso/macroscopic scales. *Proceedings of Royal Society of London A*, 460 (2045), 1477, Royal Society of London)

was a compromise between accuracy and CPU time. The number of iterations for each step of global and local iterations was about five and four respectively, and the variation in these numbers was small. The computation time can be estimated by the following information. For a simulation of the first five reversal segments of specimen $t = 267$ nm (see Figure 1.3), the three-scale multiscale analysis took about 3 minutes 40 seconds by a Dell laptop computer with processor speed of 860 MHz.

The axial strain under the applied stress along the Z-axial direction was compared with the corresponding strain under the same amount of applied stress along the X-axis direction. The two stress-strain curves shown in Figure 8.8 obtained by loadings along these two perpendicular directions are almost coincidence, indicating a satisfactory simulation for the macroscopic isotropy of the aggregate. In the following, all results presented were obtained by cyclic loading along the X-axis.

## 8.9.2 Determination of Basic Material Parameters

The elastic parameters and volume fractions used in the calculation are given in the first two rows of Table 8.4. The elastic constants were determined based on Langford's 1977 report[18] that said there are no significant differences in elastic constants between cementite and ferrite phases. Therefore, constants of steel were adopted for these two phases. The volume fraction of cementite layer over ferrite layer was assumed to be approximately constant. In fact, the formation of cementite and ferrite layers originates from the diffusion of Austenite whose composition of carbon is fixed at 0.77wt% C.[21] Because the

**Table 8.4** Material parameters for high carbon rail steel with baseline layer thickness of $t_0 = 319$ nm

| | |
|---|---|
| Elastic constants | $\mu^{(i)} = 80.12 \times 10^3$ MPa, $v^{(i)} = 0.267$ (i = 1,2) |
| Volume fraction | $V^{(1)} = 0.11$, $V^{(2)} = 0.89$ |
| Plastic parameters of the baseline material with layer thickness of $t_0 = 319$ nm | $(C^{(1)}, C^{(2)}) = (100.9, 1.355) \times 10^6$ MPa |
| | $(\alpha^{(1)}, \alpha^{(2)}) = (6803, 833)$ |
| | $(S_{y1}^{(1)}, S_{y1}^{(2)}, S_{y2}^{(2)}, S_{y3}^{(2)}) = (2500, 260, 125, 140)$ MPa |

compositions of the two resulting phases, i.e., cementite and ferrite phases, are also fixed, their volume ratio should be a constant to satisfy the balance requirements of carbon before and after the diffusion. The carbon composition, 0.78wt% C, of our specimen was very close to that of Austenite (see Table 8.2). Therefore, the approximate assumption of constant volume fraction in this work is acceptable.

Different from elastic constants, plastic parameters of each constituent phase depend substantially on the layer thickness and they are difficult to determine by experiments. Baseline material parameters of thickness $t_0 = 319$ nm were obtained by curve fitting to the cyclic testing curve. Their values are shown in the third row of Table 8.4. Plastic parameters of other thicknesses were determined based on the proposed size laws from the analysis of dislocation pile-ups given in section 8.6.2. In this work, the number n of internal variables in (8.9c) was chosen to be 1 for both cementite and ferrite layers so that the subscript under the coefficients C and $\alpha$ were dropped; but the superscript is still retained to distinguish hard and soft phases.

There are four kinematic material parameters, namely $C^{(1)}$, $\alpha^{(1)}$, $C^{(2)}$, and $\alpha^{(2)}$ and four yield stresses in the third row of Table 8.4 for material with baseline layer thickness $t_0 = 319$ nm. The first two kinematic hardening parameters are for cementite layers, and the last two for ferrite layers. Regarding the four yield stresses, we may need to pay attention to the following three points:

- First, for cementite the deviatoric yield stress $S_1^{(1)} = 2500$ MPa was chosen, which was about one order larger than $S_i^{(2)} (i = 1$–$3)$ of the ferrite layer. The much higher yield stress of the cementite layer compared to the ferrite layer is consistent with previous observations and analysis.[18]
- Second, yield stress of $S_2^{(2)} = 125$ MPa was chosen for all even reversal segments of cycling, and $S_3^{(2)} = 140$ MPa for all odd reversal segments. This choice was to simulate the cyclic deformation anisotropy of more resistance in the odd reversal segments than in the even ones, observed in Figure 1.3. This anisotropy is due to back stress and is a kind of Bauschinger effect.
- Third, yield stress of $S_1^{(2)} = 260$ MPa was chosen for the first loading reversal segment because more driving force was necessary to initiate and maintain dislocations for the "raw" material.

### 8.9.3   Determining Size Effects on Material Parameters by Size Laws

In the calculation, plastic parameters of materials with layer thickness of $t = 287$ nm and $t = 267$ nm were obtained by size laws summarized in (8.15h–j). The power index $n_1$ was analytically determined as 0.5. Power index $n_2$ and $n_3$ were determined by trial and error. The results were $n_2 = 5.69$, $n_3 = 7.8$ (or $\eta_0 = 5.19$ and $\eta_1 = 2.11$ in (8.15d–e)). Rewriting (8.15a–c) for these data and taking $t_0 = 319$ nm, we have:

$$S_{t,j}^{(m)} = S_{319,j}^{(m)} \left(\frac{t}{319}\right)^{-0.5} \qquad (m = 1, j = 1; m = 2, j = 1 - 3) \tag{8.17a}$$

$$C_t^{(m)} = C_{319}^{(m)} \left(\frac{t}{319}\right)^{-5.69} \qquad (i = 1, 2) \tag{8.17b}$$

$$\alpha_t^{(i)} = \alpha_{319}^{(i)} \left(\frac{t}{319}\right)^{-7.8} \qquad (i = 1, 2) \tag{8.17c}$$

Using these obtained material parameters, the micro/meso/macroscopic cyclic responses of the experimental curves shown in Figure 1.3 under asymmetric axial loading of $115 \pm 605$ MPa were calculated for materials with layer thickness of 319, 287 and 267 nm.

### 8.9.4   Comparison Between the Results of Three-scale Multiscale Simulation with Data of Cyclic Experiments

Numerical results obtained by the present three-scale methodology incorporating the size laws of material parameters described in this chapter are given in Figure 8.9 and Figure 8.10. In the latter figure, the

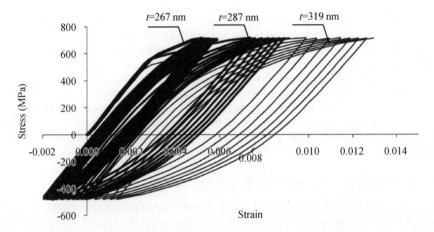

**Figure 8.9** Calculated results of three-scale simulation for cyclic creep (ratcheting) of high carbon rail steel specimens of layer thickness t = 319, 287 and 267 nm (Reproduced from Fan, J., Gao, Z., and Zeng X. (2004) Cyclic plasticity across micro/meso/macroscopic scales. *Proceedings of Royal Society of London A*, 460 (2045), 1477, Royal Society of London)

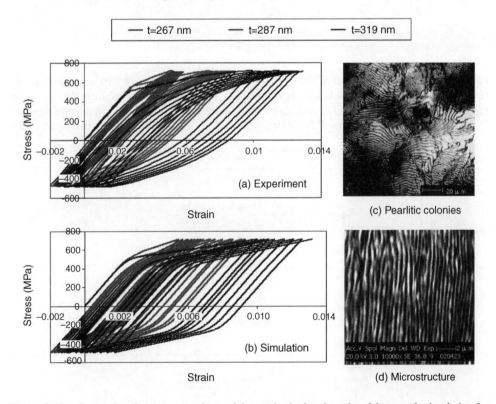

**Figure 8.10** Comparison between experimental data and calculated results of three-scale simulation for cyclic creep (ratcheting) of high carbon rail steel specimens (Reproduced from Fan, J., Gao, Z., and Zeng, X (2004) Cyclic plasticity across micro/meso/macroscopic scales. *Proceedings of Royal Society of London A*, 460 (2045), 1477, Royal Society of London)

comparison with experiments and related microstructure is also given. From the comparison between curves in Figures 1.3 and 8.9 as well as the two curves in Figure 8.10, it is evident that the present numerical multiscale analysis of size effects compares satisfactorily with the experimental curves qualitatively and quantitatively.

This is essential, because there is a lack of microscopic experiments to check the accuracy of the microscopic analysis. Hence, the result that the macroscopic counterpart matches well with the measured stress-strain curve offers certain evidence for the reasonable accuracy of the meso/microscopic analysis. However, this does not mean that the microscopic analysis for the hard and soft phases is equally accurate for the prediction of the macroscopic stress-strain curve. Direct evidence should be found in the future to verify the accuracy of the microscopic calculations.

## 8.10 Findings in Microscopic Scale by Multiscale Analysis

In the following, we will describe numerical results of effects of the microstructure size on microscopic response. This is very important since usually they cannot be observed and measured. The microscopic findings by the multiscale analysis include effective stress, plastic strain and residual stress in a microstructure whose layer thicknesses are t = 319, 287 and 267 nm. Before describing these findings at the microscopic scale, definitions of effective stress and strain are introduced. Here, the effective stress and effective plastic strain for hard phase (i = 1) and soft phase (i = 2) are defined, respectively, as follows:

$$\sigma_{eff}^{(i)} = \sqrt{\frac{3}{2} S_{mn}^{(i)} S_{mn}^{(i)}} \qquad (i = 1, 2) \tag{8.17d}$$

$$\varepsilon_{eff}^{p(i)} = \sqrt{\frac{2}{3} \varepsilon_{mn}^{p(i)} \varepsilon_{mn}^{p(i)}} \qquad (i = 1, 2) \tag{8.17e}$$

**Finding 1: The effective stress of brittle (cementite) phase can reach about 2 GPa at the peak of odd reversal, the thicker the layer the higher the stress can be, which causes brittle failure of the thick layer.** Figures 8.11 and 8.12 show size effects on effective stress evolution, respectively, in a cementite layer and ferrite layer for a meso-cell B whose Euler angles are $\theta = 288°$, $\phi = 37.4°$, and $\omega = 6.3°$. In Figure 8.11, the

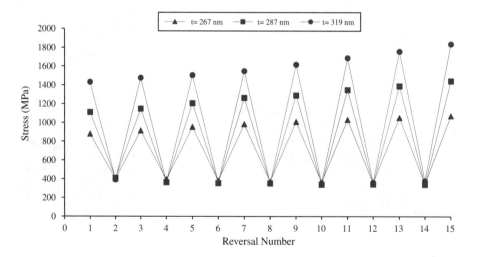

**Figure 8.11** Layer thickness effects on effective stress evolution in a cememtite layer of a meso-cell B ($\theta = 288°$, $\varphi = 37.4°$, $\omega = 6.3°$) (Reproduced from Fan, J., Gao, Z., and Zeng X. (2004) Cyclic plasticity across micro/meso/macroscopic scales. *Proceedings of Royal Society of London A*, 460 (2045), 1477, Royal Society of London)

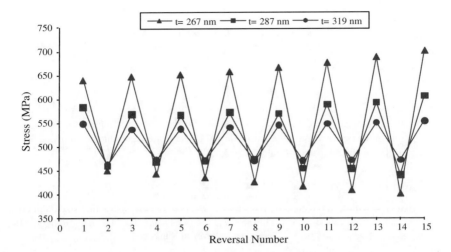

**Figure 8.12**  Layer thickness effects on effective stress evolution in a ferrite layer of a meso-cell B whose Euler angles are $\theta = 288°$, $\phi = 37.4°$, $\omega = 6.3°$ (Reproduced from Fan, J., Gao, Z., and Zeng, X. (2004) Cyclic plasticity across micro/meso/macroscopic scales. *Proceedings of Royal Society of London A*, 460 (2045), 1477, Royal Society of London)

effective stresses at the top three points at odd reversals correspond to the tensile peak stresses of the three specimens. On the other hand, the effective stresses at the bottom three points at even reversals are almost coincident, and correspond to the maximum compressive stresses of these specimens. The peak tensile stresses increased with reversals. After 15 reversals the maximum effective stress reached as high as 1.9 GPa. It merits notice that the thicker the layer, the larger the effective tensile peak stress in the cementite layer.

The findings that the maximum effective stress in the thicker cementite layer reached a value of 1.9 GPa, and the thinner cementite layer had less stress, are significant. These high stresses in thicker layers can cause fragmentation of the cementite layer so that the thicker layer has less strength. On the other hand, the thin layer has strong capability against fragmentation because much less effective stress was developed here. This judgment is consistent with experimental observations. In fact, Langford reported "there is no evidence for extensive, brittle fragmentation of cementite plates thinner than 0.01 μm."[18] He also reported "there is no evidence for gross plastic deformation of cementite plates thicker than 0.2 μm," indicating that brittle failures were observed in the thick cementite layers.

**Finding 2: The higher stress of brittle (cementite) phase in a thicker layer is caused by stress transformation cycle-by-cycle from soft phase to brittle phase. The stress transformation is caused by plastic strain accumulation in the soft (ferrite) layer cycle-by-cycle due to the cyclic loading.**
To find the underlying mechanism, the corresponding evolution of peak tensile stresses of ferrite layers was examined in Figure 8.12. From that we can see that the maximum tensile and compressive effective stresses of the thickest ferrite layer had the smallest values and those of the thinnest layer had the largest values under the same applied stress. The conjectured mechanism was that the thickest layer produced a larger microscopic plastic strain so that a larger macroscopic strain was produced as shown in Figure 1.3. This larger microscopic plastic strain can effectively accommodate stress distribution between hard and soft layers.

Specifically, this plastic strain acted as a means for transformation cycle-by-cycle of the stresses undertaken originally by ferrite layers to the cementite layers. This transformation accommodating process is a transient process, which in concert with the accompanying cycle-by-cycle plastic strain accumulation is the source for cyclic creep (ratcheting). On the other hand, the thin cementite and ferrite layers both had smaller plastic strain compared to the thicker layers, so that the stress transformation from ferrite layers to cementite layers was limited. This conjecture is verified in Figure 8.13 in which the

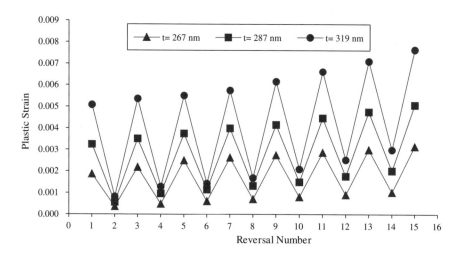

**Figure 8.13** Layer thickness effects on effective plastic strain evolution in a ferrite layer of a meso-cell B whose Euler angles are $\theta = 288°$, $\phi = 37.4°$, $\omega = 6.3°$ (Reproduced from Fan, J., Gao, Z., and Zeng, X. (2004) Cyclic plasticity across micro/meso/macroscopic scales. *Proceedings of Royal Society of London A*, 460 (2045), 1477, Royal Society of London)

maximum effective plastic strain value of a ferrite layer for the thickness of 319 nm is about 2.43 times larger than for 267 nm.

**Finding 3: Residual stress increases cycle-by-cycle and high residual stress can reach about 1 GPa at even reversal valley in the brittle (cementite) phase because of the accumulation of plastic strain in the soft phase under asymmetric cyclic loading. Therefore the asymmetric residual stress in concert with the accumulation of plastic strain plays an important part in ratcheting**.
Figures 8.14 and 8.15 also show effective residual stress evolution with reversal segments in a cementite layer and ferrite layer of a meso-cell B. The effective residual stresses in cementite layers were much

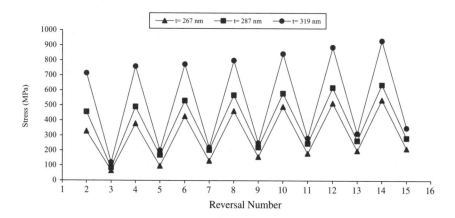

**Figure 8.14** Layer thickness effects on effective residual stress evolution in a cementite layer of a meso-cell B with Euler angles $\theta = 288°$, $\phi = 37.4°$, $\omega = 6.3°$ (Reproduced from Fan, J., Gao, Z., and Zeng, X. (2004) Cyclic plasticity across micro/meso/macroscopic scales. *Proceedings of Royal Society of London A*, 460 (2045), 1477, Royal Society of London)

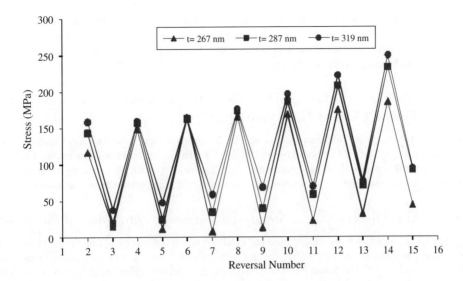

**Figure 8.15**    Layer thickness effects on effective residual stress evolution in a ferrite layer of a meso-cell B whose Euler angles are $\theta = 288°$, $\phi = 37.4°$, $\omega = 6.3°$ (Reproduced from Fan, J., Gao, Z., and Zeng X (2004) Cyclic plasticity across micro/meso/macroscopic scales, *Proceedings of Royal Society of London A*, 460 (2045), 1477, Royal Society of London)

higher than those in ferrite layers. The maximum effective residual stress occurred in the thickest cementite layer due to the larger plastic deformation; its value was as high as 0.9 GPa after 15 reversals.

Figures 8.14 and 8.15 also show that effective residual stresses at even reversal segments are much higher than at odd reversal segments in both cementite and ferrite layers. The mechanism can be explained as follows. Residual stress at even reversal segments was the stress after unloading from the maximum tensile peak stress at which a large effective plastic strain was produced. This large plastic strain caused a large residual stress because it resisted the reduction of elastic deformation of the cementite layer. On the other hand, residual stress at odd reversals, besides reversal 1, was the stress after unloading from the maximum compressive stress. At this maximum compressive stress, the effective plastic strain was much less than that produced at the maximum tensile peak. This asymmetry is due to the asymmetric loading of $115 \pm 605$ MPa.

The difference of the amount of the maximum plastic strain at the tensile peak and at the maximum compressive stress explained why residual stress at even reversal segments is much higher than at odd reversal segments. This asymmetric residual stress in concert with the accumulation of plastic strain, therefore, plays an important role for ratcheting.

## 8.11    Summary and Conclusions

In this chapter, a hierarchical multiscale methodology is developed for cyclic elastoplastic analysis that spans micro/meso/macro scales of dual-phase materials. This methodology of multiscale analysis is summarized as follows.

### 8.11.1    Methods for Bridging Three Scales

The key to this method is to use a representative meso-cell (inclusion) for bridging analysis at microscopic, mesoscopic and macroscopic scales. This bridging is effective for quantitative transitions of variables between these three scales. While the meso-macro linkage is accomplished through an improved self-consistent scheme proposed by Fan (1999),[3] the meso-micro linkage is through the development of effective constitutive equations of the meso-cell (inclusion) by satisfying approximately conditions of equilibrium, compatibility and constitutive equations of each constituent phase at the microscopic scale. This methodology is exemplified by a three-scale analysis of the cyclic behavior of high carbon rail steel that consists of thousands of pearlitic colonies with randomly distributed orientations, and where each colony includes many alternately arranged cementite and ferrite layers.

### 8.11.2    Methods in Bridging Atomistic Dislocation Analysis and the Second Class of Multiscale Analysis

The dominant characteristics of the analysis are that the size effects of a microstructure, for example the layer thickness, can be explicitly incorporated into the formulation. This is achieved through a combined analysis by dislocation pile-ups in the layers and the plasticity constitutive laws for the same shearing problem, so that the effects of layer thickness on yield stress and kinematic hardening parameters can be accounted for.

### 8.11.3    Size Effects on Yield Stress and Kinematic Hardening of Plasticity

It is found that yield stresses follow the Hall-Petch type relationship such that the yield stress of constituent phase has a reciprocal square root relation to its thickness, in which the power index of the thickness term is $-0.5$. It is found for the first time that the thickness effects of thin layers on material parameters of kinematic hardening are more serious, with the power index being $-5.69$ and $-7.8$; the thinner the layer, the larger the kinematic hardening. This is reasonable because the thinner layer develops higher concentrations and gradients of stress and plastic strain near the interface between the hard and soft phase. These concentrations and larger gradients produce strong back stress of dislocation pile-ups and material kinematic hardening.

### 8.11.4    Experimental Validation for the Size Effects on Ratcheting

A systematic experimental study of the effects of microstructure layer thickness on material cyclic behavior was conducted to verify the methodology. By changing the cooling rates of heat-treatment, specimens of high carbon rail steel with different layer thicknesses were obtained. These specimens were cyclically tested to show the layer-thickness effects on macroscopic stress-strain curves, and more specifically on the ratcheting behavior. It was found that the layer thickness has essential effects on this behavior. When the thickness reduces to 94% and 85% of the baseline thickness of 319 nm, the ratcheting strain after seven cycles reduces to about 70% and 47% respectively. A comparison between experimental curves of layers with thickness of 319, 287 and 267 nm and the numerical results obtained by the present methodology of three-scale analysis incorporating the size laws described in this chapter, shows a good agreement qualitatively and quantitatively.

### 8.11.5    Failure Mechanisms of Thicker Layer

It was found that the thicker the constituent layer, the softer (or ferrite) layer produced more microscopic plastic strain. As the microscopic strain increased, more stresses originally undertaken by the soft phase were transformed cycle-by-cycle to the hard (cementite) layer to produce a high stress that can cause fragmentation of the cementite layer. Therefore, the thin layer has strong capability against fragmentation and the thicker layer has less strength because a higher effective stress can be developed in these thick cementite layers. This finding is consistent with experimental observations.

### 8.11.6    The Formulation and Important Role of Residual Stress

Also, it was found that effective plastic strain in microstructures increased cycle-by-cycle which was a direct source for residual stress and ratcheting. The finding that residual stress and ratcheting depended very much on the size of the layered structure may offer a reasonable explanation why phenomenological models of ratcheting which do not consider the size effects have had limited success so far. This is a good example of the effectiveness of the methodology that spans the micro/meso/macro scales in understanding some notoriously difficult issues that have existed for a long time in the analysis of material behavior and fatigue.

### 8.11.7    Wide Scope of Applications of the Proposed Multiscale Methodology

The relativity of the definition of micro/meso/macro scales provides broad applications of the methodology for the second class of multiscale analysis. Because the generic meso-cell can consist of microstructures at different sub-scales and with different phase properties, it offers the possibility for extending the investigations of microscopic stress, strain and cracking to a wide range of microstructures, such as cells with domain walls, laminae, needle microstructures, dispersed microstructures, grain and grain boundaries. In the case of a complicated microstructure, an analytical approach of microscopic analysis to get an effective constitutive equation of the generic meso-cell is difficult, so a finite element method may be needed. In addition, the developed multiscale analysis should be applicable to other dissipative mechanisms and inelastic deformation such as creep and fatigue.

## Appendix 8.A Constitutive Equations and Expressions of Parameters

The elastoplastic constitutive equations of phase (1) and (2) are given in (8.10d) which can be rewritten as follows:

$$
\begin{aligned}
d\sigma_{ij}^{(m)}/2\mu_p^{(m)} &= d\varepsilon_{ij}^{(m)} + a_m d\varepsilon_{kk}^{(m)}\delta_{ij} + h_{ij}^{0(m)}dp^{(m)}/\rho_0^{(m)} \\
&= d\varepsilon_{ij}^{(m)} + a_m\delta_{ij}(d\varepsilon_{11}^{(m)} + d\varepsilon_{22}^{(m)} + d\varepsilon_{33}^{(m)}) + h_{ij}^{0(m)}dp^{(m)}/\rho_0^{(m)} \quad (m=1,2)
\end{aligned}
\tag{8A.1a}
$$

This equation can be expressed by matrix form as follows:

$$
\{d\sigma^{(m)}\} = [D^{(m)}]\{d\varepsilon^{(m)}\} + 2\mu_p^{(m)}\{h^{0(m)}\}dp^{(m)}/\rho_0^{(m)} \quad (m=1,2)
\tag{8A.1b}
$$

where

$$
\{d\sigma^{(m)}\} = \{d\sigma_{11}^{\{m\}}, d\sigma_{22}^{(m)}, d\sigma_{33}^{(m)}, d\sigma_{23}^{(m)}, d\sigma_{13}^{(m)}, d\sigma_{12}^{(m)}\}
\tag{8A.1c}
$$

$$
\{d\varepsilon^{(m)}\} = \{d\varepsilon_{11}^{\{m\}}, d\varepsilon_{22}^{(m)}, d\varepsilon_{33}^{(m)}, d\varepsilon_{23}^{(m)}, d\varepsilon_{13}^{(m)}, d\varepsilon_{12}^{(m)}\}
\tag{8A.1d}
$$

$$
\left\{h^{0(m)}\right\} = \left(h_{11}^{0(m)}, h_{22}^{0(m)}, h_{33}^{0(m)}, h_{23}^{0(m)}, h_{31}^{0(m)}, h_{12}^{0(m)}\right)^T
\tag{8A.1e}
$$

$$
[D^{(m)}] = 2\mu_p^{(m)}
\begin{bmatrix}
1+a_m & a_m & a_m & 0 & 0 & 0 \\
a_m & 1+a_m & a_m & 0 & 0 & 0 \\
a_m & a_m & 1+a_m & 0 & 0 & 0 \\
0 & 0 & 0 & 1 & 0 & 0 \\
0 & 0 & 0 & 0 & 1 & 0 \\
0 & 0 & 0 & 0 & 0 & 1
\end{bmatrix}
\quad (m=1,2)
\tag{8A.1f}
$$

where $2\mu_p^{(m)}$ and $a_m$ are given, respectively, in (8.10e) and (8.10f).

$$
dp^{(m)} = \sqrt{d\varepsilon_{ij}^{p(m)}d\varepsilon_{ij}^{p(m)}}
\tag{8.8e}
$$

$$2\mu_p^{(m)} = \frac{\rho_0^{(m)}}{1 + \rho_0^{(m)}/2G^{(m)}} \tag{8.10e}$$

$$a_m = v^{(m)}/\left(1 - 2v^{(m)}\right) + K^{(m)}/\rho_0^{(m)} \tag{8.10f}$$

By combining equations of (8.11a–d) and (8A.1c) and after a long algebraic manipulation we finally have

$$\{d\varepsilon^C\} = [M]^{ep}\{d\sigma^C\} + \{dH\} \tag{8.12a}$$

The related expressions are given as follows:

$$[M]^{ep} = \begin{bmatrix} M_N & 0 \\ 0 & M_S \end{bmatrix} \tag{8A.2}$$

$$[M_S] = \begin{bmatrix} 1/P & 0 & 0 \\ 0 & 1/P & 0 \\ 0 & 0 & 1/P_0 \end{bmatrix} \tag{8A.3}$$

$$[M_N] = \begin{bmatrix} M_{11} & M_{12} & M_{13} \\ M_{21} & M_{22} & M_{23} \\ M_{31} & M_{32} & M_{33} \end{bmatrix} = \begin{bmatrix} A/D & -B/D & -CF/D \\ -B/D & A/D & -CF/D \\ -CF/D & -CF/D & C_0 + 2C^2F/D \end{bmatrix} \tag{8A.4}$$

$$\{dH\} = \{dH_1, dH_2, dH_3, dH_4, dH_5, dH_6\}^T \tag{8A.5}$$

$$dH_1 = -\sum_{m=1}^{2} \left[ \left( Ah_{11}^{0(m)} - Bh_{22}^{0(m)} \right) - Fh_{33}^{0(m)} \chi^{(m)} \right] \Phi^{(m)} dp^{(m)} \tag{8A.6}$$

$$dH_2 = -\sum_{m=1}^{2} \left[ \left( Ah_{22}^{0(m)} - Bh_{11}^{0(m)} \right) - Fh_{33}^{0(m)} \chi^{(m)} \right] \Phi^{(m)} dp^{(m)} \tag{8A.7}$$

$$dH_3 = -\sum_{m=1}^{2} \left[ \left( 2CFa^{(m)} + \frac{D}{2\mu_p^{(m)}(1 + a_m)} \right) h_{33}^{0(m)} - CF\left( h_{11}^{0(m)} + h_{22}^{0(m)} \right) \right] \Phi^{(m)} dp^{(m)} \tag{8A.8}$$

$$dH_4 = -\sum_{m=1}^{2} h_{23}^{0(m)} V_m dp^{(m)}/\rho_0^{(m)} \tag{8A.9}$$

$$dH_5 = -\sum_{m=1}^{2} h_{31}^{0(m)} V_m dp^{(m)}/\rho_0^{(m)} \tag{8A.10}$$

$$dH_6 = -\frac{1}{P_0}\sum_{m=1}^{2} 2\mu_p^{(m)} h_{12}^{0(m)} V_m dp^{(m)}/\rho_0^{(m)} \tag{8A.11}$$

Where

$$F = A - B \tag{8A.12a}$$

$$D = A2 - B2 \qquad (8A.12b)$$

$$\Phi^{(m)} = \frac{2\mu_p^{(m)} V_m}{D\rho_0^{(m)}} \qquad (8A.12c)$$

$$\chi^{(m)} = \frac{a_m}{1 + a_m} \qquad (8A.12d)$$

$$A = \sum_{m=1}^{2} 2\mu_p^{(m)} \left( \frac{1 + 2a_m}{1 + a_m} \right) V_m \qquad (8A.13a)$$

$$B = \sum_{m=1}^{2} 2\mu_p^{(m)} \left( \frac{a_m}{1 + a_m} \right) V_m \qquad (8A.13b)$$

$$C = \sum_{m=1}^{2} \frac{a_m V_m}{1 + a_m} \qquad (8A.13c)$$

$$C_0 = \sum_{m=1}^{2} \frac{V_m}{2\mu_p^{(m)}(1 + a_m)} \qquad (8A.14a)$$

$$P = 1 \Big/ \sum_{m=1}^{2} \frac{V_m}{2\mu_p^{(m)}}, \qquad (8A.14b)$$

$$P_0 = \sum_{m=1}^{2} 2\mu_p^{(m)} V_m \qquad (8A.14c)$$

## Appendix 8.B: Derivation of Equation (8.12e) and Matrix Elements

From (8.7e) $\{\delta\} = \{1, 1, 1, 0, 0, 0\}^{\mathrm{T}}$, we have:

$$A d\varepsilon_{kk}^{ec} \{\delta\} = A \{d\varepsilon_{kk}^{eC}, d\varepsilon_{kk}^{eC}, d\varepsilon_{kk}^{eC}, 0, 0, 0\}^{\mathrm{T}} \qquad (8A.15a)$$

As we see in the text in deriving

$$[d\varepsilon^{eC}] = [M]^e \{d\sigma^C\} \qquad (8.12b)$$

where with

$$\{d\sigma^C\} = \{d\sigma_{11}^C, d\sigma_{22}^C, d\sigma_{33}^C, d\sigma_{23}^C, d\sigma_{13}^C, d\sigma_{12}^C\} \qquad (8.7c)$$

$$\{d\varepsilon^{eC}\} = \{d\varepsilon_{11}^{eC}, d\varepsilon_{22}^{eC}, d\varepsilon_{33}^{eC}, d\varepsilon_{23}^{eC}, d\varepsilon_{13}^{eC}, d\varepsilon_{12}^{eC}\} \qquad (8.7d)$$

$[M]^e$ can be derived from $[M]^{ep}$. Let us distinguish the matrix element of $[M]^e$ from that of $[M]^{ep}$ by a superscript e, thus from (8A.2) we have

$$[M]^e = \begin{bmatrix} M_N^e & 0 \\ 0 & M_S^e \end{bmatrix} \qquad (8A.15b)$$

Since $d\varepsilon_{kk}^{ec}$ only involves normal strain so only $\left[M_N^e\right]$ is introduced below. From (8A.4) we have

$$[M_N^e] = \begin{bmatrix} M_{11}^e & M_{12}^e & M_{13}^e \\ M_{21}^e & M_{22}^e & M_{23}^e \\ M_{31}^e & M_{32}^e & M_{33}^e \end{bmatrix} = \begin{bmatrix} A^e/D^e & -B^e/D^e & -C^eF^e/D^e \\ -B^e/D^e & A^e/D^e & -C^eF^e/D^e \\ -C^eF^e/D^e & -C^eF^e/D^e & C_0^e+2(C^e)^2F^e/D^e \end{bmatrix} \tag{8A.15c}$$

Any element in the above matrix can be obtained from (8A.4) by taking $\rho_0^{(m)} \to \infty$ and $\mu_p^{(m)} \to G^{(m)}$. For example, from (8A.13a), (8A.13b) and (8.10f) we can express $A^e$ and $B^e$ as follows:

$$A^e = \sum_{m=1}^{2} 2G^{(m)} \left(\frac{1+2a_m^e}{1+a_m^e}\right) V_m \tag{8A.15d}$$

$$B^e = \sum_{m=1}^{2} 2G^{(m)} \left(\frac{a_m^e}{1+a_m^e}\right) V_m \tag{8A.15e}$$

$$a_m^e = v^{(m)}/\left(1-2v^{(m)}\right) \tag{8A.15f}$$

$$D^e = (A^e)^2 - (B^e)^2 \tag{8A.15g}$$

From (8A.15c) we have

$$M_{11}^e = \frac{A^e}{D^e} = \frac{A^e}{(A^e)^2 - (B^e)^2} \tag{8A.15h}$$

Other elements in (8A.15c) can also be determined with a similar approach. From (8.12b) and (8A.15c) we have

$$d\varepsilon_{kk}^{eC} = d\varepsilon_{11}^{eC} + d\varepsilon_{22}^{eC} + d\varepsilon_{33}^{eC} = (M_{11}^e d\sigma_1^C + M_{12}^e d\sigma_2^C + M_{13}^e d\sigma_3^C) + (M_{21}^e d\sigma_1^C + M_{22}^e d\sigma_2^C + M_{23}^e d\sigma_3^C)$$
$$+ (M_{31}^e d\sigma_1^C + M_{32}^e d\sigma_2^C + M_{33}^e d\sigma_3^C) = (M_{11}^e + M_{21}^e + M_{31}^e) d\sigma_1^C + (M_{12}^e + M_{22}^e + M_{32}^e) d\sigma_2^C + (M_{13}^e + M_{23}^e$$
$$+ M_{33}^e) d\sigma_3^C = \overline{\overline{M}}_{11} d\sigma_1^C + \overline{\overline{M}}_{12} d\sigma_2^C + \overline{\overline{M}}_{13} d\sigma_3^C \tag{8A.15i}$$

If we introducing the following notation

$$\overline{\overline{M}}_{11} = (M_{11}^e + M_{21}^e + M_{31}^e) \tag{8A.15j}$$

$$\overline{\overline{M}}_{12} = (M_{12}^e + M_{22}^e + M_{32}^e) \tag{8A.15k}$$

$$\overline{\overline{M}}_{13} = (M_{13}^e + M_{23}^e + M_{33}^e) \tag{8A.15l}$$

$$Ad\varepsilon_{kk}^{ec}\{\delta\} = A\{d\varepsilon_{kk}^{eC}, d\varepsilon_{kk}^{eC}, d\varepsilon_{kk}^{eC}, 0, 0, 0\}^T$$

$$= A\{d\varepsilon_{kk}^{eC}, d\varepsilon_{kk}^{eC}, d\varepsilon_{kk}^{eC}, 0, 0, 0\}^T = A \begin{Bmatrix} d\varepsilon_{kk}^{eC} \\ d\varepsilon_{kk}^{eC} \\ d\varepsilon_{kk}^{eC} \\ 0 \\ 0 \\ 0 \end{Bmatrix} = A \begin{bmatrix} \overline{\overline{M}}_{11} & \overline{\overline{M}}_{12} & \overline{\overline{M}}_{13} & 0 & 0 & 0 \\ \overline{\overline{M}}_{11} & \overline{\overline{M}}_{12} & \overline{\overline{M}}_{13} & 0 & 0 & 0 \\ \overline{\overline{M}}_{11} & \overline{\overline{M}}_{12} & \overline{\overline{M}}_{13} & 0 & 0 & 0 \\ 0 & 0 & 0 & 0 & 0 & 0 \\ 0 & 0 & 0 & 0 & 0 & 0 \\ 0 & 0 & 0 & 0 & 0 & 0 \end{bmatrix} \begin{Bmatrix} d\sigma_1^C \\ d\sigma_2^C \\ d\sigma_3^C \\ d\sigma_4^C \\ d\sigma_5^C \\ d\sigma_6^C \end{Bmatrix} = \overline{\overline{[M]}}^e \{d\sigma^C\}$$

$$\tag{8A.15m}$$

where

$$
[\overline{\overline{M}}]^e = \begin{bmatrix}
\overline{\overline{M}}_{11} & \overline{\overline{M}}_{12} & \overline{\overline{M}}_{13} & 0 & 0 & 0 \\
\overline{\overline{M}}_{11} & \overline{\overline{M}}_{12} & \overline{\overline{M}}_{13} & 0 & 0 & 0 \\
\overline{\overline{M}}_{11} & \overline{\overline{M}}_{12} & \overline{\overline{M}}_{13} & 0 & 0 & 0 \\
0 & 0 & 0 & 0 & 0 & 0 \\
0 & 0 & 0 & 0 & 0 & 0 \\
0 & 0 & 0 & 0 & 0 & 0
\end{bmatrix}
\tag{8A.15n}
$$

# References

[1] Fan, J., Gao, Z., and Zeng, X., (2004) Cyclic plasticity across micro/meso/macroscopic scales. *Proceedings of the Royal Society of London A*, **460**, 1477.

[2] Fung, Y. C. (1994) *A First Course in Continuum Mechanics*, 3rd edn., Prentice-Hall.

[3] Fan, J. (1999) A micro/macroscopic analysis for cyclic plasticity of dual-phase materials. *Journal of Applied Mechanics*, **66**, 124.

[4] Berveiller, M. and Zaoui, A. (1997) An extension of the self-consistent scheme to plastically flowing polycrystals. *J. Mech. Phys. Solids*, **26**, 325.

[5] Fung, Y. C. (1965) *Foundations of Solid Mechanics*, Prentice-Hall.

[6] Peng, X., Fan, J., and Yang, Y. (2002) A microstructure-based description for cyclic plasticity of pearlitic steel with experimental verification. *Int. J. Solids Struct.*, **39**, 419.

[7] Valanis, K. C. and Fan, J. (1983) Endochronic analysis of cyclic elastoplastic strain fields in notched plate. *Journal of Applied Mechanics*, **50**, 789.

[8] Drucker, D. C. (1950) Some implications of work-hardening and idea plasticity. *Quarterly of Applied Mathematics*, **7**(4), 411.

[9] Valanis, K. C. (1980) Fundamental consequences of new intrinsic time measure plasticity as a limit of the endochronic theory. *Arch. Mech.*, **32**, 171.

[10] Chaboche, J. L., Dang Van, K., and Cordier, G. (1979) Modelization of the strain memory effect on the hardening of 316 stainless steel, in *Proceedings of the Fifth International Conference on SMiRT*, Berlin, L11/3.

[11] Ohno, N. and Wang, T. D. (1993) Kinematic hardening rules with critical state of dynamic recovery, Part 1: Formulation and basic features for ratchetting behavior. *Int. J. Plasticity*, **9**, 375.

[12] McDowell, D. L. (1997) An engineering model for propagation of small cracks in fatigue. *Engineering Fracture Mechanics*, **56**, 357.

[13] Jiang, Y. and Sehitoglu, H. (1996) Modeling of cyclic ratcheting plasticity. Part 1. Development of constitutive relations. *Journal of Applied Mechanics*, **63**, 720.

[14] Eshelby, J. D., Frank, F. C., and Nabarro, F. R. N. (1951) The equilibrium of linear arrays of dislocations. *Philos. Mag.*, **42**, 351.

[15] Li, J. C. M. and Chou, Y. T. (1970) The role of dislocations in the flow stress grain size relationships. *Metall. Trans.*, **1**, 1145.

[16] Hall, E. O. (1953) The deformation and ageing of mild steel: II Characteristics of the Luders deformation, *Proceedings of the Royal Society of London B*, **64**, 742.

[17] Petch, E. O. (1953) The cleavage strength of polycrystals. *Journal of the Iron and Steel Institute*, **174**, 25.

[18] Langford, G. (1977) Deformation of pearlite. *Metall. Trans. A*, **8**, 861.

[19] Hwang, K. C. and Sun, S. (1994) Micromechanical modeling of cyclic plasticity, in *Advances in Engineering Plasticity* (eds. B. Xu and W. Yang), International Academy Publishers, Beijing, 41.

[20] Peng, X. and Fan, J. (2000) A new approach to the analysis of polycrystal plasticity. *Arch. Mech.*, **52**, 103.

[21] Shackelford, F. (2004) *Introduction to Materials Science for Engineers*, 6th edn., Prentice Hall.

# 9

# Topics in Materials Design, Temporal Multiscale Problems and Bio-materials

Topics of multiscale analysis in cutting-edge areas are selected and introduced in the three parts of this chapter. Part 9.1 describes the methods and roles that multiscale analysis can take in materials design. Part 9.2 describes different concurrent and hierarchical temporal multiscale methods including principles and examples for simulation of infrequent event systems. Specifically, the nudged elastic band (NEB) method for finding minimum energy path (MEP) and saddle point is introduced and applications in investigating effects of strain rate and temperature on dislocation initiation, and in understanding large-scale separation of activation volume between plasticity and creep, and the in-depth mechanism of high strength and high ductility of nanocrystalline materials, are described and exemplified.

Part 9.3 introduces hierarchical structures of protein materials, and discusses modeling in three scales including nano (protein), mesoscale (intermediate filaments and membranes of cells) and cell-implant interaction. Possible applications in enhancing the adhesion strength between cells and implants are addressed.

## Part 9.1 Materials Design

## 9.1 Multiscale Modeling in Materials Design

In this part, functions, limitations, and choices of multiscale methods in materials design will briefly be discussed.

### 9.1.1 The Role of Multiscale Analysis in Materials Design

The words "material(s) design" or "materials by design" are terms defined to distinguish them from the traditional "trial and error" method in material development. In the traditional method, the cycle for trial and error to develop and certify a new material is usually long, say 20 years. However, with material design the goal of accelerating the insertion of new and improved materials into, say, next-generation transport vehicles and propulsion systems is beginning to be realized.[1, 2]

*Multiscale Analysis of Deformation and Failure of Materials*  Jinghong Fan
© 2011 John Wiley & Sons, Ltd

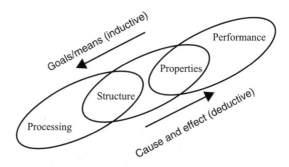

**Figure 9.1**   Olson's concept of material design (Reproduced with permission from Olson, G. B. (1997) Computational design of hierarchically structured materials. *Science*, 277 (5330), 1237, the American Association for the Advancement of Science (AAAS))

From the concept developed in Section 1.1 that material property depends on structural hierarchy, one may assemble a microstructural morphology to meet the requirements for developing or upgrading a material. This is possible because the emerging computational modeling and simulation tools and increasingly high-resolution and rapid-characterization instruments lay the foundation for the new generation of multiscale analysis.

To understand the important role of multiscale analysis in materials design, Olson's concept shown in Figure 9.1 is introduced.[3] The figure shows a deductive process which links processing to structure to properties to performance of the material. Through this bottom-up process of multiscale analysis, as discussed in Section 1.2, the underlying mechanism for processing/structure and structure/property relationship can be found.

The data point in the design space, which connects the processing and structure parameters to a ranged set of property and performance requirements, can be developed further.[2, 4] The obtained mechanism and information can offer valuable information for the material selection, which is a top-down inductive process as shown on the top left of the figure. Its task is to use the knowledge and data accumulated in the deductive process to choose composition, processing and microstructure parameters of the material to reach the required performance.

### 9.1.2   Issues of Bottom-up Multiscale Modeling in Deductive Material Design Process

Ideally, using the fast increasing power of computers and the emerging modeling and simulation tools, multiscale analysis can carry on a large amount of bottom-up intensive computation to accumulate data for design in terms of processing and structure parameters. In reality, there are some issues listed below which should be mentioned relating to the limitations of multiscale analysis.

– While multiscale modeling can play an essential role in providing valuable information for design decisions, it cannot replace the decision-making process in concurrent design of material structure hierarchy and products made of the material. Actually, the design process "requires a combination of expert estimation based on prior experience, experiments, and models and simulations at virtually every mapping."[2]
– Multiscale analysis may not give a unique answer because at this moment the uncertainty of problems, which concern processing route, stochasticity of microstructure, path dependence nature of inelasticity, etc. in multiscale analysis, is still a hot topic under intensive investigation.

– Materials design involves concurrent design of material hierarchy and product with a ranged set of performance requirements. The solution is not clear at the beginning, thus an iterative multiscale analysis is essential for bottom-up information flow combined with top-down guidance from applications and performance requirements.

While the role of multiscale analysis has a serious limitation in material design at this moment, this situation will be improved with further development of multiscale modeling and its underlying theory, including uncertainty analysis, non-equilibrium analysis, and temporal multiscale analysis. Moreover, with multimillion atom simulations coupling with the sub-millimeter continuum domain in the next decade, a more effective role for multiscale modeling in the discovery of new material concepts can be expected.

### 9.1.3   Choices of Multiscale Methods in Materials Design

Materials design involves a large range of length scales from angstrom to nanometer to micrometer to macroscopic scale. It is by now widely recognized that both natural and synthetic materials inherently have hierarchical, multiscale structures. Thus, hierarchical (sequential) multiscale methods are, in general, favorable in material design. This is consistent with the thoughts of material scientists.

Actually, in many published papers of multiscale analysis by material scientists, developed approaches at different scales that have been employed for the same material. The results obtained at different scales were used for comprehensive analysis and comparison with the goal of linking these scales. Due to a lack of quantitative analysis tools for transitions between different scales, the hierarchical multiscale analysis involved was mostly qualitative in nature. This kind of analysis, however, is effective in materials design because it is intuitive, simple, based on experience and testing the data of each individual scale, thus it can easily be used in a decision-making design process.

Recently, multiscale modeling development has paid more attention to concurrent multiscale modeling than to hierarchical modeling schemes. Its emphasis is on the atomistic-continuum coupling. This is multiscale analysis within small-length scales ranged from angstrom to nanometer to sub-micrometer dimensions. Needless to say, concurrent multiscale analysis may produce highly accurate results; it is important for nanotechnology and biotechnology to find underlying mechanisms of deformation and failure. However, it is limited to a narrow range and cannot address all the requirements of the bottom-up deductive material design process, particularly for mesostructured materials.

There is a lack of hierarchical quantitative multiscale method covering a large range of scales. Some coupled multiscale models are complex with uncertain outcomes for parameter identification and validation. This makes it difficult to conduct the bottom-up deductive process of material design. Instead, for engineering applications engineers often prefer to introduce rather more elementary, proven model concepts at each scale that lack explicit linkage between different scales via numerical or analytical means.

Multiscale modeling is still under development and there are various models and classifications at this stage. These models are creating the framework for multiscale analysis for years to come. The trend observed is that multiscale analysis will be more and more widely used, including applications in material design.

## Part 9.2 Temporal Multiscale Problems

## 9.2   Introduction to Temporal Multiscale Problems

This section introduces the significance, basic concepts and approaches of temporal multiscale analysis. Infrequent events for transition state analysis based on saddle points of the minimum energy path by nudged elastic band will be introduced in Sections 9.3 to 9.5. Other time-accelerating methods are briefly introduced in Section 9.6. The coarse-grained method using microseconds in the simulation for bio-cells is introduced in Section 9.8.

## 9.2.1 Material Behavior Versus Time Scales

There are temporal multiscale problems in the deformation and failure of materials. For instance, a material domain with molecular vibration, corrosion reaction and creep deformation involves multi-time scales and shows different aspects of material property. The former is ultrafast and needs to be described using a small time unit such as picosecond, the latter is an ultra slow phenomenon and is described in terms of hours, while the corrosion rate is in the middle and can be measured by seconds or minutes. Thus, distinctly different time scales could coexist in the same model.

Material behavior can be quite different at different time scales. There are plenty of experimental reports in viscoelasticity, viscoplasticity, rheology, and biology to show the time-scale dependent behavior of materials. It relates to the spectrum of relaxation time, which in turn is coupled to spatial scales of structure and kinetics of rearrangement processes.

Take biomaterials as an example; experiments on the actin cells collected from organs (not cells cultured in lab) show that a fresh cell exhibits fast creep in a short time ($0 \sim 0.01$ s) or in a high frequency (larger than 100 Hz) immediately after being subject to an applied force,[5] while it exhibits slow creep in a larger time scale ($0.01 \sim 10$ s or frequency smaller than 100 Hz). The short-time property is basically the same for all materials and is related to molecular thermal motion. However, the long-time property varies in different materials and environments. For instance, the temporal behavior of a cell depends on whether it is ill or healthy, and it can be changed when the cell configuration changes.

## 9.2.2 Brief Introduction to Methods for Temporal Multiscale Problems

### 9.2.2.1 Concurrent Methods by Introduction of Several Temporal Variables

One way to solve the temporal multiscale problem is to introduce concurrently (parallel) several time variables at different time scales in the analysis. This is the concurrent method. For instance, in fatigue analysis, a long time scale (low frequency) characterizing a cycle-averaged solution may be used to describe the global fatigue behavior; while a short time scale (high frequency) for a remaining oscillatory problem may be used to describe the cyclic behavior in a short period. Needless to say, the material behavior at short time scales is intrinsically related to the fatigue performance at longer time scales. A wavelet decomposition method may be introduced to integrate the two time scales and their corresponding material behavior concurrently, a method significantly reducing the computation time till fatigue crack initiation.[6]

An example of concurrent time methods is the simulation of Portevin-Le Chatelier (PLC) effect in Al and Mg alloys. Different from initial yielding, PLC is a phenomenon where stress fluctuates in a load-displacement curve during tensile tests at room temperature with small strain rates (Figure 9.2). PLC is generally considered to be the result of dynamic interactions between gliding dislocations and mobile solute atoms, i.e., the so-called dynamic strain aging (DSA).[7]

The dislocation glide is intermittent due to the pinning and unpinning competition. The unpinning process of dislocation causes the stress drop on stress-time curves; it is a fast process which may introduce a fast time scale to describe it. On the other hand, formulating a solute cloud coupling with structural evolution of forest dislocations to pin the mobile dislocation is a slow process which can introduce a slow time scale to describe it. Both time scales are used in the governing equations, thus the coupling between these two time scales can be realized.[8]

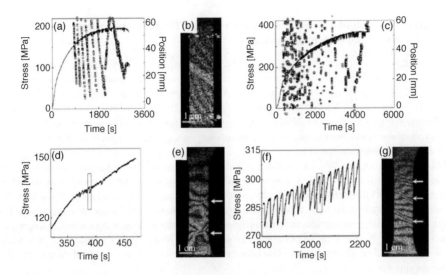

**Figure 9.2**   Plots of stress (curve) and band position (hollow dot) against time for (a) annealed and (c) solution-treated Al-4%Cu alloys at a strain rate of $10^{-4}\,\text{s}^{-1}$ (Reproduced from Jiang, H., Zhang, Q., Chen, X., *et al.* (2007) Three types of Portevin-Le Chatelier effects: Experiment and modeling. *Acta Material*, 55 (7), 2219, Elsevier)

### 9.2.2.2  Hierarchical Multiscale Temporal Methods by Introduction of a Sequence of Time Scales

One example for using hierarchical temporal scales in simulation is vapor deposition for formation of thin films (see Section 3.14.2). Vapor deposition is a multiple time-scale problem in the sense that growth of the film occurs in a reactive substrate whose time period may range from 1 to 100 seconds, while the assembly events involve a time scale in the pico- to microsecond range. In fact, atomistic assembly is more fundamentally determined by the making and breaking of chemical bonds which is described by the wavefunctions of bonding electrons with time scale of 0.1 fs or so.

In a vapor deposition system, a coating machine with vacuum process technology is capable of providing ions (e.g., oxygen ion) bombardment during the deposition process on the substrate. Several processes will then be generated after the bombardment which include reflection, resputtering and diffusion.

It is found that if the deposition process is only simulated by MD, the formed layer thickness is much thinner than the experimental data. That is understandable because MD usually can only simulate processes like bombardment, reflection and resputtering in which a short time scale is used. On the other hand, the diffusion process is important for the film formation, thus a longer time scale is necessary. One may separate the whole deposition process into many steps. In each step, first use a short time scale for MD simulation followed by much longer time scale for diffusion simulated by kinetic Monte Carlo method, which is introduced in Section 2.9. As described in (2.36), the mean time between two events is determined by the rate of adsorption (or incidence) and desorption (or reflection) as well as the size of lattice grids. The size in the Monte Carlo simulation may be adjusted to match the realistic time for the thin film growth.[9, 10]

### 9.2.2.3 Introduction of Different Characterization Time Scales to Describe Temporal Dependent Behavior

Concurrent temporal multiscale methods can also be used with different characteristic time scales related to the same time variable. Each of the introduced time scales relates to a special physical mechanism. For instance, in the analysis of tensile stress wave propagation induced by planar collision, three characteristic time scales are introduced. They describe globally and microscopically the process of material spallation or avalanche induced by the formation of large numbers of microvoids or microcracks.[11] The three characteristic time scales are the macroscopic fluctuation time $t_i$, micro-damage nucleation time $t_n$ and micro-damage propagation time $t_v$.

The Deborah number is defined as the ratio of characteristic times $t_n$ and $t_v$ over the macroscopic time $t_i$, i.e., $\text{De}^* = t_v/t_i$ and $\text{De} = t_n/t_i$ and is used to describe the competition and coupling between different damage mechanisms through these time scales. Subsequently, numerical method is employed to study the effects of the Deborah number $\text{De}^*$ and De on the evolution of the internal damage in a target plate. It was found that $\text{De}^*$, which characterizes defect propagation, is superior in describing the macroscopic failure process induced by the localization of damage. Besides, $\text{De}^*$ is also a better characteristic parameter for describing time-scale coupling effects while De is superior in describing nucleation of damage clusters in the mesoscopic space.

### 9.2.2.4 The Connection Between Spatial and Temporal Multiscale Analysis

Temporal multiscale analysis cannot completely separate from the spatial one. Limited by the characteristic time of atomic thermal vibration, the time step in MD is typically limited to the range of the scale of fs ($1\text{fs} = 10^{-15}\,\text{s}$). Such a small time step makes the description of slow processes with MD difficult, and becomes an obstacle for the development of temporal multiscale analysis. Although several time variables and characteristic time parameters can be introduced, it is still difficult to choose the system time in the analysis suitable for both atomic scale and continuum scale.

Taking wave propagation as an example, the limitation on the time step may make the increase of spatial scale difficult. In fact, the MD simulation in general covers at most several nanoseconds because it requires one million time steps to reach 1 nanosecond if time step is 1 fs. The group velocity of a sound wave is about 3000–5000 m/s, which indicates that the wave can propagate at most several micrometers in several nanoseconds.[12]

It is seen that temporal multiscale problems have become an obstacle in the development of multiscale analysis. New ideas shown in this and the next several sections are necessary to break through this bottleneck.

## 9.3  Concepts of Infrequent Events

Recent advances in temporal multiscale analysis for infrequent events have been noteworthy.[13] Examples of infrequent events include various diffusion processes, chemical reaction processes, and the reconformation process of molecules including heating event of atoms absorbed on the metal surface.

Let us first introduce the concept and definition of infrequent-event systems. As we know, during the system evolution the total energy of the system will be changed from one state to another. This functional dependence can be expressed by an energy surface over the system's configuration coordinate space. While this coordinate space consists of 3N-dimensional coordinates, the energy-state relationship may still be imagined as the simple one-to-one correspondence in the sense that one point in the generalized coordinate plane corresponds to one energy value in the vertical coordinate perpendicular to that plane. This formulates a point on the generalized energy surface. This energy surface may include several basins with local energy minima and the transition areas between these basins.

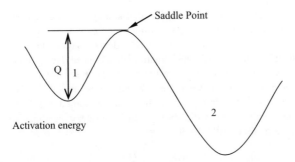

**Figure 9.3**  Energy barrier or activation energy is necessary for state transition from energy basin 1 to basin 2

Investigating energy basins and how a system moves from one basin state to another is important because an energy basin is a relatively stable configuration with lower configuration energy. The pathway between the two basins is called the reaction path. As shown in Figure 9.3 in the reaction path there is a saddle point whose energy is higher and the system is unstable so that new phenomenon such as defect (e.g., dislocation and micro-crack, vacancy) and diffusion may start at that point.

Here we just describe the basic concept of the infrequent-event system which relates to the system behavior at the basins. Specifically, when the trajectory of the system reaches an energy basin, it wanders around in the basin over a time scale of a huge number of vibrational periods. At some instant when enough energy has been localized into a reactive mode, the trajectory passes through a dividing surface of the basin and moves along the reaction pathway, entering another basin transited from the saddle point. In essence, it "accidentally" exits the original basin, beginning a new journey of vibrational wandering, with no memory of how it arrived in the original basin because it was wandering too long there.

The key feature of the infrequent-event system is that it has no memory of its history, thus its trajectory will automatically choose an appropriate escape path with no prior information. This makes simulation much simpler because one only needs to investigate phenomena occurring during the trajectory motion from the current energy basin to its neighboring one. This is technically important because we do not need to use MD to trace the system evolution in terms of its history but only consider the significant phenomena occurring on the reaction path between the two basins. We will discuss this method in detail in the next section.

## 9.4  Minimum Energy Path (MEP) and Transition State Theory in Atomistic Simulation

An important problem in material science and material mechanics is to identify a lowest energy path for a rearrangement of a group of atoms from one energy basin to another. This is also an important problem in theoretical chemistry and condensed matter physics. As discussed in the last section, the basin of energy surface over the 3N generalized atomistic space denotes local minimum of the system energy, thus it is a location for a stable configuration.

### 9.4.1  Minimum Energy Path (MEP) and Saddle Point

The lowest energy path is referred to as the minimum energy path (MEP). It is frequently used to define a "reaction coordinate" or reaction path so that infrequent events may happen along that reaction coordinate (path). Examples of the state transitions include dislocation core nucleation, changes in configuration of molecules, diffusion processes in solids, and chemical reactions.

Transition state theory (TST)[13] is the formalism that underpins all accelerating molecular dynamics simulations of these infrequent-event processes, directly or indirectly. The significance is that these acceleration methods can reach simulation times several orders of magnitude longer than direct MD while retaining full atomistic detail. In the TST approximation, the classical rate constant from energy basin 1 to adjacent basin state 2 is taken to be the equilibrium flux through the dividing surface between basin 1 and 2. The potential energy maximum point along the minimum energy path is the saddle point which gives the energy barrier Q as shown in Figure 9.3. If the saddle point along the reaction pathway on the potential energy surface can be identified, then the transition rate, $v$, becomes a very simple Arrhenius temperature dependence as follows:[13, 14]

$$v = f_v \exp\left(-\frac{Q(\sigma, T)}{k_B T}\right) \tag{9.1}$$

where $f_v$ is the system attempt frequency and $k_B T$ is thermal energy with $k_B$ as the Boltzmann constant. This energy barrier $Q(\sigma, T)$ is also called activation energy. It is seen from (9.1) that this quantity Q is of central importance for estimating the transition rate. In general, the activation energy is a function of temperature and applied stress. The simplest linear form of the temperature dependence of activation energy can be written as:[14]

$$Q(\sigma, T) = \left(1 - \frac{T}{T_m}\right) Q_0(\sigma) \tag{9.2}$$

where $T_m$ is the melting or disordering temperature. This equation satisfies the two extreme conditions: when $T = T_m$ there is no energy barrier and when $T = 0$, the activation energy equals $Q_0(\sigma)$ which is the activation energy on the potential energy surface (PES) at zero temperature and can be calculated as the energy difference between the saddle point and initial equilibrium state on the MEP of that surface (see Figure 9.3). The simplest linear form of stress dependence of activation energy can be written as:[14]

$$Q_0(\sigma) = A(1 - \sigma/\sigma_{ath})^\alpha \tag{9.3}$$

where $\sigma_{ath}$ is the maximum stress at which there is no energy barrier when there are no temperature effects. Besides the activation energy, the other important variable in the transition state analysis is the activation volume which can be determined as:

$$\Omega(\sigma, T) = -\frac{\partial Q(\sigma, T)}{\partial \sigma}\bigg|_T \tag{9.4}$$

This relationship is understandable since driving the activation volume $\Omega$ moves with the increasing stress $d\sigma$ which will reduce the activation energy $dQ$ by $\Omega d\sigma$.

### 9.4.2   Nudged Elastic Band (NEB) Method for Finding MEP and Saddle Point

#### 9.4.2.1 An Example to Show MEP and Saddle Point on the Energy Surface

The approach to use the so-called nudged elastic band (NEB) method for finding MEP and saddle point is better explained by simple examples. There are two simple models used for the introduction of the NEB method in Ref. [15]. Figure 9.4 uses the combined potential of the LEPS (London-Eyring-Polanyi-Sato) energy potential and the harmonic oscillator potential. Figure 9.5 uses the LEPS energy potential only. In the following, we will use Figure 9.4 for the general introduction of the NEB method and Figure 9.5 to show the detail for the whole determination process of the NEB method to determine the MEP and saddle points including the code development.

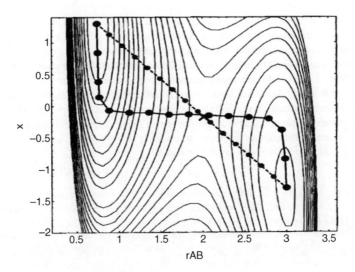

**Figure 9.4**  A contour plot of the potential energy surface for a simple NEB test problem with the LEPS + harmonic oscillator potential. The initial and final configuration and 14 images are shown (Reproduced from Jonsson, H., Mills, G., and Jacobsen, K. W. (1998) Nudged band method for finding minimum energy paths of transitions, in *Classical and Quantum Dynamics in Condensed Phase Simulations* (eds. B. J. Berne, G. Ciccotti, and D. F. Coker), World Scientific, Singapore, 385. © World Scientific Publishing Co.)

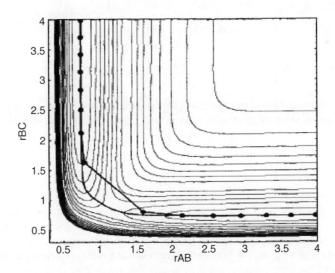

**Figure 9.5**  $V^{LEPS}$ potential surface contour (Reproduced from Jonsson, H., Mills, G., and Jacobsen, K. W. (1998) Nudged band method for finding minimum energy paths of transitions, in *Classical and Quantum Dynamics in Condensed Phase Simulations* (eds. B. J. Berne, G. Ciccotti, and D. F. Coker), World Scientific, Singapore, 385. © World Scientific Publishing Co.)

The model used LEPS potential which mimics a reaction involving three atoms A, B, C confined to motion along a line. For the case of Figure 9.4, A and C are fixed at both the left and right end of a fixed length line; atom B can form a chemical bond with either one of the two fixed atoms. An additional degree of freedom is introduced which can be interpreted as a fourth atom D which is coupled in a harmonic way to atom B. Because atoms A and C are fixed the simulation system has two degrees of freedom. When the positions of atoms B and D are determined, the distance between the four atoms are determined; after the interatomic potential function is given the total system energy can be determined. Thus the one-to-one correspondence between a point at the horizontal coordinate plane and a point at the energy surface, whose energy value is given by the contour lines, can be uniquely determined.

The potential used for Figure 9.4 is given as follows:

$$V(r_{AB}, x) = V^{LEPS}(r_{AB}, r_{AC} - r_{AB}) + 2k_c(r_{AB} - (r_{AC}/2 - x/c))^2 \tag{9.5a}$$

The first term on the right is the LEPS potential which is the function of $r_{AB}$ and $r_{BC}$ (i.e., $r_{AC} - r_{AB}$), the second term is the harmonic oscillator potential where x is used as the variable in the vertical coordinate of Figure 9.4. The LEPS potential is given as follows:

$$
\begin{aligned}
V^{LEPS}(r_{AB}, r_{BC}) = {} & \frac{Q_{AB}}{1+a} + \frac{Q_{BC}}{1+b} + \frac{Q_{AC}}{1+c} \\
& - \left[ \frac{J_{AB}^2}{(1+a)^2} + \frac{J_{BC}^2}{(1+b)^2} + \frac{J_{AC}^2}{(1+c)^2} - \frac{J_{AB}J_{BC}}{(1+a)(1+b)} - \frac{J_{BC}J_{AC}}{(1+a)(1+c)} \right. \\
& \left. - \frac{J_{AB}J_{AC}}{(1+a)(1+c)} \right]^{\frac{1}{2}}
\end{aligned}
\tag{9.5b}
$$

It can be rewritten as:

$$V^{LEPS} = Q^{term} - (J^{term})^{1/2} \tag{9.5c}$$

with

$$Q^{term} = \frac{Q_{AB}}{1+a} + \frac{Q_{BC}}{1+b} + \frac{Q_{AC}}{1+c} \tag{9.5d}$$

$$J^{term} = \frac{J_{AB}^2}{(1+a)^2} + \frac{J_{BC}^2}{(1+b)^2} + \frac{J_{AC}^2}{(1+c)^2} - \frac{J_{AB}J_{BC}}{(1+a)(1+b)} - \frac{J_{BC}J_{AC}}{(1+b)(1+c)} - \frac{J_{AB}J_{AC}}{(1+a)(1+c)} \tag{9.5e}$$

where the $Q_{term}$ represent Coulomb interactions between the electron clouds and the nuclei; the $J_{term}$ relates to the quantum mechanical exchange interactions. The forms of these functions are:

$$Q(r) = \frac{d}{2}\left( \frac{3}{2} e^{-2a(r - r_e)} - e^{-a(r - r_e)} \right) \tag{9.5f}$$

$$J(r) = \frac{d}{4}\left( e^{-2a(r - r_e)} - 6e^{-a(r - r_e)} \right) \tag{9.5g}$$

where $r_e$ is the equilibrium distance, values of parameters are:[15] a = 0.05, b = 0.80, c = 0.05, $d_{AB} = 4.746$, $d_{BC} = 4.746$, $d_{AC} = 3.445$, $r_e = 0.745$ and $\alpha = 1.942$. All parameters are normalized so only numbers are shown to make it simple and concise. With this information, we can plot the potential contour along an $r_{AB}$ and $r_{BC}$ axis to illustrate all the possible atomic configurations, as seen below in Figure 9.4.

In this simple case, the energy surface can be drawn over a 2D coordinate plane as shown in Figure 9.4. Specifically, the total energy of the system is a function of two variables, one is the distance $r_{AB}$ between atom A and atom B, the other is the coordinate, x in (9.5a). These two variables $r_{AB}$ and x are, respectively, taken as the horizontal and vertical coordinate axes in the figure.

The energy corresponding to each point in the coordinate plane is denoted by the point on the energy contours over that plane. It is easy to see that the energy value depends on the location of the point and that the same contour gives the same energy value. There are, however, three regions that need to be paid special attention. The first two, respectively, located on the top-left and the bottom-right region, denote two energy basins with local energy minima. The third one in the middle of the figure is with the maximum energy and it denotes the region near the saddle point in the transition pathway.

### 9.4.2.2 Determining MEP and Saddle Point by Nudging Elastic Forces

Take two points of energy minima as the starting and ending point of the pathway and then draw a straight line with many dot points between these two points as shown in Figure 9.4. In the figure the straight line for developing the initial images is with dashes and 14 small filled circles. These points and the two ending points define a "drag coordinate" which is used to find the MEP and the saddle point. To do so with NEB method the system energy is required to be minimized by a relaxation process along a perpendicular direction to the drag lines.

Specifically, the 14 small circles indicate the initial configuration chosen for the images in the elastic band. The larger filled circles lying close to the curved MEP show the configuration energies after the relaxation and convergence process by a minimum energy algorithm such as the so-called conjugate method which is used to find the system configuration with minimum energy.

Take note that in the NEB method the relaxation is finally completed by considering the coupling between all the images and the two ending points. These couplings are realized by springs between neighboring images. It offers forces to drag the images from the initial point to the ending point. These forces nudged by springs between band images are important for the convergence and for finding the saddling points. During the energy minimization the projections of the perpendicular component of the true force and the parallel component of the spring force to the pathway are as "nudging" forces. This may be the reason why the words "nudged" and "elastic" are used in the method's name. On the other hand, the perpendicular component of the spring force and the parallel component of the true atomistic force to the pathway are designed to be zero.

These specific force projections decouple the dynamics of the path itself from the particular distribution of images chosen in the discrete representation of the path. The spring force then does not interfere with the relaxation of the images. This is so because the relaxed configuration of the images satisfies the condition that the perpendicular force of springs to the path is zero so that they lie on the MEP (see next section). This decoupling of the relaxation of the path from the discrete representation of the path is essential to ensure convergence to the MEP. Furthermore, since the spring force only affects the distribution of the images within the path, the choice of the spring constant is quite arbitrary. In Figure 9.4, the spring constant is $k = 0.5$ near the ends of the bank and $k = 1.0$ in the middle so as to increase the spring stiffness, and in turn increase the density of the images in the most relevant region to the saddle point.

---

**Homework**

(9.1) For a model with three atoms A, B and C confined to motion along a horizontal line in which B is between A and C, and the total length of AC is limited to 4 units. Draw the energy contour of the LEPS potential shown in (9.5b) with the horizontal axis coordinate as $r_{AB}$ and vertical coordinate as $r_{BC}$ and then compare your drawing with the chart in Figure 9.5 to see whether your drawing is consistent with that figure. Please write a comparison between the saddle point that your drawing and the one identified by the figure has.

Hint: an energy contour file can be made by looping twice over the equation (9.5b) for varying 'r's and displayed using Gnuplot software with a script (see Section 10.4.1). One way by the following commands developed for drawing Figure 9.10 will help you in the drawing of this problem.

```
Gnuplot> set palette rgbformulae 10,10,10    # sets the colour ranges of a
heat map, 10,10,10 is one way to make the colours black and white
>set cbrange [-5:5]                # sets range of z values to map
>set ylabel "rBC"                  # set ylabel
>set xlabel "rAB"                  # set xlabel
>set terminal png                  # save plot as png
>set output "reNEB_1-9.png"            # set file name to save as
>plot 'contour.dat' u 1:2:3 with image, \   # plots a contour from data in
                             #'contour.dat' as a heat map
'LEPS_II_NEB001.dat' u 2:3 lt -1 pt 2, \*
'LEPS_II_NEB001.dat' u 2:3 with lines lt rgb "black", ... \**

* add data from first reiteration file, lt -1 makes the points black, pt 2
makes the points as 'x'
**\ add the same data but draw it with lines, with linetype 'lt' as an 'rgb'
colour 'black'
```

(9.2) If the first and ending points with minimum energy are given as First point: 0.7460 4.0000, Last point: 4.0000 0.7460, suppose in between these two ending points there is a straight line in which 14 images are chosen with equal spacing, determine their coordinates and the LEPS energy.

(9.3) Verify that the Coulomb Forces given in (9.5d) for the LEPS potential can be expressed as follows:

$$Q_{abTerm} = (dQab/(1+a)) + (dQac/(1+c))$$

$$Q_{bcTerm} = (dQbc/(1+a)) + (dQac/(1+c))$$

where dQab, dQbc and dQac are given in the Fortran code LEPS_II.f90 (in UNITS/UNIT2/ F90_CODES/NEB).

```
dQab=dab*alpha*exp((r0*alpha)-(2*alpha*rab(rj)))*(exp(alpha*rab(rj))-
3*exp(r0*alpha))/2
dQbc=dbc*alpha*exp((r0*alpha)-(2*alpha*rbc(rj)))*(exp(alpha*rbc(rj))-
3*exp(r0*alpha))/2
dQac=dac*alpha*exp((r0*alpha)-(2*alpha*rac(rj)))*(exp(alpha*rac(rj))-
3*exp(r0*alpha))/2
```

## 9.4.3   Mathematical Description of the NEB Method

Let us see mathematically how the above-mentioned decoupling of the spring force and true atomistic force through the designed projection of the spring and atomistic true forces is reached.

An elastic band linked by springs with N+1 image configurations of the atomistic system can be denoted by [$R_0$, $R_1$, $R_2$...$R_N$]. All bold letters here denote vectors in a generalized coordinate system. For instance, $R_I$ denotes position vector of the I image configuration. This vector uniquely determines the positions of any atom in the system at the image I state. For instance, the system in Figure 9.4 has two degrees of freedom so any system status can be described by a point which can be determined either by a position vector $\mathbf{R_I}$ (not drawn in the figure) or simply by its horizontal and vertical coordinates rAB or rBC. If the 3D system has 10,000 atoms then the basic concept is still valid. However, it is difficult to draw these position vectors because it requires 30,000 coordinate axes to describe a single position vector. Specifically, $\mathbf{R_0}$ and $\mathbf{R_N}$ are fixed end points in the generalized coordinate system by which all the atoms in the initial and final states are given. Usually, the coordinates of any atoms in the system at the energy minimum (energy basins) can be determined, thus the position vectors $\mathbf{R_0}$ and $\mathbf{R_N}$ can be uniquely determined. The N–1 intermediate images are adjusted by the optimization algorithm. The total force vector acting on the atoms of image i is the sum of the spring force along the tangent of the pathway and the true atomistic force perpendicular to the tangent. The force vector form is expressed as follows:

$$\mathbf{F_i} = \mathbf{F_i^{Spring}}|_{\parallel} - \nabla V(\mathbf{R_i})|_{\perp} \tag{9.6a}$$

where the symbols $\parallel$ and $\perp$ denote the component along, respectively, the tangential and perpendicular directions of the pathway of the images. The true atomistic force is given by the last term of (9.6a) which is the negative gradient of the potential function (see (2.1a). Here, symbol $\nabla = \frac{\partial}{\partial x}\bar{i} + \frac{\partial}{\partial y}\bar{j} + \frac{\partial}{\partial x}\bar{k}$ denotes the gradient and $\mathbf{R_i}(\bar{r}_1, \bar{r}_2 \cdots \bar{r}_N)$ denotes the set of position vectors of all atoms for image i. Obviously, the projection of the true atomistic force along the perpendicular direction to the pathway should be obtained by subtraction of the parallel force from the total force as

$$-\nabla V(\mathbf{R_i})|_{\perp} = -(\nabla V(\mathbf{R_i}) - \nabla V(\mathbf{R_i}) \bullet \hat{\tau}_i) \tag{9.6b}$$

where $\hat{\tau}_i$ is the tangent unit vector of the pathway; it can be approximated by the normalized line segment between the two images as:

$$\hat{\tau}_i = (\mathbf{R_{i+1}} - \mathbf{R_{i-1}})/|\mathbf{R_{i+1}} - \mathbf{R_{i-1}}| \tag{9.6c}$$

There are different versions of the NEB method to determine the local tangent. One way is to bisect the neighboring two unit vectors by using both $(\mathbf{R_{i+1}} - \mathbf{R_i})$ and $(\mathbf{R_i} - \mathbf{R_{i-1}})$. The other way is only to use image i and its neighbor with higher energy in the estimate.[16]

There are different kinds of NEB methods to improve the accuracy and avoid the kink of the pathway as shown in Figure 9.5, where we see that the minimum potential energy pathway does not pass the saddle point on the bottom-left corner but makes a kink in a straight line to the bottom-right energy basin. Readers interested for the detail are referred to the climbing image nudged elastic band (CINEB) method,[17] and free-end nudged elastic band (FENEB) method.[18] In the next section we will show the approach of reiteration scheme that we developed for avoiding the kink of the pathway and making the convergence of the NEB method more effective to get the MEP and the saddle point.

## 9.4.4  Finding MEP and Saddle Point for a 2D Test Problem of LEPS Potential via Implementation of the NEB Method

Here, we take the LEPS potential for a system with three atoms A, B and C along a line to exemplify the NEB method. The variable $r_{AB}$ is taken as the horizontal coordinate and $r_{BC}$ as the vertical coordinate. Different from Figure 9.4, atoms A and C are not fixed; in Figure 9.5, the range of both $r_{AB}$ and $r_{BC}$ varies from 0.0 to 4.0. The LEPS potential is given in equation (9.5b–d) with all parameters the same for Figure 9.4 except b; here b = 0.3 but for Figure 9.4 b = 0.8.

**Step 1 Determine the position of two ending points and the intermediate images**
Ending points have minimum energy, and should appear at the equilibrium position between two bonding atoms. As mentioned in 9.4.2, only one chemical bond can be formed, either between atom A and B or between B and C. In both cases, the interatomic equilibrium distance $r_e = 0.745$ which gives the minimum potential energy basin and can be taken as the ending point. It is easy to see that the coordinates of the two

**Table 9.1** Coordinates of 16 initial images

| Image | $r_{AB}$ | $r_{BC}$ |
|---|---|---|
| 0 | 0.7450 | 4.0000 |
| 1 | 0.9629 | 3.7831 |
| 2 | 1.1799 | 3.5661 |
| 3 | 1.3968 | 3.3492 |
| 4 | 1.6137 | 3.1323 |
| 5 | 1.8307 | 2.9153 |
| 6 | 2.0476 | 2.6984 |
| 7 | 2.2645 | 2.4815 |
| 8 | 2.4815 | 2.2645 |
| 9 | 2.6984 | 2.0476 |
| 10 | 2.9153 | 1.8307 |
| 11 | 3.1323 | 1.6137 |
| 12 | 3.3492 | 1.3968 |
| 13 | 3.5661 | 1.1799 |
| 14 | 3.7831 | 0.9629 |
| 15 | 4.0000 | 0.7450 |

ending points are: initial point (top-left): (0.745, 4) and ending point (bottom-right): (4, 0.745). The first one is for the bonding between atoms A and B which occurs at $r_{AB} = r_e = 0.745$, the second is for the bonding between atoms B and C which occurs at $r_{BC} = r_e = 0.745$ as shown in Figure 9.5. Their potentials are −4.51802 and −3.64840 respectively. The asymmetry is due to the different parameters chosen in the equations (9.5b–d), making the bond between A and B more stable than between B and C. Using a straight line (see Figure 9.4), connect these two ending points and make 14 intermediate images equally spaced on the line. To be specific their coordinates are listed above in Table 9.1.

**Step 2 Determination of the projections of LEPS potential force and spring force**
**(2a) Determination of the tangential direction ($\cos\theta$, $\sin\theta$) on the image pathway**
To determine the force projection for any image i (i = 1...16), it is important to determine the direction cosine or $\cos\theta$ and $\sin\theta$ of the tangential unit vector $\mathbf{e}_\parallel$ of the image pathway at image i. Figure 9.6(a)

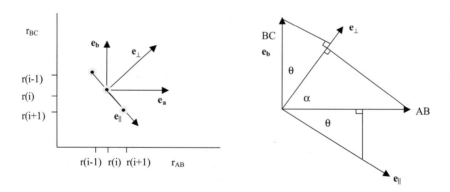

**Figure 9.6** Tangential and normal direction of the image pathway at image i

shows $e_{\parallel}$ and other unit vectors in terms of the position coordinates $r_{AB}$ (i) on the horizontal axis and $r_{BC}$ (i) on the vertical axis of the image i in relation to its neighbor image coordinates of $r_{AB}$ (i − 1), $r_{BC}$ (i − 1), $r_{AB}$ (i + 1) and $r_{AB}$ (i + 1). For a clear drawing, the subscripts "AB" on the horizontal axis and "BC" on the vertical axis are omitted. Figure 9.6(b) shows the angles $\theta$ and $\alpha$ in relation to the tangential unit vector $e_{\parallel}$ and the unit vector $e_{\perp}$ perpendicular to the pathway.

Since Figure 9.6(a) shows clearly $r_{BC}$ vs $r_{AB}$ at time steps i−1, i, and i + 1, with unit vectors parallel and perpendicular to the pathway at image i and in the AB and BC directions, the value can be easily determined below. Because the time steps are very small, we assume points i–1, i, and i+1 to be approximately in a straight line. Using trigonometry and the Pythagorean theorem, we can see that:

$$\cos\theta = \frac{r_{AB}(i+1) - r_{AB}(i-1)}{\sqrt{[r_{AB}(i+1) - r_{AB}(i-1)]^2 + [(r_{BC}(i+1) - r_{BC}(i-1)]^2}} \tag{9.7a}$$

$$\sin\theta = \cos\alpha = \sqrt{1 - \cos^2\theta} \tag{9.7b}$$

**(2b) Projection of LEPS potential force perpendicular to the image pathway**

The projection force of molecule A on B (denoted $F_{AB}$) and C on B ($F_{BC}$), derived from the LEPS potential on the perpendicular direction can be, respectively, determined as follows:

$$F_{AB\perp} = F_{AB} \times \cos(\alpha) = F_{AB} \times \sin(\theta) \tag{9.7c}$$

$$F_{BC\perp} = F_{BC} \times \cos(\theta) \tag{9.7d}$$

Summing these two together below, we get the total interatomic force, $F_{rel}$:

$$F_{REL} = F_{AB\perp} + F_{BC\perp} = F_{AB} \times \sin(\theta) + F_{BC} \times \cos(\theta) \tag{9.7e}$$

This projection force is perpendicular to the pathway and it is the relaxation force needed to draw a generic image j (j = 2...15) to the relaxation configuration of the atomistic system.

The values of $F_{AB}$ and $F_{BC}$ can be determined by the negative partial derivative of the LEPS potential with respect to, $r_{AB}$ and $r_{BC}$ (see eq.(2.1a) or (2.10c)). From (9.5c) we have

$$F_{AB} = -\frac{\partial V^{LEPS}}{\partial r_{AB}} = -\left[\frac{\partial Q^{term}}{\partial r_{AB}} - \frac{1}{2}(J^{term})^{-1/2}\frac{\partial J^{term}}{\partial r_{AB}}\right] \tag{9.7f}$$

$$F_{BC} = -\frac{\partial V^{LEPS}}{\partial r_{BC}} = -\left[\frac{\partial Q^{term}}{\partial r_{BC}} - \frac{1}{2}(J^{term})^{-1/2}\frac{\partial J^{term}}{\partial r_{BC}}\right] \tag{9.7g}$$

For the development of computer code the force perpendicular to the image pathway may be expressed with the following compact form:

$$F_{ji} = -\left(Q_{ji}^{term} - \frac{1}{2}(J^{term})^{-1/2}J_{ji}^{term}\right) \tag{9.7h}$$

where j represent AB (j = 1) and BC (j = 2); i is the image number as shown in Figure 9.6. The $Q_{ij}^{term}$ is the derivative of the $Q^{term}$ with respect to j for image i and $J_{ji}^{term}$ is the derivative of the $J^{term}$ with respect to j for image i. Using (9.5d) and (9.5e) we can determine all the terms in (9.7f) and (9.7g). Taking the calculation of $\frac{\partial J^{term}}{\partial r_{AB}}$ as an example, by using (9.5e) we have:

$$J_{1i}^{term} = \frac{\partial J^{term}}{\partial r_{AB}}\bigg|_{i=2..15} = \frac{2J_{AB}\frac{\partial J_{AB}}{\partial r_{AB}}}{(1+a)^2} + 0 + \frac{2J_{AC}\frac{\partial J_{AC}}{\partial r_{AB}}}{(1+c)^2} - \frac{\frac{\partial J_{AB}}{\partial r_{AB}}J_{BC}}{(1+a)(1+b)} - \frac{J_{BC}\frac{\partial J_{AC}}{\partial r_{AB}}}{(1+b)(1+c)} - \frac{\frac{\partial J_{AB}}{\partial r_{AB}}J_{AC} + J_{AB}\frac{\partial J_{AC}}{\partial r_{AB}}}{(1+a)(1+c)} \tag{9.7i}$$

Using (9.5g) results in

$$\frac{\partial J_{AB}}{\partial r_{AB}} = \frac{-\alpha d_{AB}}{4}\left(2\exp[-2\alpha(r_{AB}-r_0)]-6\exp[-\alpha(r_{AB}-r_0)]\right) \tag{9.7j}$$

$$\frac{\partial J_{AC}}{\partial r_{AB}} = \frac{-\alpha dC}{4}\left(2\exp[-2\alpha(r_{AC}-r_0)]-6\exp[-\alpha(r_{AC}-r_0)]\right) \tag{9.7k}$$

All the values of $r_{AB}$, $r_{AC}$, $r_{BC}$ in (9.7j, k) and in (9.5g) for $J_{AB}$, $J_{BC}$, $J_{AC}$ depend on the image i ($j = 2 \ldots 15$) involved, thus the projection force of the LEPS potential for different images can be determined.

**(2c) Projection of spring force tangential to the image pathway**
The total magnitude of these forces, acting in the parallel direction, can be derived in a similar fashion as above and is given by

$$F_{spr} = F_{sp\_AB} \times \cos(\theta) + F_{sp\_BC} \times \sin(\theta) \tag{9.7l}$$

where $F_{sp\_AB}$ and $F_{apr\_BC}$ are components of spring force along AB and the negative BC direction which can be determined by:

$$F_{sp\_AB} = k_j[r_{AB}(i+1)-r_{AB}(i)] + k_{j-1}[r_{AB}(i)-r_{AB}(i-1)] \tag{9.7m}$$

$$F_{sp\_BC} = k_j[r_{BC}(i+1)-r_{BC}(i)] + k_{j-1}[r_{BC}(i)-r_{BC}(i-1)] \tag{9.7n}$$

We may write (9.7m, n) in a compact form

$$F_{sp\_ij} = k_j[r_j(i+1)-r_j(i)] + k_{j-1}[r_j(i)-r_j(i-1)] \tag{9.7o}$$

Again j = 1 denotes $r_{AB}$ and j = 2 denotes $r_{BC}$, the difference here from (9.7h) is that there is no derivative involved and $r_{AB}$ and $r_{BC}$ of a different image should be determined by its corresponding i number.

**Step 3 Determination of position vectors with time step by Velocity Verlet algorithm**
Once the spring force and potential force at any given coordinate are determined, at each timestep the coordinates and velocities are updated to consider the coupling effects between different images. This update can be done from the coupled first-order equations of motion based on the force evaluated at the current coordinates. Specifically, Velocity Verlet algorithm described in Section 2.4.3 is used to integrate these coupling equations to find the new positions of the images. In this calculation the mass is assumed as 1. Then these new positions are used to find new forces and update positions by the obtained acceleration, different from MD, the acceleration always starts from zero for each time step, thus the MS feature is retained. A Fortran 90 code entitled LEPS_II.f90 was developed whose input file is LEPS_I.conf, readers interested can use that code to understand the formulation and use it for calculation. (Remark: the code was developed by Ross Stewart, a student of the 2010 multiscale course at Alfred University).

Specifically, the iteration is done using a do loop, namely "do rj = 1,numImages" to go from ri = 1 to the last image (ri = 16 in this case) that finds the forces at each image then stores them in an array. This example was performed with a time step of 0.001 and with 1500 steps, and a variable spring constant that increased linearly from 0.3 to 1.0 at the center.

**Step 4: Develop the reiteration scheme to solve the kink of the image pathway**
When the energy of the system changes rapidly along the pathway, but the restoring force on the images perpendicular to the path is weak, the pathway can become kinked and the convergence of the minimization energy can be difficult. This phenomenon can be seen from the straight line in the corner

of Figure 9.5 which does not pass the saddle point. Following Ref. [27] we have tried to introduce a smooth switching function $\Phi(\theta)$ that gradually turns on the perpendicular component of the spring force where the path becomes kinky

$$\phi(\theta) = \frac{1}{2}(1 + \cos(\pi\cos(\theta)))  \qquad (9.7\mathrm{p})$$

where $\theta$ is the angle between the images. $\Phi(\theta)$ is a coefficient for the perpendicular component of the spring force. When $\theta$ is close to $\pm\pi/2$, $\Phi(\theta)$ returns all of the perpendicular spring force which helps to realign the image line. When $\theta$ is closer to $\pi$ or 0 almost none of the perpendicular spring force is added.

Unfortunately, this approach does not solve the kinking problem in the example. There are other efforts to solve the kinking problem such as Ref. [16] mentioned in the last section. Here, a new method called an NEB reiteration is proposed. The difference between NEB and the NEB reiteration method, used in the code LEPS_II.f90, is that in the reiteration the elastic band or image line is prevented from relaxing fully, it is only relaxed enough for images near the two ends of the image line. The relaxation of these two regions does not stop until it becomes stable. For example a 16-image elastic band would relax until, say, images 4 and 12 are at a stable location. Here, the term "stable location" means, if images 1–4 and 12–16 are no longer moving an appreciable distance, or below a tolerance which is user defined. This distance can be found by averaging the change in distance of each image over a specified iteration step, for example every 50 steps the image's average is compared to the tolerance, if it is greater than the tolerance, it continues to iterate as in NEB. If it is less than the tolerance the image is considered to be stable.

When these images are fully stabilized the iteration process stops. At which point the reiteration begins. In the new iteration, the stable images, say, image 4 and 12 in this case, are taken for the updated end points of a new elastic band, whereby the typical NEB method is performed until the specified images are settled enough. Then another reiteration is made. The number of reiteration steps is user defined. In the code LEPS_II.f90, there is a reiteration do loop which is outside the standard NEB do loop. After each reiteration an output file is made that shows the position of each image at the iteration step where the specified images have settled. Plots of these pathways at reiteration step 3 and the last step (i.e., 9) can be found in Figures 9.7 and 9.8. The Fortran 90 code also outputs an MEP data file, consisting of the positions and energy of the stable images at the ends of the elastic band, after every reiteration. This MEP file can be used to interpolate the saddle point, see Figure 9.9 where the saddle point is at the left corner. The ensemble of image paths of all nine reiterations is shown in Figure 9.10.

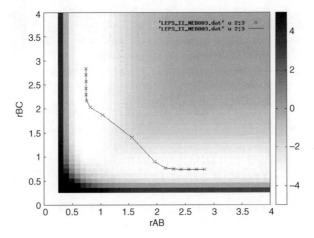

**Figure 9.7**    Reiteration step 3

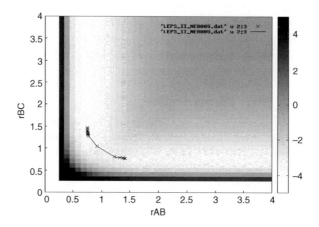

**Figure 9.8**   The last reiteration

### Step 5: Results and discussion

What this new technique provides is a way to avoid kinking in a normal NEB calculation, by taking advantage of the stable points in a normal NEB process to make a new elastic band by an updated straight line, discarding where the unsettled images were in the last iteration. The reason for discarding them is because these images tend to kink. Making a new elastic band from the previous band's stable images assures a new beginning, a shorter band, and a higher image density, all of which increases accuracy.

Recording the stable image locations and energies at every reiteration essentially maps the MEP line because the stable images are on the MEP. The image number is important for accuracy in finding the saddle point; in this example, four images on each end were allowed to settle for each reiteration, making $2 \times 4 \times 9$ images for nine reiterations, plus 16 images on the final NEB line, i.e., 88 points in total on the

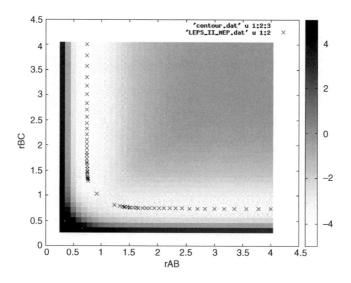

**Figure 9.9**   MEP made from stable images by reiteration method

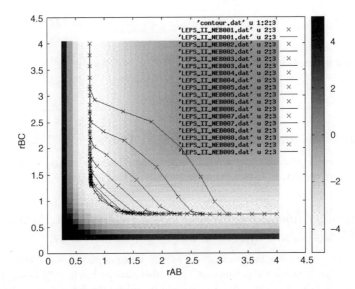

**Figure 9.10**    Image paths of all nine reiterations to show the search process for MEP and saddle points without kinking

MEP path, compared with only 16 images in the example of a typical NEB approximation. In general, three factors, namely how many images from the ends of the band are allowed to settle before reiteration, the number of reiterations, and the number of images in the NEB elastic band, determine the number of MEP points.

Since much of the code parameters are user defined, such as the amount of images allowed to settle, their tolerance and the number of iteration and reiteration steps, provides this technique with the potential to be as accurate as needed.

With this method the spring constant distribution over the elastic band is almost negligible, in fact the example above worked best when there was no change in the spring constant across the band. It was best held at 1.0 to cover as much of the saddle point as possible, as can be seen in the figures.

While the example is simple, it covers most aspects of the NEB method. A more complicated problem may involve a large amount of calculation. The computation of the potential gradient or the atomistic force for the various images of the system can be done in parallel, for example with a separate computer node in a parallel computer system handling each one of the images. Each computer node then needs to receive all atomic coordinates of adjacent images to evaluate the spring force and to carry out the force projections.

**Homework**

(9.4) Prove or disprove that the following force components determine the correct force components in (9.7e) which are perpendicular to the instant path direction. All the variables are given in the code of LEPS_II.f90.

```
Fab=-(QabTerm-0.5*(JTerm)**(-0.5)*FQMab)*(rbc(rj-1)-
rbc(rj+1))/&
      sqrt((rab(rj+1)-rab(rj-1))2.0+(rbc(rj-1)-rbc(rj+1))**2.0)
Fbc=-(QbcTerm-0.5*(JTerm)**(-0.5)*FQMbc)*(rab(rj+1)-rab(rj-
1))/&
```

```
              sqrt((rab(rj+1)-rab(rj-1))**2.0+(rbc(rj-1)-rbc(rj+1))
         **2.0)
```

(9.5) Determine the explicit expressions of spring force $F_{SP\_AB}$ and $F_{SP\_BC}$ denoted in (9.7m) and (9.7n), find the counterparts of these expressions and prove or dis-prove the consistency of your results with the code of LEPS_II.f90.

## 9.5    Applications and Impacts of NEB Methods

In this section, atomistic analysis with the NEB method is introduced to investigate the effects of strain rate and temperature on surface dislocation nucleation.[21]

### 9.5.1    Governing Equations and Methods for Considering Strain Rate and Temperature Effects on Dislocation Nucleation

Surface dislocation sources can be characterized by two quantitative measures, one is the athermal strength and the other is activation parameters. The former such as $\sigma_{ath}$ in (9.3) is the elastic limit of the surface at which dislocations nucleate instantaneously without the aid of thermal fluctuations. The activation parameters, including activation energy $Q(\sigma,T)$ and activation volume $\Omega(\sigma,T)$, characterize the probabilistic nature of dislocation nucleation by thermal fluctuation. Following (9.1) the nucleation rate at a given temperature T and stress $\sigma$ may be written as:[21]

$$v = Nv_0\exp(-Q(\sigma,T)/(k_BT))  \tag{9.8}$$

where N is the number of equivalent surface nucleation sites, $v_0$ is the atomic attempt frequency. Comparing this with (9.1), we have $f_v = Nv_0$. Taking the natural log on both sides of (9.8) gives

$$Q(\sigma,T) = -k_BT(\ln v - \ln(Nv_0))  \tag{9.9}$$

Substituting (9.9) into (9.4) we have

$$\Omega(\sigma,T) = -\frac{\partial Q(\sigma,T)}{\partial\sigma}\Big|_T = k_BT\frac{\partial\ln v}{\partial\sigma} = k_BT\frac{1}{v}\frac{\partial v}{\partial\sigma}  \tag{9.10}$$

Surface dislocation nucleation is an infrequent event, so the NEB method can be used to determine the MEP and its saddle point on the PES at Zero-T (i.e., $T=0$ K), thus the function $Q(\sigma)$ expressed by (9.3) can be determined through NEB calculations at several stress values. In turn, by (9.4) and (9.5) $Q(\sigma,T)$ and $\Omega(\sigma,T)$ can also be determined. Furthermore, one can derive the most probable nucleation stress as the load when the dislocation forms most frequently. An implicit expression was derived by Zhu et al. (see Appendix 9.A) for the nucleation stress when the nanowire is under constant temperature and strain rate.[21] It reads

$$\frac{Q(\sigma,T)}{k_BT} = \ln\left(\frac{k_BTNv_0}{E\dot{\varepsilon}\Omega(\sigma,T)}\right)  \tag{9.11}$$

### 9.5.2    Examples and Impact (1): Strain Rate and Temperature Effects on Dislocation Nucleation from Free Surface of Nanowires

An atomistic model of a copper nanowire is developed in Ref. [14] with length of $l_0 = 7.23$ nm and side length $a_0 = 5.42$ nm, the total number of atoms of the system is 18,000. EAM potential developed by

Mishin *et al.*[20] for Cu is used for the atomic interactions. Periodic boundary conditions are imposed along the axial direction and the side surfaces are free. The work applies the uniaxial load by varying the length of the simulation cell in the axial direction and allow all the atoms to freely relax.

Under a given stress and temperature, the CINEB method[17] is used to determine the MEP of dislocation nucleation by thermal activation. In CINEB calculations, an initial state is where a perfect nanowire (under a prescribed strain or stress) is fully relaxed, and a final state where an embryonic dislocation loop has formed in the wire. The activation energy Q is determined by the energy difference between the saddle point and initial states of the MEP.

The number of the equivalent nucleation sites is N = 160. This is because there are a total of 80 atoms at the four corners of the wire, and there are two equivalent slip systems at each atom site, such that the equivalent active atoms are N = 160.

Using the CINEB method for a copper nanowire with several stress values, the corresponding saddle point and activation energy are determined. By curve fitting, it is found that A = 4.8 eV, $\sigma_{ath}$ = 5.2 GPa, $\alpha$ = 4.1 in (9.3). The other parameters are determined as $v_0$ = 3.14 × $10^{11}$/s, N = 160, $T_m$ = 700 K, T = 300 K, E = 100 GPa, $k_B$ = 1.38062 × $10^{-23}$ J $K^{-1}$. Substituting these values inside (9.3), (9.4) and (9.5) results in:

$$Q_0(\sigma) = 4.8 \times \left(1 - \frac{\sigma}{5.2}\right)^4 \tag{9.12a}$$

$$Q(\sigma, T) = \left(1 - \frac{300}{700}\right) \times Q_0(\sigma) = 2.74 \times \left(1 - \frac{\sigma}{5.2}\right)^4 \tag{9.12b}$$

$$\Omega(\sigma, T) = -\frac{\partial Q(\sigma, T)}{\partial \sigma}\Big|_T = 2.1 \times \left(1 - \frac{\sigma}{5.2}\right)^3 \tag{9.12c}$$

By substituting these relationships into (9.11) we have

$$\frac{2.74 \times \left(1 - \frac{\sigma}{5.2}\right)^4 \times 1.602 \times 10^{-19} J}{1.3802 \times 10^{-23} \times 300 \times J} = \ln \frac{1.38 \times 10^{-23} \times 300 \times 160 \times 3.14 \times 10^{11} J/s}{100 \times \dot{\varepsilon} \times 2.1 \times \left(1 - \frac{\sigma}{5.2}\right)^3 \times 1.602 \times 10^{-19} \times \left(\frac{GPa}{s} \frac{J}{GPa}\right)}$$

The two sides are non-dimensional. From the above equation we can get the following approximate results. For stain rate of $\dot{\varepsilon} = 10^8$/s, $\sigma = 2.48$ GPa; for strain rate of $\dot{\varepsilon} = 10^{-3}$/s, $\sigma = 1.3$ GPa. They are consistent with the value obtained by Figure 3 in Ref. [21]. It is seen that the strain rate effects investigated by atomistic analysis can be used for strain rate as low as conducted in the laboratory.

---

**Homework**

(9.6) Given $v_0 = 3.14 \times 10^{11}$/s, N = 160, $T_m = 700$ K, E = 100 GPa, $k_B = 1.38062 \times 10^{-23}$ J $K^{-1}$, J = Nm, A = 4.8 eV, $\sigma_{ath} = 5.2$ GPa, $\alpha = 4$, determine the values of computational nucleation stress (GPa) versus the temperature at lab strain rate of $10^{-3}$ per second for the dislocation nucleation of the copper nanowires.

Hint: Use (9.2)–(9.4) and (9.11).

(9.7) Compare your stress-temperature curve obtained in homework (9.6) with the curve of Figure 3 in Ref. [21]. If the agreement is good, develop a stress-strain rate curve at temperature T = 250 K within the range of strain rate between $10^{-3}$ and $10^8$ with the following points, say $10^{-3}$, $10^{-2}$, $10^{-1}$, 1, $10^2$, $10^3$, $10^5$, $10^8$.

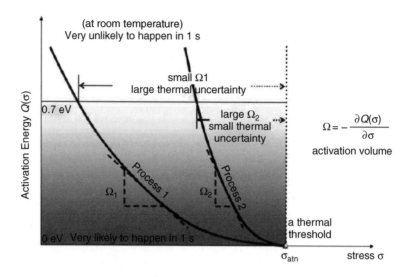

**Figure 9.11** Illustration of activation energy $Q(\sigma)$ of stress-driven activated processes (Reproduced from Li, L. (2007) The mechanics and physics of defect nucleation. *Mater. Res. Soc. Bulletin*, 32, 151. Copyright © Materials Research Society)

### 9.5.3 Examples and Impact (2): Departure Between Plasticity and Creep Based on Activation Energy and Activation Volume

As one can see from the solution process of the previous problem, the most important step is to determine the $Q_0(\sigma)$ by the NEB method to get the MEP and saddle points to determine the activation energy $Q_0$ under different stress value and 0 K, then use curve fitting to determine the three parameters A, $\alpha$ and $\sigma_{ath}$ (see (9.3) and (9.12a)). Thus, we can see the importance of the NEB method which we have discussed in the last several sections.

The parameter $\sigma_{ath}$ has clear physical meaning. It is the athermal threshold stress as when $Q(\sigma_{ath}) = 0$, in other words, it is the athermal condition corresponding to the limit where the driving force is so large that the activation barrier vanishes, so nucleation can proceed even at 0 K.

Figure 9.11 shows the curves of activation energy $Q_0(\sigma)$ versus stress $\sigma$ for stress-driven activated processes· The bottom line is with $Q = 0$, the corresponding stress at the intersection point between the curves and the bottom line is $\sigma_{ath}$ by definition. There is a horizontal line with activation energy of 0.7 eV, two curves are drawn on the figure for the two processes: one on the left with a small slope is process 1 and the one on the right with a steeper slope is process 2. Since the absolute value of the slope $-\partial Q/\partial \sigma$ of these curves denotes the activation volume $\Omega$ as shown in (9.4), the right one has much larger activation volume $\Omega$ than the left.

An activation process with larger activation volume of material may be considered more collective in motion of atoms and correspondingly there is less thermal uncertainty for the stress at which the process may happen. An example is forest dislocation cutting which usually has the volume of $\sim 10^3 b^3$ (where **b** is the Burgers vector). This large volume indicates that forest dislocation of plasticity process is a very "athermal" process, and will not happen until the applied stress $\sigma$ is high, say, 90% of $\sigma_{ath}$ (for a material with 200 MPa yield strength[25]).

On the other hand, process 1 has a small activation volume $\Omega$, so it is more thermally sensitive with a large thermal uncertainty. For instance, point defect migration of a diffusion process usually has a small activation volume $\Omega \sim (0.02-0.1) b^3$ compared to dislocation in plasticity, thus the diffusion can happen at almost any stress, but the temperature can greatly increase the diffusion rate. From the above description, it is seen that the large difference in the scale of activation volume is the physical basis for the distinction between yield and creep of the material.

### 9.5.4  Examples and Impact (3): Findings for Mechanisms of High Strength and High Ductility of Twin Nanostructured Metals

As reported by Lu and others, the introduction of coherent nano-twins in ultrafine-grained copper, typically tens of nanometers in thickness with grain size of several hundred nanometers, leads to an unusual combination of ultrahigh strength ($\sim$ 1 GPa) and high ductility (14% elongation to failure).[22–24] Zhu *et al.* developed a mechanistic framework for predicting the rate sensitivity and elucidating the origin of ductility in terms of dislocations with interfaces.[25] They used the atomistic reactions pathway calculation with NEB and found MEP, saddle points and related dislocation configuration in different conditions.

What they found is that the slip transfer reactions mediated by twin boundary are the rate-controlling mechanisms of plastic flow. Specifically, two patterns of dislocation transferring process across twin boundary are found as shown in Figure 9.12. In that figure one uses a command such as "Make Atoms Invisible" of AtomEye (see Section 10.4.3.2) to dismiss the regular crystal structure and to show only the transited local configuration such as patterns of dislocation nucleation cores, interactions between dislocation with grain boundaries, etc.

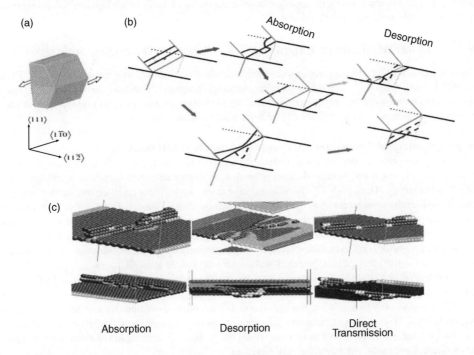

**Figure 9.12**  Atomistic modeling of interface-mediated slip transfer reactions (Reproduced with permission from Zhu, T., Li, J., Samanta, A., et al., Interfacial plasticity governs strain rate sensitivity and ductility in nanostructured metals. *Proceedings of National Academy of Sciences of the United States of America*, 104, 3031, Copyright (2007) © National Academy of Sciences, U. S. A)

**Table 9.2**  Comparison of yield stress, activation volume, and strain-rate sensitivity between experimental measurements and atomistic calculation (Reproduced with permission from Zhu, T., Li, J., Samanta, A., et al., Interfacial plasticity governs strain rate sensitivity and ductility in nanostructured metals. *Proceedings of National Academy of Sciences of the United States of America*, 104, 3031, Copyright (2007) © National Academy of Sciences, U. S. A)

|  |  | Yield stress | Activation volume $v^*$ | Strain-rate sensitivity $m$ |
|---|---|---|---|---|
| Nano-twinned copper | Uniaxial tension [LuQ4] | $\approx 1$ GPa |  |  |
|  | Nanoindentation [LuO5] | $>700$ MPa$^\dagger$ | $12$–$22b^3$ | $0.025$–$0.036$ |
|  | Atomistic calculation | $780$ MPa | $24$–$44b^3$ | $0.013$–$0.023$ |
| Diffusion-controlled processes |  |  | $\approx 0.1b^3$ | $\approx 1$ |
| Bulk forest hardening |  | $\approx \mu b\sqrt{\rho\mathrm{bulk}}$ | $100$–$1,000b^3$ | $0$–$0.005$ |

$^\dagger$ Extracted from measured hardness $H$ as $H/3$.

From the comparison shown in Table 9.2, it is seen that activation volume determined by the NEB method of atomistic simulation is $24$–$44$ $b^3$ and the test result is about $12$–$22b^3$;[23] the difference is satisfactory. It is also seen that the activation volume of the diffusion-controlled process is about $0.1$ $b^3$. The authors attribute the relatively high ductility of nano-twinned copper to the hardening of twin boundaries as they gradually lose coherency during plastic deformation. It is clear that these findings provide insights into the possible means of optimizing strength and ductility through interfacial regions to improve, say, the twin boundary structure.

## 9.5.5   Other Methods in Extending Time Scale in Atomistic Analysis

Recently, several other effective approaches have been developed to simulate infrequent events. These include hyperdynamics,[26, 27] temperature accelerating dynamics[28] and on-the-fly kMC.[29] These approaches provide a special way to accelerate the transition process near extreme points with minimum energy on the potential surface. They can be briefly introduced as follows.

### Hyperdynamics and Temperature Accelerated Dynamics (TAD) method

In the hyperdynamics method, the potential surface $V(\bar{r})$ is modified by adding a non-negative bias potential $\Delta V(\bar{r})$, which will increase the energy level of the energy basin, and so reduce the energy barrier or activation energy. Thus, an NVT classical trajectory can then be propagated on this surface with fast rate. It assumes there are no correlated events as required by the TST theory. There is also a requirement for the bias potential to be zero at all the dividing surfaces at which the bias potential does not change the energy level. In addition, the system must still obey TST for dynamics on the biased potential. Under these conditions, a trajectory on this modified surface evolves correctly from state to state at an accelerated rate. This is because the positive bias potential within the well lowers the effective barrier. Note that the evolution from state to state is correct because the bias potential does not change the relative TST rates for different escape paths from a given state.

TAD is another way to accelerate the simulation process but is more approximate than the hyperdynamics method. It relies on the harmonic TST approximation which gives an Arrhenius dependence of the rate on temperature as shown in (9.1) in which the product kT appears in the denominator, thus if temperature T increases, the transition rate increases.

### The dimer method

This method is an effective method to find the saddle points. The dimer is made up of two images of the system. These images are spatially separated by a finite distance displacement along a given vector direction. For an empirical potential, it can be small. For instance, for Al(100) it is 0.005 angstrom.[29]

Dimer can have rotation and translation. Minimizing the energy of the dimer with respect to the rotational force aligns the dimer to the right orientation. The latter corresponds to the so-called lowest curvature mode; the rotational force is the difference in the force on the two images. A first-order saddle point on a potential surface is at a maximum along the lowest curvature direction and a minimum in all other directions. Thus, to converge to a saddle point, the dimer should move up the potential along the lowest curvature mode and reduce the potential in all other directions. This is done by defining an effective force on the dimer in which the true force due to the potential acting at the center of the dimer has the component along the dimer inverted.

Minimizing with respect to this effective force moves the dimer to a saddle point. The dimer is rotated once or a few times until the rotational force is less than the specified tolerance. In the above-mentioned Al(100) calculations the maximum rotational force was set as 5 meV per angstrom. For the Al adatom on an Al(100) surface, the dimer requires on average 400 force evaluations to converge on a saddle point. Dimer searches starts from configurations close to the potential minimum. The easiest way to choose a starting position is to make a random displacement of atoms within a local region away from the minimum. This indicates that the random displacement is not for the whole atomic system but is only limited to local atoms, otherwise, the search can be biased towards finding higher energy processes. For each dimer search in the Al(100) ripening calculations,[29] an undercoordinated surface atom was chosen at random to be the center of the initial, local displacement. The displacements had a Gaussian distribution with a mean of 0.2 angstrom in each degree of freedom. The region consisted of the central atom and its first and second neighbors.

### On-the-fly kinetic Monte Carlo method

In the standard kMC method, there is no classical trajectory determined by integration of Newton's second motion law. Instead, it is determined from a list of possible transition events, one escape path is chosen randomly and the system is advanced to a new state. The time is then incremented in a way that is consistent with the average time for escape from that state (see Section 2.9.4).

The problem with this type of approach is the approximation of the list of possible escape paths, the other is the assumption that the possible escape mechanisms can be guessed in advance. It is pointed out[26] that highly concerted events can be common and important to the dynamical evolution of the system, which makes the guess sometimes questionable. A variation on kMC is proposed by combining dimer and kMC, called on-the-fly kinetic Monte Carlo method.[29] Its key is the dimer method which is essential to find the saddle points. Primary studies shows that this type of search of saddle point is so sufficient that one can start searches from many different randomly-placed configurations. The goal of the search is to find all the low-lying saddles surrounding the state, thus building an evolution path based on the MC method. In this way, there is little limitation in terms of the numbers of atoms or the spatial extent of the transition. In any state which the system evolves, the energy barriers can be found and do not need to be known before the calculation. Also, the atoms do not need to be mapped onto a lattice.

The above acceleration method and others are without sacrifice of the particular information in the atomistic scale. For example, the band acceleration approach enables use of a physical event time scale $10^6$ times larger than that used in ordinary MD simulations. The theoretical base and examples can be found in the references.[30–33]

Recently, Passerone and Parrinello proposed an interesting approach to determine the real dynamics trajectory of some non-frequent events whose initial and final configurations are known.[34, 35] They used least action theory to optimize the path and named the approach action-derived molecular dynamics (ADMA). It controls evolution using the Hamiltonian together with a penalty function in a given condition.[36]

# Part 9.3 Multiscale Analysis of Protein Materials and Medical Implant Problems

## 9.6 Multiscale Analysis of Protein Materials

### 9.6.1 Hierarchical Structure of Protein Materials

The three most abundant biological macromolecules are proteins, nucleic acids, and polysaccharides. They are all polymers composed of multiple covalently linked or nearly identical small molecules, called monomers. For a protein, the monomer is an amino acid that consists of N, H, C, and O atoms, in sequence as {-NH-CαHR-CO-}. In the brackets, the N-terminal with a positive charge is the beginning of the monomer, and the C-terminal with a negative charge is the end of the terminal. The symbol R denotes a residue which is a side chain of the amino acid. Proteins are essential building blocks of life which form a diverse group of biological material and play an important role in providing key biological functions.[37] These materials, ranging from spider silk to bone and from tendons to the skin, are different in concept from both the conventional materials and their structures, as they represent the merger of these two concepts. This merger is through hierarchical formation of structural elements that range from the nanoscale to the macroscale. Protein's hierarchical structure results in many new deformation mechanisms, whose understanding and analysis are still only beginning.

#### 9.6.1.1 Protein as Bio-structural Material

Figure 9.13 shows two examples of hierarchical multiscale structures of protein materials. The structure shown in (a) is the hierarchical structure of lamin in which α-helical protein domains assemble into dimers, which form filaments that define a lattice-like lamin network of the cell's nucleus. The one shown in (b) is a hierarchical structure of collagenous tissues from nanoscale, to mesoscale (μm), to macroscale (cm); tropocollagen triple helix assembles to form fibrils which form fibres that provide the structural basis of collagen tissues such as tendons.

The cell's hierarchical structure is also notable. The cytoskeleton of eukaryotic cells is composed of three main components, intermediate filament, globular protein tubulin and microfilament. At higher

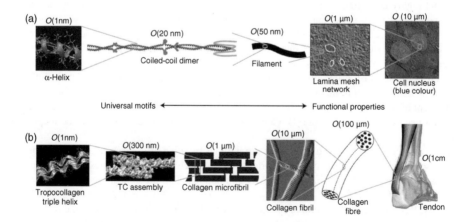

**Figure 9.13** Examples of hierarchical multiscale structures of biological protein materials (Reproduced from Buehler, M. J. and Ackbarow, T. (2007) Fracture mechanics of protein materials. *Materials Today*, 10 (9), 46, Elsevier)

scale, the structure is composed of a network of three-dimensional intermediate filaments in the cells. Its object is to strengthen the plasma membrane composed of lipid and protein. The plasma membrane wraps the cell, and thus separates it from outside environment. The cytoskeleton is responsible for shape changes in cell division, motility and phagocytosis. Besides that, it also provides a place for cell adhesion and interaction among different parts of the cell.

Intermediate filament (IF) plays a dominant role in cell structure (see Figure 9.16). Based on this understanding, its hierarchical structure and dynamic behavior will be investigated in this section, and its role in collaboration with cell membrane and nucleus membrane in cell deformation and adhesion to other materials such as implants will be discussed in the next section.

It is reported that full length filaments are about 240 nm long and consist of unit-length filaments of about 60 nm long.[38] The latter consists of eight tetramers, each of which is formed with two dimers. More detail on the hierarchical structure of the network of IF in cells will be found in the reference above. The dimer shown in Figure 9.13(a) is IF's elementary block composed of the α-helical protein. Protein is the product of polymerizing a large quantity of amino acid,[39] and will be introduced in Section 10.10.1.1.

Protein shows different functions according to the configuration of different amino acids. One of the main functions is to manage and adjust molecules, and thus enable it to control various actions of the life system. Another main function is as construction material for the biological system, bearing external force and loading induced by changes of temperature and environment. The loading condition of long fibrils is in tension. In the case of large deformation, each dimer is also subjected to tensile loading.

## 9.6.2 Large Deformation and Dynamic Characteristics of Protein Material

IF takes on different functions in different hierarchical scales. At mesoscopic scale, its particular structure enables the network of vimentin IF to act like the safety-belt in cars to protect cells. Because the flexible network of vimentin IF makes the structure very soft in small deformation and low extending velocity, under these particular conditions there is almost no resistance and change of shape during the movement of cells. However, the structure is very rigid in large deformation and high velocity, thus assuring its function at the cell scale. It is very interesting that the mechanical properties of vimentin IF are greatly affected by diseases. It has been found that degeneration of integrality and good mechanical property of vimentin IF relates closely to deterioration or mutation of intermediate filament protein.[40, 41]

When a cell is enduring large deformation, each vimentin IF undertakes tensile deformation which is distributed to a single dimer in molecular scale. Figure 9.14 illustrates the molecular geometry of the spiral helix dimer, the basic element of vimentin IF; the dimer is about 45 nm long. It comprises a head on the left end, a tail on the right end and an elongated rod-shape region. The rod-shape region is composed of four

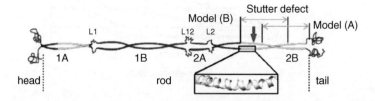

**Figure 9.14** Geometry of the molecular building block of vimentin. Model (A) is without stutter defect but Model (B) is with stutter defect as shown inside the rectangle (Reproduced from Buehler, M. J. and Ackbarow, T. (2007) Fracture mechanics of protein materials. *Materials Today*, 10 (9), 46, Elsevier)

α-helix (1A, 1B, 2A and 2B), connected by joints L1, L12 and L2. When subjected to external loading, the helix spring is going to unwind, and the hydrogen bond of amino acid is going to break. This deformation and failure process relate to whether there are defects in the dimer. A stutter defect as shown in Figure 9.14 undertakes less loading than a normal formation due to its softer nature, thus more loading is transferred to other parts of the α-helix and joint to produce debonding and fracture.

The above-mentioned functions of vimentin IF make it especially significant to study the mechanical property of the dimer when it is in small and large tensile deformation. The reason is that the structure and property in this small scale play an important role in the investigation of function and mechanism of the vimentin IF and its network in large scale. In other words, the property in large scale is determined by the property of the dimer which represents the smallest structural scale. It can be seen that the property of a biomaterial also obeys the basic concept emphasized in section 1.1, i.e., material properties are based on the atomic and microscopic structure. From the following description, this concept can be understood more comprehensively.

## 9.6.3   At Molecular (Nano) Scale: Molecular Dynamics Simulation of Dimer and the Modified Bell Theorem

### 9.6.3.1 MD Simulation for Tensile Behavior of Dimers

MD is appropriate for dimers because their structure is in the nanoscale. MD simulation for biomaterials is described in detail in Sections 10.10–12 and can be conducted with computer simulation code and examples in UNIT10–11 and UNIT12 of CSLI. Specifically, Sections 10.11 and 10.12 are for nonequilibrium MD simulation under external loading. Figure 9.15 illustrates loading curves of tensile force versus engineering strain of spiral helix vimentin dimer measured in different elongating velocity.[41] Among the curves, the ones with small fluctuation are obtained from MD. This figure provides the relation between tensile force and engineering strain in different elongating velocity for the spiral helix dimer in IF. It is seen from Figure 9.15 that the curve may be divided into three stages. The first stage shows elastic tensile strain which may be as high as 10% to 25%. The second stage is a flat region which shows that the unwinding process of the helix dimer is carried out in a constant load, while the last stage shows significant strain hardening. In this stage, the double helical structure has disappeared and the

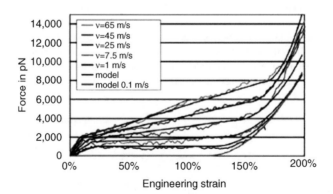

**Figure 9.15**   The curves of tensile force versus engineering strain of spiral helix vimentin dimer measured in different elongating velocity. In the figure, the solid curves are obtained by Bell theorem, the curves with small fluctuation are obtained by MD (Reproduced from Buehler, M. J. and Ackbarow, T. (2007) Fracture mechanics of protein materials. *Materials Today*, 10 (9), 46, Elsevier)

main chain of protein sequence structure has been stretched. These structural changes result in significant increase in stiffness.

### 9.6.3.2 Extended Bell Statistical Theory

In Figure 9.15, the solid curves show the response of a dimer in different elongating velocity obtained by extended Bell theorem. Statistical concept is absolutely necessary in describing the mechanical properties of protein structure, and Bell theorem is the model for describing the statistical properties of bond breaking.[42–44] In the extended Bell theory, the relations between the dimer unwinding force $F$ and the speed $v$ of bond dissociation are given as follows:

$$F(v) = a\ln v + b \tag{9.12}$$

$$a = k_b T / (x_b \cos\theta) \tag{9.13}$$

$$b = -k_b T \ln v_0 / (x_b \cos\theta) \tag{9.14}$$

$$v_0 = \omega_0 x_b \exp\left(-\frac{E_b}{k_b T}\right) \tag{9.15}$$

where $k_b$ is Boltzmann constant, $T$ is temperature (K), $\theta$ is the angle between applied force $F$ and the propagation direction of bond dissociation, $x_b$ is the distance from the position where an atom is in static balance state to the critical position where bond breaks, $\omega_0 \approx 1\times10^{13} s^{-1}$ is the natural frequency of bond vibration,[42] and $E_b$ is the energy barrier for bond breaking.

Under the same stretching mechanism where $E_b$ and $x_b$ are considered to be fixed, it can be seen from (9.1) to (9.4) that the extended Bell theory would predict that a larger angle $\theta$ corresponds to a stiffer dimer protein. Figure 9.13 illustrates a defective dimer. The defect has a large $\theta$ angle and is different from dislocation defects in crystal structure. It is generated by inserting one or two or four additional residues to the periodic structure composed of seven residues. These inserted irregular residues make the dimer softer. Therefore, the defective segment is easy to break. Also, the position of additional residues usually lies in the initial position of unwinding. It may be explained as follows.

According to the extended Bell theorem, the stretching force $F$ is smaller in the region where additional residues exist than in the other normal structural regions of protein. The angle is $\theta_{cc} = 23° \pm 10.2°$ in the spiral helix region of normal vimentin protein. However, the angle is $\theta_{st} = 16° \pm 8.5°$ in the stutter region where four additional residues exist. Because a larger angle $\theta$ induces a relatively smaller component of force in the longitudinal direction of the dimer, larger force is required for the normal vimentin protein to be stretched. When defects of additional residues appear, the angle $\theta$ is reduced, resulting in smaller resistance to the applied force.

From (9.1) and (9.2) we have

$$F_{st}(v) = \frac{\cos\theta_{cc}}{\cos\theta_{st}} F_{cc}(v) \tag{9.16}$$

where the subscript cc denotes normal spiral helix, while st represents defect region with stutter. Because $\theta_{cc} > \theta_{st}$ and $\cos\theta_{cc} < \cos\theta_{st}$, $F_{st}(v) < F_{cc}(v)$. This result has already been verified by MD simulation. It tells us that the defect region with stutter is a weak region. This defect region is the first to stretch straightly when subjected to increasing tensile loading. However, so far there is still no convincing theory to explain why defects like stutter occur and to understand their significance in biology, mechanics and physics. One thought is that the defect region with stutter provides a prescribed controllable region. It may take the lead in stretching straightly when subjected to large deformation.

## 9.6.4    Unique Features of Deformation, Failure and Multiscale Analysis of Biomaterials with Hierarchical Structure

The structure of biological protein is totally different from crystal materials, thus a number of deformation mechanisms are found which are different from those for crystals. With respect to crystal materials, the deformation mechanism is mainly described as formation and development of dislocation in ductile materials, and cleavage damage and crack nucleation/propagation in brittle materials. However, with respect to biological protein materials with hierarchical structure, the predominant deformation mechanism is stretching and slipping. In the stretching process, ruptures of bond connecting hydrogen and nitrogen atoms play a very important role by resulting in imperfect structure. This special deformation mechanism is determined by structural geometry and the arrangement of protein segments which span scales from nano to mesoscopic.

The main features of biological protein materials are their hierarchical structure and the widely-existing weak interaction among molecular structures.[39, 45, 46] Due to the particular hierarchical structure of biomaterials, the advantage of weak interaction between biological molecule and submolecule may be brought into full application. And in larger scales, it could even be more prominent. The advantage of using weak interaction is that the materials may be fabricated at room temperature with low energy consumption. The weak interaction also plays an important role in material control because these weak hydrogen bonds are capable of controlling the deformation and function of material even after the structure's larger scale is composed.

Another difference between biological protein material and traditional material is the geometrical position where defect occurs. Defects are usually distributed randomly in crystal materials. However, for biomaterials which are composed of ordered structures from nano to mesoscopic scale, one may predict the location where defect occurs with high precision in atomic or molecular scale. This feature plays a predominant role in observing material properties in meso scale. As illustrated above for defects of stutter in vimentin IF, the softening points introduced by three residues provide the location where protein starts to stretch, which is instructive for this type of analysis of material deformation and failure mechanisms.

Based on the viewpoint of hierarchical multiscale analysis, the structure and properties of biological protein materials in a smaller scale influence the structure and properties of complex hierarchical structures in a larger scale, and therefore affect the behavior and characteristics of the whole biological system. This is to say that the concept of multiscale analysis, that atomistic and microscopic structures determine macroscopic properties, is also applicable for biomaterials. For crystal materials, there is a size effect whereby reducing grain size increases the strength of metal. The same mechanism also works in biomaterials. For example, changing the molecule length of tropocollagen in collagen filament would directly control the deformation mechanism and material strength.[47–48]

There is a significant difference between the analysis of biomaterial and traditional materials, i.e., it is not applicable for biomaterials to use the approach which derives relative information in a larger scale by averaging relative variables in a smaller scale. This approach calculates effective material constants by averaging variables related to microstructures. It is applied in many multiscale approaches, but may fail when applied to biomaterials. One of the reasons is that it is not capable of taking into account enough relative elements in the lower scale. Another reason is that the averaging may result in loss of some critical information at a lower scale that can be important to the material properties at larger scales.

It can be seen from the above discussion that there are many new deformation mechanisms in biomaterials. It is a great challenge to understand these mechanisms. It is required to know how variables at different scales are coupled and how the phenomena related with molecular process influence material behavior and properties in a larger scale. People may benefit from understanding these multiscale and multi-physics problems in designing and fabricating new materials with complex structure, using the viewpoint of integrating from microscopic to macroscopic scale. Some questions of long standing in biology and biomedicine could also be solved, so that multiscale analysis can provide an attractive approach for multidisciplinary and multiscale investigation.

## 9.7 Multiscale Analysis of Medical Implants

Multiscale analysis of biocompatibility and design of medical implants such as hip joint implants which interact with bones are significant. This challenging topic is chosen for illustration of the features of multiscale analysis in biomaterials in which cell deformation behavior plays a central role. Thus it is necessary to have substantial knowledge of cell molecular biology, and readers are directed to Ref. [48]. In this section we will first introduce the background of this problem and some thoughts on methods for analysis at different scales, then mention some complexities of the problem.

### 9.7.1 Background

As longevity increases in human populations, degenerative diseases are a prevalent and critical issue. As organs, joints and other critical body parts degenerate, modern medical science offers transplant and/or implant solutions. Despite the success of surgical implants such as artificial hip and knee joints, the materials used in these procedures still do not satisfy the demands of a durable functioning joint. Current synthetic materials, such as stainless steel, titanium alloys, polymers and ceramic composites experience degradation after 10 to 15 years of use in the human body. Longer lives of patients may lead to several replacement surgeries in their lifetimes.[49]

The common failures of joints are severe. The primary cause of failure is implant loosening by mechanical failure, corrosion and poor biocompatibility between the materials and the body. While the wear resistance and corrosion of implant materials have been improved, evaluating biocompatibility and quantifying the adhesive strength of new materials remains a challenging task.

The cell's adhesive strength is the major concern in the development of implant materials. High cell density, strong adhesion and cell growth are pivotal factors to providing sufficient lubrication and self-renewal capacities, and the functional qualities of biocompatibility. Fundamental understanding of adhesion of bio-cells to the implants is critical to the development of new materials and toward reliable implants. Improving cell adhesion involves the choice of material chemical composition at the atom/nano/submicron scale, cell deformation and failure mechanism to resist external action at the mesoscopic scale, and the design of implant surface morphology at the macroscopic scale. In the following subsections, the characteristics and relevant methods of analysis at these multiscales will be introduced.

There are a serious of efforts in developing microstructure-based models of cells focused on cell mechanics, cell adhesion and cell migration. This includes a paper using a micropipette to attach and then detach a red blood cell on a flat substrate mediated by receptor-ligand binding,[50] and a multiscale modeling effort for a complex cellular mechanism that involves key mechanical and biochemical events at multiple scales.[51]

### 9.7.2 At Atom-nano and Submicron scale: Selection of Implant Chemical Composition Based on Maximum Bonding Energy

From the discussion in Section 1.1 it is seen that material property depends on the atomistic and microscopic structures. The same concept is applicable to the selection of chemical composition for implant composite material. In other words, among various factors which affect the adhesion strength between bio-cells and the implant, the chemical compositions and their related atomistic and microscopic structure is dominantly important.

#### 9.7.2.1 Basic Assumption for Selection of Implant Compositions

It is important for adhesion enhancement to choose chemical composition of implants with high adhesion energy between proteins (or bio-cells) and the implant. Thus, it is reasonable to assume the following criterion for preliminary selection of implant composition: Among all possible compositions of implant materials, the one with the largest bonding energy between proteins and the implant is the best candidate for the implant.

### 9.7.2.2 Method

Using the schemes and NAMD software described in detail in Sections 10.10–12 and their corresponding CSLI lab units for simulation of protein-water systems, investigate two arrangements of the same protein-implant system, A. One is in the normal adhesion condition with water molecules in between. The other separates implant and protein-water by a large distance. Both arrangements may use idealized settings of a protein-water system with periodic conditions in the implant plane along the x- and y-direction and non-periodic conditions in the z-direction perpendicular to the plane. The energy difference of these two settings approximately denotes their bonding energy $E_A$. Repeat this process by changing the composition of implant material but keeping all others the same to get $E_B$, $E_C$, $E_D$; the one with the maximum binding energy is the best candidate for the implant material.

### 9.7.2.3 Rationale of Using Idealized Protein-Water System

It may be questioned why proteins not cells are used in the idealized setting for the composition selection. In fact, it is reported that soluble proteins in body fluids adsorb rapidly to the surface of an implant material – more rapidly than and prior to the adhesion of living cells to the surface, because living cells are much larger than proteins and thus move slowly.[52] Living cells do not typically contact the molecular structure of the material surface. Instead, the adsorbed ECM protein layers through a cell-membrane receptor interact with the ECM ligand leading to adhesion.

In reality, protein adherence is a complicated process. Proteins initially adsorb to random areas of the surface, possibly in reversible fashion, then reorient and spread out on the surface to become irreversibly adsorbed; the process continues until most of the available surface is covered with irreversibly adsorbed proteins. But while tracing the detailed moving process is a complicated one, for the purpose of pre-selection of composition we can use the idealized protein-water system setting because the comparison is based on the relevant bonding energy value under the fixed constant condition.

## 9.7.3 At Mesoscopic Scale (μm): Cell Adhesion Strength is Calculated and Characterized

At this stage, cell deformation and failure in terms of their interaction with an implant should be investigated and cell adhesion strength under the external shear force may need to be calculated and characterized. Cell deformation and interaction with an implant are controlled by its cytoskeleton, because the cytoskeleton provides cells with structure and shape. Therefore, a mesoscopic cell model considering the cytoskeleton structure is the key to investigating cell deformation and its interaction with an implant. In the next section, we will introduce the mesoscopic cell model proposed by Bertaud *et al.*[53] Based on this model, a consideration to use that model for investigation of the interaction between cells and implant will be discussed.

### 9.7.3.1 Coarse-grained Mesoscopic Cell Model

Coarse-grained models in the sense of reductionist models have been around for a long time. Voth and his colleagues have shown for the first time how to rigorously construct such models bottom-up from the underlying potential energy function and statistical mechanics.[54–55] Specifically, a new approach is presented for obtaining coarse-grained (CG) force fields from fully atomistic MD trajectories. The coarse-graining of the interparticle force field is accomplished by an application of a force-matching procedure to the force data obtained from an explicit atomistic MD simulation of the biomolecular system of interest. The theoretical foundations were then rigorously proved and the full methodology defined.[56–58] Recently Li *et al.* investigated the nature of the coarse equation through microscopic simulations.[59]

In the following, a mesoscopic cell model based on a coarse-grained bead which satisfies the above requirements will be introduced. This kind of bead is similar to the generalized particle introduced in Chapter 5 in the sense that it represents a group of atoms. The bead, however, is different from the generalized particle because it does not keep the same material structure as the latter does. In fact, the introduction of a bead is related to the introduction of a different bead-spring elementary model whose parameters are determined by fitting the model's behavior with results of geometrical analyses and experimental and atomistic simulations; thus it is an empirical model rather than a model based on a rigorous hierarchical multiscale approach. It can, however, be used to identify generic structure-property relationships of the mechanical roles for a broader range of cell types.

## A. Bead-spring chain model for IF and membranes

As introduced in Section 9.2.1, cytoskeleton of eukaryotic cells is composed of three main components, intermediate filaments, globular protein tubulin and microfilament. Among these, it has been suggested recently that IF plays a crucial role in providing mechanical stability to cells. Based on this, a mesoscale model is proposed which offers the route to investigate the deformation of cells and its interaction with implants. This is a simple empirical coarse-grained computational model of the IF network in eukaryotic cells. This model is shown on the right of Figure 9.16A, it consists of nuclear membrane and cell

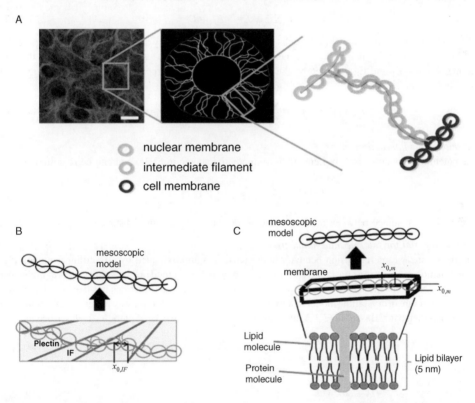

**Figure 9.16** A coarse-grained cell model. (A) Schematic of the cell model geometry inspired by experimental pictures. (B) Schematic of the 1D bead-spring chain model of IF. (C) Schematic of the bead-spring chain model of membranes (Reproduced from Bertaud, J., Qin, Z., and Buehler, M. J. (2010) Intermediate filament-deficient cells are mechanically softer at large deformation: A multi-scale simulation study. *Acta Biomaterialia*, 6, 2457, Elsevier)

membrane on the two extremes, and IF in between to connect the two membranes. This model is inspired by experimental pictures shown on the left of Figure 9.16A for the geometry of Madin-Darby canine kidney (MDCK) epithelial cells, in which IF extend from the nucleus to the cell membrane.[60]

IF is a hierarchical arrangement of alpha-helical proteins that has a one-dimensional shape, thus it can easily be described by a bead chain with $x_{0.IF}$ as its bead equilibrium distance (Figure 9.16B). Membranes are lipid bilayers with embedded membrane-associated proteins as shown in the lower part of Figure 9.16C. In the mesoscale model, a membrane is also represented by a one-dimensional bead chain, as shown in the top of Figure 9.16C. To reduce the approximation, the equilibrium bead distance $x_{0.IM}$ also determines the width of the membrane that is represented by one bead as shown in the middle figure of Figure 9.16C.

## B. Potential energy for the coarse-grained method

All beads in the cell mesoscale model interact according to a specific intermolecular multibody potential. The potential is developed to reflect the key physical properties of IF and membranes, including stretching and bending. Under this framework, MS and MD software can be used for the calculation of the coarse-grained method.

The mathematical expression for the total energy of the system is given by[53]

$$E(X') = E_T + E_B + E_I \tag{9.17}$$

where $X'$ denotes the positions of all beads, $E_T$ the total tensile energy and $E_B$ the total bending energy. $E_I$ is the total intermolecular interaction energy which is introduced to avoid IF bead penetration through the membranes. $E_T$ and $E_I$ are given by the sum of all pairwise interactions, and $E_B$ by the sum of all three-body interactions, i.e.,

$$E_T = \sum_{pairs} \phi_T(x), \quad E_I = \sum_{pairs} \phi_I(x), \quad E_B = \sum_{triples} \phi_B(\theta) \tag{9.18a, b, c}$$

*Intermolecular interaction energy*

The potential $\phi_i$ is described by Lennard-Jones potential shown in (2.3a) and can be rewritten as:

$$\phi_I(x) = 4\varepsilon \left[ \left( \frac{\sigma}{x} \right)^{12} - \left( \frac{\sigma}{x} \right)^{6} \right] \quad x < x_c \tag{9.19}$$

where x is the distance between the mesoscale beads and $x_c$ is the cutoff distance.

*Energy potential for intermediate filaments*

The IF nonlinear force-extension behavior under tension is approximated by a multilinear model. It is a combination of six spring constants $K_T^{(i)}$ ($i = 1, \ldots, 6$) which are turned on at specific values of molecular stretch.[53] A similar model with simple piecewise harmonic potential functions has been used in earlier studies of fracture in crystalline model materials[61] and provides an effective way to describe the nonlinear constitutive behavior. Based on (2.1c), the tensile force between two-bead particles is described as:

$$F_T(x) = -\frac{\partial \phi_T(x)}{\partial x} \tag{9.20a}$$

where

$$\frac{\partial \phi_T}{\partial x}(x) = H(x_{break} - x)$$

$$\times \begin{cases} K_T^{(1)}(x - x_0) & x < x_1 \\ R_2 + K_T^{(2)}(x - x_1) & x_1 \leq x \leq x_2 \\ R_3 + K_T^{(31)}(x - x_2) + K_T^{(32)}(x - x_2)^2 + K_T^{(33)}(x - x_2)^2 & x_2 \leq x \leq x_3 \\ R_4 & x \geq x_3 \end{cases} \tag{9.20b}$$

In this equation, $H(x_{break}-x)$ is the Heaviside function $H(a)$ which is defined as 0 for $a < 0$ and 1 for $a \geq 0$. This function makes the calculation effective only before the break bond point (i.e., $x \leq x_{break}$); in fact, after the break point the force transformation is discontinued and $H = 0$ to make all the forces zero. Here, $x_0$ is the equilibrium point, which is $x_{0,IF}$ and $x_{0,M}$, respectively for IF and membrane in Figure 9.16B and C.

The bending energy of a triplet of three-bead particles is given by

$$\phi_B(\theta) = \frac{1}{2} K_B(\theta - \theta_0)^2 \tag{9.21}$$

where $K_B$ is the bending stiffness parameter relating to the IF bending stiffness EI through $K_B = 3EI/x_0$. The IF bending stiffness EI is related to the IF persistence length $L_P$ through $EI = L_p k_B T$, where $k_B$ is the Boltzmann constant and T denotes the temperature.[37]

*Energy potential for cell membrane and nuclear membrane*
Modeling a 2D membrane as a 1D bead chain is more approximate. At least, we need to adjust the 1D spring constant $K_T$ to match the area expansion modulus $K\varepsilon$ so that both can produce the same potential energy. For the 1D case, the energy is

$$\phi_T(x) = \frac{1}{2} K_T(x - x_0)^2 \tag{9.22a}$$

The tensile energy of the membrane in the 2D equal tension is given by

$$\phi_T(\Delta A) = \frac{1}{2} K_\varepsilon \left(\frac{\Delta A}{A_0}\right)^2 A_0 = \frac{1}{2} K_\varepsilon (2\varepsilon)^2 x_0^2 \tag{9.22b}$$

In deriving the last equation the relations $\Delta A/A_0 = \varepsilon 1 + \varepsilon 2 = 2\varepsilon$ and $A_0 = (x_0)^2$ are used. Denoting $d = x - x_0$ as the elongation of $x_0$ segment we have $d = \varepsilon x_0$. Substituting these two relations, respectively, into (9.10a) and (9.10b) and then making the energy of $\phi_T(x) = \phi_T(\Delta A)$, the required stiffness relationship is given as follows:

$$K_T = 4K\varepsilon \tag{9.22c}$$

The bending energy of a triplet of three-bead particles of the mesoscale model is also given by

$$\phi_B(\theta) = \frac{1}{2} K_B(\theta - \theta_0)^2 \tag{9.22d}$$

where $K_B$ relating to the bending stiffness $k_b$ of the membrane through $K_B = 4\pi k_b$. This relation is derived in the following by identifying the bending energy of a three-bead mesoscale model to the bending energy of a round membrane with radius $x_0$ during a bending experiment for small deformation.[53] The bending energy of the membrane with an initial area $A_0 = \pi(x_0)^2$ is given by

$$\phi_B(\theta) = \frac{1}{2} K_b \left(\frac{2}{R}\right)^2 A_0 \tag{9.22e}$$

Where 1/R is the curvature caused by the deflection d and the two principal curvatures are identical. For small deflections, $1/R = 2d/(x_0)^2$ and $\theta - \theta_0 = 2d/x_0$, substituting these two equations, respectively, into (9.22e) and (9.22d) then equating the bending energy expressed by (9.10d, e), the following relationship is established:

$$K_B = 4\pi K_b \tag{9.22f}$$

### C. Model parameter identification for intermediate filaments
In Ref. [53] the equilibrium bead distance $x_0 = 0.2492$ μm per bead is chosen, which provides significant computational speedup while maintaining a sufficiently fine discretization for IF. The latter is rational

because $x_0$ is much lower than the IF persistence length, which is about 1 μm. It leads to a bead particle mass m = 8,723,077 amu. The five parameters in (9.20b) are then fitted to the force strain curve behavior of stretching a single IF obtained using experimental and computational results.[12, 31, 32] The fitting result is given in Table 1 of Ref. [53] which includes numerical values of critical distances $x_1$, $x_2$, $x_3$; tensile stiffness parameters in linear region $K_T^{(1)}$, $K_T^{(2)}$; tensile stiffness parameters in nonlinear region $K_T^{(31)}$, $K_T^{(32)}$, $K_T^{(33)}$; tensile stiffness parameters $R_2$, $R_3$, $R_4$; bond breaking distance $x_{break}$; equilibrium angle $\theta_0$, bending stiffness parameter $K_B$ and mass of each mesoscale particle.

*Cell membrane*
The equilibrium bead distance $x_0 = 0.1571$ μm per bead is chosen, which provides significant computational speedup while maintaining a sufficient fine discretization for the membrane. This value, combined with data on the membrane mass composition,[53] leads to mass m = 171,880,665 amu.

It is noted that the thickness of a cell or nucleus membrane is approximately 5 nm. Typical values of membrane extension stiffness $K_\varepsilon$ determined by experiments are in the range of 0.1 to 1 Nm⁻¹ for various types of lipid bilayer. For instance, $K_\varepsilon$ is about 0.45 Nm⁻¹ for red blood cell (RBC) membranes.[53] Based on the average value and (9.22c), $K_T = 1.44$ kcal mol⁻¹Å⁻² can be determined. Of the other two parameters, one is the equilibrium angle $\theta_0 = 180.00°$ which corresponds to a relaxed membrane structure with no curvature. The other is bending stiffness parameter $K_B$ which can be determined by $K_b$ based on (9.22f). Typical $K_b$ values lie in 10⁻¹⁹ Nm for RBC or lipid bilayers and (1–2) × 10⁻¹⁸ Nm for other cell types (e.g., neutrophils, endothelial cells) that possess a more extensive cytoskeleton network. In the work, $K_b$ is taken as 2 × 10⁻¹⁸ Nm which produces $K_B = 288$ kcal mol⁻¹rad⁻².

*Nuclear membrane*
The same method is used as for cell membrane. The result is given as follows:[57]

| | |
|---|---|
| Equilibrium bead distance $x_0$ (in Å) | 1571 |
| Tensile stiffness parameter $K_T$ (in kcal mol⁻¹Å⁻²) | 1.44 |
| Equilibrium angle $\theta_0$ (in degrees) | 180.00 |
| Bending stiffness parameter $K_B$ (in kcal mol⁻¹rad⁻²) | 288 |
| Mass of each mesoscale particle m (in amu) | 227,312,178 |

It is seen that the nuclear membrane is heavier than the cell membrane because its membrane mass composition data is different from cell membrane.

## D. Parameters of model geometry
Cell model geometry used in Ref. [53] is an approximation of the geometry of a eukaryotic cell which is shown in the top-left part of Figure 9.17 in which IF extends from the nucleus to the membrane. The

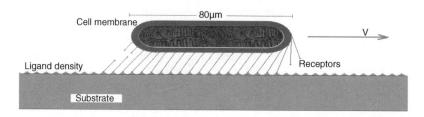

**Figure 9.17**   Using coarse-grained mesoscale filament model to connect the membrane in the cell receptors with substrate ECM ligands

proposed model represents the main geometrical feature. The cell is designed with the size of circular cell membrane of diameter 10 μm and that of nuclear membrane of diameter 4.6 μm. The design peripherally distributes radial coarse-grained IF that are attached from one extremity to the nucleus and from the other one to the membrane. The tensile bonds between IF extremities and membranes are chosen to be the same as the intra-IF tensile bonds. There are no angular bonds defined for these junctions. Each IF is made up of 12 beads. All IF are given a slightly slack behavior after equilibration by setting the initial interbead distances to a smaller value than the bead equilibrium distance.

### E. Simulation process and the results
The right part of Figure 9.16 depicts a schematic of the cell geometry after equilibrium. This system has 40 IF and 1000 particles in total. All simulations are carried out at 300 K in an NVT ensemble. The equilibrium process takes 4 μs under constraining longitudinal motions along the vertical direction of the top and bottom extremities. Then the cell model was pulled at a pulling velocity of 0.0001 Å ps$^{-1}$ (0.01 m per second), and used to measure the force-strain curve. Since each bead of the cell mesoscale model represents a large group of atoms, their large masses enable the choice of a relatively large time step, thus a time step of 10,000 fs is chosen. This step is below the characteristic time constant $\tau = (m/k)^{0.5}$. It is reported that the simulation typically runs for less than an hour on a multi-CPU Linux workstation, providing a rapid approach for testing the cell model mechanical behavior under varying IF densities. The results of the simulation seem encouraging to catch the main characteristics of the cells.[53]

### 9.7.3.2 Using the Mesoscale Cell Model for Analysis of Interactions Between Cells and Implants

There are some models for investigating the interaction between the cells of implants such as the focal adhesion complex model.[62] These models are frequently oversimplified because they do not consider the interactions between implant and cell cytoskeleton structure in detail. The coarse-grained mesoscale cell model brings new hope since it can consider the interaction between membranes and IF. We may extend the model to investigate the interactions as follows.

It is realized that the interaction between cells and implant usually does not imply a direct contact between the cell membrane and the implant. The first effect that occurs post-implantation is the deposition or adsorption of proteins onto the surface of the implanted device, usually known as biofouling. This will depend on the surface properties of the device, with hydrophobic surfaces being particularly affected. As a result, the adsorbed extracellular matrix (ECM) protein layers through a cell-membrane receptor interact with the ECM ligand leading to adhesion. In other words, "The adhesive interaction is mediated by cell adhesion receptors binding specifically with their counter-receptors or ligands to form noncovalent bonds."[52]

There are phenomenological models to simulate the receptor-ligand binding such as the work using micropipette.[50] Here, we would recommend accounting for the biomaterial nature of both ligand and receptors in the simulation. A ligand is a group of atoms or molecules which can combine with other materials, while a receptor is molecular material which may be located on cell membrane or enzyme. Thus, it may be reasonable to simulate the receptor-ligand binding through IF-type elements to link with cell membrane and the membrane in the bio-layer attached to the implant.

Related to the bond mechanism, two important issues have not found a good solution for a long time; one is how to simulate the binding filaments which connect the receptors and ligands, the other is how the filaments connect the membranes. These two issues are important because both the constitutive behavior of filament and membrane and the connection condition between filament and membranes affect strongly the mechanical behavior of tissues and the resistance for cell migration. Since the mesoscopic cell model offers a relatively good solution to these issues, a step toward the understanding of the cell-implant interaction may be to use the above coarse-grained bead-spring mesoscopic model to solve these two difficult problems.

Figure 9.17 shows a schematic of this thought. In this figure, the coarse-grained cell model is adopted to simulate the constitutive behavior of tension, bending of nuclear membrane, cell membrane and the IF. Furthermore, the IF mesoscale model with six spring constants is also proposed to investigate the nonlinear behavior of the filaments linking the receptors in the cells and the ligands in the substrates as shown in Figure 9.17.

More specifically and approximately, the filaments in linking the cell receptors of cells and ligands could also be modeled by the six-constant model of (9.20b). Their two ends link to the membranes in cells and the bio-layer on the substrates. The Lennard-Jones potential can also be used to avoid the filament bead penetration through the membranes in the receptors and the ligands. While the setting may not necessarily be accurate to meet the realistic condition, it can be used to investigate the effects of the roughness, morphology versus the cell size and property on the shearing strength and bonding energy because the obtained data for the comparison is under the same approximate conditions.

## 9.7.4  Discussion

It can be seen from foregoing that multiscale analysis is inevitable if one hopes to design biocompatible implants and break through the life limits of implanted material. Without analysis at atomistic and nano-submicron scale, it is difficult to determine the important effects of implant composition on surface adherence between proteins and the implant. Without mesoscopic analysis based on coarse-grained cell model and cell-plant interaction model, it is difficult to have in-depth understanding of the cell adhesion strength and cell resident condition in terms of the surface morphology, roughness and waviness, and to develop guidance for the design of implants. Without macroscopic analysis it is difficult to design the implant to match multiple requirements related to the bone and surrounding environments. Due to the complexity of the implant work condition, several points are noted here.

(1) Multiscale analysis should combine with experimental study to verify the result. In the microscopic and mesoscopic scale, the direct measurement using an atomic force microscope (AFM) is essential. The AFM probe will be functionalized by molecules that are active in cells. The outcome of this approach will be the in-depth understanding of the nature of adhesion forces, such as van der Waals, ionic, or otherwise. This knowledge is critical to determine the chemical and physical properties of building blocks for the biomaterials. At the meso- and macroscopic scale, the effects of surface morphology on adhesions of multiple cells can be investigated by a rotating rheometer which can measure the shear strength and then make comparison with the predictions of multiscale analysis.

(2) A very interesting phenomenon of cell residency on implants is the so-called contact guidance which indicates that cell motion and orientation is guided by the topography of the substrate they are in contact with.[63–64] This contact guidance of cells, although proven to occur readily, is very complicated and has different factors that affect it. It may be a good area where multiscale analysis can contributen to explaining the behavior of cells.

(3) The implant process is a very complicated biological process that we should pay close attention to. Damage to vascularized tissue at the point of implant leads to the release of materials and cells that form what is known as a provisional matrix around the device. This begins the process of inflammation and wound healing. The initial response of the immune system leads to an accumulation of immune cells, which will attach to the surface of the implant. In the acute inflammatory stage the mobile white blood cells or leukocytes known as neutrophils will move to the foreign body, which is typically too large for them to attack through the normal process of phagocytosis. Instead they may produce products that will try to degrade the implant. As time goes on, larger immune cells known as monocytes and lymphocytes will migrate to the implant where they may divide into macrophages which will again attempt to engulf the implant. If they are unable to do so they may release signaling chemicals or cytokines, which will encourage the recruitment of more macrophages. These can then

fuse to form giant foreign body cells with multiple nuclei. If the implant is too large for these cells to engulf by phagocytosis they can remain on the surface and produce enzymes and other products in an attempt to break down the device. At the same time as the immune cell response there will also be a wound healing response characterized by the proliferation of fibroblasts and vascular endothelial cells. In the first few days after implantation they will begin to produce granulation tissue and cause neovascularization/angiogenesis. New capillaries or blood vessels will be formed and the fibroblasts may begin to produce fibrous tissue that will eventually form a capsule around the implant. This is known as fibrosis or fibrous encapsulation and is the usual end point in the body's response to an implant.

## 9.8 Concluding Remarks

Chapter 6 emphasizes that long-term multiscale method has a vital role in the resolution of the discrepancy between theory and experiment for Al twinning. Since it is a time-scale problem, we would like to briefly introduce how the long-term simulation in Reference [65] was done. The concurrent multiscale method was used to minimize the amount of atoms without affecting the physical results and to reduce the time in the computation. The loading was along the [111] direction with the crack penetrating along the [11-2] direction. This crack setting is the most favorable for twin emission under mode I loading. The simulations were all held at 300 K and the dimensions of the model were $0.2 \times 0.2 \times 0.002 \ \mu m^3$. The atomistic region with 15141 atoms was $84 \times 147 \times 20 \ \mathring{A}^3$ around which were 8,282 finite element nodes in a plane strain linear elastic region. Loading was applied by displacements to the perimeter of the model with the mode I stress intensity factor, $K_I$.

To run the long-time simulation, the parallel replica method[27] was used. This method is the simplest and most accurate method within the catalog of accelerated dynamics techniques of infrequent events (see Sections 9.3 and 9.5.5). The only assumption is that the infrequent events obey first-order kinetics (exponential decay), that is, the probability distribution function $p(t)$ for the time of the next escape from the energy basin is given by Equation (2.33a) or rewritten as

$$p(t) = re^{-rt}$$

where $r$ is the constant rate for escape. The equation arises as a natural example for the ergodic, chaotic exploration of an energy basin. The method is clearly illustrated in Figure 4 of Reference [32]. In practice[65], the method runs many nominally identical simulations on independent CPUs. When the first nucleation event of interest occurred on any processor, all simulations were stopped and their individual simulation time was summed up to be the total time of the simulation. In other words, "the time of the nucleation event was considered to be the sum of the simulation time on all the processors."[65] On the basis of this technique of long-time multiscale simulation ($t > 10$ ns), it is found that "as the load decreases, instances of trailing dislocation nucleation become dominant at the lowest loads and long times $t > 10$ ns" while "twin formation dominates at high loads $k_I > 0.185$ eV $\mathring{A}^{-2.5}$ and very short times $t < 100$ ps." A conclusion for the transition from twinning to full dislocation emission was obtained to resolve the mentioned discrepancy of theory and experiment.

## Appendix 9.A Derivation of Governing Equation (9.11) for Implicit Relationship of Stress, Strain Rate, Temperature in Terms of Activation Energy and Activation Volume

Defining f as the survival probability of initially elastic nanowire, the supplement materials of Ref. [14] proposed the following rate equation:

$$\frac{df(t)}{dt} = -vf(t) \tag{9A.1}$$

where $v$ is the nucleation rate of defects given by (9.8). This proposed equation may be explained as follows. At the beginning time ($t=0$), the survival probability may be taken as one, thus the survival probability will be reduced with time, thus the negative is used in (9A.1). The rate is assumed to be proportional to $v$ because the life reduces quickly if the defect nucleation rate is high. On the other hand the survival rate is assumed to be proportional to existing life f which indicates a high life reduction rate at the beginning when f is high and low reduction rate when f is small, a phenomenon observed in many failure processes. Integration of (9A.1) gives $f = f(0)\exp(-vt)$ which indicates the large decaying probability reduction at the beginning and almost zero reduction when the existing probability is very low.

Because the load is applied by constant strain rate, one has $\sigma = E\dot{\varepsilon}t, dt = d\sigma/(E\dot{\varepsilon})$, so

$$\frac{df(\sigma)}{d\sigma} = -\frac{v(\sigma)}{E\dot{\varepsilon}}f(\sigma) \tag{9A.2}$$

and

$$\frac{d^2f(\sigma)}{d\sigma^2} = \frac{d\left(\dfrac{-v(\sigma)}{E\dot{\varepsilon}}f(\sigma)\right)}{d\sigma} = -\frac{v(\sigma)}{E\dot{\varepsilon}}\frac{df(\sigma)}{d\sigma} - \frac{f(\sigma)}{E\dot{\varepsilon}}\frac{\partial v(\sigma)}{\partial \sigma} \tag{9A.3}$$

Substituting (9A.2) into (9A.3), we obtain

$$\frac{d^2f(\sigma)}{d\sigma^2} = \left[\frac{v(\sigma)}{E\dot{\varepsilon}}\right]^2 f(\sigma) - \frac{f(\sigma)}{E\dot{\varepsilon}}\frac{\partial v(\sigma)}{\partial \sigma} \tag{9A.4}$$

The most probable nucleation stress is defined by the peak of $df(\sigma)/d\sigma$. This requires

$$\frac{d^2f}{d\sigma^2} = 0 \tag{9A.5}$$

Substituting (9A.4) into (9A.4), one obtains:

$$v = \frac{E\dot{\varepsilon}}{v}\frac{\partial v(\sigma)}{\partial \sigma} \tag{9A.6}$$

Combining this equation with (9.8), one obtains:

$$\frac{E\dot{\varepsilon}}{v}\frac{\partial v(\sigma)}{\partial \sigma} = Nv_0 \exp\left(-\frac{Q(\sigma,T)}{k_BT}\right) \tag{9A.7}$$

$$\frac{-Q(\sigma,T)}{k_BT} = \ln\left(\frac{E\dot{\varepsilon}}{Nv_0 v}\frac{\partial v(\sigma)}{\partial \sigma}\right) \tag{9A.8}$$

Combining this equation with (9.10) results in:

$$\frac{Q(\sigma,T)}{k_BT} = -\ln\left(\frac{E\dot{\varepsilon}}{Nv_0 k_BT}\Omega(\sigma,T)\right) \tag{9A.9}$$

$$\frac{Q(\sigma,T)}{k_BT} = \ln\left(\frac{k_BTNv_0}{E\dot{\varepsilon}\Omega(\sigma,T)}\right) \tag{9.11}$$

# References

[1] McDowell, D., Panchal, J., Choi, H.-J., *et al.* (2009) *Integrated Design of Multiscale, Multifunctional Materials and Products*, Butterworth-Heinemann.

[2] McDowell, D. L. and Olson, G. B. (2008) Concurrent design of hierarchical materials and structures. *Scientific Modeling and Simulation*, **15**, 207.

[3] Olson, G. B. (1997) Computational design of hierarchically structured materials. *Science*, **277**(5330), 1237.

[4] McDowell, D. L., Choi, H., Panchal, J., *et al. (2007) Plasticity-related microstructure-property relations for materials design, Key Engineering Materials*, **340-341**, 21.

[5] Deng, L., Trepat, X., Butler, J. P., *et al. (2006) Fast and slow dynamics of the cytoskeleton. Nature Materials*, **5**(8), 636.

[6] Anahid, M., Chakraborty, P., Joseph, D. S., and Ghosh, S. (2009) Wavelet decomposed dual-time scale crystal plasticity FE model for analyzing cyclic deformation induced crack nucleation in polycrystals. *Modelling Simul. Mater. Sci. Eng.*, **17**, 064009.

[7] Jiang, H., Zhang, Q., Chen, X., *et al. (2007) Three types of Portevin-Le Chatelier effects: Experiment and modeling. Acta Mater.*, **55**(7), 2219.

[8] Fressengeas, C., Beaudoin, A. J., Lebyodkin, M., *et al. (2005) Dynamic strain aging: A coupled dislocation–Solute dynamic model. Materials Science and Engineering A*, **400-401**, 226.

[9] Jacobsen, J., Cooper, B. H., and Sethna, J. P. (1998) Simulations of energetic deposition: From picoseconds to seconds. *Phys. Rev. B*, **58**, 15847.

[10] Lou, Y. and Christofides, P. D. (2003) Estimation and control of surface roughness in thin film growth using kinetic Monte-Carlo models. *Chemical Engineering Sciences*, **58**, 3115.

[11] Bai, Y. L., Wang, H. Y., Xia, M. F., *et al. (2005) Statistical mesomechanics of solid, linking coupled multiple space and time scales. Appl. Mech. Rev.*, **58**(6), 372.

[12] Wagner, G. J. and Liu, W. K. (2003) Coupling and continuum simulations using a bridging scale decomposition. *J. Comput. Phys.*, **190**, 249.

[13] Voter, A. F., Montalenti, F., and Germann, T. C. (2002) Extending time scale in atomistic simulation of materials. *Ann. Rev. Mater. Res.*, **32**, 321.

[14] Zhu, T., Li, J., Samanta, A., *et al. (2008) Temperature and strain-rate dependence of surface dislocation nucleation. Phys. Rev. Lett.*, **100**, 025502.

[15] Jonsson, H., Mills, G., and Jacobsen, K. W. (1998) Nudged band method for finding minimum energy paths of transitions, in *Classical and Quantum Dynamics in Condensed Phase Simulations* (eds. B. J. Berne, G. Ciccotti, and D.F. Coker), World Scientific, Singapore, 385.

[16] Henkelman, G. and Jonsson, H. (2000) Improved tangent estimate in the nudged elastic band method for finding minimum energy paths and saddle points. *J. Chem. Phys.*, **113**, 9978.

[17] Henkelman, G., Uberuaga, B. P., and Jonsson, H. (2000) A climbing image nudged elastic band method for finding saddle points and minimum energy paths. *J. Chem. Phys.*, **113**, 9901.

[18] Zhu, T., Li, J., Samanta, A., *et al. (2007) Interfacial plasticity governs strain rate sensitivity and ductility in nanostructured metals. Proc. Nat. Acad. Sci. USA*, **104**, 3031.

[19] http://www.itap.physik.uni-stuttgart.de/~imd/download/imd-20010-10-06.tgz

[20] Mishin, Y., Mehl, H. J., Papaconstantopoulos, D. A., *et al. (2001) Structural stability and lattice defects in copper: Ab initio, tight-binding, and embedded-atom calculations. Phys. Rev. B*, **63**, 225106.

[21] Zhu, T., Li, J., Samanta, A., *et al. (2007) Interfacial plasticity governs strain rate sensitivity and ductility in nanostructured metals. Proc. Nat. Acad. Sci. USA*, **104**, 3031.

[22] Lu, L., Shen, Y. F., Chen, X. H., *et al. (2004) Ultrahigh strength and high electrical conductivity in copper. Science*, **304**, 422.

[23] Ma, E., Wang, Y. F., Lu, Q. H., *et al. (2004) Strain hardening and large tensile elongation in ultrahigh-strength nano-twinned copper. Appl. Phys. Lett.*, **85**, 4932.

[24] Dao, M., Lu, L., Shen, Y., and Suresh, S. (2006) Strength, strain-rate sensitivity and ductility of copper with nanoscale twins. *Acta Mater.*, **54**, 5421.

[25] Li, L. (2007) The mechanics and physics of defect nucleation. *Mater. Res. Soc. Bulletin*, **32**, 151.

[26] Voter, A. F. (1997) Hyperdynamics: Accelerated molecular dynamics of infrequent events. *Phys. Rev. Lett.*, **78**(20), 3908.

[27] Voter, A. F. (1998) Parallel replica method for dynamics of infrequent events. *Phys. Rev. B*, **57**, R13985.

[28] Sorensen, M. R. and Voter, A. F. (2000) Temperature-accelerated dynamics for simulation of infrequent events. *J. Chem. Phys.*, **112**, 9599.

[29] Henkelman, G. and Jonsson, H. (2001) Long time scale kinetic Monte Carlo simulations without lattice approximation and predefined event table. *J. Chem. Phys.*, **115**, 9657.

[30] Jonsson, H., Mills, G., and Jacobsen, K. W. (1998) Nudged elastic band method for finding minimum paths of transitions, in *Classical and Quantum Dynamics in Condensed Phase Simulations* (eds. B. J. Berne, G. Ciccotti, and D._F. Coker), World Scientific.

[31] Miron, R. A. and Fichthorn, K. A. (2003) Accelerated molecular dynamics with the bond-boost method. *J. Chem. Phys.*, **119**, 6210.

[32] Voter, A. F., Montalenti, F., and Germann, T. C. (2002) Extending the time scale in atomistic simulation of materials. *Ann. Rev. Mater. Res.*, **32**, 321.

[33] Jacobsen, J., Cooper, B. H., and Sethna, J. P. (1998) Simulations of energetic beam deposition: From picoseconds to seconds. *Phys. Rev. B*, **58**(23), 15847.

[34] Passerone, D. and Parrinello, M. (2001) Action-derived molecular dynamics in the study of rare events. *Phys. Rev. Lett.*, **87**(10), 108302.

[35] Passerone, D. and Parrinello, M. (2003) A concerted variational strategy for investigating rare events. *J. Comput. Phys.*, **118**(5), 2025.

[36] Lee, I. H., Kim, S. Y., and Jun, S. (2004) An introductory overview of action-derived molecular dynamics for multiple time-scale simulations. *Computer Methods in Applied Mechanics and Engineering*, **193**(17–20), 1633.

[37] Buehler, M. J. and Yung, Y. C. (2009) Deformation and failure of protein materials in physiologically extreme conditions and disease. *Nature Materials*, **8**, 175.

[38] Buehler, M. J. and Ackbarow, T. (2007) Fracture mechanics of protein materials. *Materials Today*, **10**(9), 46.

[39] Lodesh, H., Berk, A., Matsudaira, P., *et al. (2004) Molecular Cell Biology*, 5th edn., W. H. Freeman and Company, New York.

[40] Kiss, B., Karsai, A., and Kellermayer, M. S. Z. (2006) Nanomechanical properties of desmin intermediate filaments. *Journal of Structural Biology*, **155**(2), 327.

[41] Omary, M. B., Coulombe, P. A., and McLean, W. H. I. (2004) Intermediate filament proteins and their associated diseases. *New England Journal of Medicine*, **351**(20), 327.

[42] Bell, G. J. (1978) Models for the specific adhesion of cells to cells. *Science*, **200**(4342), 618.

[43] Evans, E. and Ritchie, K. (1997) Dynamic strength of molecular adhesion bonds. *Biophys. J.*, **72**(4), 1541.

[44] Dudko, O. K., Hummer, G., and Szabo, A. (2006) Intrinsic rates and activation free energies from single-molecule pulling experiments. *Phys. Rev. Lett.*, **96**, 108101.

[45] Lakes, R. (1993) Materials with structural hierarchy. *Nature*, **361**(6412), 512.

[46] Ahl, V. and Allen, T. F. H. (1996) *Hierarchy Theory – A Vision, Vocabulary, and Epistemology*, Columbia University Press, New York.

[47] Buehler, M. J. (2008) Nanomechanics of collagen fibrils under varying cross-link densities: Atomistic and continuum studies. *Journal of the Mechanical Behavior of Biomedical Materials*, **1**(1), 59.

[48] Alberts, B., Johnson, A., Lewis, J., *et al. (2002) Molecular Biology of the Cell*, 4th edn., Garland Science.

[49] Steven, K. M. (2004) *The UHMWPE Handbook: Ultra-High Molecular Weight Polyethylene in Total Joint Replacement*, Elsevier Academic Press, San Diego.

[50] Cheng, Q. H., Liu, P., Gao, H. J., and Zhang, Y. W. (2009) A computational modeling for micropipette-manipulated cell detachment from a substrate mediated by receptor-ligand binding. *J. Mech. Phys. Solids*, **57**, 205.

[51] Rangarajan, R. and Zaman, M. H. (2008) Modeling cell migration in 3D: Status and challenges. *Cell Adhesion & Migration*, **2**, 106.

[52] Latour, Robert A., Jr., (2005) Biomaterials: protein-surface interaction, in *Encyclopedia of Biomaterials and Biomedical Engineering*, Taylor & Francis.

[53] Bertaud, J., Qin, Z., and Buehler, M. J. (2010) Intermediate filament-deficient cells are mechanically softer at large deformation: A multi-scale simulation study. *Acta Biomaterialia*, **6**, 2457.

[54] Izvekov, S. and Voth, G. A. (2005) A multiscale coarse-grain method for biomolecular systems. *J. Phys. Chem. B*, **109**, 2469.

[55] Izvekov, S. I. and Voth, G. A. (2005) Multiscale coarse graining of liquid-state system. *J. Chem. Phys.*, **123**, 134105.

[56] Noid, W. G., Chu, J.-W., Ayton, G. S., and Voth, G. A. (2007) Multiscale coarse-graining and structural correlations: Connections to liquid state theory. *J. Phys. Chem. B*, **111**, 4116.

[57] Noid, W. G., Chu, J.-W., Ayton, G.-S., *et al. (2008) The multiscale coarse-graining method I. A rigorous bridge between atomistic and coarse-grained models. J. Chem. Phys.*, **128**, 244114.

[58] Noid, W. G., Chu, J.-W., Ayton, G.-S., *et al. (2008) The multiscale coarse-graining method II. Numerical implementation for coarse-grained molecular models. J. Chem. Phys.*, **128**, 244115.

[59] Li, J., Kevrekidis, P. G., Gear, C. W., and Kevrekidis, I. G. (2007) Deciding the nature of the coarse equation through microscopic simulations: The baby-bathwater scheme. *SIAM Review*, **49**(3), 469.

[60] Alon, R., Hammer, D. A., and Springer, T. A. (1995). Lifetime of the P-selectin-carbohydrate bond and its response to tensile force in hydrodynamic flow. *Nature*, **374**, 539.

[61] Kreplak, L. and Fudge, D. (2007) Biomechanical properties of intermediate filaments: From tissues to single filaments and back. *Bioessays*, **29**, 26.

[62] Liu, W. K., Karpov, E. G., and Park, H. S. (2006) *Nano Mechanics and Materials*, John Wiley & Sons, Ltd.

[63] Curtis, A. S. G. and Clark, P. (1990) The effects of topographic and mechanical properties of materials on cell behavior. *Crit. Rev. Biocomp.*, **5**, 343.

[64] Harrison, R. G. (1911) On the stereotropism of embryonic cells. *Science*, **34**, 279.

[65] Warner, D. H., Curtin, W. A., and Qu, S., (2007) Rate dependence of crack-tip processes predicts twinning trends in f.c.c. metals. *Nat. Mater.*, **6**, 876.

# 10

# Simulation Schemes, Softwares, Lab Practice and Applications

The focus of this chapter is on study methods, skills and applications of atomistic simulation. Accompanying the chapter is a Computational Simulation Laboratory Infrastructure (CSLI) which can be downloaded from http://multiscale.alfred.edu and used for lab practice. CSLI contains 14 UNITS which correspond one-to-one with the 14 sections of this chapter. The chapter is in three parts.

Part 10.1 (Sections 10.1 to 10.6) introduces the basics of computational simulations. It starts with an introduction to UNIX, analysis of a simple MD code, and basic knowledge of Fortran 90. It introduces GULP software used for optimization of atomistic static lattices, calculation for defects and determination of potential parameters. It also introduces the visualization tools of Gnuplot, VMD and AtomEye as well as a public MD software DL_POLY.

Part 10.2 (Sections 10.7 to 10.9) covers simulation applications in metals and ceramics by MD.

Part 10.3 (Sections 10.10 to 10.14) introduces the NAMD software that can be used for simulation of a protein-water system. A brief introduction to the LAMMPS MD package and the GP multiscale methods is also given.

Each UNIT in CSLI contains files for computer simulation and the corresponding section describes the simulation processes operated in that UNIT. This chapter is written in a style which can be used both in a laboratory class of computational simulation and by self-learning readers at different levels.

## Part 10.1 Basics of Computer Simulations

## 10.1 Basic Knowledge of UNIX System and Shell Commands

Due to its stability, flexibility and wide range of applications, the UNIX operating system will be used as the basic system in this chapter. Readers are advised to read "UNIX Tutorial for Beginners" on the website http://info.ee.surrey.ac.uk/Teaching/Unix/. The following sections will provide a brief introduction to the UNIX system based on that website's documentation.

### 10.1.1 UNIX Operating System

The UNIX operating system is made up of three parts: the kernel, the shell and the programs. The kernel of UNIX is the hub of the operating system. It allocates time and memory to programs, as well as handles

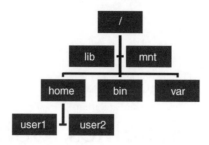

**Figure 10.1**  A UNIX directory tree at upper levels (Directories under the UNITS directory will be shown in several sections)

the storage of files and communications in response to system calls. The shell acts as an interface between the user and the kernel. When a user logs in, the login program checks the username and password and then starts the shell. The shell is a command line interpreter (CLI), which interprets the commands the user types in and arranges for them to be carried out as programs. When the process has finished running, the shell then returns the UNIX prompt $ to the user, indicating that it is waiting for further program commands (in some UNIX versions the prompt symbol is %, not $).

By typing part of the name of a command then pressing the [Tab] key, the shell will complete the rest of the name automatically. If the shell finds more than one name beginning with the typed letters, it will beep in some machines, prompting for more letters to be typed before pressing the [Tab] key again.

The shell keeps a list of the commands that have been typed in. If a command is repeated after it has been typed previously, use the cursor keys to scroll up and down to find the desired command or type "history" for a list of previous commands.

All the files are grouped together with a structure shown by a directory tree as follows. The top of the hierarchy is traditionally called the root (written as a slash /), with the lower levels under the upper level also separated by a slash/. A typical GNU/Linux directory tree is shown in Figure 10.1. Below user1 and user2, there are several levels of directories (not shown) such as user1/UNITS/UNIT2. The details of directories below each UNIT are different between different sections. Figures 10.5, 10.6, 10.10 and 10.13 show various details of structure within the UNITS directory.

## 10.1.2   UNIX Shell Commands

Several basic shell commands are introduced in this section. When first logging in, the working directory will be the home directory named after the username selected. This home directory will be where all personal files and directories are saved. Note that after each computation required by a command is completed, a new command line should appear, prompting for the next command.

### 10.1.2.1  Common Commands

**ls:** Use the "ls" command to list the files within the current directory.

```
~$ ls
```

Note that the symbol "~" before the prompt symbol $ denotes the home directory. After the symbol "~" may appear the path to the current directory, for instance:

```
~UNITS/UNIT2/MD_CODE $ls
Crystal_M_simple.f90 multi1.f90 Model_M_simple.in mt1.in MD_CODE_out
```

The listed files are in the current directory MD_CODE which belongs to the UNIT2 directory of the computer directory tree; that UNIT is associated with the text content in Section 10.2. The first two files in the MD_CODE directory are Fortran 90 codes developed, respectively, for creating a simulation model and carrying on MD simulation; the third and fourth files are their corresponding input data files. The fifth file "MD_CODE_out" is a directory. All directories ending "_out" are directories with input, processing and output files during the simulation process, which are listed for readers to learn the simulation skills.

**cd:** Use "cd" to change directories. For example, to change from the home directory to the directory UNITS, type "$cd UNITS". The open directory will then be UNITS, shown as the command line ~/$UNITS to designate that the open UNITS directory is within the home directory. Once the new directory is open, the files within the new directory can be listed by typing "ls", which should appear as:

```
~/UNITS$ ls
```

in the command prompt screen. This command will show the following subdirectories:

```
UNIT1  UNIT2  UNIT3  UNIT4  UNIT5  UNIT6  UNIT7  UNIT8  UNIT9  UNIT10_11
UNIT12, UNIT13, UNIT14, APPENDIX, DL_POLY_2.18 and NAMD_2.6
```

As an exercise, continue to use the "cd" and "ls" commands to change between directories to review the files within each directory.

**mkdir:** Use "mkdir" to make a new directory. For example, two directories called "HOMEWORK" and "unixstuff" can be created inside the shell directory ~/UNITS/UNIT1. Type "mkdir HOMEWORK" and "mkdir unixstuff" after the prompt symbol, i.e.,

```
~/UNITS/UNIT1$ mkdir HOMEWORK
~/UNITS/UNIT1$ mkdir unixstuff
```

Note that, while in the shell, the path of the current directory before the prompt symbol is shown; for simplicity in the remainder of this chapter, the above command lines will be written as:

```
$ mkdir HOMEWORK
$ mkdir unixstuff
```

**Directories (.) and (..):** There are two special directories called (.) and (..) in all directories of the UNIX system. To verify, type "$ ls -a", and the directory listing will include all files and subdirectories within the current directory as well as the directories (.) and (..). In UNIX, a single dot (.) means the current directory and a double dot (..) indicates the "parent" of the current directory. Typing "cd ." with a space between the cd and the dot, is the command to remain inside the current directory. This may not seem very useful at first, but using (.) as the name of the current directory will save a lot of typing, as shown later in the tutorial. Also try typing "$ cd .." which will move up one level above the current directory in the hierarchy. Furthermore, by typing "cd" with no other text, the current directory will return immediately to the home directory, which is very useful.

**vi:** Only a brief overview of "vi" is given here for creating and editing files. An in-depth view of its extended features and reference information can be found in Appendix 10.C. Typing the command "vi xxxxx" in shell will open an empty file with the name "xxxxx". Then by typing the letter "i", the word "insert" will be shown at the bottom of the screen, indicating insert mode. Under this mode you can type and edit text files. After completing the typing and editing, hit ESC to exit the insert mode. Typing "w" will save the file, "q" will exit the file, and "wq" will save and then exit the file in one step. The "xxxxx" file with the saved content will be in the directory that was open at the time of using the "vi" command.

**cp:** The first function of "cp" is to copy files. The command "cp file1 file2" creates the file2 as an exact copy of file1 in the current working directory. The "cp" command can be used to copy files across directories. To show the usage of cp at different directory levels, take a file stored in the UNIT2 directory, and use the cp command to copy it to the UNIT2/unixstuff directory. First, create a text file in the ~/UNITS/UNIT1 directory by typing:

```
~/UNITS/UNIT1$ vi example.txt
```

The prompt will disappear and a clean shell will open. Type "wq" and press Enter. Once a new prompt command line is ready, type "ls -lrt" to verify that the example.txt file has been created. If the file was created correctly, it should be the last file on the list, as it was created last; and should have the current date and time. Then change the current directory to the unixstuff directory using the "$cd unixstuff" command as follows:

```
~/UNITS/UNIT1/unixstuff $cp ../example.txt.
```

This will copy the file "example.txt" in the UNIX0 directory (up one level) to the current directory. Do not forget the dot symbol "." at the end. The above command will copy the file example.txt at the upper level to the current directory while keeping the name the same as the original file.

The second function of cp is to copy directories. To copy a directory, use "-r" after the command (i.e., "cp -r path1 path2"). Here path 1 belongs to directory 1 and path 2 belongs to directory 2. Suppose that "jason" and "fanj" are two directories that belong to the home directory. To copy a directory "CRYSTAL1" in jason to fanj with the same name, type:

```
"~/Jason$cp -r CRYSTAL1/.. cp - r CRYSTAL1/fanj
```

**mv:** To move a file from one directory to another, use the "mv" command. This has the effect of actually moving rather than copying the file. This command can also be used to rename a file, by moving the file to the same directory, but giving it a different name. For example, the command "mv file1 file2" moves (or renames) file1 as file2. As a tutorial example, rename the file "example.txt" to "example_old.txt". First, change the active directory to the unixstuff directory. Then, inside the unixstuff directory, type:

```
$ mv example.txt example_old.txt
```

**rm:** To delete (remove) a file, use the "rm" command. For example, create a copy of the example.txt file, and then delete it. Open the unixstuff directory and type:

```
$ cp example_old.txt tempfile.txt
```

to create the file. Use "$ ls -lrt" to verify that the file exists, and then type:

```
$ rm tempfile.txt
```

to remove the file. Use "$ ls -lrt" again to verify that the file has been deleted. The command "rm -r" can also be used to remove a directory.

### 10.1.2.2 Other Important Commands

**ls -lrt:** Use "ls -lrt" to review additional information about the current directory. For example, by typing "$ ls -lrt", the shell will list the files present within the directory with additional information about the date and time created. Each file will be listed as a long string format in reverse time order, i.e., the last line is the newest file created, thus by using this command the origins of files and directories can be monitored. By typing this command in the current open directory, the two new directories "HOMEWORK" and "unixstuff" in UNIT1 should be listed. The string format should look similar to the following:

```
drwxr-xr-x 2 mauro mauro   4096 2010-10-17 17:17 unixstuff
```

The first character is "d" for the directory. After this there are nine characters, respectively, for the user, group and the others. Here, "r" denotes reading, "w" writing and "x" execution. If there is a "rwx" for the file, this indicates that the user has all the rights to read, write and execute the file. However, if a symbol "-" is used to replace one of the characters, this denotes limited permission to the file. For instance, suppose the nine characters are written as:

```
rw-rw-rw-
```

This string means that the "user", "group" and "others" have read and write permission but no execute function. The string:

```
rw-r--r--
```

means that the file will allow the user to have read and write permission, while the "group" and "others" will only have permission to read the file.

The third field in the long string is the number of link, the fourth is the owner of the file, and the fifth is the group of the file. Note that the default setting will make the fourth and fifth value the same. The last two fields are the date and the file name.

**chmod:** Use "chmod" to change the permission of a file. For example, the following command will allow the file "sfmakedepend" in the QC software to be readable, writable and executable:

```
$ chmod a + rwx sfmakedepend
```

There are also other expressions for this command, such as:

```
$chmod 755 filename (e.g.,job.bat)
```

The permissions for the file (e.g., job.bat) will now be rwxr-xr-x so all users will be able to execute the batch file.

**pwd:** Use "pwd" (print working directory) to locate where you are in relation to the whole file system. For instance, type "$ pwd" and the full pathname will be shown as ~/UNITS/UNIT1/unixstuff.

### 10.1.2.3 How to Transfer Files Between Gnu/Linux and Windows Operating Systems

WinScp is a free software that can transfer files between a Gnu/Linux remote machine and Microsoft Windows of a local machine. The details of WinScp can be found on the website http://winscp.net/eng/index.php. The procedure for operating WinScp is as follows:

(1) Install winscp code in a directory.
(2) Open winscp by opening the winscp407.exe file in the specified directory.
(3) A window will open. The following information should be filled into the blanks:
    **Hostname:** The remote computer name (e.g., fanlab10@alfred.edu or fanlab10 if both machines are in the same location).
    **Username:** The desired username at the remote computer with/Linux.
    **Password:** The desired password in the remote computer.
    **Login:** Yes.

After clicking "yes", a new window with left and right parts will appear on the screen. The right portion is the Linux system of the remote computer; and the left portion is the Windows system in the current computer. Files can be transferred from left to right or right to left as needed. For example, files in the Linux system, such as example1.gnu and 8000a.cfg files in the UNIT4/GNUPLOT_ATOMEYE_EXAMPLES directory, can be transferred into a Windows directory, such as My Documents, using the following three steps:

(1) Browse the right portion of the window to find and open the UNITS/UNIT4/GNUPLOT_ATOMEYE_EXAMPLES directory and select the file: example1.gnu.
(2) Browse the left portion of the window to find and open the C:/My Documents directory in the local computer.
(3) Click F5 copy in the bottom of the window. The two files should copy to the desired directory in Windows.

Files in the Windows system can also be moved to the Linux system in the remote computer using the same commands as above.

---

**Homework**

(10.1) Create a directory in UNIT1 called UNIT1_yourname. Open the new directory, and create a text file called Shell_command using "vi". Use the file to list and describe the UNIXshell commands learned in Section 10.1.

(10.2) Create a directory in UNIT1 called HOMEWORK_yourname. Copy the directory UNIT1_yourname into the HOMEWORK_yourname directory.

(10.3) If problems (10.1) and (10.2) were completed successfully, the "Shell_command" file should be in the directory: home/Homework_yourname/UNIT1_yourname. Copy this file to HOMEWORK_yourname of your home directory.

(10.4) In ~/UNITS/UNIT2/F90_CODES there are nine files including example1.f90 through example5.f90, as well as example1.dat through example4.dat. Make a directory UNIT1/F90_Codes_1 and copy these Fortran codes and data files to this new directory.

(10.5) Move the MD_CODE directory in/UNITS/UNIT2/to the UNITS/UNIT1/F90_Codes_1/ directory.

---

## 10.2   A Simple MD Program

In this section two MD simulation codes will be introduced and can be found in the directory UNITS/UNIT2/MD_Code:

(1) Source code: crystal_M_simple.f90; input file: Model_M_simple.in
(2) Source code: multi1.f90; input file: mt1.in and conf.in.dat (The second input file includes the x, y, z coordinates of atoms in the model; this input file is created only after running crystal_M_simple.f90)

The first group (1) creates an initial configuration of an atomic system which is not an MD code, but rather a code to develop the geometric model for the MD simulation. The second group (2) is the simple MD program. The MD code is written in the Fortran 90 language. In this section, several commands of Fortran 90 will be introduced through typical lines of the simple MD code to illustrate the code structure. Concepts of a "Module" and a "Subroutine" for Fortran 90 will also be introduced. The last section will introduce additional knowledge about the MD code including its compiling and running. The mathematical formulation of numerical algorithms used in this section can be found in Section 2.5. Readers interested in additional details on Fortran 90 should see Appendix 10.B.

## 10.2.1 Five Useful Commands of Fortran 90

In this section, five basic commands of Fortran 90 are introduced along with their functions. In Appendix 10.B, there are corresponding sections with five simple codes to explain these items in more detail. These five simple examples are located in the UNIT2/F90_CODES directory. It is beneficial to read, compile and run these examples to get more practice in addition to the introduction in this section. While the way used in this book for readers to learn Fortran 90 is not a systematic way as usually done, the results seem to be positive, see Fortran codes developed by students for homework (2.5) and NEB methods (Section 9.4.4) which are also placed in UNIT2/F90_CODES.

### 10.2.1.1 Write Command

The first command below listed in multi1.f90 will open the file "output.out" with file location ID = 15, while the second and third commands will write parameters within the "output.out" file. The second command specifically lists six parameters "step", "temperature", etc. which will be at the top line of the output.out file. The third command gives the format to write the numerical values of these parameters. These comments are listed, respectively, at lines of multi1.f90 as shown after the symbol "!" below.

```
open(15,file='output.out')                          ! line 7
write(15,*) "step temp ene_kin ene_pot ene_tot pres" ! line 98
write(15,'(i4,1x,5f16.5)'),step,temp, &             ! line 181
    ene_kin, ene_pot,ene_tot, pres                  ! line 182
```

Note: (1) When variables in line 182 are related to the command in line 181, a continuation symbol "&" is necessary. (2) In line 181, the parenthesis gives the writing format, "i4" indicates to write an integer variable such as "Step" with four columns, "1x" indicates one blank column to separate the integer variable from the succeeding real variables. "5f16.5" is the rfw.d-type of writing descriptor. In this format, r is the integer number indicating how many times the descriptor is to be used; r = 5 denotes writing five real variables such as temp, ene_kin etc. with the format "f16.5". Here, "f" indicates a type-real decimal form (without an exponent), "w" establishes the writing field width, "w = 16" indicates 16 output columns, "d" specifies the number of digits following the decimal point, "d = 5" specifies five decimal places.

### 10.2.1.2 Do Cycle

The "Do cycle" is a repeatable set of lines between the command line "do" and the line "enddo". The "do" line includes the words "do", the variable name and its range. Commands inside the "do cycle" will repeat as the variable increases in value from its initial value to the maximum value, then the "do cycle" stops.

Two "do cycle" examples are given below. Their functions together are to calculate displacement, velocity and acceleration for each atom at each time step of the whole simulation process. The first example uses the variable "i" which covers all atoms in the system (i = 1, 2. . .natms), where "natms" denotes the total number of atoms in the system. The second example uses the variable "step", which ranges from the initial time step to the ending time step (step = 1, 2. . .Nsteps), where "Nsteps" denotes the last step of the simulation. The first example starts from line 166 and ends at 177, while the second example starts at line 164 and ends at line 192. The larger range of the second example indicates that the second "do cycle" contains the first "do cycle" within it. Therefore, for each time step of the second do cycle, the first group must be completed. Physically, it indicates that in each time step the calculations of position, velocity and acceleration must be completed for each atom in the system. The following shows the code developed to create the two do cycles:

1. Lines 166 to 177 of the multi1.f90 file show the first "do cycle" that will calculate position (pos), velocity (vel) and acceleration (acc) for each atom of the atomistic system.

```
do step=1,nsteps                                              ! line 164
  write(*,*) step
  do i = 1,natms
  pos(1,i) = pos(1,i) + dt*vel(1,i) +0.5d0*(dt**2)*acc(1,i)   ! line 167
  pos(2,i) = pos(2,i) + dt*vel(2,i) + 0.5d0*(dt**2)*acc(2,i)
  pos(3,i) = pos(3,i) + dt*vel(3,i) + 0.5d0*(dt**2)*acc(3,i)
  vel(1,i) = vel(1,i) + 0.5d0*dt*acc(1,i)                     ! line 170
  vel(2,i) = vel(2,i) + 0.5d0*dt*acc(2,i)
  vel(3,i) = vel(3,i) + 0.5d0*dt*acc(3,i)
  call compute_forces                                        ! line 173
  vel(1,i) = vel(1,i) + 0.5d0*dt*acc(1,i)
  vel(2,i) = vel(2,i) + 0.5d0*dt*acc(2,i)
  vel(3,i) = vel(3,i) + 0.5d0*dt*acc(3,i)                     ! line 176
  enddo !i = 1,natms
  call compute_temp(temp)                                    ! line 178
  ene_tot= ene_kin + ene_pot
  pres = density*temp + virial/volume
  write(15,'(i4,1x,5f16.5)'),step,temp, &
                  ene_kin,ene_pot,ene_tot,pres               ! line 182
! print configuration xyz format
  write(45,*) natms
  write(45,*) "timestep", step
  do i = 1,natms
   write(45,'(a4,3f10.5)') "Ar", pos(1,i), pos(2,i), pos(3,i)
  enddo ! i = 1,natms
! end print configuration

enddo !step=1,nsteps                                         ! line 192
```

2. Lines 164–192 show the second large "do cycle" which covers the "do cycle" of the first. This indicates that the total number of cycles to calculate the position (pos), velocity (vel) and acceleration (acc) will equal natms multiplied by Nsteps.

3. Lines 167–169 are used to calculate the position vector $\bar{r}_i$ (t), and lines 170–176 are used to calculate the velocity $\bar{V}_i(t)$.
4. Lines 167–176 show the Fortran program for the "Velocity Verlet" numerical integration of Newton's differential equation of particles (second law). See Section 2.4.3 and equations (2.16h, i) for more detail. Specifically, the results of force and acceleration are calculated by calling the subroutine "Compute_Force" (line 173). The calculation is based on equation (2.16i) in order to compute velocity and atom position.
5. After position and velocity are determined for the whole atomistic system, the temperature, energy and pressure can be determined as shown in lines 178–180, and then printed out as shown in lines 181–182.

### 10.2.1.3 Arrays and Allocation

"pos(1,i)" of line 167 shows an array of x coordinates when atom variable i varies from 1 to ntam. If natms = 100, there are 100 values in the array arranged sequentially. The code is written as an array to make the Fortran 90 more concise. The first value "1" in "pos(1,i)" indicates the x coordinate of the position vector $\bar{r}_i$. This value can be changed to 2 or 3 to denote the y or z coordinates of the position vector $\bar{r}_i$. Thus an array of "pos(m,i)" can be formed, where the maximum value of m is Dim = 3. The parameters "Dim" give the dimensions of the array. The exact values of Dim and natms are typically given at the beginning, but may be determined during calculation if needed. This latter case is stated by using the "allocation" statement as shown on line 31 of multi1.f90 below:

```
real(kind=8), dimension(:,:), allocatable :: pos          ! line 31
```

The "natms" value is given on line 85 by reading the data file "conf_in.dat". This data file is developed by "crystal_M_simple.f90". Only after this code runs can the variable "natms" be determined. Thus, before reading "natms", open or insert the file "conf_in.dat" in a defined Fortran 90 location, such as the location "UNIT=10" described in line 82 as follows:

```
open(UNIT=10,file="conf_in.dat",action='read')          ! line 82
read(10,*) natms                                         ! line 85
```

### 10.2.1.4 If Then Else

This statement shows that if a given condition is satisfied, the command will perform the desired task; otherwise it will do something else or simply do nothing. The latter depends on whether an "else" statement is in between the line "If" and "endif". The following "if then if" statement of multi1.f90 between lines 230 and 242 is used to determine the acceleration of atom i caused by atom j. The "if" condition is given on line 230 and states that if "dist" is less than the square of "rcut" then the acceleration value of atom i will be calculated by lines 232–234; otherwise the acceleration of atom i caused by atom j will be zero. Here, the "dist" variable denotes the square of the distance between atom i and atom j, and "rcut" is the cutoff radius of atom i.

In the following program, variables accx, accy, and accz are proportional to the atomic acceleration components along the x, y, and z directions. Variable "coeff" is determined based on the LJ potential[*] and denotes a coefficient related to force and acceleration. Command line 230 indicates that if atom j is located inside the sphere with atom i as the center where the radius is "rcut", then its effect on atom i should be accounted for. Otherwise, the distance is larger than the cutoff radius and the force on atom i is negligible, so atom j's effect on atom i's acceleration is not considered.

```
dist = dx**2. + dy**2. + dz**2.                          ! line 228

   if(dist.le.(rcut**2.)) then                           ! line 230
      coeff=4.*((12.)/(dist**7) - (6.)/(dist**4))
      accx=coeff*dx                                       ! line 232
      accy=coeff*dy
      accz=coeff*dz
      ene_pot = ene_pot + 4.*((1.)/(dist**6) &
           -(1.)/(dist**3)) - pot_cutoff
      virial = virial + (accx*dx) + (accy*dy) + (accz*dz)
   else
      accx = 0.d0
      accy = 0.d0
      accz = 0.d0
   endif !(dist.le.(rcutg2))                              ! line 242
```

* The multi1.f90 program uses LJ units, i.e., the atom mass is 1, the unit of length is $\sigma$, and the unit of energy is $\varepsilon$, where $\sigma$ and $\varepsilon$ are the typical LJ parameters (see Section (2.3.1) and equation (2.3a)).

### 10.2.1.5  Commands for Read of a File and Call a Subroutine

In this section, the program "example5.f90" that opens the "example4.dat" file is described. This program is designed to calculate the mean values of the second and third column of the "example4.dat" file, which can be found in the UNITS/UNIT2/F90_CODES directory. Open the UNITS/UNIT2/F90_CODES directory and then use the following two commands to compute and run example5.f90:

```
$ gfortran example5.f90 -o example5.exe
$ ./example5.exe
```

After the commands have processed, the mean value of the second and the third columns, x and sin(x), respectively should be outputted. The program was developed with a "read" statement similar to the "write" statement, as shown in line 22 of example5.f90 as follows:

```
read(15,'(i4,2x,2f10.5)') nn, one_on_x(ix), sinx_on_x(ix)   ! line 22 in ex. 5
```

Compare it with the write statement used in example4.f90:

```
write(15,'(i4,2x,2f10.5)') ix, one_on_x(ix), sinx_on_x(ix)  ! line 31 in ex. 4
```

Note that the format is the same. After the file is read, the same two statements can be found in example5.f90:

```
call mean_value(one_on_x,nx,mean1)
call mean_value(sinx_on_x,nx,mean2)
```

These commands call a "subroutine mean_value". The subroutine mean_value is a subprogram which is given at the end of the example5.f90 file as follows:

```
!ccccccccccccccccccccccccccccccccccccccccccccccccccccccccccccc
subroutine mean_value(vect,n,mm)
integer :: i,n
real(kind=8),dimension(1:n) :: vect
real(kind=8) :: mm
mm = 0.
do i = 1,n
  mm = mm + vect(i)
enddo !i = 1,n
mm = mm/float(n)
end subroutine mean_value
!ccccccccccccccccccccccccccccccccccccccccccccccccccccccccccccc
```

Note that the subroutine declaration is outside the program, i.e., after the "end program" statement. The following points are features of subroutines:

1. A subroutine does not share the same variables as the program that calls the subroutine. Therefore if a subroutine must operate with the same variables as the main program, these variables must be included in the subroutine. For example, the subroutine "mean_values" calculates the mean value of an array "vect(i)" of a given dimension and then stores this mean value as a real variable. Three variables must be passed, including the real array, the array dimension (integer value) and a real variable on which the subroutine puts the calculated value.
2. A subroutine can be used more than one time in a program. In example5.f90 the mean values of the two arrays "one_on_x" and "sinx_on_x" are calculated using the same subroutine with different variables. After compiling and running the code, the mean values of the two arrays will be displayed on the screen as two separate values from the same program code.

## Exercise 10.1

Change the name of the file "example4.dat" by using the command:

```
mv example4.dat example4_test.dat
```

Run the example5.exe file again. An error message will appear. Why does this occur? Try to correct the problem. Hint: Change the open statement in the example5.exe file to contain the new file name: "open (15, file=....)".

### 10.2.2  Module and Subroutine

Subroutine and module are two basic structural blocks of Fortran 90. In a complex code such as MD programming, the variable management can be difficult. For this reason Fortran 90 and other programming languages use subroutine and modular structure to make the program concise. An executable program consists of one main program and any number (including zero) of other program units, including subroutines and modules. In Fortran 90 a module is a set of variables and subroutines, and can be included in the main program, subroutines and other modules.

To show the basic usage of modules, consider the code "crystal_M_simple.f90" found in the UNITS/UNIT2/MD_CODE directory. Open it using vi then type ":100" on the bottom line and press

Enter, which puts the cursor on line 100, where the main program starts. The first command in the main program is

```
use data                                                            ! line 103
```

"data" is a Fortran 90 module defined in lines 2–21 of the crystal_M_simple.f90 file. "use data" means that the variables defined in the "data" module can be used by the program "main".

Then the main program calls several subroutines, including "const" at line 110, "read_in" at line 111 and "basic_cell" at line 121. Each subroutine starts with the command "use data" (see lines 27 and 53). This means that every subroutine will use the "data" module variables. Otherwise, the variables would need to be passed to the subroutine. For the latter case, see example5.f90 in the UNITS/UNIT2/F90_CODES directory.

A similar way to use modules is shown in the multi1.f90 file found in the UNITS/UNIT2/MD_CODE directory. This program is a simple but inefficient molecular dynamics program.

### 10.2.3    Using crystal_M_simple.f90 to Create Initial Configuration

The crystal_M_simple.f90 code creates a simple initial configuration of atoms on a face centered cubic (FCC) lattice. This running must be done to produce the conf.in.dat file before running MD simulations in the next section. Open its input file "Model_M_simple.in" found in the UNITS/UNIT2/MD_CODE directory using vi. The file is a text file with three lines of code as follows:

```
1.6  1.6  1.6
4  4 4
0. 0. 0.
```

The first line contains the basic cell dimension (value 1.6 in each direction), the second line shows the number of basic cells in each direction (value 4), and the third line shows the coordinates x, y, and z of the bottom-left corner of the configuration, the position of the first atom. It is easy to see that the model has 64 unit cells ($4 \times 4 \times 4 = 64$) and 256 atoms because for FCC crystal each unit cell has four atoms. This file is read by crystal_M_simple.f90 using the subroutine "read_in". To compile crystal_M_simple.f90, open the UNITS/UNIT2/MD_CODE directory and type the following:

```
$ gfortran crystal_M_simple.f90 -o crystal.exe
```

"gfortran" denotes the free Fortran compiler software, which can be installed as shown in Appendix 10.A. The first file, crystal_M_simple.f90, is the source file. The last file, crystal.exe, is the name of the executable file to be produced after the compiling process is complete. Verify that the program completed successfully using "ls -lrt". Run the executable file by typing the following:

```
$ ./crystal.exe
```

Check the output using "ls -lrt". Two files have been created, including conf_in.dat (the input file for the molecular dynamics program "multi1.f90") and conf.xyz (that will be used during the following sections to show the initial configuration with VMD).

### Exercise 10.2

Open the "conf_in.dat" file and review the file's structure. Check the write(30,*) command in the crystal_M_simple.f90 file.

## 10.2.4   Use multi1.f90 to Run a Molecular Dynamics Calculation

After we have developed the conf_in.dat file it is time to review the first MD code, multi1.f90. This code is written in a very simple way, which will allow for easy understanding of the file content, but cannot be used for a large system because of its low efficiency. The main program starts from the beginning and contains only three commands for calling subroutines as follows:

```
call read_input                    ! line 10
call read_conf                     ! line 11
call evolve                        ! line 12
```

The input files, "mt1.in" and "conf_in.dat", are the input data, respectively, of the read_input and read_conf subroutines, and are introduced below.

The first subroutine reads the "mt1.in" file which contains only three lines as follows:

```
test              !Name of simul.
30                ! of time steps
0.003             !Time step (ps)
```

The second subroutine reads the file "conf_in.dat". The latter is produced by running the crystal_M_simple.f90 file for model structure formulation as introduced in the last section; thus it stores the number of atoms, the box dimensions, and the list of initial position of all atoms.

The third subroutine contains the time evolution code. In this subroutine it is very easy to recognize the Velocity Verlet scheme (see the do cycle from lines 166–177).

Compile and run the multi1.f90 file using the following commands:

```
$ gfortran multi1.f90 -o multi1.exe
$ ./multi1.exe
```

The program will run for approximately 10 minutes. Use this time to analyze the structure of the code.

The only output is the file "output.out". The variables and their format in the output file are given by the following command in the multi1.f90 file:

```
open(15,file='output.out')                  ! line 7
write(15,'(i4,1x,5f16.5)'),step,temp, &            ! line 181
ene_kin,ene_pot,ene_tot, pres       ! line 182
```

The above write statement will arrange the data from left to right as follows: Step, temperature, kinetic energy, potential energy, total energy, and pressure. The symbol "&" in line 181 is a continuous symbol of the command, which includes the contents on line 182 into the command in line 181.

---

**Homework**

(10.6) The interatomic force determined by the LJ potential is given by (2.5) as follows:

$$f_{ij}(r) = 24\frac{\varepsilon}{\sigma}\left[2\left(\frac{\sigma}{r}\right)^{13} - \left(\frac{\sigma}{r}\right)^{7}\right]$$

Suppose that the parameters $\sigma$, $\varepsilon$, and atom mass m are assumed to be 1. Verify that the acceleration component, $a_x$, of atom i along the x-direction is consistent with the Fortran command from lines 228–232 of multi1.f90 listed as follows:

```
dist=dx**2.+dy**2.+dz**2.                              ! line 228

If(dist.le.(rcut**2))then                              ! line 230
coeff=4.*((12.)/(dist**7) - (6.)/(dist**4))                    ! line 231
accx=coeff*dx                                    ! line 232
```

Hint: (1) The variables dx, dy, dz are component differences along the x, y, z directions between atom i and j; and rcut is the cutoff radius which defines the spatial domain for using the interatomic potential. (2) The acceleration component, $a_x$, along the x-direction is equal to the acceleration multiplied by the cosine of the angle between the $r_{ij}$ direction which connects atom i and j and the x-axis.

(10.7) Verify that the recursion formulas (2.16h) and (2.16i) of the Velocity Verlet numerical algorithm are consistent with commands from lines 166–177 in the multi1.f90 code. In other words, based on the commands in these lines, the related computer calculation can produce velocity and position vectors of atoms following these formulas.

(10.8) Use the multi1.f90 code to run a simulation with 300 time steps. (Hint: Change the number of time steps by opening the "mt1.in" data file and modifying the value of time step with the vi command; it is important that you check whether you have produced conf.in.dat. It should include x, y, z coordinates of 256 atoms.)

(10.9) Write a code that opens the "output.out" file and calculates the mean values of temperature, kinetic energy and pressure in that file obtained from running 300 steps in (10.8).

Hints: (1) Verify that the output data from the file "output.out" in the UNIT2/MD_CODE directory contains a value for the step, temperature, kinetic energy, potential energy, total energy and pressure at each time step. These variables should follow the write statement of lines 181–182 listed above. Then change the write statement to a read statement to get all the information. Do not try to write a code from scratch. (2) Use the source file example5.f90 in the UNITS/UNIT2/F90_CODES directory as the sample code to write the required mean values. (3) Since the "output.out" has five lines for letters in the beginning, you may need to delete these five lines so that you can use, say, the following sentence to read the numerical data of step, temperature, etc. listed in the 300 lines for the 300 time steps.

```
Read(15, "(i4, 1x, 5f16.5)') STEP, temp(ix), temp(ix),ene_kin(ix), ene_-
pot(ix), ene_tot(ix), Pres(ix).
```

If one wants to keep one line with letters such as "STEP Temperature . . .", one may write the following line before the above line to read that line by

```
Read (15,*)
```

Here, the symbol "*" indicates a free format to read the letters.

## 10.3    Static Lattice Calculations Using GULP

GULP (General Utility Lattice Program), developed by Julian D. Gale, Nanochemistry Research Institute, Department of Chemistry, Curtin University of Technology, Australia,[1,2] is a program that can perform a variety of simulation tasks including fitting potential parameters, static lattice calculation, molecular dynamics, Monte Carlo and so on. In the following sections, GULP will be used to show how to perform static lattice calculation and potential parameter fitting. GULP is free of charge to academics but users should register through its website https://www.ivec.org/gulp/ to obtain the latest package and install it in a suitable directory of their computer.

Two directories for installation of softwares in this chapter are suggested. One is the home directory, such as for GULP to be installed in this section. The other is under the UNITS directory such as for DL_POLY_2 and NAMD_2.6 (see Figure 10.5). These kinds of installation will have different paths. The first one is easier because its path can be expressed as "$~/GULP", no matter where the current directory is. This is because the symbol "~" denotes the home directory. The second one depends on where the current directory is, see the difference of the two installation paths for running jobs in Section 10.3.2 for GULP and Section 10.3.4 for DL_POLY_2.18.

## 10.3.1   Installation and Structure of GULP

After unpacking the software package, the user should have a directory with a structure shown in Figure 10.2. The Src directory contains the source files which need to be compiled into the GULP executable by the following steps:

1. Open the shell script "getmachine" in the Src directory.
2. In the getmachine file, find the section corresponding to the user's operation system, select a set of proper flags with comment symbol #. These include four parts:
   – Compiler (RUNF90=), three sets of optimization flags (OPT=, OPT1=, OPT2=)
   – Fortran flags 'FFLAGS=)
   – Libraries (LIBS)
   – Math libraries (BLAS=, LAPACK=, CDABS=)
     If not sure, try to use default value or consult with the computer administrator.
3. Save "getmachine" then run "make" in the Src directory. The objects will be compiled into a directory in Src. The final executable "gulp" will be placed in Src.
4. After successful compilation, move the executable "gulp" to the objective directory. Here we put the executable directory into the "GULP" directory.

Other directories of the GULP package include documentations (Docs), example files (Examples), force field libraries (Libraries) and utilities (Utils).

## 10.3.2   Input File Structure and Running GULP

From examples offered by the GULP package we can see the basic structures of the input file. In the following, example 6 will be used to introduce each part of the structure. This example is a defect simulation using a shell model for MgO (see Section 3.3 for shell model).

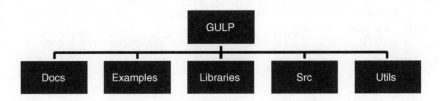

**Figure  10.2**   GULP directory structure

### 10.3.2.1  GULP Input File

The GULP input file consists of the following six subsections: Keywords, Title, Structure, Species, Defects and Potentials.

#### Keywords

The first line contains the keywords that define what kind of simulation is going to be performed. Example 6 is a defect calculation with Li substituting for Mg in MgO. Its first line contains the following keywords:

#### opti defect regi conp

These keywords mean: optimize (minimize the energy of the lattice), defect (perform defect calculation; on the concept of "region 1" in defect analysis see Figure 3.4 and Section 3.5), regi_before (output region 1 coordinates before defect calculation), conp (constant pressure). Only the first four letters are needed if there is no confusion. For an example, "prop" is the same as "property".

#### Title

The second part is the title, starting with a line of "title" and ending with a line of "end". In this example the title is:

#### title
#### Example of defect impurity calc: Li substituted into MgO
#### End

#### Structure

A lattice structure can be defined by parameters of the lattice and atoms in the unit cell. An example of lattice parameters is given as follows:

```
cell
4.212 4.212 4.212 90 90 90
```

Using symbol "frac" denoting fractional coordinates and using one line for each atom in the unit cell, the expressions for core and shell coordinates as well as electric charges can be expressed as follows:

```
frac
O  core 0.5 0.5 0.5  0.86902
Mg core 0.0 0.0 0.0  2.00000
O  shel 0.5 0.5 0.5 -2.86902
```

Here, the last value of each line denotes the charge of the corresponding core and shell. Charges can also be defined in the next "species" section, in this case, there are only three fraction coordinates after the atom (or ion) name. In addition, space group can be used to input crystal structure (see Section 2.5 for space group of crystal structure). In this example MgO has a rock salt structure, and its expression with the space group symbol is written as follows:

#### space
#### 225

#### Species

This subsection defines the charges of atoms in the simulation, for example if Li has a charge 1.0:

```
species
Li 1.0
```

**Defects**

The defect subsection is to input the information needed for a defect calculation, an example for impurity Li replacing Mg is given:

```
centre Mg
impurity Li Mg
size 4.0 10.0
```

Here, the first number 4.0 is the size of region 1 and the second number 10.0 is the size of region 2.

**Potentials**

The last subsection in the input file is to define potentials. It includes potential name followed by a line with the sequence of "Atom Symbol" "core/shell" "Atom Symbol" "core/shell" "parameters" and "cutoff". For example here we have:

```
buck
Mg core O shel 1280.0 0.3 0.0 0.0 8.0
```

This defines the short-ranged interaction between $Mg^{2+}$ and $O^{2-}$ as Buckingham potential with $A = 1280\,eV$, $\rho = 0.3\,\text{Å}$, $C = 0\,eV\text{Å}^6$, and interaction range is between 0.0 and 8.0 Å. Different simulation tasks will need different corresponding subsections which will be introduced, respectively, in detail in Sections 10.3.3 to 10.3.6.

### 10.3.2.2 Running GULP

To run GULP, enter the following in the shell:

```
/GULP path/gulp <"input file" >"output file"
```

For example, if GULP is installed in "/home/GULP" directory and the executable file "gulp" is under GULP, to run example 6 one should first enter the directory with the input file of example 6, that is the directory "GULP/examples". Enter that directory by typing:

```
$ cd /home/GULP/Examples
~/GULP/Examples $ ~/GULP/gulp <example6.gin >example6.gout
```

It is important to note that in <example6.gin > there is a space between <example6.gin and the symbol >.

## *10.3.3   Structure Optimization and Output File Structure*

A simple simulation of MgO structure will be performed in this section. This so-called rocksalt structure is shown in Figure 3.2. First, create a new input text file named "MgO_structure_1.gin" in the UNITS/UNIT3/structure directory and edit it as:

```
opti conp prop
title
static lattice calculation of MgO
```

```
end
cell
4.212 4.212 4.212 90 90 90
frac
O 0.5 0.0 0.0
O 0.0 0.5 0.0
O 0.0 0.0 0.5
O 0.5 0.5 0.5
Mg 0.0 0.0 0.0
Mg 0.5 0.5 0.0
Mg 0.5 0.0 0.5
Mg 0.0 0.5 0.5
species
Mg 2.0
O -2.0
buck
Mg O 1280.0 0.3  0.0  0.0 10.0
buck
O 1280.0 0.3 27.88 0.0 10.0
```

In the first line, we ask GULP to perform a structure optimization (opti) under constant pressure (conp), and calculate the material properties (prop). Title, structure and potentials are followed. MgO has a simple rocksalt structure, which can be easily inputted. But if the user is familiar with crystallography, the structure can also be input using space group. Buckingham potentials are used for both $Mg^{2+}$-$O^{2-}$ and $O^{2-}$-$O^{2-}$ interaction and the cutoff is set to 10 Å.

**Running**
After saving the input file, run the following command:

```
/GULP path/gulp <MgO_structure_1.gin >MgO_structure_1.gout
```

For the setting in UNIT3, GULP is installed in "/home/GULP" directory and the executable file "gulp" is under GULP, first enter the structure directory and then give the command as:

```
~UNITS/UNIT3/structure$ ~/GULP/gulp <Mgo_structure_1.gin
 >MgO_structure_1.gout
```

The calculation should be done in a short moment, and the output information will be saved in the text file "MgO_structure_1.gout".

**Output file**
Now open and read through the output file. The first part of the output file is a repeat of the input information. User should carefully check it to make sure the input has been correctly read by GULP. Following the input information is the energy calculation and minimization. After several calculation cycles, the optimized structure is obtained and the final lattice energy per unit cell is printed. Be sure to look for the line "Optimization achieved". The optimization is successful only when this line is presented. If no valid optimization can be achieved, the result is meaningless and the input should be checked for errors or unphysical conditions. The optimized structure parameters are printed along with the energy information. If "prop" is presented in the keywords, material properties such as elastic constants and dielectric constants will be calculated and printed in the output. At the end of the output is the computational information such as the job CPU time.

From the output it can be seen that the equilibrium lattice parameter of MgO is calculated to be 4.26294 Å. At this lattice parameter, the force on each atom becomes zero and the system energy is minimized to be $-163.195726$ eV.

To better understand this structure optimization process, rename the input file as "MgO_structure_2.gin" and change the initial lattice parameter from 4.212 Å to 10 Å and run the job again. In the new output of MgO_structure_2.gout, although the initial lattice parameter has been changed to 10 Å, the same optimized result 4.26294 Å is achieved. From an initial structure, GULP can optimize the structure to minimize the energy and find a stable structure close to the initial one. However, it should be noticed that more optimization cycles and CPU time have been taken for the optimization of the initial lattice parameter of 10 Å. Therefore a reasonable initial structure is always preferred.

---

**Homework**

(10.10) Optimize lattice constants a, b, c and elastic constants $C_{11}$, $C_{12}$, $C_{44}$ with initial constants $a = b = c = 4.57$ Å by GULP through modifying input file "MgO_structure_1.gin" and running it in the directory UNIT3/structure. Compare the optimized lattice and elastic constants with the single crystal properties found in the literature. What are the possible reasons for the difference?

Hint: Get property data from the literature, for instance, *Landolt-Bornstein – Group III Condensed Matter*, Springer-Verlag, vol. 41B: 1, and K. Marklund and S. A. Mahmoud (1971) Elastic Constants of Magnesium Oxide, *Phys. Scr.*, 3: 75. In this case references are given as follows: lattice parameter: 4.212 Å, $C_{11} = 306.7$ GPa, $C_{12} = 93.71$ GPa, $C_{44} = 157.6$ GPa.

(10.11) MgO can have two types of structure. The first one is rocksalt as shown in Figure 3.2 and discussed in the text of this section. The coordinates of the 4 Mg and 4 Cl ions in the unit cell are given in the input file "MgO_structure_1.gin" in UNIT3/structure directory. The second type is with CsCl (caesium chloride) structure with 1 Mg and 1 O ion which can be described in GULP input format as follows:

```
cell
2.57   2.57   2.57   90   90   90
frac
O  0.5   0.5   0.5
Mg 0.0   0.0   0.0
```

(10.12) Use the same potentials and condition listed in the text of this section by running GULP to find the optimized lattice parameter for the second type of structure and the lattice energy per mole MgO. Then compare the lattice energies per mole MgO of this type with those for the rocksalt type obtained from the output file of the text. Which one is more stable under the same simulated condition?

Hint: The results obtained for rocksalt type are for 4 Mg and 4 O, thus the energy should divide 4 to compare with the result of the CsCl type of MgO which only has 1 Mg and 1 O.

(10.13) Adding the following line GULP input file MgO_structure_1.gin in the directory UNIT3/structure to simulate the structure under 350 GPa:

**pressure 350**
Simulate both structures of MgO under 350 GPa pressure, compare the results and find which structure is more stable. Then try to find approximately under what pressure the two structures become equally stable. Compare the result to the literatures.

Hint: The one with lowest energy is more stable.

## 10.3.4   Determining Potential Parameters by Fitting Calculations

Appropriate interatomic potential models are essential for the accuracy of atomistic simulations of materials. One way to obtain these potential parameters is fitting them to empirical data. From the last section we can see that GULP can calculate observable material properties with given potentials. Conversely, GULP can also vary potential parameters to find the best match to the given properties. This section will introduce how to determine the parameters of potential functions by GULP's fitting function.

**Input file**
In an input file for the potential parameter fitting, initial parameters of the potential and the experimental data such as elastic constants should be given. Create a new GULP input file "MgO_fitting.gin" in the UNITS/UNIT3/fitting directory and edit it as below:

```
fit conp opti prop
title
potential fitting for MgO
end
cell
4.262936 4.262936 4.262936 90 90 90
frac
O  0.5 0.0 0.0
O  0.0 0.5 0.0
O  0.0 0.0 0.5
O  0.5 0.5 0.5
Mg 0.0 0.0 0.0
Mg 0.5 0.5 0.0
Mg 0.5 0.0 0.5
Mg 0.0 0.5 0.5
species
Mg 2.0
O -2.0
observables
elastic
 1 1 342.5550
elastic
 1 2 180.2137
elastic
 4 4 180.2137
buck
Mg O 1300.0 0.25 0.0 0.0 10.0 1 1 0    ! the 4th and 5th are the reading of
          ! the range (0-10) inside the cutoff-radius. "1 1" denote the first two
          !parameters (i.e., A and ρ) need to be determined.
buck
O  O 1280.0  0.3  27.88  0.0  10.0  0  0  0
```

In the first line, the keyword "fit" asks GULP to perform potential fitting. Optimization and property calculation will be performed after fitting with the fitted potential; it is realized by the keywords "opti" and "prop" listed. In this input, structure parameters and three observables (elastic constants $C_{11}, C_{12}, C_{44}$) are provided for potential fitting. The values of these properties are obtained from the calculation in the last section. Initial values of potential parameters should be provided and followed by same number of flags. Flag of "1" means corresponding parameter can be varied and "0" means it is fixed. Here we only fit the first two parameters of $Mg^{2+}$-$O^{2-}$ Buckingham potential, not change the potential parameters of $O^{2-}$-$O^{2-}$ at all. Since MgO is ionic material, it can be assumed that there is no attraction between $Mg^{+2}$ and in the term of short-ranged potential, see (3.5) based on assumption (3) of Section 3.4.1.

### Running
After saving the input file, run it with the following command:

```
/GULP path/gulp <MgO_fitting.gin >MgO_fitting.gout
or
~UNITS/UNIT3/fitting$ ~/GULP/gulp <Mgo_fitting.gin >Mgo_fitting.gout
```

### Output file
In the output file, the fitting process is shown after the input information. One important output content is the fitting quality expressed by "Final sum of squares"; the smaller the sum, the better the quality. Since the properties used here are derived from the calculation, they are perfectly self-consistent. Therefore a "Final sum of squares" $=0$ can be achieved. Compare the fitted potential parameters with the ones used for the structure optimization in the last section. They should be the same, since the properties calculated in the last section were used to fit the potential. Also compare property calculation results after the fitting with the ones calculated in the last section.

### Fitting strategy and procedures
Parameters in the potential function are correlated, so a good strategy and related sequence in the fitting process may be important. As an example, change the second parameter of $Mg^{2+}$-$O^{2-}$ Buckingham potential from 0.25 to 0.1 as follows:

```
Mg O 1300.0  0.1  0.0  0.0  10.0  1  1  0
```

and then run the fitting job again to see if the same results can be achieved. It can be seen that when the parameter is too far from the appropriate value, the fit quality becomes worse. This is because the first two parameters of Buckingham potential are correlated. Therefore it is better to use a "one variable" strategy: fix the second parameter sequentially as 0.2, 0.29, 0.31 and 0.4 and fit the first parameter respectively. Comparing the "Final sum of squares" for each run, we can see that fitting quality increases when the second parameter changes from 0.2 to 0.29 but decreases with the change from 0.31 to 0.4. From such a trend, it can be seen that 0.3 could be a good value for the second parameter.

Another suggestion for achieving good fitting quality with high speed is to use reasonable initial values. These values can be obtained from an earlier study or a likewise potential. Keep in mind that the potential fitting could be much more complicated when using actual experimental data for a more complex material system. When a new set of parameters is obtained, always compare the simulation results with experimental data to check if it can reproduce different types of material properties.

---

### Homework
(10.13) Fit the first two parameters of $Mg^{2+}$-$O^{2-}$ Buckingham potential to the elastic constants determined by experiments as shown below (see 'Hint' of homework (10.10); you may use the data: $a = b = c = 4.212$ Å, $C_{11} = 306.7$ GPa, $C_{12} = 93.71$ GPa, $C_{44} = 157.6$ GPa).

## 10.3.5   Shell Model

In the previous simulations, all the ions are assumed to be rigid. For more sophisticated simulations, the charge redistribution inside the ions must be considered. This can be included by using the shell model (see Section 3.3). In the shell model, an ion consists of a "core" and a "shell" connected with a harmonic spring. Below is the modification of the input file from 10.3.2, applying the shell model to oxygen ions.

```
opti conp prop
title
static lattice calculation of MgO, with shell model for oxygen ions
end
cell
4.212 4.212 4.212 90 90 90
frac
O core 0.5 0.0 0.0
O core 0.0 0.5 0.0
O core 0.0 0.0 0.5
O core 0.5 0.5 0.5
Mg 0.0 0.0 0.0
Mg 0.5 0.5 0.0
Mg 0.5 0.0 0.5
Mg 0.0 0.5 0.5
O shel 0.5 0.0 0.0
O shel 0.0 0.5 0.0
O shel 0.0 0.0 0.5
O shel 0.5 0.5 0.5
species
Mg 2.0
O core 0.86902
O shel -2.86902
buck
Mg O shel 1280.0 0.3 0.0 0.0 10.0
buck
O shel O shel 1280.0 0.3 27.88 0.0 10.0
spri
O 74.92
```

In this input file, all the oxygen ions in the unit cell consist of cores and shells. Core and shell charges are defined in the "species" section. The Buckingham potentials act only between shells, and a harmonic spring potential for the oxygen ion is added to connect its core with the shell. Now complete the following steps:

1. Save the input file as "MgO_shellmodel.gin" in UNITS/UNIT3/shell_model and then run

   ```
   /GULP path/gulp <MgO_shellmodel.gin >MgO_shellmodel.gout
   ```

   Specifically for UNIT3, the command is

   ```
   ~UNITS/UNIT3/shell_model$~/GULP/gulp<Mgo_shellmodel.gin>Mgo_
   shellmodel.gout
   ```

2. Compare the output file with the one obtained in Section 10.1.2 and find what properties become different and find whether there are new properties calculated.
3. Try to answer the following questions. What is the physical meaning of these properties? How does the shell model affect these properties when the Buckingham potentials are still the same? In what situation can the shell model produce substantially more accurate results?

The shell model includes the ion polarizability, therefore it would be important when polarization was present in the simulated materials. Such situations include simulating material under electric field, low symmetry structure, defects and so on. When fitting the shell model parameters from experimental data, dielectric constants are usually used because of their sensitivity to the polarizability. Other properties can also have strong correlation with the polarizability, therefore typically the short-ranged potential models and shell models are determined from one set of experimental data. Users should pay attention to maintain the correspondence between short-ranged potentials and shell models.

### 10.3.6 Defect Calculation

One important capability of GULP is the defect calculation, which can provide insights for complicated problems such as material doping and ion transportation. GULP uses the two-region strategy to calculate the energy of material structure with defects (see Mott-Littleton approximation model of Section 3.5). Below is an input file for a calculation of one oxygen vacancy in MgO:

```
opti conp prop defe
title
defect calculation of an Mg vacancy in MgO
end
cell
4.262936 4.262936 4.262936 90 90 90
frac
O1 core 0.5 0.0 0.0
O2 core 0.0 0.5 0.0
O3 core 0.0 0.0 0.5
O4 core 0.5 0.5 0.5
Mg1 0.0 0.0 0.0
Mg2 0.5 0.5 0.0
Mg3 0.5 0.0 0.5
Mg4 0.0 0.5 0.5
O1 shel 0.5 0.0 0.0
O2 shel 0.0 0.5 0.0
O3 shel 0.0 0.0 0.5
O4 shel 0.5 0.5 0.5
species
Mg 2.0
O core 0.86902
O shel -2.86902
centre Mg1
size 6.0 12.0
vacancy Mg1
buck
Mg   O shel 1280.0 0.3 0.0 0.0 10.0
```

```
buck
O shel O shel 1280.0 0.3 27.88 0.0 10.0
spri
O 74.92
```

The input file is based on the shell model simulation of MgO in the last section. The keyword "defe" is added in the first line for a defect calculation. Dielectric constants are required to calculate defect energy, so property calculation should be performed or reloaded. To easily define and indentify the defect, all the ions in the unit cell are labeled with numbers. Three new lines are added for defect information:

```
centre Mg1
size 4.0 10.0
vacancy Mg1
```

The first two lines indicate the center "Mg1" for the two regions (i.e., the two regions have the same center) and the sizes of region 1 and region 2 are given, respectively as 4 and 10 Å. The third line asks GULP to create a vacancy by removing Mg1. Region 1 should include the most disturbed ions around the defect, therefore the region center is placed at the defect. When more than one defect is simulated, the centre of regions should be placed at the centre of the defect cluster. But users should try to choose the center so that the structure maintains a high symmetry with respect to it. This will help GULP to use symmetry to reduce the calculation time. The larger the regions, the more accurate the result should be. But large region size will cause a great increase in the calculation time. For a serious simulation, the size of region should be tested so that an acceptable approximation can be achieved. Now finish the following steps:

Save the input file as "MgO_VMg.gin" in UNITS/UNIT3/defect and then run

```
/GULP path/gulp <MgO_VMg.gin >MgO_VMg.gout
```

    or

```
~UNITS/UNIT3/defect$ ~/GULP/gulp <MgO_VMg.gin >MgO_VMg.gout
```

Carefully read the output file, check the input information for the defect, make sure the expected defect is created.

In the output file, the defect calculation follows the perfect lattice optimization and property calculation. The total charge on defects and the numbers of ions are listed and should be verified. Here only one $Mg^{2+}$ is removed, so the total charge is −2, which represents a theoretical situation. Users should pay close attention to the actual coordinates of the defect since GULP treats different ways of defect input with different methods: when atom symbol or fractional coordinates (default) are used, GULP will utilize the structure symmetry and use its nearest image to the region center as the defect; when Cartesian coordinates are used for the input of defects, GULP will use the absolute coordinates without any transformation.

Once the input is confirmed and valid optimization of defect structure is achieved, the defect energy can be analyzed. The defect energy is calculated as the energy difference between the lattice with and without defect:

$$E_{defect} = E_{defected\ lattice} - E_{perfect\ lattice}$$

Comparing defect energy of different defect configurations provides information about defect formation and association. The final coordinates of the defective lattice are also provided in the output file, so users can compare with the perfect lattice if interested.

**Homework**[*]

(10.14)  Use GULP to calculate the defect energy of rocksalt type of MgO in the following three cases:
(a) Remove Mg1 only as shown in the input file MgO_VMg.gin in the directory of UNIT3/ defect; here Mg1 or O1 are element ID in that file; (b) Remove O1 only, in other words, calculate the defect energy of one oxygen vacancy in MgO (hint: "vacancy O1" in the input file will remove the core and shell together); (c) Try to calculate one $Mg^{2+}$ vacancy and $O^{2-}$ vacancy at the nearest neighbor site of each other by removing simultaneously both Mg1 and O1, then compare the total defect energy with the sum of the energy obtained from (a) and (b). Discuss the reason for the energy difference between the sum and the coupled defect energy obtained in (c).

Hint: when calculating two vacancies at each other's nearest neighbor site, the region center can be placed on one of the vacancies.

[*]In the real material, charge neutrality is maintained when defects are generated. One way to keep charge neutrality is by generating "Schottky defect". Schottky defect is where the vacancies of cation and anion are formed in stoichiometric units. In MgO, it means one $Mg^{2+}$ vacancy will be accompanied by one $O^{2-}$ vacancy.

# 10.4   Introduction of Visualization Tools and Gnuplot

In this section three common graphical tools are presented. They are Visual Molecular Dynamics (VMD), AtomEye, and gnuplot. The first two are codes for visualization and analysis of molecular dynamics configurations. They have different advantages. AtomEye has several spatial functions for data processing, but VMD is more easy to use in the Linux system. Beginners may choose either software to study how to make visualizations, while gnuplot is a standard GNU plotting program, which will be used to plot data obtained from the MD simulations. More functions of VMD in developing MD models for biological materials such as protein will be introduced in Sections 10.10 and 10.11.

It is important to note that gnuplot and other graphical programs (e.g., VMD) will not work when using PUTTY software to connect remote computers. The computers must use UNIX to operate gnuplot. For Windows systems, use the KNOPPIX software to access the UNIX system. As shown in Appendix 10.B, a KNOPPIX CD is required to install GNU/Linux system.

## 10.4.1   Gnuplot

Gnuplot software can be used to plot different variables in different files. The software installation process can be found in Appendix 10.A. Once installed, open the program by typing the following command in the directory where the data file is located:

```
$ gnuplot
```

The following gnuplot prompt will appear on the screen:

```
gnuplot>
```

This prompt symbol > indicates that gnuplot is running and ready to take commands for plotting.

### 10.4.1.1  Plot a Single Function

Type the following command:

```
>plot sin(x)
```

A graphical window reporting the sine function will appear. The x axis will range from −10 to 10 as the default x range for gnuplot. To change this interval, close the graphical window and type the following command in the gnuplot prompt:

```
>plot[-2:15] sin(x)
```

The graphical window will appear again with the sine function, but now the x interval ranges from −2 to 15. The y interval is determined by the minimum and the maximum of the plotted function (in this case −1 and 1). To change the range of the y interval, close the graphical window and type the following command in the gnuplot prompt:

```
>plot[-2:15][-1.5:2] sin(x)
```

The first [#:#] represents the x range, and the second [#:#] the y range.

### 10.4.1.2 Plot Multiple Functions on One Graph

In gnuplot, it is possible to plot several functions on the same graph using the same "plot" command. Separate the two functions with a comma, for example:

```
> plot sin(x), 0.2*x
```

This command plots the two functions sin(x) and 0.2*x on the same graph.

## Exercise 10.3

Plot the function $0.2*x^2$ (typed on gnuplot as 0.2*x**2), 2*sin(0.2*x) and the square root of x (typed on gnuplot as x**0.5) on the same graph. Use −2 to 10 for the x interval. Explain why the last function is only shown in part of the graph.

### 10.4.1.3 How to Plot Data from a File

To use the import function of gnuplot, open the output.out file in the UNITS/UNIT4/MD_CODE_vmd directory. The first line contains the names of the quantities reported in each column. Each line from the first to the last contains six numbers: the first is an integer and the other five are real numbers. To use gnuplot to plot the data reported in this form, add the character # to the first line in position 1,1 (i.e., first character of the first line). The # character is a gnuplot comment character. It means that the rest of the line will be ignored by gnuplot. Now open gnuplot using the command:

```
$gnuplot
```

and then type the following command:

```
>plot "output.out"
```

A graph of the output.out file appears. In this graph, each point corresponds to a variable. The x value corresponds to the first column (in this case step number) and the y value is the second column (in this case temperature). Note: the difference between plotting a function and plotting a file is that when plotting a file put quotation marks in the file name such as "output.out" and "STATUS_ela2.gnu", but when only plotting a function, such as sin(x), there is no need for quotation marks.

To select different columns as values for the x and y axis, type "using" as follows:

```
>plot "output.out" using 1:3
```

This command tells gnuplot to plot values of the third column versus the first column. The "using" option also allows gnuplot to plot functions of column variables. For instance, the product of column 1 (step) times 0.003 denotes the simulation time, because 0.003 (ps) is the time interval per step. This product can be used as the x axis; any algebraic operation of columns can be taken as the y axis, such as the sum of columns 3 and 4. The corresponding command is as follows:

```
>plot "output.out" using ($1*0.003):($3+$4)
```

This command plots the sum of the kinetic and potential energies (see columns 3 and 4 of the output.out file) as a function of the time. Note: the symbol $ followed by the column number indicates the value of that column; therefore the expression ($1*0.003) means the value of column 1 times 0.003.

Also in the case of data plotting, it is possible to plot more than one data set in the same graph, for example:

```
>plot "output.out" using 1:4, "output.out" using 1:3
>plot "output.out" using 1:2, using 1:4
```

## Exercise 10.4

Use gnuplot to verify that the temperature and kinetic energy are mutually related. Hint: plot the two corresponding data sets on the same graph, and use the "using" function.

### 10.4.1.4  Creating .png, .jpg, and .eps Files

The following example shows how to produce .png files:

```
gnuplot> plot "output.out" using ($1*0.003):($2)
gnuplot> set xlabal "Time t (ps)"
gnuplot> set ylabel "Temperature T(K)"
gnuplot> set terminal png
gnuplot>set output "T_t_plot.png"
gnuplot> replot
gnuplot>exit
```

After the .png file has been created, use "winscp" to move that file to a directory of Windows and then use "Paint" to print the figure. From "All Programs" go to Accessories, Paint and then File; open the file in the saved directory by browser, then you can look at and print that figure. If you want to produce a .jpg file, you just need to change the lines gnuplot>set terminal png to gnuplot>set terminal jpeg and gnuplot>set output "T_t_plot.png" to gnuplot>set output "T_t_plot.jpg".

The best way to create a PostScript image file in gnuplot is to use a "script", i.e., a file with a list of statements and commands. Enter the UNITS/UNIT4/GNUPLOT_ATOMEYE_EXAMPLE directory and open the example1.gnu file using vi (vi example1.gnu). This file is an example of the script, which reads:

```
reset
set terminal postscript enhanced eps
set output "example1.eps"
#
set xlabel "t"
set ylabel "E"
#
set xrange[0:1]
set yrange[0:7000]
#
plot "output.out" using ($1*0.003):3,"output.out" using ($1*0.003):4
#
reset
set terminal X11
reset
```

The first three lines of the code are used to meet the two basic requirements in creating a PostScript image file in gnuplot: (1) Change the previous "terminal" and (2) define the name of the output file. Therefore, line 1 resets any previous specifications, line 2 sets a new terminal, and line 3 sets the name of the output file.

Lines 5 and 6 specify the name of the axes, in particular the x axis label is defined as "t" and the y axis label is defined as "E". Lines 8 and 9 specify the x and y interval. Line 11 states the command to plot the file. In this particular case two lines are plotted, including the third and the fourth columns of the "output. out" file as functions of the first column times 0.003 (i.e., the time).

Lines 13–15 reset the terminal and the option. These should not be altered. Copy the output.out file in the MD_CODE_vmd directory to the current directory using the command:

```
...../GNUPLOT_EXAMPLE$ cp ../MD_CODE_vmd/output.out.
```

Then open gnuplot and type the command:

```
>load "example1.gnu"
```

Close gnuplot (exit or q) and find the new file, "example1.eps", created by the third line of the "example1.gnu" file in the shell window. This is a PostScript file. If the computer has "gv" installed then one can open it by using a PostScript viewer, for example:

```
$gv example1.eps
```

Note that the x and y ranges and labels match those specified in the "example1.gnu" file.

## Exercise 10.5

Create a plot called "regime.eps" with an x interval from 0.4 to 0.9 and a y interval consisting of column 3 of the "output.out" file. Hint: make a copy of "example1.gnu" and modify it.

### 10.4.1.5  Producing .pdf and .jpg Files from .eps and .tga Files on a Linux System

File formats created by VMD and gnuplot, such as .eps and .tga files, can be converted to .pdf and .jpg formats using "epstopdf" (or "ps2pdf") and "convert" commands as follows:

```
$ epstopdf example2.eps            ! Transform. eps file to. pdf file

$ ps2pdf example2.eps              ! Transform. epd file to. pdf file
$ convert snap.tga snap.jpg         ! Transfer snap.tga file from VMD
                                    output to. jpg file

$ convert example2.eps example2.jpg  ! Transform. eps file to. jpg file
```

In many cases, .pdf file format is favorable.

### 10.4.1.6  Abbreviated Commands and Additional Features

All gnuplot commands have an abbreviation that can be used to simplify typing when creating gnuplot images. For example, instead of using the "plot" command, "p" can be used to perform the same operation. The script example3.gnu is identical to example2.gnu, except several of the commands have been used in the abbreviated form. Open both files and compare the differences. Additional gnuplot features can be added to the image as well, such as line style, titles, and interval marks.

## Exercise 10.6

What is the meaning of "w l ls 1" in line 19 of the example3.gnu file? Hint: compare with the example2.gnu file.

## 10.4.2  Visual Molecular Dynamics (VMD)

VMD will be used to plot instantaneous configurations of the MD simulations. See the website http://www.ks.uiuc.edu/Research/vmd/ for more details. Open the UNITS/UNIT4/MD_CODE_vmd directory and check whether there are crystal_M_simple.f90 and multi1.f90 files along with their input files Model_M_Simple.in and mt1.in.

### 10.4.2.1  Procedure to Develop xyz-type File for Visualization

Xyz-type files such as conf_in.xyz and conf.xyz are files including information of atomistic configuration at different time steps to show the evolution of the model with time. It includes the atom name and the xyz coordinates of all atoms in the simulation system and at each required recording time step. The total step number of the MD simulation in the multi1.f90 code is controlled by the input file "mt1.in" as discussed in Section 10.3.4. Modify the step number in the second line of "mt1.in" to be 10 using the "vi mt1.in" command so that the simulation lasts for only 10 increment steps to save time. For the time being, the geometric model which has been produced in Section 10.3.3 should not be changed; it can be reproduced in the current directory by source file crystal_M_simple.f90 with the data file Model_M_simple.in, or directly copy the conf_in.dat file in Section 10.3.3 by the following steps:

- Open the UNIT4/MD_CODE_vmd directory.
- Copy the conf_in.dat file from the UNIT2/MD_CODE/MD_CODE_OUT directory using the following command:

```
~ UNIT4/MD_CODE_vmd$ cp ../../UNIT2/MD_CODE_MD_CODE_out/conf_in.dat.
```

- Compile the code using the following command:

```
$ gfortran multi1.f90 -o multi1.exe
```

- Run it using the following command:

```
./multi1.exe
```

- Check the output using "ls -lrt". A new file should be created called "conf.xyz".
- Open this file with vi. The first line shows the number of atoms (e.g., 256). The second line shows the timestep i which corresponds to the block with 256 lines followed. These lines show the 256 atoms in the simulation at that time step. Each line provides information of each atom's position. The format of the structure is determined by the write command of line 188 of multi1.f90 as shown above. Specifically, among the four columns, the first one is the atom name, and the other three denote the atom's x, y, and z coordinates. These coordinates of all the atoms give the configuration of the simulation system at the given step or time.
- The above block structure is repeated many times during the simulation process. In this particular case, the configurations are printed out for each time step, thus there are 10 blocks for 10 configurations, which show how the atomic system evolves with time. The total lines of the conf.xyz can be calculated using 10 configurations with 256 atoms, or $10 \times (256 + 2) = 2580$.

### 10.4.2.2 VMD Windows/Menus and Procedures for Visualization

This kind of file is called an "xyz" file (e.g., conf.xyz and conf_in.xyz), and can be read by VMD. Note: VMD has a very good tutorial and a well-written user manual which can be found on the VMD website. To use VMD:

1. Type "vmd" in the terminal where these files are located (e.g., MD_CODE_vmd) to open the VMD program.
2. Several windows will appear, one of these is called "VMD main", where several menus can be found including file, molecule, graphics, and so on at the top of the window.
3. In the menu "file", select the "new molecule" command. A new window called "molecular file browser" will appear.
4. Click the "browse" button on this window and select the file "conf.xyz" (or conf_in.xyz).
5. Click the "load" button.

Two things should have happened. First, the window with a black background and many green lines should have appeared; and second, the string like the following should have appeared in the VMD main window:

```
0 T A D F conf.xyz  256  10  0
```

The values in the string correspond to the operations performed: "conf.xyz" is the name of the file that was loaded, the number 256 is the number of atoms, and the number 10 is the number of frames (configurations) or blocks in the conf.xyz file. The first number in the string, in this case 0, indicates the reference ID of the molecule.

VMD visualizes the last frame which shows the configuration of the last step within the black background window. Each intermediate step in the 10-step series can be viewed by moving the horizontal scrolling bar in the bottom part of the VMD main window left or right. Try this and note the very slight differences in the image as the steps progress.

The visualization will be changed by the following operations:

1. On the graphics menu in the VMD main window select the "representation" command. A new window will appear.
2. This window is very well described in the VMD tutorial and user manual. First, select a different kind of representation style. Change the drawing method from "lines" to "VDW". Note: VDW requires more memory to compute. It is better to start with "lines" for the whole model. Then, if it is desired to show a more detailed partial view of the model, use VDW or other more time-consuming representation styles.
3. To see all atoms in the simulation, change the sphere scale from 1.0 to 0.2 in the text box at the right of the graphic window.
4. Close the graphical representation window.
5. On the display menu select "orthographic" instead of "perspective".
6. Now use the mouse on the black background simulation window to rotate the crystal image to view the structure for the current configuration.

### 10.4.2.3 Periodicity Check with VMD

An important feature of vmd is the ability to create periodic images of the loaded molecule. It also allows the user to check if the box dimensions used in the MD simulation are compatible with the initial configuration. In particular, one can check if some atoms overlap due to the periodicity. This feature will be used several times in the next sections. To plot periodic images of the loaded molecule, perform the following steps:

1. Load a molecule. Follow the steps described in the previous section and load the molecule "conf.xyz". If it is already done, you do not need to start from the very beginning.
2. On the "extension" menu of the VMD main window, select the Tk console command. A terminal will appear which will allow the user to perform a wide variety of operations. At this point only a few commands will be introduced. For further instruction see the VMD User's Guide "8 Tcl Text Interface."
3. Write the following command in the Tk console: "molinfo 0 set a 6.8". Press "Enter". Molinfo 0 denotes the molecule "0" which is the ID of the currently loaded molecule. As said before, this ID is specified by the first number of the string on the VMD main window. The remainder of the command including "set a 6.8" sets the cell dimension along the a-direction to be 6.8 Å. Do the same for b and c to be 6.8 Å with the other two comment lines such as "molinfo 0 set b 6.8". In VMD these three variables are the periodic box dimension for the x, y and z axes. These values are a little larger than the model size of 6.4 Å ($=4 \times 1.6$). This important and necessary gap is discussed in detail in Section 10.9.2.2 and Figure 10.9.
4. Now, use VMD to plot the periodic image. Open the representation window ("representation" command under the "graphics" menu of the VMD main window) and select the "periodic" tabs in the middle of the window.
5. Inside the periodic tab select the desired periodic image, for instance $+X$ means the periodic image along the positive x axis.

## Exercise 10.7

Try to change the value of the variable a (for instance "molinfo 0 set a 7.5"). What are the changes you find in your image?

### 10.4.2.4  Exporting and Changing the Default Color of the VMD Image

To export an image, select the "render" command on the "file" menu of the VMD main window. Within the render window, several options can be modified; however, the default option (tga) is sufficient to view a good image. The image file is created in the same directory in which VMD was opened. Click "start rendering" and then check the directory using the "ls -lrt" command to see if the tga file was created. To open the image, use the "display snap.tga" command on the UNIX terminal. One may also change it to a .jpg file (see Section 10.4.1.5).

Another very common procedure to make visualization easier and allow for printing is to change the background and atom default colors. Typically, a black background with light atoms is appropriate for visualization, but a white or light gray background with dark atoms is preferred for printing. To change the default colors select the "colors" option on the "graphics" menu in the VMD main window. A window called "color controls" will appear. Select "display" in the "categories" subwindow located at the top-left corner of the window, and select the desired background and atom colors by clicking "background" in "names" and then clicking colors, say "white", then the background will change color from black to white.

## 10.4.3   AtomEye

Among the numerous MD visualization software packages available, AtomEye[3,4] is the free, preferred software that will be used throughout this section. AtomEye software can be found along with a detailed manual and image gallery at the http://mt.seas.upenn.edu/Archive/Archive/Graphics/A/ website.

### 10.4.3.1  File Format and Operations

AtomEye supports the following formats: PDB, Standard CFG and extended CFG. There are four example files including, 2a.cfg, 8000a.cfg, 30000a.cfg, and 60000a.cfg, in the UNITS/UNIT4/GNUPLOT_ATOMEYE_EXAMPLES directory. The number in the file name denotes the number of loading steps.

If using windows, the software "cygwin" should be downloaded from the website: http://x.cygwin.com. Roughly speaking, cygwin is a Linux-like system for Windows. It will allow one to use some basic Linux stuff such as terminals (or shell). Xterm is the name of a Linux terminal. After downloading is completed, put the .cfg file in a consistent directory in the path. One also needs to create an Xterm window, then type the command:

```
$ A filename.cfg    (example:A 8000a.cfg)
```

A text window and a screen (visualization) window will appear.

In order to have a good visualization, the object (model) should be rotated or shifted. The object can be rotated using keyboard arrow keys or by dragging the mouse. Press the left, right, up, and down arrow keys to rotate the object as if the object was rolling inside a ball. For in-plane rotation, use Shift + Up (clockwise) and Shift + Down (counter-clockwise). Shifting of the object in-plane can be done by simultaneously pressing Ctrl + Left, Ctrl + Right, Ctrl + Up, or Ctrl + Down. The object can be pushed away from the viewpoint or zoomed out using Ctrl + Shift + Up, while Ctrl + Shift + Down will pull it closer or zoom in.

As we discussed in Chapter 2 (also see Sections 10.6 and 10.7), periodic boundary condition (PBC) is important for simulation. To shift objects under PBC, press the Shift button while dragging the mouse.

Alternatively, when Caps Lock is on, the left, right, up, down arrow keys, Shift + Up, Shift + Down will perform equivalent operations, such as shift the object described above, except now with the PBC. Press Z to recover the initial PBC state, where there is no shift.

### 10.4.3.2 Obtaining Information for Atomistic Simulations with AtomEye

The following functions that AtomEye offers are used frequently.

#### Find basic information for an individual atom

This is realized by clicking that atom on the screen or typing the number of the individual atom in the bottom line of the accompanying window. Specifically, by selecting an atom, relevant information about that atom will appear in the accompanying text box. The information shown will include the atom's ID number, its x, y, z coordinates of position vector and shear strain, and its coordination number (CN) with its neighbors. To find an atom if one only knows its number but not its location, pressing F on the keyboard and then typing the number of the desired atom will show the atom's location and other information and allow one to find it, since the model will move around it.

#### Coordination number (CN) coloring

This is the most useful method to show the atomistic configuration around a defect, such as a dislocation. The CN is the number of the nearest neighboring atoms of the selected atom. The definition of nearest neighbors can be changed by the "cutoff control" command in AtomEye. The CN depends on the size and structures of adjacent atoms, therefore the value of the CN can reflect the density of the adjacent atom region. For the normal FCC and HCP structures inside the body, the CN is 12, while on the surface the number is 8. For BCC inside the body, the CN is 10. The CN can also be used effectively to detect atom defects, because when defects occur the CN at the core positions of the defects will be larger, say, 13, 14, or 15, compared to that of a normal atom.

When the AtomEye screen is opened using the command "A filename", a table will appear which shows the atom number in the system which has a certain CN value. In the table its corresponding color in the configuration is also given. This is very helpful to find the defect initiation step or strain value. Using an FCC and HCP crystal as examples, if the CN values all equal or are less than 12, that configuration is normal, and no defects have occurred. If the CN value is larger than 12, it denotes that the defect occurs. Following this method, it is possible to determine the loading step or strain that will cause the defect to start. It is then possible to watch the development of the defect by monitoring the increase in atoms with high CN.

Press K on the keyboard to toggle the coordination number coloring. To use the CN for the detection of defects, use the "make atoms invisible" command to make certain kinds of atoms invisible. These atoms which will not be seen on the screen can be atoms with regular CN (e.g., CN equals 12 or less for FCC and HCP). For example, Figure 10.3 shows how to use the CN to investigate the defect initiation and evolution. Specifically, Figure 10.3 (a) shows that other than some atoms at the right and left edges, most atoms have $CN \leq 12$ at 280,000 steps (strain = 15.6%). Figure 10.3 (b) shows the result of the command "make atoms invisible" to make these atoms ($CN \leq 12$) disappear, leaving only atoms with CN > 12 in the figure. This figure shows that the dislocation starts at the edges of the nanoscale bar. Figure 10.3 (c) shows the dislocation evolution to a large area at the two side edges at 320,000 steps. In UNIT2/F90_CODES is a code CFG_convrt_cmd.f90 which can be used to develop .cfg files for visualization from the output files of atomistic simulations.

#### Make atoms invisible

To clearly see defect (e.g., dislocation) cores, it is often best to remove the perfectly coordinated atoms as shown in Figure 10.3 This is done using the "make atoms invisible" function in AtomEye. Specifically, Ctrl + Shift + Right-click will make a certain species (e.g., atoms with a low CN value) invisible. The coloring of atoms can be restored by pressing the O key on the keyboard.

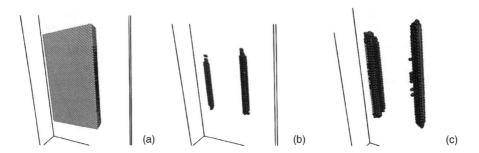

**Figure 10.3** Using the coordination number (CN) to explore locations of defect (dislocation) initiation and evolution

**Central symmetry parameter coloring**

There are some important defects, such as twinning and stacking fault, which cannot be demonstrated by the CN because the CN does not change when these defects occur. In order to demonstrate these defects, a central symmetry parameter $c_i$ surrounding relevant atoms is adopted.[5] Figure 10.4 shows a high $c_i$ region, which denotes the stacking faults. Thus, twinning and stacking fault can be found by AtomEye's function of central symmetry parameter coloring. To do so, pressing Meta(Alt) + H will color-code the atoms according to their central symmetry parameter. In Figure 10.4, using the "make atoms invisible" function, atoms with $c_i$ less than 0.00376 are not shown.

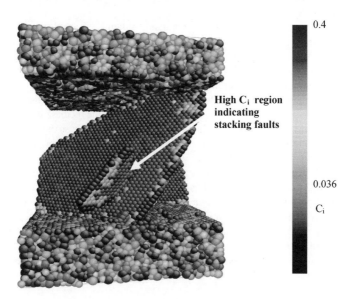

**Figure 10.4** Using central symmetry parameter $C_i$ to find the stacking faults of Cu/Zr nanolayer (Reproduced with permission from Wang, Y. M., Li, J., Hamza, A. V., and Barbee, T. W. (2007) Ductile crystalline-amorphous nanolaminates. *Proceedings of the National Academy of Sciences of the United States of America*, 104, 11155. Copyright © National Academy of Sciences)

**Atomistic local von Mises shear strain invariant coloring**
Typing Meta(Alt) + G will color-code the atoms according to their local von Mises shear strain invariant. Shift + G will toggle whether the system-averaged strain tensor should be included or disregarded when computing the invariant. AtomEye assumes that the default is no.

**Making jpeg, png, and eps screenshot files**
Press J on the keyboard to make a .jpg screenshot file. Press P to make a .png screenshot file. Press E to make a high-resolution Encapsulated PostScript Screenshot .eps file.

---

**Homework**

(10.15)  Use the crystal_M_simple.f90 code to develop a Lx × Ly × Lz = 5a × 4a × 4a FCC atomistic model where Lx, Ly, Lz are the model sizes along the x, y, and z direction, and a = 1.6 Å is the lattice constant of the basic cell. Run a 300 time step MD simulation by the multi1.f90 code for the model created, found in the UNITS/UNIT2/MD_CODE directory.

Prepare a two-page report containing two images.

(1)  The first figure must contain two curves: (a) the temperature time evolution T(t) where T is the instantaneous temperature and t is the time (t = $\Delta t \times n$; n steps, $\Delta t$ time step described in multi1.f90), and (b) for comparison, a line expressing the mean value of the temperature during the simulation period. The mean value is a constant, therefore it should be a straight horizontal line. Hint: Use the code developed in the previous UNIT homework to calculate the mean value of the temperature of the simulation period. Discuss the meaning of the temperature versus time evolution.

(2)  The second figure is a snapshot of the first and the last configuration developed in VMD. Hint: The source file is the crystal_M_simple.f90 file and the data file is the Model_M_-simple.in file, found in the UNIT2/MD_CODE/directory. It is important to modify the size of the data file to satisfy the values of the unit cell. The code is for an FCC structure. Modify the data in the input file and then run the code to produce the atomistic model.

---

# 10.5 Running an Atomistic Simulation Using a Public MD Software DL_POLY

In Section 10.3 a simple MD code, multi1.f90, is introduced for readers to develop a basic understanding of the MD code structure. However, this type of code is not efficient and accurate for a model with more than 500 atoms. This section will introduce how to perform an MD simulation using various MD software packages.

## 10.5.1 Introduction

There are several high-quality MD programs in the public domain, including:

- LAMMPS:[6] Developed by the Sandia National Laboratory, USA http://lammps.sandia.gov.
- DL_POLY:[7,8] Developed by CCLRC Daresbury Laboratory, Cheshire UK, http://www.cse.clrc.ac.uk/msi/software/DL_POLY.
- Moldy:[9] A portable molecular dynamics simulation program for serial and parallel computers.
- NAMD:[10,11,13] Focuses primarily on biomolecular simulation, developed by the University of Illinois and Beckman Institute, USA.
- IMD:[12] Developed by University of Stuttgart, Germany.

In the following sections, DL_POLY_2 and NAMD will be used to show how MD software packages can carry on atomistic simulations. DL_POLY_2 is a package of subroutines, programs and data files, designed to facilitate MD simulations. This package has been developed under the support of the Engineering and Physical Science Research Council, UK, and is the property of the Council for the Central Laboratory of the Research Councils. DL_POLY_2 will be used in Sections  for simulation of one-phase and two-phase metals. NAMD is a powerful public MD software, which will be used to simulate a protein-water system under tensile loading in Sections 10.10–10.12.

While both DL_POLY_2 and NAMD are an academic free package, readers must receive an academic license for DL_POLY_2 and register for NAMD/VMD through their websites before using these packages. Licensing and registration is easy to complete by following the instructions of these websites.

## 10.5.2   Installation and Structure of DL_POLY_2

After obtaining a DL_POLY_2 license, copy the corresponding DL_POLY_2 package into the UNITS/ dl_poly_2 directory as shown in Figure 10.5. Initially, the dl_poly_2 directory will be empty, awaiting a copy of the copyrighted files. (Due to copyright restriction, no directories or specific input or output files from DL_POLY software could be included. Therefore UNIT5 will not include any "_out" directory for readers to check the running process.) The version of DL_POLY_2 described in this chapter is DL_POLY_2.18, which may differ from the latest versions. Usually, the difference between versions is small, so any version after 2.18 can be used. After installation, the dl_poly_2 directory will have eight subdirectories as shown in the directory tree of home/UNITS/dl_poly_2.

The first frequently used directory is "srcmod" which includes many primary subroutines including the main subroutine dlpoly.f; the compiling process is carried out in this directory. The second frequently used directory is "execute" used to execute the MD simulation. The directory "data" has 31 MD examples including input and output files, covering metals (Al, Cu, Ni-Al), ceramics (K-Na disilicate glass), DNA, carbon nanotubes, and carbon diamond with applications of different potentials. Each of these examples has the corresponding directory labeled from TEST1 through TEST31. In each directory, the LF and VV subdirectories can be found. The first one uses a leap-frog algorithm and the second one uses a Velocity Verlet algorithm (see Sections 2.5.2 and 2.5.3). One may refer to the DL_POLY_2 manual for the functions of other directories.

## 10.5.3   General Features of DL_POLY_2 Files

In DL_POLY the initial configuration is given in a formatted file called "CONFIG"; details of the interatomic potential are defined in the "FIELD" file (sometimes, a Table file is needed for potentials); the simulation control mode and parameters such as $\Delta t$, f, temperature T, total time steps, etc. are defined in the "CONTROL" file. Therefore the user must prepare these three files for the simulation before running DL_POLY (sometimes one needs to prepare the TABLE file if potential is prepared by tables). Within this section, the CONFIG, CONTROL, FIELD and Table input files used in dl_poly_2.18/data/TEST1 will be discussed. Starting from Section 10.4, input files of different problems will be developed. In Section 10.4, argon is taken as an example for both simulations under ensembles nve and npt. Readers who want to use these files can go directly to Section 10.4 after gaining a basic knowledge of DL_POLY through reading this section.

Section 10.5.4 will introduce several output files including OUTPUT, REVCON, STATIS and HISTORY files. Both OUTPUT and REVCON files are mandatory, the first one summarizes the simulation conditions, the main results or error message; the REVCON file is a re-start file including all the information about atoms at the last step of simulation. Thus it can be changed to the CONFIG file for the next simulation.

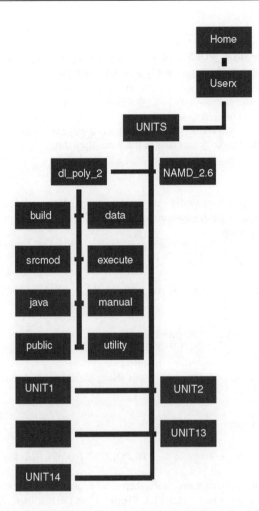

**Figure 10.5** Directory tree for UNITS directory with emphasis to the structures of DL_POLY_2. (The dl_poly_2 and NAMD_2.6 directories are empty until readers put the codes in)

STATIS and HISTORY files contain a lot of data but they are non-mandatory. The user can use command in the CONTROL file to state whether and how these files should be printed. HISTORY is a huge file which one must be careful to deal with. By using two softwares, ela_STATIS.f90 and ela_history_2009.f90, developed in this book, quantitative data processing of these two files can be much easier to carry out (see Sections 10.6–10.9).

## 10.5.4 Compile and Run

To compile a Fortran code connected with many files, the standard "gfortran…" compiling command used in the previous UNITs cannot be used. Each program package should include a file called Makefile. There are two types of Makefiles. The first is called "MakeSEQ" for sequential operation, and the other is "MakePAR" for parallel operation of the computer cluster. They are stored in the subdirectory "build" of dl_poly_2.

## 10.5.4.1  Compiling

For compiling a simulation in DL_POLY_2 perform the following steps in the srcmod directory. Clean all unnecessary files that may have been previously produced by typing

```
~srcmod$ make clean
```

Look what the compiling systems are, which have been previously set in DL_POLY_2. This information can be found by typing the "make" command in the same shell, i.e.,

```
~srcmod$ make
```

Then a series of existing target settings will be shown in the screen. If you can find one of the previous sets that match the user's computer cluster system, say "gfortran" then type the following command in the shell srcmod directory for compiling:

```
~/.../srcmod $make gfortran
```

For some cluster settings with the ifort command, the target may match "intel-linux-ifc"; in this case use the following "make" command:

```
~/.../srcmod $make intel-linux-ifc
```

These commands will require about one minute to complete because the commands compile all related Fortran files in the srcmod directory. When the process is completed the following message will appear:

```
$mv DLPOLY.X../execute/DLPOLY.X
```

The message indicates that the executable file "DLPOLY.X" has moved into the execute directory of DL_POLY_2.18.

If no changes to the source codes are necessary, there is no need to compile the software again. In other words, if only input files change (e.g., CONFIG, FIELD, CONTROL), there is no need to run a new compilation. In fact, in Section 10.10 for the software NAMD we only offer compiled files such as namd32 and namd64 for users to run simulations, no compiling will be conducted.

## 10.5.4.2  Execution

Computations can occur in the execute directory or in any directory containing the CONFIG, FIELD and CONTROL input files. The details for the two cases are described as follows.

### Running in a directory that does not contain the executable file

Suppose that it is desired to run the TEST1 example of DL_POLY_2 for the simulation of K-Na disilicate glass at 1000 K in the UNIT5/Simu_4 directory, the procedure is as follows.

Enter the UNITS/UNIT5/Simu_4 directory. Move the CONFIG, CONTROL, FIELD and Table files of TEST1 from the DL_POLY_2 package into this directory. These files can be found in the dl_poly_2.18/data/TEST1/LF directory, accessed using the command:

```
...~UNIT5/Simu_4 $ cp ../../dl_poly_2.18/data/TEST1/LF/CONFIG.
```

Repeat this command to move other files (CONTROL, FIELD, and Table files) into the directory. Note: (1) The last period "." in the command is to copy the file to the current directory "Simu_4". (2) If using a newer version, such as DL_POLY_2.20, then replace dl_poly_2.18 with dl_poly_2.20 in the command. (3) The table file is required to provide potentials between oxygen-oxygen ions and oxygen-silicon ions described in the FIELD file.

Run DLPOLY.X in the directory using the following command:

```
$..Simu_4$. /../../dl_poly_2.18/execute/DLPOLY.X
```

This process will use the copied CONFIG, FIELD, CONTROL, and Table input files in the UNIT5/ Simu_4 directory to run the simulation in the execute directory of DL_POLY_2.18. Because the DL_POLY_2.18 directory is one level higher than the Simu_4 directory, the command should include the appropriate "./../../dl_poly_2.18/" to change to the correct directory. After the simulation is complete, type:

```
$ls -lrt
```

Several output files have been created. Compare the output files obtained during this running process with those listed in the LF or VV directory offered by the DL_POLY_2 package to see whether they are consistent.

**Running the simulation in the execute directory**
Move the CONFIG, CONTROL, FIELD and Table files directly to the execute directory using the command:

```
~ execute$ cp ../data/TEST1/VV/CONFIG.
```

Just like the previous commands, the last period "." in the above line is important to transfer the file to the current directory.

After all four files have been moved into the execute directory, start the simulation by typing:

```
$. /DLPOLY.X
```

## 10.5.5    Units of Measure

Internally, all DL_POLY_2 subroutines and functions assume the following molecular units as defined:

- Time: $1 \times 10^{-12}$ seconds, i.e., picoseconds (ps)
- Length: $1 \times 10^{-10}$ meters, i.e., angstroms (Å)
- Mass: $1.6605402 \times 10^{-27}$ kilograms (i.e., atomic mass units)
- Charge: $1.60217733 \times 10^{-19}$ coulombs (i.e., unit of proton charge)
- Energy: $1.6605402 \times 10^{-23}$ joules ($10 \, \text{J mol}^{-1}$)
- Pressure or stress: $1.6605402 \times 10^{7}$ pascal (163.882576 atm)
- Force: $0.16605402 \times 10^{-12} \, \text{N}$ ($0.16605402 \, \text{pN}$)

In the CONTROL and OUTPUT files, however, the pressure is given in units of kilo-atmospheres (katm); and the unit of energy is either the DL_POLY units specified above, or other units specified in the second line of the FIELD file. The energy can be in eV for electron-volts, kcal for kilocalories $\text{mol}^{-1}$, kJ for kilojoules $\text{mol}^{-1}$ or K for Kelvin$^{-1}$.

Frequently, it is necessary to convert the units in the FIELD file to a more appropriate unit scale. For example, when converting eV to kcal mol$^{-1}$, the following basic data is necessary for the translation:

- 1 eV $= 1.60219 \times 10^{-19}$ J (see Section 2.3.4)
- 1 kcal $= 4184$ J
- Avogadro's Number (atoms per mole) N $= 6.02217 \times 10^{23}$

Therefore:

$$1\,\text{eV} = 1.60219 \times 10^{-19} \times \frac{1}{4184} \times 6.02217 \times 10^{23} = 23.06085\,\text{kcal/mol} \tag{10.1}$$

$$1\,\text{eV} = 1.60219 \times 10^{-19} \times 6.02217 \times 10^{23} = 96.4866\,\text{kJ/mol} \tag{10.2}$$

## 10.5.6   Input Files of DL_POLY

In this section, the input files of `glass structure` described in dl_poly_2.18/data/TEST1 are introduced.

### 10.5.6.1  Structure and Parameters of the CONFIG File

Open the CONFIG file of the TEST1 directory with "vi". The first nine lines in the CONFIG file contain the following information:

```
DL_POLY TEST CASE 1: K Na disilicate glass structure
      2    3
  24.1790000000    .0000000000    .0000000000
  . 0000000000  24.1790000000    .0000000000
  . 0000000000    .0000000000  24.1790000000
Na+         1
  -10.18970354   -11.14553975    2.950816701
  -10.92491513   -11.32922344   -1.683043107
   80710.967958    7831.492182   14290.88665
K_+      2
   4.203354201   -6.599949388   11.67055019
  -.4336920163  -10.629860244    .5802665381
   14372.08258    98010.543805   4104.320538
```

**Line 1:** The simulation name, which can be altered by the user to any name desired.
**Line 2:** The first number, 2, is the "levcfg". If levcfg is 0, only the position coordinates (x, y, z) of each atom are shown. If levcfg $= 1$, the velocity of each atom will also be shown; and if levcfg $= 2$, then position, velocity, and force of each atom will be shown. The second number is the "IMCON", which can be any one of seven numbers that indicate different periodic boundary conditions. See Appendix 10.B of DL_POLY_2 manual for more information. For the present case, IMCON $= 3$ indicates that the boundary condition is the parallelepiped periodic boundary condition.
**Lines 3–5:** There are nine real numbers with three per line. Each line shows, respectively, the lattice axis vector $\bar{a}, \bar{b}, \bar{c}$ with respect to the global x, y, and z axis. If only diagonal values are not zero, then these diagonal values denote the model size along the x, y, and z direction, respectively.

**Line 6:** Shows the atom's name and ID number.

**Line 7:** Shows the atom's x, y, z coordinates.

**Line 8:** When levcfg is not zero, line 8 presents velocity components. The initial velocity can be given in the CONFIG file by previous calculation. DL_POLY can also randomly generate initial velocities for a given temperature.

**Line 9:** If levcfg = 2, line 9 presents the force components of the atom.

Lines 6–9 are then repeated for K. Note: The CONFIG file is a formatted file, ensure that there are no changes to the position of the periods between the integer part and floating part of the values.

### 10.5.6.2 FIELD File

A FIELD file contains detailed information about the molecular type, molecular numbers of each type, and parameters of potentials. For each molecule, the information about its atoms including atom number, weight and charges are described. In the case of the TEST1 example, the first nine lines of the FIELD file are given as follows:

```
DL_POLY TEST CASE 1: K Na disilicate glass forcefield
UNITS
MOLECULES 3
Potassium Sodium
nummols 120
atoms 2
Na+      22.9898   1.000
K_+      39.1000   1.000
Finish
```

The first line is the simulation title, and the second line is the unit of energy as mentioned above. If there is no specified unit in this line, it indicates that the default unit of DL_POLY_2 described in Section 10.5.5 will be used.

The third line contains the number of molecule types, which in TEST1 includes three types. For each molecule there is a block constituted by several lines, in this case lines 4–9. The first line of each block contains the molecule name (potassium sodium here), followed by a line that contains the number of molecules (120 here), while the third denotes the number of atoms in each molecule (2 here). Then for each atom there is a line in which the atom name, mass and charge is reported. The "finish" statement closes the block. There are three blocks in the TEST1 example. The other two are for silicon and oxygen.

The FIELD file is strictly related to the CONFIG file. The number of atoms of each type of molecule in the CONFIG file must be the same as described in the FIELD file. For instance, in the corresponding CONFIG file of the above FIELD file, initially there must be 120 potassium sodium molecules with alternatively changed Na and K atoms. If the position order or the number of Na and K atoms is different from the FIELD file, an error message will appear.

After listing all molecule blocks, the file will contain interatomic interactions. In the TEST1 example, there are ten pair potential interactions where eight are given by Buckingham (Buck) pair analytical potential, and the remaining two pair potentials are given by Tables (tab) with numerical values. There are also two three-body potentials (TBP) with the bvs2 type. The detailed expressions of these potentials can be found in the manual (e.g., Table 4.7 and 4.17), including the values that correspond to the desired parameters.

```
VDW 10
K    K    buck    0.00    1.0000
.........................................................................................
O    Si    tab
O    O     tab
TBP 2
O_2-    Si4+    O_2-    bvs2    4824.2644 109.4666667    1.2    2.6    3.45
Si4+    O_2-    Si4+ bvs2    2412.1322 144.0000000    4.0    2.6    3.45
```

## Exercise 10.8

DL_POLY "internal" units are defined in 10.5.5. Compare the value of the epsilon parameter of the Argon LJ potential with the value found in the literature.

### 10.5.6.3 CONTROL File

The CONTROL file contains information for the simulation, including the number of time steps, the cutoff radius, the ensemble in which the simulation is performed, etc. Specifically, the CONTROL file in TEST1 of DL_POLY_2, found in the data/TEST1/LF/CONTROL directory, includes the following data: Steps = 500, temperature = 1000 K, pressure = 0, cutoff radius = 12, and the ensemble is nve.

## 10.5.7   Output Files

### 10.5.7.1 OUTPUT File

The OUTPUT file contains a summary of the simulation conditions described in the CONTROL and FIELD files. It briefly shows the results of the initial and final data for the position, velocity and force of selected atoms, as well as 30 parameters at various steps which are the statistical averaging values of parameters over a large number of time steps. If the simulation has errors, the error number can be found in the OUTPUT file, which can be used to determine the source of the error as described by the code troubleshooting portion of Appendix 10.C: "DL_POLY Error Messages and User Action".

If in the CONTROL file the lines of "rdf" and "print rdf" are written, at the last part of the OUTPUT file, radial distribution function will be given. For each function, a head line states the atom types ('a' and 'b') represented by the function, and the configuration number for averaging will be given followed by three columns of r, g(r) and n(r) in a tabular form.

Function g(r) is defined by $g(r) = \rho(r)/\rho_0$. Function $\rho(r)$ is the local density at r which is calculated in a small annular region between $(r + \Delta r/2)$ and $(r - \Delta r/2)$, and $\rho_0 = N/V$ is the averaged atomic volume density. n(r) is the average number of atoms of type 'b' within a sphere of radius r around an atom of type 'a'. Note that a readable version of the rdf data is provided by the RDFDAT output file. The rdf distribution data is important since it enables the performance of quantitative analysis of atomistic structure to see whether the system is in solid crystal structure, or at the liquid or gas state. Specifically, for crystal structure, it should show some kind of periodic peaks and valleys in the distribution curve versus the radius due to the periodic structure of crystal. On the other hand, a gas state may show a continuous curve without many peak and valleys.

### 10.5.7.2 HISTORY File

The HISTORY file contains the time evolution of the molecular configuration. It is defined in the CONTROL file by the command "traj i j k". The variable "i" is the starting time step for dumping configurations, "j" is the interval between configurations, and "k" is the data level which is similar to "levcfg" in the previous sections: 0 prints position; 1 prints position and velocity; and 2 prints position, velocity and force. For example, the command "traj 0 100 0" indicates that the HISTORY file will be written from the beginning of the simulation and writes the atom positions for every 100 time steps.

Open this file with VMD by the following steps:

1. Open VMD.
2. Select the File menu, then select "new molecule". The "molecular file browser" window will appear.
3. Select the HISTORY file in the "filename" tab.
4. Select DL_POLY History in the "determine file type tab" and click "load" in the window.

After the HISTORY file has been loaded, change the visualization image using the representation window. Note: The HISTORY file is not overwritten by the code; if two simulations are run in the same directory, only one HISTORY file is produced. The first part of the file is related to the first simulation, and the last part is related to the second simulation. If the first part is not desired, delete the HISTORY file when starting the new simulation.

## Exercise 10.9

Type the command "traj 10, 50, 2" in the control file to run TEST1 again. Compare the first and the last frame of the image using VMD for the obtained HISTORY file.

### 10.5.7.3 STATIS File

The STATIS file contains the time evolution of all calculated variables such as temperature, kinetic energy, configuration energy, chemical bonding energy, electrostatic energy, total energy, volume, pressure, etc. The first 27 variables are fixed; however, the total number "mxnstk" of variables is based on the following formulas:

```
mxnstk > 27 + ntpam (number of atom types in the model)
       + 9 (if stress tensor is calculated)
       + 9 (if constant pressure simulation is requested)
```

The location of each variable and its definition are described in the DL_POLY_2 user manual. A sample block of variables is given below:

```
 1  5.000000E-03   37
-2.374721E+06 1.953518E+02 -2.376398E+06 -2.376398E+06 0.000000E+00
 0.000000E+00  0.000000E+00 0.000000E+00  0.000000E+00 -3.958073E+06
 0.000000E+00  4.753407E+06 4.753407E+06  0.000000E+00  0.000000E+00
 0.000000E+00  0.000000E+00 0.000000E+00  1.164243E+05  0.000000E+00
 0.000000E+00  0.000000E+00 9.000000E+01  9.000000E+01  9.000000E+01
 0.000000E+00 -9.325196E+02 3.048230E-04 -9.324903E+02 -10.716020E-03
 1.214737E-03 -10.716020E-03 -9.325775E+02 -5.956594E-02 1.214737E-03
-5.956594E-02 -9.324910E+02
```

The first integer number on the first line of the STATIS file denotes the current MD time step "nstep". The second real number is the elapsed simulation time which equals nstep × Δt. The third integer number denotes the variable number (or array elements number) to follow. In the present case, there are 37 variables after the first line. This is the case in which there is only one type of atom and no stress tensor component being produced. The values of these variables are printed into the STATIS file after every n steps where "n" is the pre-set value given in the CONTROL file (after the "stats" statement; in the present case n = 1). Therefore, the number of total blocks is equal to the total step number divided by n. Before the first block, there are two lines to start the STATIS file. The first line contains the simulation name and the second line describes the energy units used in the simulation.

## 10.5.8  Data-Processing for Variable Evolution Versus Time by the ela_STATIS.f90 Code

The above standard form of the STATIS file cannot be plotted by gnuplot, therefore it is not possible to draw the time evolution of the variables. To do this use an ad hoc code called ela_STATIS.f90 within the Simu_4 directory. The following is the procedure to use this code for drawing the parameters of STATIS:

1. Open the Simu_4 directory.
2. Open the ela_STATIS.f90 file with vi and go to line 29 (by typing ":29" at the bottom of the screen).
3. Assign the value 500 to nstep_t (i.e., nstep_t = 500).
4. Assign the value 1 to stat_every (i.e., stat_every = 1).
5. Assign the value 37 to nstpval (i.e., nstpval = 37).
6. Compile the code using the command:

```
$gfortran ela_STATIS.f90 -o ela_STATIS.exe
```

7. Create the directory POST_PROC using the command:

```
$mkdir POST_PROC
```

8. Run the code using the command:

```
$./ela_STATIS.exe
```

9. Enter into the POST_PROC directory.
10. Type "ls -lrt". Three new files have been created in the "POST_PROC" directory, including STATIS_ela.gnu, STATIS_ela2.gnu and statistic.dat. Open each file and review the contents using the "vi" command.

**statistic.dat**
Contains mean values, variance and standard deviation of the variables reported in the STATIS file. The two .gnu files contain exactly the same data reported in the STATIS file, but in a format that can be plotted using gnuplot. Specifically, the structure and usage of STATIS_ela2.gnu is different from STATIS_ela.gnu as follows.

**STATIS_ela2.gnu**
Each line contains 38 numbers, the first is the time and the other 37 are the variables reported in STATIS file as shown in the above sample block. Because nstep_t = 500 steps and stat_every = 1 are chosen, there are 500 lines or blocks. To plot this data with gnuplot, open gnuplot and type the command:

```
>plot "STATIS_ela2.gnu" using 1:3
```

This command plots the time evolution of temperature using the columns for time (column 1) and temperature (column 3). If stat_every = 10 the curve may not be quite good, because it only has 50 data points. That is the reason we use stat "1" in the control file and use Stat_every = 1 in the ela_STATIS.f90 to increase the data point from 50 to 500. Note: While temperature is the second variable after the energy column in the STATIS block, it is the third column for plotting because time is the first column. Similarly, since the pair potential (or VDW) and the volume are, respectively, the 4th variable and 19th variable, the following two commands, respectively, plot the time evolution of pair (VDW) energy and volume:

```
>plot "STATIS_ela2.gnu" using 1:5      !Pair potential versus t
>plot "STATIS_ela2.gnu" using 1:19                 !Volume versus t
```

**STATIS_ela.gnu**
There are only 37 blocks sequentially arranged according to the order of the 37 variables. Each block has 500 lines, with each line consisting of two columns. The first column, time, is the same for all blocks; and the second line shows the corresponding variables at different times. When using gnuplot, there is little advantage using the STATIS_ela.gnu file. However, it may be desired to transfer the file to the windows file format through winscp and use excel to develop the plot.

Note that the STATIS file is not overwritten by the code, therefore if two simulations are run in the same directory, only one STATIS file is present, similar to the HISTORY file. The first part of the file is related to the first simulation, and the last part to the second simulation.

## 10.5.9   Useful Tools for Operating and Monitoring MD Simulations

The command "nohup" can be used to run a long MD simulation along with "top" that can be used to monitor the simulation.

**Nohup**
Molecular dynamics simulations can require a great deal of time (the time required to run a single simulation may be days even on parallel machines). To run long simulations, it is not convenient to use the processes listed in the previous sections, because if for some reason the shell is closed (for example if the simulation is running on a remote machine and it is desired to logout of the current terminal) the simulation will stop.

To avoid this problem, run the simulation with the nohup command. This command allows a code to run "outside" the shell, which means that if the shell is logged out, the program will continue. To understand how nohup works, enter the UNIT5/Simu_4 directory. Within this directory, open the "test.f90" file with vi. It is a simple program that can be used to show how the nohup command works. Compile the code using the command:

```
~UNIT5/Simu_4$gfortran test.f90 -o test.exe
```

Run the code using the command:

```
~UNIT5/Simu_4$./test.exe
```

Running will require several seconds and a great deal of numbers will appear on the screen. Then type the following command:

```
~UNIT5/Simu_4$nohup./executable_file > outputname &
```

where "outputname" is the name assigned to the resulting txt file. Other names can also be assigned, an example may be:

```
~UNIT5/Simu_4$nohup./test.exe > test.out &
```

After several seconds, a message will appear that states:

```
[1] + Done       nohup./test.exe >test.out
```

This means that the program is finished and that the output is the file "test.out". Open the OUTPUT file with vi and verify that the numbers match those in the previous run without the nohup command.

## Top

If the simulation ran successfully "outside" the shell (i.e., if the nohup command worked correctly), a special program can be used to verify that the simulation is running. This program is called "top". Run the simulation again with nohup using the command:

```
$nohup./test.exe > test.out &
```

and type:

```
$ top
```

The shell will change and a set of lines should appear similar to the following:

```
2790 mauro  25 0 2564 676 560 R 910.3 0.1 0:06. 41 test.exe
```

Take a few moments to review the first number, called the PID (or process ID); the first string stating the user that ran the process (in this case mauro); and the last string stating the name of the process (in this case "test.exe"). To exit the top program, type "q". Note: Do not enter any other character other than "q", otherwise errors can occur and exiting top can become difficult.

## Kill

To kill or stop a process, the PID is required. For instance, run the test simulation:

```
nohup./test.exe > test.out &
```

Use "top" to check the PID of the process and then in the "top" screen type the following command to kill the job:

## kill pid

where "pid" is the PID number of the process. Note: No message will appear if the kill program is completed. However, if the PID number is incorrect, an error message will occur.

## Less

If the program is running and it is desired to check the output files to avoid error, do not use "vi". Instead, use the program "less". The command allows the user to read a file but not to modify it. The commands are similar to vi. For example the command:

```
$ less test.out
```

will open the test.out file. The cursor can be moved inside the file as in vi (for example: use the command "gg", ":n" and "shift G", respectively to move the cursor to the first line, line n and the last line). To exit from the "less" program, type "q".

**Head**
This command can be used to view the headings (the first several lines) of a desired file, such as a HISTORY file using the command:

```
$ Head HISTORY
```

This command and the "less" command can be useful when dealing with a large file such as a HISTORY file, when it can be difficult and unnecessary to open the whole file.

---

**Homework**

(10.16) There are CONFIG, CONTROL and FIELD files for a nanoscale coating of iron nitride ($Fe_4N$) on iron substrate (Fe) in UNIT5/HW11. The atomistic system has already experienced equilibration process with three-dimensional periodic condition (IMACON = 3) with npt ensemble. Run two DL_POLY simulations with the same initial configuration but with different temperatures, one at 50 K, the other at 300 K under the given ensemble nvt in the control file and with IMCON = 6 in the CONFIG file. The boundary condition is therefore changed from parallelpiped periodic boundaries to slab (x,y periodic, z nonperiodic) condition.

(1) Write a report in which the differences between the two simulations are clearly shown (select appropriate variables such as energy, volume, temperature, pressure to do it).

(2) In the report, images created by VMD or AtomEye must be included. For instance, use VMD for visualization of HISTORY file to see what is the initial configuration and what is the difference of the last configuration from these two cases.

Hint: The only change that you need is the temperature in CONTROL file. Because the initial configuration is not changed you do not need to change the FIELD and CONFIG files. Then you can use Section 10.3.4 for compiling and running. You may use the "mv" command to change the names of output files with, say, "STATIS_50" or "STATIS_300" to distinguish two kinds of results. Remark: When no change of source files in srcmod occurs, one does not change the executable file DLPOLY.X again, thus two runs can use the same DLPOLY.X.

(10.17) Use the "Find" atoms (F key) command and the "print window" (Ctrl + left click) command in AtomEye to investigate the evolution of the right x-boundary surface using the files in the UNIT4/GNUPLOT_ATOMEYE_EXAMPLES directory: 2a.cfg, 8000a.cfg, 30000a.cfg, 60000a.cfg. Select 10 random points on that surface or use the following suggested points: 12279, 12319, 12359, 12399, 12439, 12479, 12519, 12559, 12373, 12371, 12369, 12367, 12365, 12363, 12361, 12359, 12357, and 12355. The investigation is intended to focus on the x-displacements of each of the points selected.

---

## 10.6   Nve and npt Ensemble in MD Simulation

In this section DL_POLY will be used to perform simulations in the "nve" and "npt" ensembles described in Section 2.7. In the "nve" ensemble the total number of atoms (n), volume (v), and energy (e) of the system remain constant. For the "npt" ensemble the number of atoms (n), temperature (t), and pressure (p) are kept constant, which therefore requires a thermostat and barostat to maintain the desired properties.

Pressure and volume (v) are conjugate variables, as explained by the ideal gas law: npv = RT where p and v are inversely related if the total number of atoms and the temperature remains constant.

Thus, for the npt ensemble, the pressure is controlled by adjusting the volume of the system, which will be controlled using a barostat or piston. The temperature is controlled by a thermal reservoir (bath) as described in Section 2.7. In the directory UNITS/UNIT6, there are two directories, NVE_Ar and NPT_Ar. Both files use argon under different pressures (1 and 1000 atm), and different temperatures (100 and 310 K), to show how to use nve and npt ensembles for simulation. In each of the simulations, the following four files in each directory are needed in order to run: crystal_structure.f90 with the data file Model_input to produce the model CONFIG file, and CONTROL and FIELD file for the nvt and npt simulation.

Use of the term "equilibrium" and "non-equilibrium" indicates thermodynamic equilibrium and thermodynamic non-equilibrium, respectively. For non-forced simulations of an isolated system, some well-defined statistical mechanical ensembles reach a thermodynamic equilibrium after a transient process. For instance, "nve" is called a microcanonic ensemble in statistical mechanics; its equilibrium is reached at the maximum entropy of the system after the transient process. For a canonical ensemble (for instance nvt with a Nose-Hoover thermostat) the equilibrium is reached at the minimum Helmholtz free energy value.

Note that the term "equilibration" is used for any simulation that uses a thermostat and barostat to reach an equilibrium state, without any external loading. "Non-equilibrium" is used for forced MD simulations or steering MD simulations in which external loading is applied, causing the system to reach a stationary state, not an equilibrium state.

## 10.6.1   Nve Simulation with DL_POLY

The first task is to develop a simulation model which can serve as the CONFIG file, and then use VMD for visualization to check whether all necessary requirements to the model are satisfied.

### 10.6.1.1  Developing a CONFIG File for a Solid with a FCC Crystal Structure

Open the UNITS/UNIT6/NVE_Ar directory; a CONFIG file will be developed using the source file "crystal_structure.f90" and data file "Model_input". Open the Model_input file using vi. This file contains six lines:

```
1 nphase
arg
fcc
5.75 5.75 5.75      cell dimension
8 9 10         ncell
0 0 0      r0
```

**Line 1:** Indicates the number of phases of the crystal structure; 1 indicates only one phase, 2 indicates two phases, and so on. For each phase, five lines must be specified. In this case there is only one phase, so lines 2–6 contain the information necessary to generate the whole structure.

**Line 2:** States the name of the phase as "arg", but can be changed to any name desired. For this case, "arg" represents argon which has a crystal structure under high pressure.

**Line 3:** Defines the crystal structure, in this case FCC.

**Line 4:** Contains the basic cell dimensions, i.e., the values of a, b, c of the primary unit cell.

**Line 5:** Provides the number of cells in each direction. In this case, 8, 9, and 10 are the cell numbers along the x, y, and z direction, respectively. If they are not these numbers, change them to the required ones.

**Line 6:** Contains the x, y, z coordinates which denote the position from the bottom left corner of the structure, which is the reference point of the model. Thus, the above six lines define the "arg" model, which has $8 \times 9 \times 10$ basic cells of an FCC structure, starting from the position 0 0 0.

Now compile the crystal_structure.f90 code using the command:

```
: ~NVE_Ar$ gfortran crystal_structure.f90 -o crystal.exe
```

Run it using the command:

```
:~NVE_Ar$./crystal.exe
```

Check the output using "ls -lrt" to verify the existence of the files. Several output files have been created, including CONFIG and conf.xyz. The CONFIG file is the input file for DL_POLY_2, and the conf.xyz file is used in the next section for the VMD visualization.

### 10.6.1.2  Verify CONFIG File Using VMD Visualization

The conf.xyz file contains the configuration in the xyz format of VMD. Open it with vi and scroll the cursor to the last line (after using vi to open the conf.xyz file, move the cursor to the last line by pressing Shift + G). The file should contain 2882 lines, as created using the crystal_structure.f90 file to generate a $8 \times 9 \times 10$ of FCC basic cells; each cell contains four atoms, thus producing $8 \times 9 \times 10 \times 4 = 2880$ lines. This number plus the first and second lines makes a total of 2882 lines.

Now go to the first line ("gg" command in vi) and replace the string "number of atoms" with "2880". Save and close the file (:w and:q). Open VMD by typing "vmd" in the UNIX shell, and load the conf.xyz molecule. Select a VDW representation by selecting Representation under the Graphics menu, as shown in Section 10.4.3.2.2. Select an orthography view by selecting the orthographic option under the Display menu. Open the Tk console by selecting the Extension menu, and type the following commands:

```
molinfo 0 set a 46.000
molinfo 0 set b 51.750
molinfo 0 set c 57.500
```

The three values are assigned to the periodic box dimensions at 8, 9, and 10 times the basic cell dimension (i.e., 5.75), respectively. In the periodic tabs on the representation window, periodic images can be added or removed from the visualization. This check is very useful because it can allow the user to search for overlaps in the initial configuration that would cause the MD simulation to work improperly.

After the VMD check, open the CONFIG file to check for issues as described in Section 10.6.1.1. Go to the end of the CONFIG file to verify that the total number of atoms is 2880. The most important check, however, is whether the box sizes of 46.00, 51.75, and 57.5 are correct in the diagonal positions of lines 3–5. In addition, verify that the periodic boundary condition described by the "imcon" number in the second number of the second line is correct. In the present case, three-dimensional periodic condition is used, therefore the "imcon" number should be 3.

### 10.6.1.3  FIELD File for Argon by LJ Potential

A FIELD file contains the details of the molecular interaction parameters. This file is strictly related to the CONFIG file, i.e., the number and sequence of atoms in the CONFIG file must be the same as the FIELD file. The example below is a FIELD file for Argon:

```
DL_POLY argon
UNITS internal
MOLECULES 1
Argon
nummols 2880
atoms 1
Ar    39.95    0.000
finish
VDW 1
Ar   Ar   lj   99.5581   3.405
CLOSE
```

**Line 1:** Defines the name of the simulation.

**Line 2:** Indicates the unit of measurement for energy. If there are no units specified or if "internal" is stated, DL_POLY_2 will use the internally accepted units (see Section 10.5.5).

**Line 3:** Contains the number of molecule types, in this case one. For each molecule there is a block consisting of several lines, which in this case is only one block, lines 4–8 with five lines.

**Line 4:** The first line of the block, which contains the name of the molecule, in this case Argon.

**Line 5:** The second line of the block, which contains the number of molecules, in this case 2880.

**Line 6:** The third line of the block, which contains the number of atoms for each molecule, in this case 1.

**Line 7:** The fourth line of the block. For each atom there is a line in which the atom's name, mass, and charge are reported.

**Line 8:** The "finish" statement line of the block, which must be the last line of each block to close the block. After all the molecule's blocks are completed, there are lines for the interatomic interactions.

**Line 9:** The first line for the potential description. In this case, there is only one VDW (pair potential) interaction.

**Line 10:** Shows the detail of the VDW potential, which is the LJ potential where the parameters are $\varepsilon = 99.5581$ and $\sigma = 3.405$ (see Section 2.4.1).

**Line 11:** The "close" line.

### 10.6.1.4 CONTROL File

The CONTROL file contains the information for the simulation. Specifically, the CONTROL file in the NVE_Ar directory contains the commands to perform a 500 time step simulation (i.e., step 500) with an nve ensemble (no thermostat). After all the input files are ready, compile and run the simulation following Section 10.5.4 as follows. Open the UNITS/dl_poly_2.18/srcmod directory and type:

```
$make gfortran
```

to produce the DLPOLY.X executable file in the execute directory. If the source file in srcmod is identical to the file used in other simulations, with only changes to the input files, there is no need to compile it again. Open the NVE_Ar directory and run DL_POLY using the command:

```
~/UNITS/UNIT6/NVE_Ar/$. /../../dl_poly_2.18/execute/DLPOLY.X
```

### 10.6.1.5 Error Message

The simulation may take several minutes to run. If there is an error, the simulation will stop. Use the "vi" command to open the OUTPUT file and review the error message at the end of the file. The message is usually given by the error index. Appendix 10.C of the DL_POLY_2 manual lists all error messages based on the reported number, and briefly shows the probable cause and recommendations to solve the problem. A common error for beginners may involve an improper number of molecules in the FIELD file. For instance, the number of molecules in the nve simulation is 2880; but for the npt simulation, the number of molecules is 2016 (see below). Thus, if the FIELD file in the NVE_Ar directory is used for the NPT_Ar simulation, the CONFIG file will not match and the error message will appear. Another probable cause is an incorrect box dimension, which should be 46.0, 51.75, and 57.5 for the diagonals of the third to fifth lines of the CONFIG file. If problems persist, review the UNITS/UNIT6/NVE_Ar_out directory on the http://multiscale.alfred.edu website for the files produced in the simulation which can be used to check one's own simulation.

## 10.6.2  Npt Simulation with DL_POLY

The npt simulation will consist of liquid argon in equilibrium using a thermal bath at a constant temperature of 100 K and a "piston" that imposes a constant pressure of 1 atm. To achieve this result, both the thermostat and barostat must be used. First, prepare the initial configuration (CONFIG) and interaction potential parameter (FIELD) files.

### 10.6.2.1  CONFIG File Preparation

Open the UNITS/UNIT6/NPT_Ar directory. Use the following commands to create the CONFIG and conf.xyz file using the crystal_structure.f90 code.

Open the Model_input file and insert the values 7, 8, and 9 on line 5 (i.e., the "ncell" line). These numbers will create a structure constituting 7 cells along the x, 8 along the y, and 9 along the z direction. The cell size is 5.75, therefore the model size (40.25, 46.0, 51.75) is less than the model in the NVE_Ar simulation. Compile the crystal_structure.f90 file using the command:

```
$gfortran crystal_structure.f90 -o crystal.exe
```

Run the code using the command:

```
$./crystal.exe
```

Check the output using "ls -lrt". The CONFIG and conf.xyz files have been created. Open the CONFIG file and modify lines 3, 4, and 5 to obtain the following result:

```
40.2500000000    0.0000000000    0.0000000000
0.0000000000    46.0000000000    0.0000000000
0.0000000000     0.0000000000   51.7500000000
```

The first non-zero number is seven times the x cell dimension ($7 \times 5.75 = 40.25$). The second value in the diagonal is seven times the y cell dimension ($7 \times 6.57 = 46.00$), and the third value is seven times the z direction ($7 \times 7.39 = 51.75$), as described in the last section. Now the CONFIG file is ready for the simulation.

Note: It may be beneficial to re-verify the box dimensions of the conf.xyz file using VMD as shown in the previous sections. In order to use the conf.xyz file for visualization, change the first line of the file to match the number of atoms, which in this case is 2016. This check will help to avoid trivial errors due to box periodicity, for example, overlapping of periodic images. Test the initial configuration with VMD. Use section 10.6.1.2 to assist with the verification.

### 10.6.2.2 FIELD File

The FIELD file is exactly the same as that used in the NVE_Ar simulation. The only change is the atom number; which should be changed from 2880 to 2016 ($7 \times 8 \times 9 \times 4 = 2016$).

## Exercise 10.10

Run two simulations with an incorrect number of atoms; the first with an atom number greater than the number of atoms in the CONFIG file (for example 3000 atoms in line 5 of the FIELD file); and the second with a smaller value (for instance 1000 atoms in line 5 of the FIELD file). Review the results for each case. Hint: Use the user manual to explain the error messages.

### 10.6.2.3 CONTROL File

The instructions to perform the npt simulation are as follows. Change the pressure value to 0.001. The pressure units are in katm, therefore the given value equals 1 atm, the standard atmosphere pressure. Compared to the pressure of 1 katm in the NVE_Ar control file, the npt control file is 1000 times smaller; therefore it is possible to have a liquid or gas phase depending on the controlled temperature (see Homework 10.14).

Type the following command for an npt ensemble with a Berendsen thermostat:

```
ensemble npt ber 0.5 5.
```

where "ber" denotes a Berendsen thermostat and barostat (see Section 2.7). The first number $\tau_T = 0.5$ is the time constant of the thermostat, and the second number $\tau_P = 5$ is the time constant of the barostat. These constants determine the rate of the control process. The thermostat controls the velocity of the atoms to maintain a constant total kinetic energy k, therefore causing the temperature to remain constant (see (2.18a)). Increasing the value of $\tau_T$ will reduce the rate for velocity adjustment because it reduces the change rate of the friction coefficient as seen in (2.18h) and (2.18i). This can reduce the dissipative work while slowing down the control process, therefore this can be a reasonable choice if these constants are necessary. Normally $\tau_T$ is in the range of 0.5 to 2.0 ps. The same concept applies to the barostat in adjusting the size and shape of the simulation cell to maintain a constant pressure. Its adjusting rate is controlled by the time constant $\tau_P$ for pressure fluctuation.

Change the number of time steps to 1000 and the temperature value to 100 K. DL_POLY is ready to run. Again, if only the input files have changed, not the code or environment, there is no need to compile the code again to obtain the executable DLPOLY.X file. Therefore, open the NPT_Ar directory and run DL_POLY using the command:

```
~..npt_Ar$./../../dl_poly_2.18/execute/DLPOLY.X
```

## 10.6.3 Data Post-processing via STATIS and HISTORY Output Files

### 10.6.3.1 Methods

When the simulation has finished, use "ls -lrt" to check the output files which should include the HISTORY and STATIS files. These two output files will be used for data processing.

### 10.6.3.2  Watching Volume Change Using the HISTORY File

First, open the HISTORY file with vi. If the HISTORY file is too large to open, use the "less" command. Lines 2, 3, and 4 contain the box dimensions of the configuration printed at time step 10. Check whether the box volume has changed by searching for the word "timestep" using the "/timestep" command. The cursor is now on line 4037, i.e., the first occurrence of the word "timestep". Lines 4038, 4039, and 4040 state the box dimension for time step 20. Note that the box dimension has changed. Continue to search the file for dimensional changes by pressing the "n" key after "/timestep". After completing the check, close the file by pressing "q".

### 10.6.3.3  Watching Phase Change Using the HISTORY File

Open VMD and view the HISTORY file using the "New molecule" window. Click the "determine file type" tab, and select DLPOLY History. Since there are 1000 steps in total and writing will occur once every 10 steps, there are 100 frames in the HISTORY file. Adjust the bold vertical short bar in the bottom of the main VMD window from left to the right to view the structure as it changes throughout the simulation. The system passes from the crystal state to an amorphous state (i.e., liquid structure).

### 10.6.3.4  Quantitative Analysis Using ela_STATIS.f90

For a quantitative analysis, use the ela_STATIS.f90 file to read the STATIS file in the NPT_Ar directory as described in Section10.5.8. First, open the STATIS file using vi. At line 3, the last number is 46, which means that for each block of the STATIS file, 46 variables are printed. The value of 46 originates from the number of variables in the NVE_Ar directory (37), and the box parameter number (9). Therefore in the STATIS file the time evolution of the box parameters are reported. The increase in the value of the box number (9), is automatically produced if a constant pressure is required when the npt ensemble is given (see Section 10.5.4.3).

Now open the ela_STATIS.f90 file with vi, then change the number in line 29 to match the number of blocks in the STATIS file. This number can be obtained by opening the STATIS file. Otherwise, the value can be determined by dividing the total steps "n_step_t" (in this case 1000) by the number "stat_every" listed after the command "STATS" in the CONTROL file (in this case 10). The latter denotes printing frequency to the STATIS file, therefore the number should be 100 (1000/10) in this example.

Go to line 32 and assign the variable nstpval the value 46. Create a directory called POST_PROC using the command:

```
$ mkdir POST_PROC
```

Compile and run the ela_STATIS.f90 file using the command:

```
$gfortran ela_STATIS.f90 -o ela_STATIS.exe
```

Then use the command:

```
./ela_STATIS.exe
```

Open the POST_PROC directory and verify that the correct output files were generated using "ls -lrt". As in Section 10.5.8, use gnuplot to visualize the time evolution of the variables reported in the STATIS file found in the POST_PROC directory. Open gnuplot, and type the command:

```
>p"STATIS_ela2.gnu" u 1:3
```

Remember that the gnuplot abbreviation "p" stands for plot and "u" for using. The graph reports the time evolution of the temperature. The horizontal axis is the time (ps) and the vertical axis is the temperature. The CONTROL file requires the temperature to be 100 K. The gnuplot curve shows that after about 2.5 ps transition time, the temperature varies around the range of 97–101 K.

Note that the variables used in the above comment should have 1 added to the location number in the STATIS file. This is because the first column is occupied by the time step. For example, the temperature variable is 2 in the STATIS file; therefore the number 3 is used in the above commands. Similarly, the pressure and box dimension variables are 27, 38, 42, and 46, therefore the command for plotting these variables should be given as follows:

```
>p"STATIS_ela2.gnu" u 1:28
```

This command will plot pressure, while the command:

```
>p"STATIS_ela2.gnu" u 1:39, "" u 1:43, "" u 1:47
```

will plot the three box dimensions. These values do not reach a designated value, due to the low number of time steps. In most cases, a barostat needs much more time than a thermostat to bring the system to a stationary state.

Pressure fluctuations are very large. This is normal; instantaneous pressures (and stresses) have large fluctuation because they are related to atomic forces which fluctuate greatly as well. In the UNIT6/ NPT_Ar/NPT_Ar_out/POST_PROC directory there are six png files which respectively show the curves of pressure, energy and temperature versus time (P-t, T-t) and the box size along x, y, and z directions versus time (Lx-t, Ly-t and Lz-t). Compare the obtained result with these curves.

### 10.6.3.5 Running a Long Simulation Using nohup

Use the nohup command to run a 100,000 time step simulation. Before running, delete any previous STATIS and HISTORY files within the NPT_Ar directory using the rm command. Change the time step to 100,000 in the CONTROL file using the command:

```
steps    100000
```

Make appropriate changes to the CONTROL file to allow DL_POLY to print the HISTORY file every 200 steps using the following command in the CONTROL file:

```
traj 0 200 0
```

In general, the command "traj i j k" in the CONTROL file writes the HISTORY file which consists of:

  i. Starting time step for dumping configurations. Here the value is 0. In other simulations, if the information at the beginning is not needed, change this value to the desired time step. For instance, if the value is set at 200, the configuration would begin recording from step 200.
 ii. Time step interval between configurations; here it is 200.
iii. Data level (i.e., variable keytrj see Table 4.3 of DL_POLY manual). In this case the value is 0 because only the coordinates are to be recorded, not velocity or force.

Change the reading data in the STATIS file for every step using the command:

```
stats 10
```

Use a large value for the simulation time (ps units) in the CONTROL file, for example 10,000,000, otherwise the simulation could stop. Then run DL_POLY with the following command:

```
$ nohup ./../../execute/DLPOLY.X > test.out &
```

**Homework**

(10.18) The simulation in Section 10.6.2 is with npt ensemble, thus the total number of variables in STATIS output file is 46, specifically, variables 38, 42, and 46 are the x, y, z components of the cubic cell. After the simulation by the "nohup" command with 100,000 steps required in Section 10.6.2.2 is finished, plot the time evolution of the box dimension using gnuplot.

Hint: (1) Verify that the simulation results satisfy the requirements for the "steps 100000", "traj 0 200 0", and "stats 10" commands in the CONTROL file. In other words, verify that the simulation reaches 100,000 steps, writes the x, y, z coordinates for each atom in the HISTORY file every 200 steps, and writes variables every 10 steps in the STATIS file.

(2) Remember to change lines 29, 31, and 32 of the ela_STATIS.f90 file and then run it for data analysis.

(10.19) Perform the same simulation at a temperature of 50 instead of 100 K. The easiest way to achieve this is to perform the following steps:

1. Make a copy of the NPT_Ar directory (from UNIT6) to NPT_Ar_solid using the command:

```
cp -r NPT_Ar NPT_Ar_solid
```

2. Open the NPT_Ar_solid directory.
3. Remove the output files of the previous simulation, in particular the STATIS, HISTORY, and OUTPUT files.
4. Open the CONTROL file with vi and insert the value 50 after the "temperature" command.
5. Check that the number of time steps is small (for example 1000).
6. Insert the value 1 after the "stats" command and 0,200,0 after the "traj" command.
7. Run DL_POLY using the command:

```
./../../dl_poly_2.18/execute/DLPOLY.X
```

Analyze the HISTORY file with VMD and the STATIS file with ela_STATIS.f90 and gnuplot, and then answer the following questions.

(10.19a) Is the system obtained in the solid or liquid state? Give a physical explanation of what happened during the simulation and compare the configuration energy (i.e., total potential energy) and volume of the two simulations at 50 and 100 K for Argon.

(10.19b) If the simulation ran correctly, the fusion temperature of Argon is smaller than 100 K and larger than 50 K. Perform some simulations with different temperature values and try to estimate the fusion temperature of Argon.

# Part 10.2: Simulation Applications in Metals and Ceramics by MD

## 10.7 Non-equilibrium MD Simulation of One-phase Model Under External Shearing (1)

In this section and Section 10.8, iron film behavior under external shearing rate will be simulated. These two sections involve only one phase of the material. In Section 10.9, the behavior of a nanoscale coating layer composed of iron nitride ($Fe_4N$) on an iron substrate will be simulated. In addition, the features of a two-phase atomistic simulation will be emphasized.

The key issue in all non-equilibrium simulations under external loading is to obtain the equilibration configuration of the atomistic system before loading. This is a complicated process; however this process must be completed to ensure the simulation runs accurately. In this section the model development and the first equilibration process called 3D npt equilibration with npt ensemble will be discussed. In the next section, another equilibration process called 2D nvt equilibration with nvt ensemble will be discussed.

## 10.7.1   Features and Procedures of MD Simulation Under Shearing Strain Rate

The basic features and simulation procedure of the layer under a given shearing for the case discussed here are as follows. The lower layer of the crystal structure is fixed. The layer will sustain a double shearing parallel to the xy plane. The positive and negative shearing plane has the same distance to the xy plane; however the shearing direction will be opposite. The top layer moves in the positive direction, while the bottom layer moves in the negative direction. This external shearing is given by DL_POLY_2.18's external potential, as explained in detail in Section 10.8.3.

The procedure for the simulation and its related subdirectory information are described as follows.

- The first step is to develop a CONFIG file in UNIT7/INI_CONF by crystal_structure.f90 and Model_input in that directory.
- 3D npt equilibration: This equilibration process involves triple-periodic boundary conditions, including the parallelepiped periodic boundary condition (IMCON = 3) to reach the pre-simulation temperature and pressure. This is needed because the potentials used may not be accurate. All equilibrations will allow the user to check for problems caused by the potential used. These files can be found in the UNIT7/EQUI_npt directory.
- 2D nvt equilibration: This equilibration process involves bi-periodic boundary conditions, including the slab periodic conditions along the x and y direction, but not the z direction, which is non-periodic (IMCON = 6). (See the DL_POLY_2 manual.) The equilibration process will reach the pre-simulation temperature and volume before the shearing is started along the xy plane along the x-direction. The second equilibration is necessary because the boundary conditions have changed from IMCON = 3 to IMCON = 6, also the volume must be held constant during the shearing process. Note: Pressure p is not constant.
- After the equilibration is completed, find the reference position of the crystal atoms before shearing using the file MEAN.xyz produced by data-processing of the HISTORY file, using ela_history_2009. f90 code, found in the UNIT8/2D_EQUI_nvt/POST_PROC directory.
- The bi-periodic nvt simulation under the applied shearing rate of the layers is simulated. For simplicity, the related IMCON = 6 simulation is called the bi-periodic nvt simulation, or 2D nvt simulation; therefore the directory is called 2D_nvt (see Figure 10.6). Consequently, sometimes the term "3D simulation" is used to denote that the simulation is under the full three-dimensional periodic condition (IMCON = 3).
- Information from the HISTORY file from the ela_history_2009.f90 code is obtained to estimate the mean displacement of the deformed configuration at different shearing times. These files including the input file histo.inp can be found in the UNIT8/2D_FORCED/POST_PROC directory.

Non-equilibrium simulations involve multiple-stage equilibration processes; therefore it is convenient to put files at different stages in different directories. Table 10.1 shows a flowchart for the process, connections, and functions of each stage. Figure 10.6 shows the directory tree and the files in each directory. As described in the last several sections the file names containing "_out" include detailed input and output files to compare with simulation results. In addition, the POST_PROC directory includes more post-processing files; the file names and functions of the ID listed in each box of the directory tree are given in Table 10.2.

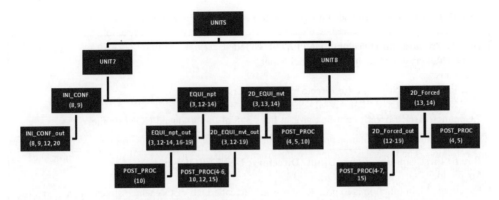

**Figure 10.6**   Directory trees of UNIT7 and UNIT8. The bottom subdirectories are simulation results so readers can compare their simulation with these results

## 10.7.2   Preparation for Input Files and Running 3D npt Equilibration

### 10.7.2.1   CONFIG File

In this 3D simulation to establish equilibrium conditions, there is no need to add any forces or constraints to the atoms. To develop the corresponding model, perform the following steps:

1. Open the UNIT7/INI_CONF directory and open the Model_input file. In this file there are six lines. The first line is the number of phases, in this case "1" phase. The second line is the name for the element, in this case "iron". The third line is the basic cell crystal structure, in this case "bcc". The fourth line is the cell dimension in the three directions, in this case "2.87 2.87 2.87". The fifth line is the number of cells in the three directions x, y, and z; in this case "16, 14, and 20", which produces 8960 atoms (i.e., the number of cells times the number of atoms for each cell, in this case 2 for the bcc structure). The sixth line is the coordinates, $ro_x; ro_y, ro_z$ of the position vector of the first atom "o" (the one in the left bottom corner with the lowest values of x, y, and z).

2. These values must be chosen carefully because, in general, the origin of the system should be located at the center of the initial configuration. Note that the simple models used in Section 10.2.3 and 10.6.1.1 assume these values to be (0 0 0) which is a simplification, but not a common reference. Use the following experience rule for the coordinates $ro_x; ro_y, ro_z$ along the x, y, and z direction:

$$ro_x = -(0.5 \times (\text{number of cell in x direction}) - 0.25) \times \text{cell dimension in x} \qquad (10.3)$$

   Use the same formula structure for $ro_y$ and $ro_z$. In this example $ro_x; ro_y, ro_z$ are calculated to be: −22.2425, −19.3725, −27.9825.

3. Compile the program using the command:

```
$gfortran crystal_structure.f90 -o crystal.exe
```

   The last part of the command line, crystal.exe, is the name given to the produced executable file. It is up to the reader to determine an appropriate name by replacing crystal.exe.

4. Verify that the executable file was created using the "ls -lrt" command. Now run the code using the command:

```
./crystal.exe
```

**Table 10.1**   Directory files and functions for non-equilibrium simulation

---

**Box A: Preparation Directory: UNIT7/INI_CONF**
Source file: crystal_structure.f90, Model_input
Functions:
   1. Produce CONFIG file for directory EQUI_npt (see next box)*
   2. Produce conf.xyz file for VMD check
* Common error: Forget to put box size in lines 3–5 of the produced CONFIG file

---

**Box B: First Equilibration (3D-npt) Directory: UNIT7/EQUI_npt**
Source file: CONFIG, CONTROL, FIELD*
Functions: Main function is to produce a re-start file called REVCON which stores all data after the equilibration process in this directory. REVCON file will be sent to UNIT8/2D_EQUI_nvt directory and be transformed into a new CONFIG file for the second equilibration.

* no HISTORY file output is needed, thus drop the command "traj" in CONTROL file

---

**Box C: Second Equilibration (Bi-Periodic) Directory: UNIT8/2D_EQUI_nvt**
Source file: the same as box B but with the following changes:
CONFIG: Produced by the command "mv REVCON CONFIG", then change the IMCON number from 3 to 6 in line 2 of the produced CONFIG file.
CONTROL: Using "ensemble nvt Berendsen 0.5" to replace "ensemble npt Berendsen 0.5 5.0"
FIELD: Increase one phase (molecules) with fixed boundary condition.
Functions:
   1. Produce a new REVCON file and then input it to the ../2D_FORCED directory for external shearing on the top layer.
   2. Produced MEAN.xyz and then input it to 2D_FORCED as a reference configuration to get displacement.
   3. Mean.xyz is obtained via HISTORY file by using function 1 of ela_history_2009.f90 and the histo.inp data file.

---

**Box D: Loading Directory: Unit8/2D_FORCED**
Source file: CONFIG file is developed by the command "mv REVCON CONFIG".
FIELD file: increase 3 lines for external shearing field.
Functions:
   1. Obtain deformation patterns through developing the "disp_distr.dat" file by using function 2 of ela_history_2009.f90 & histo.inp.
   2. Using VMD and AtomEye visualize the deformation pattern and find defect (e.g., voids, dislocations) distributions.

---

Several files have been created:

1. conf.xyz: This contains the position coordinates for all atoms. Replace the first line containing the string "number of atoms" with the actual number of atoms, in this case 8960. This change will allow the user to visualize the file with VMD.
2. CONFIG file: This contains each atom's position in the same format as the standard DL_POLY CONFIG file. However, ensure that the correct box dimensions replace lines 3, 4, and

**Table 10.2**   File ID, name and basic functions

| ID | File name | Functions |
|----|-----------|-----------|
| 1 | Multi1.f90 | A simple code for MD simulation |
| 2 | mt1 | Input file for multi1.f90 |
| 3 | ela_STATIS.f90 | Code for processing data with block structure such as STATIS file |
| 4 | ela_history_2009.f90 | Code for processing data with configurations of all atoms such as HISTORY file. Function 1: Producing average reference position after equilibration. Function 2: Determining distributions of average displacement ux, uy and uz along the global X, Y, and Z axis of the model after loading. |
| 5 | histo.inp | The input file for "4" |
| 6 | MEAN.xyz | Reference file for calculating average displacement |
| 7 | disp_distr.dat | The output file of average displacement |
| 8 | crystal_structure.f90 | A simple model developing code such as BCC and FCC |
| 9 | Model_input | The input file for "8" |
| 10 | STATIS_ela.gnu | The output file for, the same for STATIS_ela2.gnu (see Section 10.5.8) |
| 11 | example.gnu | Example files for plot |
| 12 | CONFIG | Input file for MD simulation by DL_POLY (see Section 10.5) |
| 13 | FIELD | Input file for MD simulation by DL_POLY (see Section 10.5) |
| 14 | CONTROL | Input file for MD simulation by DL_POLY (see Section 10.5) |
| 15 | HISTORY | Output file after MD simulation by DL_POLY (see Section 10.5) |
| 16 | STATIS | Output file after MD simulation by DL_POLY (see Section 10.5) |
| 17 | OUTPUT | Output file after MD simulation by DL_POLY (see Section 10.5) |
| 18 | REVCON | Output file after MD simulation by DL_POLY (see Section 10.5) |
| 19 | REVIVE | Output file after MD simulation by DL_POLY (see Section 10.5) |
| 20 | conf.xyz | Output file after running code (8) |
| 21 | monte.f90 | Fortran 90 code for $\pi$ value calculation, homework (2.5) |
| 22 | absorb.f90 | Fortran 90 code for Monte Carlo dynamics calculation, homework (2.6) |
| 23 | LEPS_II.f90 | Fortran 90 code for NEB calculation (see Section 9.4.4) |
| 24 | MaterNew_2010.Comb.f90 | For developing Model.MD file for DL_POLY_2 |
| 25 | CONFIG.f90 | For finally developing CONFIG file of DL_POLY_2 |
| 26 | Multi_2010_4.f90 | For producing multiscale Model.MD file |
| 27 | CFG_convrt_cmd.f90 | For producing .cfg files from .MD output files |

5 that are currently all zeros. Specifically, in the diagonal, the values should be equal to the basic cell dimension times the number of basic cells. In this case, they are: 45.92, 40.18, and 57.4 along the x, y, and z direction. After checking the CONFIG file, move it into the EQUI_npt directory.

# Exercise 10.11

Use VMD to visualize the conf.xyz file. Check the periodicity of the cell with VMD. Hint: As shown in UNIT4 of Section 10.4.2.3, load the conf.xyz file, open the tcl console, and type:

```
molinfo top set a 45.92
```

Open the representation window and select "$+X$" and "$-X$". Repeat for y and z.

### 10.7.2.2 FIELD File

Enter the UNIT7/EQUI_npt directory and open the FIELD file with "vi" command. Check whether the number of atoms is consistent with the number of atoms listed in the last several lines of CONFIG file generated with the executable file "crystal.exe". The number of atoms should equal the product of the number listed on the line "nummols" and the number on the line "atoms" in the FIELD file. Taking the FIELD file for the simulation of $Fe_4N$ coating layer on iron substrate as an example (see Section 10.9), if there are 1000 $Fe_4N$ molecules, then the number on the line "nummols" is 1000, and the number on the line "atoms" is 5 because each molecule has 4 Fe and 1 N, thus the atom number for $Fe_4N$ coating layer is 5000. If the substrate iron layer has 2880 atoms, then the total atom number checked in the CONFIG file should be 7880.

In this case there is only one type of molecule and the number of atoms is 8960 as calculated in section 10.7.2.1. The involved pair potential is only 1 (VDW = 1) and is listed in the second to last line of the FIELD file. It is only between iron and iron atoms (Fe-Fe), and is written as:

```
VDW 1
Fe Fe mors 0.4172  2.845  1.389
```

Note that the Morse potential is used and the three parameters were taken from Table 2.2 in Chapter 2. In addition, the energy unit given in line 2 of this FIELD file is eV.

# Exercise 10.12

Use the DL_POLY user manual to interpret the FIELD file. What is meant by the "mors" potential? What is the analytic expression for the potential?

### 10.7.2.3 CONTROL File

The CONTROL file is in the UNIT7/EQUI_npt directory. The file shows that an npt simulation is designed by the command (see line 6 of CONTROL file):

```
ensemble npt berendsen 0.5 5
```

The required temperature is 300 K and pressure is 1 atm (or 0.001 katm). The total simulation steps are 30,000 with time step = 0.005 ps, so the total equilibration time is 150 ps. The cutoff radius is 6 and the STATIS file is printed, respectively, every 50 steps (see line 14). Refer to the DL_POLY manual for further information on the CONTROL file. The "equilibration 1000" command uses the velocity-scaling method of (2.20c) discussed in Section 2.7 to control the temperature.

### 10.7.2.4 Running the Simulation

To run the code perform the following steps:

1. Open the UNIT7/EQUI_npt directory.
2. Copy the CONFIG file generated in the INI_CONF directory to the EQUI_npt directory using the command:

```
$cp ../INI_CONF/CONFIG .
```

3. Run DLPOLY.X. The simulation will require one or two hours for the 30,000 steps, depending on the processor. Use nohup by typing the command:

```
$ nohup ./../../dl_poly_2.18/execute/DLPOLY.X > equi_run_date &
```

### 10.7.3  Post-processing Analysis for Equilibration Data

In order to verify that the equilibration phase was performed correctly, analyze the results of the simulation assigned in the previous paragraph. The simulation will take approximately an hour to complete. In the meantime, work in the UNIT7/EQUI_npt/EQUI_npt_output directory that contains the output files of the npt simulation.

## Exercise 10.13

Compare the result of the simulation with the one reported in the UNIT7/EQUI_npt_output directory.

Open the EQUI_npt_output directory and use "ls -lrt" to verify that the standard DL_POLY output files (HISTORY, STATIS, OUTPUT, REVCON) are present.

Now, use ela_STATIS.f90 to print the behaviour of temperature and volume over time. If a stationary state is reached, the equilibration phase is sufficient and no further equilibration is needed. Use the procedure to develop the temperature vs. time and volume vs. time relationships:

1. Copy the ela_STATIS.f90 file from the UNIT6/NPT_Ar directory to the EQUI_npt/EQUI_npt_out directory.
2. Open the EQUI_npt/EQUI_npt_out directory.
3. Open the ela_STATIS.f90 file and insert the appropriate values for the nstep_t (line 29), stat_every (line 31), and nstpval (line 32) variables (in this case, they are, respectively, 30000, 50 and 46).
4. Compile the ela_STATIS.f90 code using the command:

```
$ gfortran ela_STATIS.f90 -o ela_STATIS.exe
```

5. Create a directory called POST_PROC.
6. Run the executable file using the command:

```
./ela_STATIS.exe
```

The program will create output files in the POST_PROC directory. Use gnuplot to print the time evolution of temperature and volume. Open gnuplot by typing "gnuplot" in the EQUI_npt/ POST_PROC shell. Gnuplot must be opened in the directory in which the relevant files, such as STATIS_ela2.gnu, are located. Use the following command:

```
>p"STATIS_ela2.gnu" u 1:3
```

to plot the temperature. In this plot, the horizontal axis will lie over the interval from 0 to 150 ps (see the CONTROL file for total steps of 30,000 and 0.005 ps per each step). The temperature distribution varies randomly between 296 and 305 K, while the control value is 300 K. Then use the following three commands, respectively, for the evolution of box size along the x, y, and z directions.

```
>p"STATIS_ela2.gnu" u 1:39
>p"STATIS_ela2.gnu" u 1:43
>p"STATIS_ela2.gnu" u 1:47
```

After a very short transient period a stationary state is reached. Note that the volume has changed, for example, the simulation started with box dimensions 45.92, 40.18, and 57.4 and reached 46.797, 40.947, and 58.497 at 30,000 steps. The initial change is very fast. If it is assumed that the iron lattice distance is 2.7, then the box size in the x-direction is 43.2 ($16 \times 2.7 = 43.2$). The box size after 30,000 steps will reach 46.8, almost the same if the lattice size was assumed to be 2.87 and the box initial size to be 45.92. This indicates that both cases reach the stable structure, therefore regardless of the equilibration process, the final result is the same.

## Exercise 10.14

Use VMD to compare the conf.xyz file with the HISTORY file. Hint: Use VMD to load more than one molecule in the same graphical window. The other file, conf.xyz, can be loaded by selecting "New molecule" on the "File" menu. In this way, it will be easier to compare the initial configuration with the HISTORY file configuration at different loading steps.

---

**Homework**

(10.20) Write a detailed report of the npt equilibration conducted in EQUI_npt directory followed by the text in Section 10.7.3. The report must contain the proof that equilibrium has been reached (Hint: report graphs of time evolution of pressure, temperature and volume).

---

## 10.8   Non-equilibrium MD Simulation of a One-phase Model Under External Shearing (2)

After the first equilibration in the EQUI_npt directory, it is important to perform another equilibration in the EQUI_2D_nvt directory as illustrated in the third box of Table 10.1.

### 10.8.1   Bi-periodic nvt Equilibration in 2D_EQUI_nvt

The simulation process for the second equilibration includes the following features. It is periodic in the x and y direction, and the lower plane in the z direction is made of fixed atoms, indicating that this equilibration is constrained within the shearing of the upper layer only. Open the UNITS/UNIT8/ 2D_EQUI_nvt directory and move the REVCON file from the EQUI_npt directory using the command:

```
~/......2D_EQUI_nvt$ cp ../../UNIT7/EQUI_npt/EQUI_npt_out/REVCON.
```

#### 10.8.1.1  CONFIG File

The CONFIG file in the second equilibration is obtained from the transformation of the REVCON file from the EQUI_npt simulation, as shown in box C of Table 10.1. The REVCON file contains all of the data for atomic positions, velocities and forces at the last time step of the previous EQUI_npt equilibration. Use this data as the initial state for the new EQUI_2D_nvt simulation. To do so, change the name of the REVCON file to CONFIG file using the command:

```
$ mv REVCON CONFIG
```

Open the new CONFIG file and change the periodic condition from tri-periodic to bi-periodic by changing the value of the imcon key on the second line from 3 to 6. The value IMCON = 6 indicates that the slab periodic conditions are only applied in the x and y direction. (See the DL_POLY user manual for additional information.)

### 10.8.1.2 CONTROL File

Open the CONTROL file and change the command for the ensemble to:

```
Ensemble nvt berendsen 0.5
```

This command indicates that the nvt ensemble uses the Berendsen thermostat with a time constant of 0.5; as well as uses a volume control instead of pressure control. Also, remove the equilibration phase (equilibration 0 instead of 1000) because the temperature has been controlled since the first equilibration. This simulation is the last one before the external shearing, therefore the HISTORY file is needed to obtain a MEAN.xyz file for the average reference position. To produce the HISTORY file, use the following command in the CONTROL file:

```
Traj 0,100,2
```

This command indicates that the writing of the HISTORY file starts from the beginning, and then writes again after 100 steps. Due to the last number 2 in the command, the produced HISTORY file will include data for position, velocity, and force. If velocity and force data are not needed, the last number may be changed to 0 to save time and hard drive space.

### 10.8.1.3 FIELD File

Open the FIELD file in the EQUI_2D_nvt directory. This file differs from the FIELD file in the EQUI_npt simulation due to the addition of a new molecule called "Fe fixed" as follows:

```
Molecules 2                                            ! line 3
Fe fixed
Nummols 448
Atom 1
Fe 55.85 0.000 1 1                                     ! line 7
finish
```

The fixed atom is steel, therefore the first three numbers of line 7, including weight, charge, and repeat counter number, do not change. The fourth number is the "ifrz" number, which establishes whether the atom is free to move or is frozen in place. In this case, ifrz>0, therefore the atom is a frozen atom.

The Fe fixed atom is located in the lower plane and makes up the first 448 atoms of the CONFIG file. For this reason, this type of molecule is listed in the FIELD file as the first molecule. Note: The first of the 448 atoms are located at the bottom-left corner (see Sections 10.6.1.1 and 10.7.2.1). The value 448 was obtained using the following: $16 \times 14 = 224$ unit cells in the bottom (see 10.7.2.1 for the cells along the x and y directions) and each cell contains 2 fixed atoms, therefore $224 \times 2 = 4410$.

#### 10.8.1.4 Running

Now run the simulation using the command:

```
$nohup ./../../dl_poly_2.18/execute/DLPOLY.X > equi_run_date &
```

The calculation should take roughly one hour to complete. After completion, the simulation should have developed an OUTPUT, STATIS, and HISTORY file. The REVCON file will move to the 2D_FORCED directory to become a new CONFIG file as described in BOX D of Table 10.1. Use ela_STATIS.f90 to plot the curves of the 37 variables over time from the STATIS file, as described in Section 10.7.3. The following will introduce how to use function 1 of the ela_history.f90 file to produce a MEAN.xyz file for the average reference position of all atoms in the system after it reaches the equilibrium state, before external action.

### 10.8.2   Reference Position Calculation via Producing MEAN.xyz

Use the output files to read the HISTORY file and calculate the mean positions of the atoms. The ela_history_2009.f90 code requires its own input file called histo.inp. An example of histo.inp is included in the 2D_EQUI_nvt/POST_PROC directory. The basic line structure is as follows:

```
1                    ! control, indicating the function 1 of ela_history_2009.f90
                       will be used.
CONFIG                 ! reference position
8960                   ! number of atoms
HISTORY                ! the name of the file for data processing, if the name
                       changes this needs to be changed.
300                    ! number of configurations in the file (e.g., HISTORY)
1, 300               ! first and last configurations of the file (e, g, HISTORY)
                       between which the average is carried on.
2                     ! "keytrj" key to write HISTORY file in the CONTROL file.
3                     ! The key for the produced displacement file (1=x, 2=y,
                       3=z)
-28.0 28.0 56         ! r0, rmax, number of intervals.
```

**Line 1:** Contains an integer that ranges from 1 to 2 that controls the two functions of the ela_history_2009. f90 file. A value of 1 calculates the atom reference position, and 2 determines the displacement compared to the reference position.
**Line 2:** Not used in this function; however, it will be discussed in Section 10.8.4.
**Line 3:** Contains the number of atoms.
**Line 4:** States the name of the file that must be read. This file is typically the HISTORY file, but if the file has been renamed, change the line string accordingly.
**Line 5:** Contains the number of configurations that are reported in the HISTORY file.
**Line 6:** Contains the range of configurations in which the statistical analysis is desired. For function 1, the entire range of time steps may be used. For function 2 (see Section 10.8.4 and Figures 10.11 and 10.12), it

may be beneficial to perform the involved statistical analysis for only a portion of the configuration interval, for example 1–20, 31–40, 41–50, etc. to calculate the displacements at different instantaneous times. In addition, large displacement values after long periods of time may not be accurate due to the negative displacement of image atoms involved in the average.

**Line 7:** Establishes the information to be printed in the history file. A value of 1 means that only the position is reported, 2 means that position and velocity are reported, and 3 means that position, velocity, and force are reported.

**Line 8:** States the information needed to calculate the displacement profile of ux, uy, and uz along the global X, Y, and Z-axis of the model. The displacement variation ux(Z) of the shearing problem in this section determines the shearing strain $\gamma_{xz}$, although uy and uz are listed in the third and fourth column of the same output file "disp_ditri.dat". However, if this line contains the number 1 or 2, the output file "disp_disti.dat" will be different. It will show the displacement profile, respectively, along the X and Y.

**Line 9:** This line shows the coordinates of the starting and ending points for the displacement profile. The last number is an integer which denotes the number of slices for averaging. This line should be consistent with line 10. In the present case, the displacement profile is for the z-axis interval from $z0 = -28$ Å to $Zmax = 28$ Å with 56 slices, each with a thickness of 1 Å. The interval between z0 to zmax should be larger than the model dimension; otherwise, an error message will appear.

The HISTORY file in the EQUI_2D_nvt directory has 300 configurations. This section will perform statistical analysis from 1 to 300. To achieve this result, use "300" on line 5 and "1,300" on line 6 to perform statistic averaging from configuration 1 to configuration 300. Also, because position, velocity, and force are reported on the HISTORY file, line 7 contains the value 2. Starting in Section 10.9, only the value of 0 will be used to print the HISTORY file, which will reduce the time required to run the simulation.

To use ela_history_2009.f90, perform the following steps:

- Open the UNIT8/2D_EQUI_nvt/POST_PROC directory. This directory should include ela_history_2009.f90, histo.inp, and the HISTORY file that was just produced.
- Compile the ela_history_2009.f90 file using the command:

```
$gfortran ela_history_2009.f90 -o ela_history_2009.exe
```

- Open the histo.inp file and verify that the values reported in lines 5, 6, and 7 match the HISTORY file. It may be necessary to open the CONTROL file of the simulation that generates this HISTORY file.
- Run the simulation using the following command:

```
$./ela_history_2009.exe
```

The following output files will be obtained:

- histo.out: This file contains the same information as the histo.inp file, and is useful only for check and debugging.
- check_histo: This file contains all of the configurations obtained using data from the HISTORY file, and is only for debugging.
- MEAN.xyz: This file contains the average atom position for each atom. It has 8962 lines including 8960 lines for the 8960 atoms and 2 lines for setup. It is important to note that the first 448 lines in the file are all zero because they are fixed atoms.
- msd.dat: This file contains the mean square displacement of the atoms' positions in the MEAN.xyz file. This file also has 8962 lines.

Now use VMD to load the MEAN.xyz file and compare it with the HISTORY file. Open both files on the same graphical window and use the orthographic view with "point" representation.

### 10.8.3 MD simulation Under Shearing Rate on the Top Layer

As shown in the last box of Table 10.1, an MD simulation will be run in which a constant shear strain rate is applied at the upper layer of the atoms. The last configuration of the second equilibration performed in EQUI_2D_nvt will be used to develop the new CONFIG file as the initial configuration. To run the simulation, perform the following steps.

Open the UNIT8/2D_forced directory and copy the following files from the UNIT7/EQUI_2D_nvt_out directory into REVCON, CONTROL, and FIELD using the following command:

```
~...../2D_FORCED$ cp ../2D_EQUI_nvt/2D_EQUI_nvt_out/REVCON.
~..../2D_FORCED$ cp ../2D_EQUI_nvt/2D_EQUI_nvt_out/FIELD.
~/..../2D_FORCED$ cp ../2D_EQUI_nvt/2D_EQUI_nvt_out/CONTROL.
```

The CONTROL file is identical to the EQUI_2D_nvt_out file REVCON. The new CONFIG file can be obtained by changing the obtained REVCON file to CONFIG using the command:

```
$mv REVCON CONFIG
```

The only difference is the FIELD file, which is discussed in detail as follows.

In DL_POLY_2 simulations, it is possible to include the effects of external fields, such as electric fields, gravitational fields, magnetic fields, oscillating shears, continuous shears, and repulsive walls (harmonic). These external functions are described in detail in the manual (e.g., Table 4.17 for DL_POLY_2.18). Use the following three commands before the "VDW" line in the field file to apply the continuous shear:

```
Extern
Shrx
A z0
```

where the first two lines denote that the command is for external shear determined by the following formula:

$$v_x = (1/2)A(|z|/z) \qquad \text{for } |z| > z0 \tag{10.4}$$

The first symbol A of the third line denotes the strain rate with units of Å/ps; and the second number denotes z0 with units of Å, or the absolute z0-coordinate of the two cutting interfaces. One is positive z0 at the top, while the other is the negative z at the bottom. The area between the two interfaces is the shearing area. Note that the shearing direction is opposite to that above, the top portion is moving in the positive x direction, while the bottom portion is moving in the negative x direction. Set the shear rate to 0.25 Å/ps, therefore $A = 0.5$ Å/ps. If the value for z0 is assumed to be 16.43, then the values "5 16.43" are used in place of "A z0" in the above FIELD file.

Now run the simulation using the shear rate as follows:

```
~/UNITS/UNIT8/2D_FORCED$nohup ./../../dl_poly_2.18/execute/DLPOLY.X >
equi_run_date &
```

The simulation will require approximately one hour to complete.

## 10.8.4    Data Analysis Using ela_history_2009.f90 for Shearing

Use the files listed in the 2D_FORCED_out directory to analyze the resulting output files of the previous simulation. The procedure is as follows:

Step 1: Open the UNIT8/2D_FORCED/POST_PROC directory.

Step 2: Copy the MEAN.xyz file from where it is located, say, the file is in the UNIT8/2D_EQUI_nvt/ POST_PROC directory. To move that file to the current directory use the command:

```
~/...../2D_FORCED/POST_PROC$cp ../../2D_EQUI_nvt/POST_PROC/MEAN.xyz .
```

and then move the produced HISTORY file from the 2D_FORCED (or 2D_FORCED_out) directory to the POST_PROC directory.

Step 3: Open the histo.inp file in the POST_PROC directory with vi. The file should appear as follows:

```
2
MEAN.xyz
8960
HISTORY
30
21, 30
2
3
-35.0 35.0 70
```

Review the following descriptions for the modification of each line, as well as Section 10.8.5 tips on how to reduce the possibility for errors when using ele_history_2009.f90.

**Line 1:** Replace the value of "1" with "2" on this line. This change indicates that we change the function 1 to function 2 of the ela_history_2009.f90 to produce the displacement data based on the MEAN.xyz. The latter file was produced by function 1 of that code after the equilibration process in, say, the directory 2D_EQUI_nvt.

**Line 2:** Replace "CONFIG" with "MEAN.xyz". The latter file contains reference positions of the atomic system obtained from Section 10.7.2. This string is not used in function 1 but it is important for function 2 of the ela_history_2009.f90 code. Function 2 calculates the mean displacement relative to the reference position reported in "MEAN.xyz".

**Line 3:** Insert the total atom number.

**Line 4:** Insert a file, such as HISTORY or HISTORY_4000, for data processing.

**Line 5:** Insert the total configuration number as stated in the HISTORY file. This value must equal the total step number divided by the number described by traj in the CONTROL file. Suppose that the total number is reduced from 30,000 to 3000 while the value of the interval number in the traj command is 100, line 4 should be 30 (not 300) there. For equilibration, more steps are required to obtain a relaxed equilibration configuration, therefore 30,000 steps in the 2D_nvt equilibration is used. For investigating displacement immediately after the shearing rate is applied, however, small steps (e.g., 3000) may be used in the shearing process in the 2D_FORCED directory.

**Line 6:** States the interval for data averaging. The first number denotes the starting number; which should not be 0, but some other real number value. The second number is the ending number. Do not include a comma between the two values and do not develop an interval that is too large; otherwise, the mean value in that domain will not be accurate.

**Line 7:** Establishes the information to be printed in the history file. A value of 1 means that only the position is reported, 2 means that position and velocity are reported, and 3 means that position, velocity, and force are reported.

**Line 8:** States the index number for the printing character of the displacement profile.
**Line 9:** In order to create the displacement profile for the z-axis with the index of "3", the model domain must be divided into slices parallel to the xy plane. In the histo.inp file of Section 10.8.2, the values "−28.0 28.0 56" were used for line 10. In this case, the numbers (−35.0 35.0 70) are used. This is due to boundary atoms on the top layer that may move beyond the top boundary Zmax. If so, the code will return a series of error messages that will request that the intervals be increased to cover all the atoms in the simulation.
Step 4: Compile and run the code using the following commands:

```
~/...../2D_FORCED$gfortran ela_history_2009.f90 -o ela_history_2009.exe
./ela_history_2009.exe
```

Several output files have been generated including:

– histo.out: Used only for checking and debugging.
– disp.dat: Contains the displacement of each atom at each time step. This file may be useful when debugging during further analysis.
– check_histo: Used only for checking and debugging.
– DISP.dat: Includes the mean displacement and standard deviation of each atom.
– disp_distr.dat: States the displacement profile of the model along the z-direction. The first column is the z coordinate and the other three are the displacement components along the x, y, and z-directions. The profile is averaged over time.

Figure 10.7 shows the result of the average displacement u(z) between steps 1100 and 2000. The average time is 1.55 ps (1550 × 0.001 = 1.55 ps). From the curves it is easy to determine the shear strain of the shear layer as:

$$\bar{\gamma}_{xz}(1.55\,ps) = \frac{\partial u}{\partial z} \approx \frac{u(16.43) - u(-16.43)}{2*16.43} \approx \frac{0.4 - (-0.4)}{32.86} = 0.02435 \tag{10.5}$$

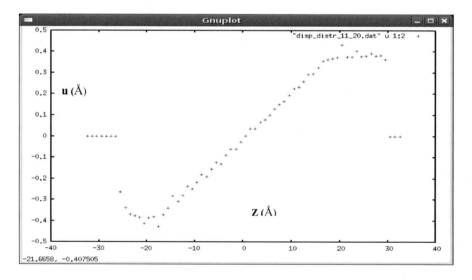

**Figure 10.7** Plot of the x-averaging shearing displacement u(z) versus the z-coordinates using gnuplot. Averaging was taken between configurations 11 and 20, or 1100–2000 steps with $\Delta t = 0.001ps$

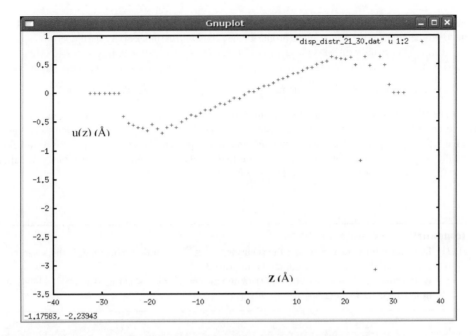

**Figure 10.8**    Gnuplot plot of the x-averaging shearing displacement u(z) versus the z-coordinates. The averaging is between configurations 21 and 30, or 2100–3000 steps with $\Delta t = 0.001$ ps

This equation can be used to calculate a series of shearing strain versus time to investigate the visco-elasticity behaviour of the layer. Note: The displacement is more random beyond the double-shearing middle area. If the range of the z coordinates is beyond the object, the displacement must be zero. Periodic boundary conditions are taken along the x and y direction, therefore displacements along these directions must be very small; otherwise, the image will enter from the opposite boundary and cause a negative displacement although the strain rate is in the positive direction. This will cause the average value to be calculated incorrectly, which will limit the strain rate A or create a shear force with a small value.

Figure 10.8 shows the displacement profile along the Z-axis from 2100 to 3000 steps with an average time of 2.55 ps. The displacement range for the shearing area increases from 0.8 (e.g., −0.4 to 0.4 Å) to about 1.4 Å (e.g., −0.7 to 0.7 Å). This is reasonable due to the increased time that will cause more displacement. However, the points with large negative displacements at positive z-coordinates (e.g. the point at $z = 27$) cause the data to concentrate at the top and should be deleted because they are actually the image points which enter the box from the negative periodic boundary along the x-direction (see Figure 2.7).

## 10.8.5    Tips to Reduce Error When Using ela_history_2009.f90

If the code receives an input that is incorrect (for example the wrong number of configurations, the wrong range for the z coordinates, etc.), the code will crash without any type of error message for troubleshooting. Therefore, note the following points.

(1) Frequently, error messages will occur due to a misunderstanding of the above descriptions of each line. The other typical reason is due to the HISTORY file itself. Ensure that the original HISTORY file is deleted before attempting to develop a new HISTORY file.

(2) Sometimes, at the beginning of the HISTORY file, there are two additional lines denoting the file name, etc. These two lines should be found using "less HISTORY" (do not use vi, the HISTORY

file is too large). If the file contains these two addition lines, use the following commands to delete them:

```
$ awk '(NR>2)' HISTORY >HISTORY2
```

If this command is used, change the file name in line 4 from HISTORY to HISTORY2.

(3) The HISTORY file is too large. Be careful to keep the size of the HISTORY file to a minimum. If there is no need for configuration data processing, do not produce it. If only the position value is needed, change the third value after the "traj" command line to 0 to prevent the velocity and force from being printed. Likewise, if the position and velocity are needed, but the force is not, change the third value after the "traj" command line to 1 to prevent the force being printed.

---

**Homework**

(10.21) Draw an approximate curve for the shearing $\gamma_{xz}\left(\equiv \frac{du(z)}{dz}\right)$ strain versus time. Is this relationship linear? How can the accuracy be improved?

Hint: From the obtained curves of u(z), obtain the approximate shearing strain $\gamma_{xz}\left(\equiv \frac{du(z)}{dz}\right)$ at times 0.55, 1.55, and 2.55 ps.

---

## 10.9 Non-equilibrium MD Simulation of a Two-phase Model Under External Shearing*

In this section, the procedure presented in Sections 10.7 and 10.8 for the one-phase problem is extended to the two-phase problem. Specifically, the two phases used in this section are iron nitride ($Fe_4N$) and iron (Fe), where $Fe_4N$ will be the coating layer and Fe the substrate. The crucial step will be the preparation of the initial configuration; therefore this section will concentrate on the preparation of the tri-periodic equilibration with an npt ensemble.

### 10.9.1 Dimensional Equilibration of the Individual Phase

Before preparation of the two-phase simulation, the cell dimensions for each phase at the equilibrium state must be determined. For BCC iron, the cell dimensions were determined from the tri-periodic npt equilibration performed in Section 10.7. The same calculation must be performed for the $Fe_4N$ coating.

#### 10.9.1.1 Developing the Initial Configuration for $Fe_4N$

Create the initial configuration using the crystal_structure.f90 file found in the UNIT9/INI_CONF_fe4n/ directory. This configuration file will include all the atomic initial coordinates in the $Fe_4N$ geometric model. Modify part of the "const" and "basic" cell subroutine in the crystal_structure.f90 file using the following procedure:

1. Open the UNIT9/INI_CONF_fe4n directory.
2. Run the npt MD simulation to allow the system to relax to its equilibrium configuration. This equilibration can be used to determine the lattice equilibrium value for the given potential. Note that

the lattice value determined by the equilibration may differ slightly from the literature, because the potential parameters are obtained by considering all the experimental data. Therefore the values from the information in this section may have used different source data to produce a different result.

3. Open the Model_input file found in the UNIT7/INI_CONF directory, and change the following lines:
   - Line 2. Change the name to fe4n to match the layer material.
   - Line 3. Replace "bcc" with "f4n".
   - Line 4. Replace the values of the cell dimension with 3.6 3.6 3.6.
   - Line 5. Replace the original cell numbers 16 14 20 along the x, y, and z direction with the new numbers 8 8 10.
   - The equilibration will only be used to find the correct volume of the basic cell; therefore the information in line 6 is not necessary and the model size can be much smaller than the realistic model.

4. Run crystal_structure.f90 with the Model_input file to produce the CONFIG and conf.xyz files using the command:

```
$ gfortran crystal_structure.f90 -o crystal.exe
```

Run the file created using the command:

```
$. /crystal.exe
```

Check that the crystal file contains 2560 atoms. This number was determined using $8 \times 8 \times 8$ cells in x, y, and z directions, respectively; times 5 because each unit cell contains 5 atoms (4 atoms for iron and 1 for nitrogen). The box size equals 28.8 ($8 \times 3.6 = 28.8$). Insert this value in the diagonal of lines 3–5 in the CONFIG file.

Use VMD to visualize the output conf.xyz file to verify the quality of the produced model. Ensure that the first line of the conf.xyz file contains the correct number of atoms or VMD will not run properly.

### 10.9.1.2 Run an npt MD Simulation for Fe$_4$N Alone

Perform a 3D npt equilibration in the UNIT9/INI_CONF_fe4n directory, using the following procedure:

1. Open the CONTROL and FIELD files and review the differences between them with their corresponding files in UNIT7. For the FIELD file, the VDW lines are increased to three, due to the three interactions between Fe-Fe, N-Fe (Morse potential), and N-N (LJ potential). In addition, the description of the 2560 atoms of Fe$_4$N should look similar to the following format:

```
FeN
Nummols 512
Atom 5
Fe 55.85 0.000 4 0
N 14.01 0.000 1 0
finish
```

2. Check the last line of the CONFIG file to verify that the total number of atoms is 2560, and the order of arrangement is consistent with that above; i.e., after every four iron atoms there is a nitrogen atom. Repeat 512 times. In addition, check the diagonal of lines 3–5 again to verify that the values of the diagonal are 28.8.

3. Run DL_POLY using the following command:

```
$nohup ./../../../dl_poly_2.18/execute/DLPOLY.X > equi_run_date &
```

Use ela_STATIS.f90, ela_history.f90 and gnuplot to perform post-processing analysis for the STATIS and HISTORY files. Note: The produced STATIS file indicates that the box dimensions reach a stationary state with a box length of about 28.8. This means that the interatomic potential used corresponds to a basic cell dimension of $28.327/8 = 3.5409$, which is slightly smaller than the given lattice constant and we will take 3.46 for model development (the value 8 originates from the initial configuration with eight basic cells).

## 10.9.2   Developing the Initial Configuration for the Two-phase Model

The two-phase simulation can now be developed. This will be the most difficult part in the MD simulation due to the time required to build a good initial configuration for the $Fe_4N$-Fe two-phase non-homogeneous system. The interatomic distances of the single-phase problem are known, because these values were obtained through the equilibration process. However, how to build the interface is not known, because the distance required between the two phases is unknown.

### 10.9.2.1  Choose a Compatible Model Size

The calculated lattice constants in the equilibrium status of the 3D_npt processes for the two phases are:

- BCC iron: From UNIT7 the basic cell for the given interatomic potential is 2.925. This value was determined using the final value of the z box dimension after the npt simulation, which equals 58.4967 and contains 20 cells; therefore $58.4967/20 = 2.925$. This value is higher than the value in Table 2.2. List some possible reasons in the homework.
- Iron Nitride ($Fe_4N$) coating layer: From the previous step it is shown that the basic cell dimension is 3.56.

Use a suitable number for the cell for the two-phase model to generate a compatible structure. A reasonably good choice is to use 12 cells of BCC iron ($2.92 \times 12 = 35.04$) and 10 cells of $Fe_4N$ ($3.56 \times 10 = 35.6$).

Create the CONFIG by opening the UNIT9/INI_CONF_coating directory and verifying that the Model_input file appears as follows:

```
2 nphase
iron
bcc
2.92 2.92 2.92    cell dimension
12 12 10         ncell
-17.0 -17.0 -31.625  ro-Fe_
fn
f4n
3.56 3.56 3.56    cell dimension
10 10 10         ncell
-17.0 -17.0 -2.000   ro-Fe4N
```

The key to develop the geometric model is to choose the position coordinates of the first atom of each phase (i.e., atom ro-Fe and ro-$Fe_4N$). Their Cartesian coordinates, $r_{oz}, r_{oy}, r_{0z}$, are listed in the last line of each phase in the Model_input data file (see above). Note that they are near the bottom-left corner, inside the model as approximately estimated by (10.3) in Section 10.7.2.

The coordinate $r_{ox}$, $r_{oy}$ values are calculated by (10.3) based on $Fe_4N$ as follows:

$$r_{ox} = r_{oy} = -(0.5*10-0.5)*3.56 = -16.05 \qquad (10.6)$$

We will take this value as $-17$. The box size along the z-direction is complicated because it involves the combination of the two phases. It will be described in the next section.

### 10.9.2.2 Determining Gap Lengths Between the Two Phases

#### Concept of two gaps

Note that while there is a good estimation of the cell box dimensions of each phase of the two-phase system, the distance between the phases is unknown. Specifically, it is difficult to estimate the exact value of $r_{oz}$, the third number on the last line of the Model_input file.

Figure 10.9 shows two types of gaps. The first, $\delta_1$, is between the two primary phases (i.e, iron substrate and Fe4N coating layer with solid boxes); and the second, $\delta_2$, is between the primary phase (solid box) and the imaging phase (dashed box). These gaps can be controlled, respectively, by changing the $r_{oz}$ value in the Model_input file and the box size. The variable $r_0$ denotes the first $Fe_4N$ atom; hence changing its position will change the gap $\delta_1$ between the two primary phases. In addition, the box dimension along the z-direction given in the CONFIG file regulates the gap, $\delta_2$, between the primary phase and the imaging phase. This is because the periodic boundary conditions are repeated by the box size.

#### Strategy for determining gap values

There is no general rule or procedure to estimate the gap values. The method used here includes the following:

1. Test several possible values for the Model_input to produce the CONFIG file. Use VMD to verify that the atoms between the two phases are not too close. The typical value of the interatomic distance for a

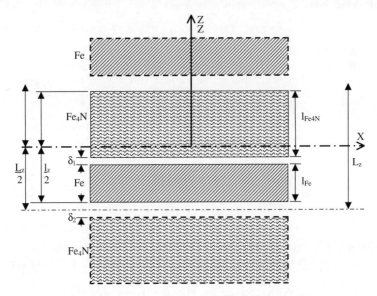

**Figure 10.9** Schematic for two types of gaps. The first is between the primary phases (solid boxes); and the second is between the primary phase (solid box) and the image phase (dashed box)

single phase can be used for the initial selection of the gap. The length of the first kind of gap can then be set equal to the second. Therefore after choosing the gap between the two phases (solid boxes) the gap between the solid phase and its image phase can be determined.

2. The selected model expressed in the CONFIG file is used for the npt equilibration, which will be discussed in Section 10.9.3. Note: There is a possibility that after certain steps the box dimension becomes very large, for example 10^3, due to the high repulsive force between overlapping atoms that imposes a large variation of volume in the barostat equation. In this case, no errors will be present in the OUTPUT file, however the simulation will be incorrect because the final state of equilibration is not a crystalline solid, but a gas. This is caused by a few particles in a large volume, hence the large interatomic distances. If this issue occurs, enlarge the box by 0.1 to 0.2 Å, which will increase the gap between the phases.

3. When changing gap values, it is important to change the parameters of the Model_input file to produce an initial configuration for the entire system that is a box with -Lx/2 Lx/2 in x, -Ly/2 Ly/2 in y, and -Lz/2 Lz/2 in z. Use VMD to check that this condition is satisfied.

After several trial-and-error simulations, a reasonable result should be obtained for the Model_input of the two-phase coating $Fe_4N$-Fe system. The determination process can be described as follows. The box size should be slightly longer than the model size, $l_z$, which is the sum of the height of the coating, $l_{Fe4N}$, plus the height of the substrate, $l_{Fe}$:

$$l_z = l_{Fe4N} + l_{Fe} = 10 \times 3.56 + 10 \times 2.92 = 64.8 \text{ Å} \tag{10.7}$$

The box size, Lz, along the z-direction should therefore be 65.8 Å. $\delta_1$ is 1 Å longer than the model size, $l_z$. This extra 1 Å length will be used to consider the gap between the model and its images at the top and bottom along the z-direction. There is no unique way to choose the coordinate $r_{oz}$ for the first atom of each phase; the only requirement is to keep the origin of the coordinates at the center of the model. Therefore the total height from the first coating atom to the model top is given as $(l_z/2 - r_{0z})$. For the current example, one may choose $r_{0z\_Fe4N} = -2$ Å for the $Fe_4N$ coating. In this way, the location of the first atom of Fe4N is 1.2 Å inside of the model as calculated by

$$r_{oz\_Fe4N} = l_{Fe4N} - (l_z/2 - r_{0z}) = (10*3.56 - l_z/2 - 2) = 1.2 \text{ Å} \tag{10.8}$$

Note: The obtained value in (10.6) consists of the gap $\delta_1$.

For the iron substrate, $r_{oz\_Fe} = -31.65$ Å. Based on this choice, determine the distance between the location of the first atom of the iron substrate and the nominal bottom of the iron model as follows:

$$r_{oz\_Fe} = l_z/2 - 31.65 = 64.8/2 - 31.65 = 0.75 \text{ Å} \tag{10.9}$$

The obtained values in (10.6) contribute to the gap $\delta_2$ minus the contribution from the difference, $(Lz-l_z)/2 = 0.5$ Å, between the box size and the model size; therefore the total second gap will be approximately:

$$\delta_2 = r_{0Z\_Fe} + (Lz - l_z)/2 = 0.75 + 0.5 = 1.25 \text{ Å} \tag{10.10}$$

## 10.9.3 Run the 3D_npt Equilibration in the INI_CONF_coating Directory

In the UNIT9 directory create a directory called COATING_npt. The tri-periodic equilibration for the two-phase configuration will be run inside this directory. Now perform the following steps:

1. Remain inside the INI_CONF_coating directory for the 3D_npt equilibration.
2. Open and review the CONTROL and FIELD file found in the directory. The CONTROL file indicates that there will be 30,000 steps with a time step of 0.002 ps. Therefore the simulation will span 60 ps.

3. Open and review the FIELD file, which describes two molecules. The first is iron, which has 2880 ($12 \times 12 \times 10 \times 2 = 2880$) atoms. The second is the iron nitride coating, consisting of 5000 atoms with 1000 molecules, where each molecule consists of 4 iron atoms and 1 nitrogen atom.
4. Verify that the CONFIG file is correct. Go to the last line, and verify that there are 7880 atoms and that the separation between the iron and $Fe_4N$ molecules starts from atom 2881. Also check the order of the atoms of $Fe_4N$. There should be 4 iron atoms followed by 1 nitrogen, repeated 1000 times. Also check lines 2–5 to verify that the diagonal values are 35.6, 35.6, and 65.10.

Run DL_POLY. At the end of the simulation carefully check the time evolution of box dimension using ela_STATIS.f90 and the STATIS file. If the simulation is reasonable, the difference between the obtained box size and the initial box size should vary only by a few percent. In the last box, the box size should be 35.4858, 35.4858, and 65.58898; which differs from the initial size (35.6, 35.6, and 65.8) by roughly 0.3%.

## 10.9.4   Non-equilibrium Simulation of the Coating Layer Under Top Shearing Strain Rate

The work listed in Boxes A and B of Table 10.1 has now been completed. As a result, an equilibrium configuration under tri-periodic boundary conditions and an npt ensemble has been formed. The next step is to prepare and simulate the non-equilibration simulations as shown in Boxes C and D of Table 10.1. To make the operation more clear, the two simulation directories, 2D_EQUI_nvt and 2D_FORCED, will be grouped inside of the 2D_Non_EQUI directory. In addition, the operation answers can be found in the subdirectories, respectively, as 2D_EQUI_nvt_out and 2D_FORCED_nvt_out. Use these files to find and check the procedures and results for any possible errors.

### 10.9.4.1  Bi-periodic nvt Equilibration for Two-phase Model in 2D_EQUI_nvt

All the procedures for the two-phase simulation for the 2D_Non_EQUI are the same for the one-phase case described in Section 10.8.1; therefore only the basic procedure will be reviewed as follows:

1. REVCON file: Open the UNIT9/2D_Non_EQUI/2D_EQUI_nvt directory and move the REVCON file in the INI_CONF_coating directory to this directory using the following command:

```
$ 2D_EQUI_nvt$ cp ../../INI_CONF_coating/REVCON .
```

2. CONFIG file: Change the name of the REVCON file to CONFIG using the command:

```
$ mv REVCON CONFIG
```

3. Open the new CONFIG file and change the periodic condition from tri-periodic to bi-periodic by changing the value of the imcon key on the second line from 3 to 6.
4. CONTROL file: Open the CONTROL file and change the command for the ensemble to:

```
Ensemble nvt berendsen 0.5
```

5. HISTORY file: Generate the HISTORY file using the following command in the CONTROL file:

```
Traj 0,100,0
```

This command indicates that the HISTORY file will be written from the beginning, and then written again after 100 steps. This HISTORY file will print only the position to save space.

6. FIELD file: This file will have three types of molecules: 288 fixed iron atoms listed first in the CONFIG file, 2592 normal iron atoms listed in the middle of the CONFIG file, and 5000 atoms of 1000 $Fe_4N$ molecules listed in the last part of the CONFIG file. The fixed atom is controlled by the "ifrz" index (ifrz>0) which is the fourth number of line 7 in the FIELD file.

Now run the simulation using the command:

```
$ 2D_EQUI_nvt$nohup ./../../../dl_poly_2.18/execute/DLPOLY.X >
equi_run_date &
```

Any file name can be used in place of "equi_run_date", which will be the output name for the simulation. The calculation should take roughly one hour to complete. After completion, the simulation should have developed an OUTPUT, STATIS, and HISTORY file. The REVCON file will move to the 2D_FORCED directory to become a new CONFIG file as described in Box D of Table 10.1. Use ela_STATIS.f90 to plot the curves of the 37 variables over time from the STATIS file, as described in Section 10.5.8. For the nvt ensemble, the box size will not appear in the STATIS file; therefore the variables will be reduced from 46 for the npt ensemble to 37.

### 10.9.4.2 MD Simulation Under Shearing Rate on the Top Layer

As shown in the last box of Table 10.1, run the MD simulation for a constant shear strain rate applied at the upper layer of the atoms. Open the 2D_Non_EQUI/2D_FORCED directory and verify that the input files are correct.

CONFIG file: The last configuration of the second equilibration performed in EQUI_2D_nvt will be used to develop the new CONFIG file for the initial configuration. Copy the REVCON file from the EQUI_2D_nvt_out directory to this directory and then rename it as the CONFIG file.

CONTROL file: This file will be the same as that in the 2D_EQUI_nvt simulation. Only the position vectors are printed to save space in the HISTORY file.

FIELD file: There is little difference between this field file and that used in the 2D_EQUI_nvt simulation, except the continuous shear as described in Section 10.8.3. After several trial-and-error attempts, it was found that $A = 0.1$ Å/ps and $|z0| = 24$ Å. This will prevent a large displacement that can cause image problems, etc. The values "0.1 24" were then inserted before the "VDW" pair potential of the FIELD file.

Use the following command to run the simulation:

```
. $ nohup ./../../dl_poly_2.18/execute/DLPOLY.X > equi_run_date &
```

## 10.9.5   Post-data Processing to Determine the Displacement of the Coating Layer Under a Given Shearing Rate

To determine the displacement for the shearing strain under a given shearing rate at different times, t, the same procedure used to produce Figure 10.9 should be followed. Specifically, move the HISTORY file produced by the 2D_nvt equilibration process in the 2D_EQUI_nvt directory to the POST_PROC directory. In that directory, the "MEAN.xyz" file, which references the position of atoms using function 1 of the ela_history_2009.f90 code, will be produced (see Section 10.8.2), and then moved to the 2D_FORCED/POST_PROC directory. In the latter directory, the HISTORY file produced from the non-equilibrium shearing process will first be moved from the 2D_FORCED directory, and then the "disp_dist.dat" file with the displacement information of u(z), will be produced by function 2 of the ela_history_2009.f90 (see Section 10.8.4).

It is important to note that the displacement is developed using statistical time averaging (see (2.21a) in Section 2.8.2). The interval for the time averaging can be altered on line 6 of the "histo.inp" data file. If the interval is too small, there will not be sufficient data for accurate averaging. On the other hand, if the interval is too large, the average result will not accurately represent the displacement at the mean time of that interval. For example, each configuration is printed in the HISTORY file by 100 steps and each step takes 0.001 ps, created, respectively, using the commands:

"Traj 0,100,0" and "timestep 0.001"

in the CONTROL file. The corresponding time and time interval for the input of "81 100" on line 6 is (8100, 10000) steps and (8.1, 10) ps. The obtained displacement u(z, ti) distribution should correspond to the mean time, 9.05 ps, of that interval (i.e., ti = 9.05 ps).

To determine a series of values for ti, u(z, ti), develop a series of reference files (e.g., MEAN_81_100. xyz, etc.), and then insert the name of the files on line 2 of the histo.inp file to match the interval (e.g., "81 100") on line 6. After running ela_history_2009, change the produced "disp_dist.dat" file to "disp_dist_81_100.dat" to distinguish it from the data obtained at the other time interval.

Figures 10.11 and 10.12 show the average displacement u(z, t), respectively, at time 5.05 ps and 9.05 ps. These displacements are due to the double shearing rate of 0.05 Å/ps with the top shearing along the positive direction at z = 24 Å and the bottom shearing along the negative direction at z = −24 Å (see the FIELD file with |z| = 24 = Å). Compared to UNIT8, this double shearing is for an $Fe_4N$ coating layer on an iron substrate, where the interface between the two phases is at the coordinate $z = -3.63$ Å. From these

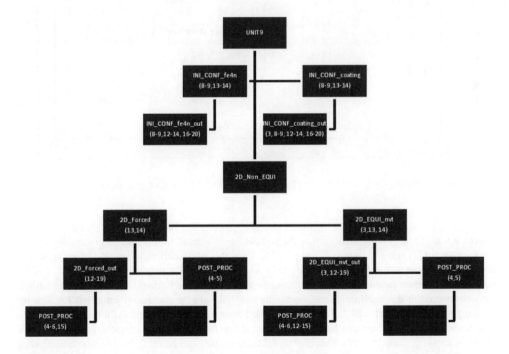

**Figure 10.10** Flowchart of non-equilibrium simulation of coating layer under shearing strain rate

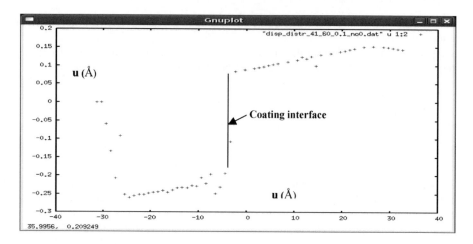

**Figure 10.11**   Displacement u(z) for the interval (41 60) with averaging time 5.05 ps

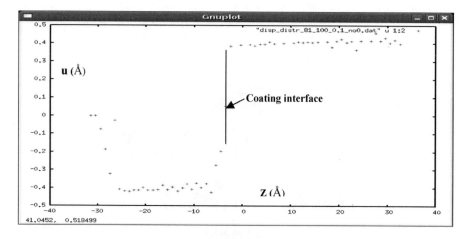

**Figure 10.12**   Displacement u(z) for the interval (81 100) with averaging time 9.05 ps

figures, it is seen that at the interface where the vertical line is located, the displacement increases rapidly by roughly $0.35\,\text{Å}$ (e.g., $0.1-(-0.2)=0.35$). Far from the interface the values of deformation u(z) increase until about $0.16\,\text{Å}$ for $z = 24\,\text{Å}$ and $-0.27\,\text{Å}$ for $z = -24\,\text{Å}$. For the outer region of $|z| > 24\,\text{Å}$, the deformation is more random until $u = 0$ due to the fixed iron at the bottom. Also, from Figure 10.12, the deformation at the interface increases rapidly to $0.6\,\text{Å}$, and the maximum value $\pm0.4\,\text{Å}$ is at the shearing plane location $|z| = 24\,\text{Å}$. This is reasonable because more deformation will be developed for a longer time at the given shear rate. Note: The increase at the coating interface indicates that the bonding at the interface is not strong enough. This may be considered a physical bonding, but not atomistic bonding or metallurgical bonding.

**Homework**

(10.22) In DL_POLY_2 MD codes, the boundary condition is controlled by an index called IMCON which appears in the position of the second value of the second line of the CONFIG file. Go inside to UNIT9/INI_CONF_coating directory and UNIT9/2D_Non_EQUI/2D_FORCED directory (see Figure 10.10) and find the IMCON number in each of the CONFIG files. Based on the IMCON number and the manual of DL_POLY_2, answer the following questions:

(a) What is the periodic boundary condition used in these directories?

(b) Based on the introduction below for the different functions of the simulation conducted in these two directories, explain why they use different boundary condition for the simulation. Remark: The INI_CONF_coating directory carries on equilibration of the coating model of iron-nitride ($Fe_4N$) on the steel substrate; 2D_FORCED carried on external shearing along the XOY plane for the layers of that coating model.

(10.23) In Table 10.1 there are four boxes with four directories to take different roles for the simulation of a thin iron layer under shearing. The three directories with CONTROL files are EQUI_npt in UNIT7, 2D_EQUI_nvt and 2D_FORCED in UNIT8. In each of the CONTROL files the ensemble used for the related simulation is given. Based on this information answer the following questions:

(a) What are the ensembles used in the simulation conducted in each of these directories?

(b) Why are these ensembles used in these directories?

(10.24) Change the value of the number of slices (or thickness of slice) in the histo.inp file in UNIT9/2D_Non_EQUI/2D_FORCE/2D_FORCED_out/POST_PROC to see their influence on the profile of displacement distribution u versus the z-coordinate.

# Part 10.3: Atomistic Simulation for Protein-Water System and Brief Introduction of Large-scale Atomic/Molecular System (LAMMPS) and the GP Simulation

## 10.10    Using NAMD Software for Biological Atomistic Simulation

NAMD is an MD code designed for the simulation of biological systems, developed by the Theoretical Bio-Physics Group at the University of Illinois at Urbana.[13] This group also developed VMD (see Section 10.4). This section will introduce the fundamental aspects of NAMD along with a simple example used in the UNIT10_11/FIRST_RUN directory. The next section will discuss how the initial configuration of a fibronectin module of a protein surrounded by water can be developed by VMD and equilibrated using a thermostat and barostat with NAMD. Section 10.12 will discuss how the equilibrated fibronectin module can be used for a non-equilibrium simulation in which the protein will be stretched.

Readers must register on the NAMD website in order to download the software; which is available at no charge. To get the codes and additional information including a NAMD/VMD user guide, tutorials, and command listings, visit the NAMD website at http://www.ks.uiuc.edu/Research/namd/.

### 10.10.1    Introduction

In this section, the basic features of biological macromolecules and the parts of NAMD that account for these features will be briefly introduced.

#### 10.10.1.1 Compositions and Polymeric Structures of Biological Macromolecules

As explained in Section 9.6.1, protein is a product of polymerizing a large quantity of amino acids. Depending on the construction of the R-side chain, there are 20 types of residues of the α-amino acid. Some residues are hydrophobic (e.g., Ala, Val, lle, Leu, Met, Phe, Tyr, Trp), and therefore attempt to remain on the protein surface. Some are hydrophilic residues (Lys, Arg, His, Asp, Glu, Ser, Thr, Asn, and Gln), which are usually buried within the protein's core.[14] Three special amino acids which are classified as either hydrophobic or hydrophilic residues are Cysteine (Cys), Glycine (Fly), and Proline (Pro).

Proteins are linear polymers containing more than ten to several thousand α-amino acids of the above 20 types, linked by peptide bonds. Peptide bonds link the amide nitrogen atom of one amino acid with the carbonyl carbon atom of an adjacent atom in the linear polymers. For a nucleic acid and polysaccharide the monomer is, respectively, nucleotide and monosaccharide, while the linkages between different monomers are still through peptide bonds. These covalent bonds between monomer molecules are usually formed by dehydration reactions in which a water molecule is lost.

### 10.10.1.2  Four Basic Files of NAMD to Conduct Atomistic Simulation of Bio-materials

In order to run any MD simulation for biomaterials, NAMD requires at least the following four files:

- PDB (Protein Data Bank) file: This file stores atomic coordinates and/or velocities for the system. PDB files may be generated by hand, but are also available via the internet for many proteins at http://www .pdb.org. Water.pdb in Section 10.10.2.1 and 1fnf.pdb in Section 10.10.1 are examples. Specifically, some of the 20 kinds of residues of α-amino acids and the connection sequences will appear in these files (see 1fnf.pdb and solvate.pdb file in Sections 10.10 and 10.11). To some extent, this file corresponds to a CONFIG file in DL_POLY_2.
- PSF (Protein Structure) file: This file stores structural information of the protein, such as charges, atomic masses, and various bonding and angle structures. Water.psf in Section 10.10.2.1 and fibr.bsf in Section 10.10.1 are examples. The latter is developed by VMD's script called guess_coord.tcl. To some extent, this file corresponds to the first part of a FIELD file in DL_POLY_2.
- Force field parameter file: A force field is a mathematical expression of the potential between atoms in the system. CHARMM, X-PLOR, AMBER and GROMACS are four types of basic force fields used for biological materials. NAMD uses all of them. The file sets bond strength, equilibrium lengths, and parameters listed in the mathematical expressions. The par_all27_prot_lipid.inp file of GROMACS used in this and next two sections is an example. This file corresponds to the last part of a FIELD file in DL_POLY_2.
- Configuration file: This file sets all the options such as temperature, pressure, time step, and the total steps that will be used when running a simulation. The file "first_run.conf" in this section and "fibr.conf" in the next section are examples. To some extent, this file corresponds to a CONTROL file in DL_POLY_2. The difference is that this file will write specific PDB and PSF file names for the simulation.

## *10.10.2   A Simple Simulation Using VMD and NAMD*

### 10.10.2.1  Input Files for a Water Simulation System

After registering at the NAMD website, download namd32 and namd64 into the directory NAMD_2.6 shown on the directory tree in Figure 10.13. Note that before these files are imported from the website, these directories are empty due to copyright protection. These two files are precompiled NAMD executable files for 32 and 64 Linux architecture and cannot be read by vi. Use namd32 for 32-bit machines and namd64 for 64-bit machines. All examples in this book will be based on the 32-bit namd32 file. If using a 64-bit machine, replace namd32 with namd64 in all commands herein.

All other files are included in the directory UNITS/UNIT10_11. Open the FIRST_RUN directory. In this directory there are four files including water.pdb, water.pdf, par_all27_prot_lipid.inp, and first_run. conf. These four files, respectively, are the four basic types of NAMD files listed in the last section. These files are introduced to simulate a periodic box of water containing the molecules of interest. An introduction to these input files is given as follows:

```
water.pdb
```

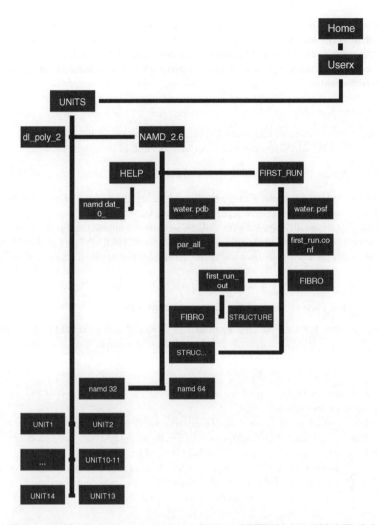

**Figure 10.13**  Directory tree with emphasis on UNIT10-11 and NAMD_2.6

Used for the initial configuration. Open with vi using the command:

```
$ vi water.pdb
```

This file contains 2478 lines, one line per atom. The lines are structured using column alignment to separate information. It is important to note that each line has 78 column positions. The column number can be found on the bottom line of the vi screen as the last number in the vi command footer. The number before the column number is the row (or line) number. Specifically, columns 14–17 represent atom type, which defines the order of atoms in the file (in the example, OH2 is listed for the first atoms, H1 for the second and so on); and columns 34–54 represent atom position (x, y, z). This file format is considered the standard PDB file format for these types of molecule files.

```
water.psf
```

Contains addition atom information including charge and mass in lines 9 to 2486, as well as the intra-molecular potential for the system. For example, in the case of water, there is a section of the file containing information on bonds labeled BOND beginning at line 2488, as well as ANGLES beginning at line 2903. The detailed structure of this code can be created by VMD or by using a utility called psfgen that will be presented shortly.

```
par_all27_prot_lipid.inp
```

Contains all the interatomic interactions (bond, spring angle, dihedral for the force field) of the molecule, and is related to the CHARMM force field.[13] No modification of this file is required.

```
first_run.conf
```

Text file describing all of the features of the simulation including the number of time steps, the thermostat and barostat constants, and the name of input and output files.

The first two files are usually created with VMD, due to the complexity and delicacy of these files in the simulation process. The third file is the standard file for the CHARMM potential and it must not be modified. The fourth file allows the user to change the simulation parameters and will be described in detail below.

### 10.10.2.2 Detailed Description of the first_run.conf File

Open the first_run.conf file with "vi" and go to line 13. Lines 13 and 14 should look as follows:

```
structure          water.psf
coordinates        water.pdb
```

These lines will force NAMD to read the molecule structure on the water.psf file and the initial configuration on the water.pdb file.

Now go to line 17, which contains the name of the output file, first_run_water. This output file name will form the base name for all output files from the simulation, including first_run_water.dcd, first_run_water.coor, etc. Later in the section the output files of NAMD will be discussed in more detail.

Now go to line 28, which contains the name of the file to be used as the molecular interaction parameter (i.e., par_all27_prot_lipid.inp). Ensure that the block of lines following this line is not changed, otherwise, errors will occur. These lines include crucial simulation commands such as setting the time step at 2 fs, etc.

Go to line 50 that states the following command:

```
langevin    on
```

The langevin command determines whether the system is coupled to a langevin thermostat (on), or is not coupled by a langevin thermostat (off).

Lines 57–60 contain the box dimensions, set at 32 by 32 by 32 Å; and the box center, set at the point 16, 16, 16.

Lastly, see line 77 that works exactly like line 50 but for a barostat, and line 107 which states the number of time steps for the simulation.

### 10.10.2.3 Running the Simulation

Open the UNIT10_11/FIRST_RUN directory. Open a second terminal for this simulation, one for running the code and the other for checking the output files created by NAMD. Run NAMD using the following command for namd32:

```
~/...../FIRST_RUN$./../../NAMD_2.6/namd32 first_run.conf
```

or

```
~/....../FIRST_RUN$ nohup ./../../NAMD_2.6/namd32 first_run.conf
>first_run.log &
```

The last part of the command (>first_run.log) means that NAMD will record some of the output in a text file called first_run.log. The simulation will run for several minutes. Open the second terminal and use the "ls -lrt" command to view the files as they are formed.

The simulation will develop many files, each with the name first_run_water, followed by a dot and three or four characters. The first_run_water.coor and first_run_water.vel files contain the coordinates and the velocities of the atoms during the simulation, and can be useful if the simulation is restarted. Both files are binary fields, therefore the file contains a list of strange symbols if opened using vi. The first_run_water.dcd file contains the history of the system time evolution, and is also a binary file.

When the simulation is finished, the terminal used to run the code will be clean. Open vmd using the command:

```
$vmd
```

and perform the following steps.

Load the water.psf file by selecting "new molecule" in the file menu of VMD, then "Browse". Select water.psf in the "Molecule File Browser" window, click "OK", and then click the "load" button. Nothing will appear on the black display, but on the VMD main window a line will appear referring to the molecule water.psf. Close the "Molecule File Browser."

Left click on the water.psf line, and right click; a drop-down menu will appear in the main window. Select "load data into molecule". This will re-open the "Molecule File Browser" window. Click "Browse", select the "first_run_water.dcd" file, click "OK", and then click the load button. The window will now show the atomic configuration on the black display. The VMD main window will show the molecule containing 10 frames. Scroll through the frames to view the evolution over time.

Note that the simulation in this section is a simple example for NAMD. It is an equilibration using a thermostat and barostat (see the first_run.conf file); therefore the configuration will not present an apparent change. During this initial simulation, the liquid molecules will move only slightly from their initial configuration.

### 10.10.3   Post-processing Data Analysis

Using the simulation output files, it is possible to observe instantaneous changes in temperature, pressure, and volume over time. This information can be found in the first_run.log file. Open it with vi using the command:

```
$vi first_run.log
```

The first part of the file (typically lines 150–200) contains the information about the simulation, such as the number of atoms, the total charge of the system, the parameters of the thermostat and barostat, etc.

Starting from line 135, the statistics of the simulation are reported with the following typical structure:

```
ETITLE:  TS   BOND   ANGLE   DIHED   IMPRP
      ELECT  VDW  BOUNDARY   MISC  KINETIC
      TOTAL   TEMP  TOTAL2   TOTAL3  TEMPAVG
    PRESSURE GPRESSURE   VOLUME  PRESSAVG GPRESSAVG
```

```
ENERGY:    0  7710.7081  170.8783   0.0000   0.0000
   -7949.3358  701.6210   0.0000   0.0000   0.0000
   -62910.1183   0.0000  -62910.1183  -62910.1183   0.0000
   -6355.4795  -2394.2538  327610.0000  -6355.4795  -2394.2538
```

The first four lines are the heading for the quantities reported in the four lines below them. For example, the variable BOND has a value of 7710.7081, and the variable PRESSURE has a value of –6355.4795. These values can then be extracted and organized in a form that can be plotted with gnuplot.

To extract data from this file, use a script called namddat_gnuplot.csh, which is a modification of a standard script that is used to print data in a format compatible with gnuplot. The namddat_gnuplot.csh file can be found in the UNIT10_11/HELP directory; or a similar version of the file with the same function can be found on the NAMD website. The script is called to extract one or more file values, such as in the command to extract temperature:

```
$./../HELP/namddat_gnuplot.csh TEMP first_run.log
```

The first part of the command (./../HELP/namddat_gnuplot.csh) opens the executable file namddat_gnuplot. The second part of the command (TEMP) is the list of variables to be extracted from the file, which in this command is the first_run.log file. The command will create an output file called data.dat. Open it with vi. The first line (#TS TEMP) is the row of variable names reported in the file. Each variable corresponds to a column of the file. In the reported example, the data.dat file is formed by two columns, the first being the time step, and the second being the temperature.

The script can also be used to extract multiple variables simultaneously. For example, both temperature and pressure can be extracted using the following command:

```
$./../HELP/namddat_gnuplot.csh TEMP PRESSURE first_run.log
```

Once completed, the shell window should provide a list of the names of different variables, called ETitles; as well as a confirmation, such as "calculating average of TEMP PRESSURE", displayed in the shell.

## Exercise 10.15

Use the namddat_gnuplot.csh script to extract temperature, pressure and volume. Use gnuplot to plot the time evolution of these three quantities.

---

**Homework**

(10.25) Use namddat_gnuplot.csh script to extract temperature, pressure and volume from the first_run.log file used in section 10.10.1.3. Use gnuplot to plot volume as a function of time in ps. Plot the image in an .eps file.

---

## 10.11   Stretching of a Protein Module (1): System Building and Equilibration with VMD/NAMD

This section will prepare the initial configuration for the fibronectin protein, and then run its first equilibration using NAMD. The next section will discuss how to develop the non-equilibrium simulation. Along with collagen and elastin, a fibronectin protein is one of the important structural components of the extracellular matrix (ECM), which contributes to the mechanical behavior of tissues. The modular structure and detailed compositions of a fibronectin protein can be found by searching for the PDB file on the PDB website.

## 10.11.1    Preparation of the Initial Configuration with VMD

The preparation of the initial configuration to be used by NAMD is a complex procedure. A step-by-step description for the 11 steps will be introduced. Simultaneously, subtitles to group these steps based on their functions are also used below.

### 10.11.1.1  Preparing the Working Directory and Downloading the PDB File

1. Open the UNIT10_11/FIRST_RUN directory. Create a directory called "FIBRO" using the command:

   ```
   $mkdir FIBRO
   ```

   Open this directory and create a directory called STRUCTURE using the command:

   ```
   $mkdir STRUCTURE
   ```

2. Download the structure for the protein. Go to the web page http://www.rcsb.org, which contains a database of resolved protein structures. Search for the file 1fnf.pdb, by typing "1fnf" in the search bar. Download the "pdb text" file and save it to the STRUCTURE directory.
3. Open the file with vi. The file contains many lines starting with "REMARK". These lines contain information about the protein and the experimental procedure used to obtain the structure, which are not necessary for this simulation. The first line with coordinates is line 344; use ":344" to reach this line. The four characters from column 23 to 27 are the residue number. For the first atom the residue number is 1142, while the second residue, 1143, starts at line 353, and so on. The last portion of the file, starting from line 3788 to the end of the file, contains the water molecules of the system. If for any reason the 1fnf file is not available through the web site, the 1fnf.pdb file can be found in the UNIT10_11/HELP directory.
4. Open VMD and load the 1fnf.pdb file. Note that the protein is formed by four modules. Also note the water molecules present as created by the PDB file as described above. Several hydrogen atoms on the protein structure are not present in the PDB file, because the experimental procedure used to calculate the structure often cannot resolve hydrogen atoms.

### 10.11.1.2  Extracting Necessary Information by Editing the PDB File

Only one module of the fibronectin protein will be simulated, therefore the PDB file must be edited as follows:

5. Extract one of the four modules from the PDB file. In particular, extract the module that consists of lines 1418–1509. To do this, open the graphical representation window in VMD by clicking the representation command in the Graphics menu, and write the string:

   ```
   $ protein and resid > 1417
   ```

   in the Selected atoms box. Note: The selection "protein and resid > 1417" means "select the entire protein residue that has a residue number larger than 1417".

6. Review the remaining module and then close VMD. Edit the 1fnf file to remove the entire residue with the exception of the residue from 1418 to 1509. To do this, first make a copy of the 1fnf.pdb file using the command:

```
$ cp 1fnf.pdb mod.pdb
```

Open the mod.pdb file with vi. Remove the first 343 lines that contain the molecule information using the following command on the first line:

```
343 dd
```

Now remove all lines with the exception of residual 1418 (at the new line number 2602 after deleting the first 343 lines) to residual 1509 (line 3444). First delete the lines after line 3444 and then delete the lines before line 2602; use the "dd" command above (see also Section 10.C.2 in Appendix 10.C). If problems occur, use the file mod.pdb in the UNIT10_11/HELP directory to complete the steps thus far.

### 10.11.1.3  Using VMD Script for Model and File Development

7. Copy the files guess_coord.tcl and top_all27_prot_lipid.inp from the UNIT10_11/HELP directory into the STRUCTURE directory. The file guess_coord.tcl is a script for VMD that will serve several functions including: (1) Read topology files such as the mod.pdb file; (2) Add the atoms that are not present in the mod.pdb file (usually hydrogen), and (3) Create a PSF (fibr.psf) and a PDB (fibr.pdb) file. The files fibr.psf and fibr.pdb attain their respective names from the configuration file used.

8. Use VMD to produce the PDB and PSF file using the command in the shell (not in the graphical window):

```
~/...../FIBRO/Structure$vmd -dispdev text -e guess_coord.tcl
```

The portion of the command "-dispdev text" means that VMD does not have to produce a graphical output but instead produces a text output on the terminal. The portion of the command "--e guess_coord.tcl" forces VMD to execute the list of commands within the file "guess_coord.tcl". After running, two output files fibr.psf and fibr.pdb will be obtained.

9. Open VMD and load the fibr.pdb file to verify that the file is correct. Note that the hydrogen atoms (in white) are now present. If problems have occurred and the file will not open or does not contain the hydrogen atoms, correct copies of the fibr.pdb and fibr.psf file can be found in the UNIT10_11/HELP directory.

### 10.11.1.4  Determining Model Geometric Parameters

10. Use VMD to check the dimensions of the protein in the fibr.pdb file. Open the TK console by selecting "TK Console" from the "extension menu", and write in the new terminal the following command:

```
$set fib [atomselect top all]
```

This command will create a selection of all atoms of the molecule called "fib". Use the string $fib to select this group of all atoms in the molecule. Measure the minimum and the maximum of the x, y, and z values of the protein using the command:

```
$measure minmax $fib
```

The output is formed by two vectors, for example {2.63800001144 11.937000247 −68.4509963989} and {38.21900177, 41.9169998169, −30.875}. The first contains the minimum values in the x, y, and z direction, and the second contains the maximum values. Take note of these numbers, as they will be important in the next step.

11. Use VMD to add a water box to surround the protein. Open the "Add Salvation Box" window by selecting the "Modeling" submenu from the "extension menu". A graphical window will appear. At the top of the window, unselect the "waterbox only" checkbox, which should then allow the "Browse" button to be selected for the PSF and PDB boxes. Click "Browse" and select the fibro.psf and fibro.pdb files for the PSF and PDB boxes, respectively. In the box size text boxes in the middle of the window, insert the values of the box dimensions with some additional space for clearance. In this particular example, allow 5 Å of clearance in each direction for the minimum/maximum values found in step 10. Therefore, subtract 5 Å from each of the minimum values (e.g., −2.362, 6.937, −73.4509963989), and add 5 Å to each of the maximum values (e.g., 43.21900177, 46.9169998169, −25.875). Therefore the box size for the salvation box in this example should be: 45.581 Å, 39.98 Å and 47.576 Å. Determine the center of the model by finding the average of the minimum and maximum values including clearance. Therefore the coordinates of the center of the model in this example are: 20.4285, 26.927 and 49.663. Click on the "solvate" button. Close VMD and verify that the files have been updated using "ls -lrt". The solvate.pdb and solvate.psf file have been created. Use vi and VMD to view the file structures. If problems have occurred, use the solvate.pdb and solvate.psf files found in the UNIT10_11/HELP directory.

## 10.11.2   Preparation of the NAMD Input File

Now that the initial configuration files have been created, the ".conf file" of NAMD can be developed for the equilibration as follows:

1. Prepare the input files for NAMD. Open the FIBRO directory. Initially this directory will be empty. Copy the par_all27_prot_lipid.inp and the first_run.conf files from the FIRST_RUN directory to this directory.
2. To avoid confusion, change the first_run.conf file to fibro.conf using the command:

```
$mv first_run.conf fibro.conf
```

3. Open the fibro.conf file with vi. Find the "structure" line (line 13) and "coordinates" line (line 14) and replace the "water.psf" and "water.pdb" files with "solvate.psf" and "solvate.pdb" files. These files are in the STRUCTURE directory; therefore at lines 13 and 14, respectively, the file names should be written as STRUCTURE/solvate.psf and STRUCTURE/solvate.pdb.
4. Change the "outputname" in line 17 by replacing first_run_water with any desired name, for example "fibro_equilibration".
5. Insert the correct box dimensions in lines 57–59. Perform an NPT simulation using a barostat to change the volume of the system. Therefore, use a box size that is slightly larger, such as 2 Å, than the box dimension created by the solvate process. For this example, the dimensions are 45.581 Å, 39.98 Å and 47.576 Å; therefore the values of 47.581, 41.98, and 49.576 Å in the X, Y, Z directions (cellbasicvecto1, 2, 3 in the code) are used.
6. Insert the values of the center of the model created by solvate as the cell origin values on line 60, i.e., 20.4285, 26.927 and 49.663.

Review the PME command lines beginning on line 66. PME is short for Particle Mesh Ewald method (see page 51 of NAMD user guide) and is used for the efficient calculation of the electrostatic force. The PME requires a grid size along the x, y, z direction in the following three lines. For this example, the maximum value of the model produced by solvate was 47.576 Å, therefore the lines 67–69 must be changed to 48. This editing is necessary to produce a grid size that is large enough for the simulation. It is important that the PME dimensions are integer values, otherwise error messages will result.

7. Remove the symbol "#" in line 104 and enter 100 after "Minimize". This value is the number of steps for energy minimization.
8. Go to the last line, line 107, and enter 400 for the number of time steps.
9. Use ":wq" to save and exit the file.

### 10.11.3   Run the NAMD Simulation

In order to run the NAMD simulation, the code will need a PSF, PDB, force field, and configuration file. The solvate.psf and solvate.pdb files are noted in lines 13 and 14 of the fibro.conf configuration file; therefore the simulation will only need to check the fibro.conf configuration file and the par_all27_prot_lipid.inp force field file to obtain all four files. Suppose that the desired output file name is fibro.log, the "nohup" command to run the simulation is as follows:

```
~/..../UNIT10_11/FIRST_RUN/FIBRE$nohup
././../../NAMD_2.6/namd32
fibro.conf > fibro.log &
```

### 10.11.4   Error Messages and Recommended Action

The above simulation will require several minutes to finish. If it stops immediately, this should indicate that an error has occurred. The NAMD log file should contain information about the current errors. If an error should occur, open the "fibro.log" file with vi and check all errors using the command ":/ERROR" followed by ":n" in the bottom line of the screen (see Appendix 10.C.2). For example, if a non-integer value (e.g., 48.2) is used for the PME dimension on lines 67–69 of the fibro.conf file, the following error message will appear on lines 17–19 and the last two lines of the "log" file (see the fibro.log file in the: ~UNITS/UNIT10_11/FIRST_RUN/FIRST_RUN_out/FIBRO2 directory):

```
ERROR: Expecting only a number for 'PMEGridSizeX' input, got: 48.2
ERROR: Expecting only a number for 'PMEGridSizeY' input, got: 48.2
ERROR: Expecting only a number for 'PMEGridSizeZ' input, got: 48.2
Application called MPI-Abort (MPI_COMM_World,1)-Process 0.
```

Here, MPI is short for message passing interface among parallel-computing systems with many computer processes working together. However, the error here occurs for the processor that reads the fibro.conf file. It is not an "mpi" error but a "conf" error because PME parameters must be integers on lines 67–69 of the fibro.conf file. The problem can be solved by replacing 48.2 with the integer value 48 on these lines (see the correct file in the First_Run_out/FIBRO directory, and wrong file and error messages in the First_Run_out/FIBRO2 directory).

The MPI abort error in the last two lines may have many causes. The most logical reasoning is that one processor sends the error message to stop all processes to the other machines, creating an error that spreads to all machines in the system, triggering the MPI message. For more information related to MPI and its error message, review the following website: http://en.wikipedia.org/wiki/Message_Passing_Interface.

After the simulation has completed, follow the same procedure used in Section 10.10.3 of data processing to analyze the fibro.log file and use namddat_gnuplot.csh to extract information from these files.

---

**Homework**

(10.26) Perform an equilibration simulation using 10,000 time steps. Use namddat_gnuplot.csh script to extract temperature, volume and pressure from the fibro.log file. Write a brief report for the equilibration procedure. Include in the report gnuplot images representing the time evolution of pressure and volume. Was the 10,000 time step equilibration enough? Why?

---

## 10.12   Stretching of a Protein Module (2): Non-equilibrium MD Simulation with NAMD

This section will introduce an example of a non-equilibrium MD simulation related to several recent scientific studies of an unfolding pathway of a protein.[15] The protein used in the last section, fibronectin, will be used again in this section. One end of the fibronectin protein, equilibrated in the previous unit, will be fixed, while a force is applied to the other end to allow for an atomic force microscopy (AFM) measurement. This type of non-equilibrium simulation is referred to as Steered Molecular Dynamics (SMD). Further information on SMD can be found in Chapter 3 of the NAMD tutorial on the NAMD website.[13]

The SMD simulation will start from the equilibration performed in the previous unit, and will be performed as follows:

– Use VMD to determine the atoms that will be fixed and the ones that will be pulled.
– Create the files for a NAMD non-equilibrium simulation.
– Run the simulation.
– Extract useful data from the NAMD output files.

### 10.12.1   Preliminary Steps

This non-equilibrium simulation can only be performed if the equilibration in the previous section was completed correctly. Use the completed homework from the last section for the following procedure. If problems were experienced while running the equilibrium simulation, use the files in the FIBRO_equi directory in the UNIT12/HELP directory. The procedure for preparation is as follows:

1. Create a directory called NON_EQUILIBRIUM inside the UNITS/UNIT12 directory (see Figure 10.14) using the command:

```
$ mkdir NON_EQUILIBRIUM
```

2. Open the NON_EQUILIBRIUM directory and create a directory called STRUCTURE. Open it, and copy the solvate.psf and solvate.pdb files from the UNIT12/HELP/FIBRO_equi/STRUCTURE directory into it. Also copy the fibro_equilibration.restart.coor from the HELP/FIBRO_equi directory into it. Review the content of its counterpart directory, NON_EQUILIBRIUM_out/STRUCTURE. If difficulties arise during the simulation procedure, use the NON_EQUILIBRIUM_out and HELP directories which contain additional simulation information and files to compare and use.

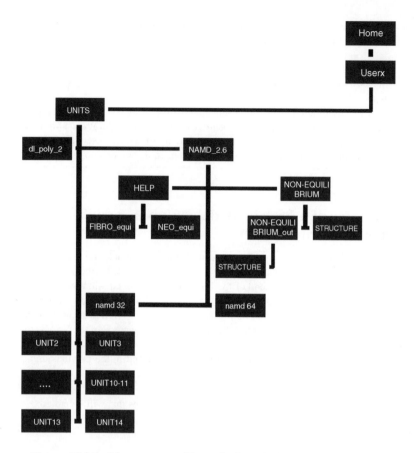

**Figure 10.14**   Directory tree with emphasis on UNIT12 and NAMD_2.6.

3. Open the solvate.psf file with VMD as a new module and click load. There will be no image in the black window. Left and then right click on "solvate.psf" in the main window. A drop-down menu will appear; select "load data into module", which will cause a new load molecule window to appear. Select the "fibro_equilibration.restart.coor" file and click load. This process is described in detail in items 1 and 2 of Section 10.10.2.3.

4. After the new file has been loaded, the fibronectin module surrounded by a water box should be visible on the screen. Use VMD to select two specific atoms on the structure: the Cα of the first and the last amino acid of the protein chain. Do this using the Graphical Representation window found in the "Graphics" menu. Note: For simplicity in codes and related descriptions, the symbol "CA" will be used to represent the Cα carbon atom.

5. Type "protein" in the "selected atoms" box, press Enter, and then select "tube" in the "Drawing style" menu. The water molecules will disappear, and the protein will appear as a green tube. Now click on the "create rep" button at the top of the window. In the green part of the Graphical Representation window a new line will appear. Select the "selection tab" on the right of the Draw Style tab. Remove the string "protein" in the "selected atoms" box and double click on the "resid" string in the "keyword" menu, and then click "1418" in the "value" box. The string "resid 1418" will appear in the "selected atoms" box. Add to this string "and name CA" to produce the string "resid 1418 and name CA". Open the "draw style" tab and select VDW in the Drawing Method menu.

The image should appear. Repeat the procedure for the last residue: Click "create rep" and replace the 1418 with another number, for instance 1500. Another green sphere will appear on the screen, based on the new location along the polymer chain as selected by the new "create rep" number. Note: The last residue of the protein is 1509. If the procedure was performed correctly, a tube with two spheres at the two ends should be visible. For clarity, change the color of the two spheres using the Draw style tab, select the "color ID" option, and select a color.

In order for the simulation to run properly, one end of the protein will be fixed, therefore it will not change location with time; while the other end moves as a result of a force applied. The CA atom of residue 1418 will be fixed; while the CA atom of residue 1509 is pulled. VMD must be used to obtain this value. Open the Tkconsole in the Extension menu, and type the following command:

```
set ca1 [atomselect top "resid 1418 and name CA"]
```

This command is similar to the one used in the Tkconsole command in Section 10.10.1.4 to select the fibronectin molecule. However, this command differs in the fact that instead of selecting the entire molecule as in the previous section, this command will only select the CA atom of resid 1418.

Now, as in the previous section, determine the minimum/maximum of the selection values using the command:

```
measure minmax $ca1
```

The selection consists of a single atom; therefore the minimum and the maximum values are equal and correspond to the atom's position. The results are: 16.3646450043 19.7755908966 −35.3768692017. If these results do not match the output, check files in UNIT12/HELP/FIBRO_equi or NON_EQUILIBRIUM_out/STRUCTURE directory files.

## Exercise 10.16

Repeat the procedure above for the CA atom of residue 1509. What is the command for the atom selection?

### 10.12.2   Preparation of the NAMD Input Files

The NAMD input file required for the simulation can now be created. The non-equilibrium simulation will have a fixed atom and a pulled atom. This section will develop a new PDB file with fixed and pulled atoms that is consistent with the current file structure.

#### 10.12.2.1  PDB Files for the Fixed Atom

Open the NON_EQUILIBRIUM/STRUCTURE directory and make a copy of the solvate.pdb file using the following command:

```
$ cp solvate.pdb fixed.pdb
```

Now open the fixed.pdb file with vi. Essentially the most important columns in the file include the atom name and index (columns 10–16), the corresponding residue name and index that the atom belongs to (columns 10–26) and the atom position coordinates (columns 33–54). In addition, columns 110–120 show the Asp amino residue which appears in lines (1–14) and (69–80); there is also information for other residues Val (lines 15–30), Pro (lines 31–44), Arg (lines 45–68), Asp (lines 69–80), etc. This information is important because we know protein consists of 20 α-amino acid residues (see Section 10.10.1.1); for a specific protein and how these residues are arranged is described in these lines.

Note that for all atoms, columns 57–60 contain the string "1.00". This column is not typically used for MD simulations. However, this column can be used to add specific features to the simulation, if desired. For example, in the present case these columns will be used in the following way:

This column will control which atoms will be free to move, using the value 0.00; as opposed to those atoms which will be fixed, using the value 1.00 (see NAMD user guide for additional details). To achieve this result, use vi to insert the value 0.00 on all the lines using the following substitution command:

```
:%s/ 1.00 0/ 0.00 0/
```

This command will substitute (s) the string "1.00 0" with "0.00 0" in the entire file (%). Note: The % symbol is a percent sign, and there is no space between the % symbol and the letter "s". Also note that there are two blank spaces before and after the strings "1.00" and "0.00", for a total of nine characters for each string.

Go to the line of the atom to be fixed (ATOM 5, i.e., the CA atom of residue 1418) and insert the value "1.00" in columns 57–60. In addition, the current x, y, z position 49.8, 19.878 and −35.382 of ATOM 5 in columns 33–54 should be replaced by the values 16.364 19.775 and −35.376 found with VMD. The resulting line should appear similar to:

```
ATOM   5 CA ASP A1418   16.364  19.775  -35.376  1.00  0.00   FIBR C
```

Note that it is crucial to maintain proper line spacing when altering values in PDB files. The produced fixed.pdb file should be copied to the NON_EQUILIBRIUM directory for the simulation. If problems arise during this procedure, use the fixed.pdb file in the UNIT12/HELP directory.

### 10.12.2.2 PDB File for the Atom to be Pulled

Perform the same procedure used for the pulled atom (i.e., the CA atom of residue 1509). However, instead of copying the solvate.pdb file, copy the fixed.pdb file using the command:

```
$ cp fixed.pdb smd.pdb
```

Open the smd.pdb file and replace the "1.00" of ATOM 5 with "0.00" and the "0.00" of ATOM 1366 with "1.00". Note ATOM 1366 is the CA (or Cα) of the residue 1509 as shown on line 1367 (see below). It is not necessary to insert the value obtained from VMD for the position. Therefore, the line should appear similar to:

```
ATOM 1366 CA THR A1509 32.264 23.769 -59.906 1.00 0.00 FIBR C   ! line 1367
```

The produced smd.pdb file should be copied to the NON_EQUILIBRIUM directory for simulation. If a problem occurs during this process, use the smd.pdb file in the UNIT12/HELP directory.

### 10.12.2.3 Other Input Files

The configuration (.conf) file can now be prepared for the simulation. This file can also be prepared using the NAMD tutorial and the NAMD user guide.[13] The necessary files have been prepared in the UNIT12/HELP directory. Open the NON_EQUILIBRIUM directory to obtain the "fibro_nonequi.conf" file from the HELP/NEQ_equi directory.

The other three files generated by the equilibrium simulation, including "fibro_equilibration.restart .coor", "fibro_equilibration_restart.vel", and "par_all27_prot_lipid.inp" will be copied from the HELP/FIBRO_equi directory. The first two files contain the initial position and velocity for the new simulation. The last file includes all the potentials introduced in Section 10.10.

## 10.12.3    Explanation of Important Lines in the fibro_nonequi.conf File

As explained in Section 10.10, the most important file to control the atomistic simulation in NAMD is the configuration (.conf) file. Open the fibro_nonequi.conf file with vi. The following lines are important configuration commands used in the simulation. Additional command descriptions can be found in the NAMD user guide.

Line 9 "set inputname fibro_equilibration": The "set" command creates a variable that can be used throughout the .conf file. For example, the line 9 command creates the inputname as the variable "fibro_equilibration". This means that any time the variable "inputname" appears in the .conf file (in particular the string $inputname), NAMD will interpret it as the string "fibro_equilibration".

Lines 11 and 12 "Bincoordinates $inputname.restart.coor" "Binvelocities $inputname.restart.vel": These lines indicate that NAMD will read the initial coordinates and initial velocities from the two files "fibro_equilibration.restart.coor" and "fibro_equilibration.restart.vel", respectively. Note that the variable "inputname" is used, as defined by line 9 above.

Line 21 "set outputname fibro_nonequi": As in line 9, the code creates a variable called "outputname".

Lines 59–63: These lines contain information for the periodic box. In this simulation, NAMD will perform a simulation without periodicity.

Line 68 "FullDirect yes": Without a periodic box condition, it is impossible to use a PME algorithm; therefore the Coulomb force must be computed in the standard way indicated by "FullDirect".

Lines 69–85: These lines indicate that PME and a piston will not be used, because both conditions require periodic boundary conditions.

Lines 102–104: If there are fixed atoms in the simulation, these three lines indicate the appropriate files that contain information about the fixed atoms in the simulation. These lines indicate the file where the atoms are located (fixedAtomsFile STRUCTURE/fixed.pdb) and in which column (fixedAtomsCol O). Note that the fixedAtomsCol line will indicate the number of columns after the atom position columns.

Lines 106–111: These lines provide information for the SMD simulation. A new potential is added to the total energy of the system as shown:

$$U(\vec{r}_1, \vec{r}_2, \ldots, t) = \frac{1}{2} k \left[ vt - (\vec{R}(t) - \vec{R}_0) \cdot \vec{n} \right]^2$$

where r1, r2, ... are the atoms indicated in the "SMDFile" (line 107). R(t) is the center of mass of the atoms. $R_0$ is the initial value calculated from the SMDFile positions. V is the velocity value reported in the SMDvel file (line 109), with units of Å per time step. The direction of the force is the unit vector in the SMDdir command (line 110). SMDOutputFreq (line 111) is a parameter that controls the frequency of output printing. For example, the number 10 indicates that after every 10 steps the data related to SMD should be printed out.

Line 122: The last line contains the number of time steps. For this simulation, use 5000 steps.

## 10.12.4    Run NAMD Simulation and Data Processing

### 10.12.4.1   Running

Before running the simulation, verify that the following six files are in the directory:

- fibro_nonequi.conf
- fibro_equilibration.restart.coor
- fibro_equilibration.restart.vel
- par_all27_prot_lipid.inp
- smd.pdb
- fixed.pdb

The command for running NAMD is as follows:

```
~/UNITS/UNIT12/NON_EQUILIBRIUM $nohup. /../../NAMD_2.6/namd32 fibro_
nonequi.conf > smd.log &
```

The simulation will require several minutes to finish. Follow the same procedure used in Section 10.10.3 to analyze the fibro.log file and use namddat_gnuplot.csh to extract information from the files.

## 10.12.4.2 Data Processing

The data processing includes two parts. The first part is qualitative analysis and the second part is quantitative analysis. They are described as follows:

**Part A: Qualitative analysis using VMD**
Open the UNIT12/HELP/NON_EQUILIBRIUM directory. Use the following procedure to develop the time evolution of the configuration using VMD:

1. Open VMD.
2. Load the STRUCTURE/solvate.psf file.
3. Load fibro_nonequi.dcd from the HELP/NEQ_equi directory into the solvate.psf file. Right click on the molecule record in the VMD main window and select "Load Data into Molecule" (see Section 10.11.1.3 as reference).
4. The protein structure will now be surrounded by water. By default the z axis is normal to the screen, therefore the protein elongation may not be obvious. Rotate the protein to view the z axis parallel to the screen.
5. Use the frame bar in the main window to view each frame.

**Part B: Quantitative analysis using VMD**
The information related to the SMD simulation is printed in the smd.log file. Open it with vi using the command:

```
$ vi smd.log
```

Go to line 164. The line should start with the string SMD, similar to the following:

```
SMD  10 15.6105 25.2805 -53.696 0 0 -2661.82
```

The seven numbers printed after SMD have the following meaning:

6. The first is the current time step as an integer. In line 164, this number is 10; in line 165, this number is 20, and so on. The printing frequency is given by the SMDOutputFreq parameter in the .conf file. In this case, as noted above, the value is 10.

7. The following three numbers are the components of the position vector of the center of mass of the steered atoms. In this simulation, only one atom is steered, therefore the values are components of the residue 1509 CA atom.

8. The last three numbers are the force components, in pN, acting on the steered atoms. In this simulation, the direction is given in the SMDdir command parallel to the z axis.

To extract this information from the smd.log file, use the VMD script "smd_extract.tcl" found in the UNIT12/HELP directory. Open the file with vi. Check that the second line contains the following:

```
set file [open smd.log r]
```

This will ensure that the script will open the smd.log file. To run this script, use the following command:

```
$ vmd -dispdev text -e smd_extract.tcl
```

This command will cause the script to create a file called "smd.dat" where the lines of the smd.log file, starting with SMD, will be reported. The file can then be plotted with gnuplot.

---

**Homework**

(10.22) Answer the following questions related to the atomistic simulation of the protein-water system.
  (a) What is the typical value of the time step of a MD simulation for a biological system?
  (b) Does an NPT simulation change the periodic box volume size?
  (c) In an NPT equilibration which quantities do you check in order to understand if the equilibration time was enough?
  (d) What kind of information is present in the parameter file (e.g., par_all27_prot_lipid.inp) and topology file (e.g., mod.pdb) of the CHARMM force field?
  (e) Which information is present in the dcd file generated by NAMD?
  (f) How can you use VMD to see the time evolution of the system configuration?

---

## 10.13   Brief Introduction to LAMMPS

### 10.13.1   General Features of LAMMPS

LAMMPS stands for Large-scale Atomic/Molecular Massively Parallel Simulator. It was developed at Sandia National Laboratories. It is a freely available open-source code, distributed under the terms of the GNU Public License (GPL), which means you can use or modify the code however you wish. LAMMPS is a classical MD simulation program designed for parallel computers, and also runs efficiently on single-processor desktop or laptop machines. One can download it directly from the website http://lammps .sandia.gov, which has more information about the code and its uses.

LAMMPS can model systems with only a few particles up to millions or billions. It models atomic, polymeric, biological, metallic, granular, and coarse-grained systems using a variety of force fields and boundary conditions for an ensemble of particles in a liquid, solid, or gaseous state with lots of commands. The current version of LAMMPS is written in C++. LAMMPS is designed to be easy to modify or extend with new capabilities, such as new force fields, atom types, boundary conditions, or diagnostics. In the most general sense, LAMMPS integrates Newton's equations of motion for collections of atoms, molecules, or macroscopic particles that interact via various force fields with a variety of initial and/or

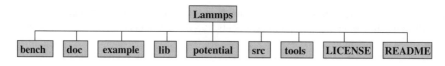

**Figure  10.15**   LAMMPS directory

boundary conditions. For computational efficiency LAMMPS uses neighbor lists to keep track of nearby particles.

## 10.13.2    Structure of LAMMPS Package

LAMMPS is still releasing new versions. The following introduction is based on the version of 2010-1-10. When you download LAMMPS you will need to unzip and untar the downloaded file with the following command:

```
$ tar zxvf lammps*.tar.gz
```

where * is the version of LAMMPS. Place the file in an appropriate directory such as in the home directory as GULP does; the path is similar to GULP as described in Section 10.3. LAMMPS directory contains two files, README and LICENSE, and several subdirectories, as shown in Figure 10.15. The "bench" directory includes benchmark problems; the "doc" directory documentation and manual; the "examples" directory some simple test problems; the "lib" some additional packages that LAMMPS can be linked with, such as MEAM, POEMS, REAX, gpu, etc; "potentials" includes various interatomic potential files; "src" all the source files; "tools" some pre- and post-processing tools for LAMMPS.

## 10.13.3    Building  LAMMPS  and Run

LAMMPS will be used on the Unix or Linux platform and the download file is the source code; you should build it on your machine, though it will be non-trivial. The src directory contains all the C ++ source files and header files for LAMMPS. It also contains a top-level Makefile and a MAKE subdirectory with low-level Makefile.* files for several machines. Within the "src" directory, type a command like:

```
$ make or $gmake
```

You should see a list of available choices. If one of those is the machine and options that your system has, you can type a command like:

```
$make linux or $gmake mac
```

If you get no errors and an executable file like lmp_linux or lmp_mac is produced, it is done. However, the first time you may not be successful. If this happens, please refer to the following processes.

### 10.13.3.1  Make LAMMPS by Serial

LAMMPS is designed for parallel computers, but it also runs efficiently in serial. In the previous step, when you enter the make command, you will find an option for serial. The corresponding explanation says that neither an MPI library (Parallel Library) or FFT is used, only a C ++ compiler is needed.

Therefore, if you carry out the serial compilation, for each related parallel compiler option, you must specify a parallel pointer to the virtual function file by typing the command in the src/STUBS directory:

```
..src/STUBS$ make
```

This command will allow you to generate the connection file of virtual functions, i.e., libmpi.a. If this build fails, you will need to edit the STUBS/Makefile for your platform. Otherwise, return to the src directory, typing

```
.. src$ make serial
```

This will require several minutes to complete because the command compiles all related C++ files in the src directory. If you get no errors and an executable like lmp_serial is produced, it is done and you can find it in the src directory. Then you can copy this file to your own directory for the calculation.

### 10.13.3.2  Make LAMMPS by Parallel

If you want LAMMPS to run in parallel, you must have an MPI library installed on your platform. If you use an MPI-wrapped compiler, such as "mpicc" to build LAMMPS, you can probably leave these three variables (MPI_INC, MPI_PATH and MPI_LIB) blank. If you do not use "mpicc" as your compiler/linker, then you need to specify where the mpi.h file (MPI_INC), the MPI library (MPI_PATH) and its name (MPI_LIB) are found.

If you are installing MPI yourself, MPICH 1.2 or 2.0 is recommended. In this case, you need to set the appropriate MPI_INC, MPI_PATH and MPI_LIB variables in the Makefile.foo file like:

```
MPI_INC = -DMPICH_IGNORE_CXX_SEEK -I/opt/mpich/include
MPI_PATH = -L/opt/mpich/lib
MPI_LIB = -lmpich -lpthread
```

Furthermore, if you want to use the particle-particle particle-mesh (PPPM) option in LAMMPS for long-range Coulomb potential via the kspace_style commands, you must have a 1d FFT library installed on your platform. The FFTW 2.1.X is a fast, portable library that should work on any platform. When you finish building FFTW for your box, you need to set the appropriate FFT_INC, FFT_PATH and FFT_LIB variables in the Makefile.foo file like:

```
FFT_INC = -DFFT_FFTW -I/usr/local/include
FFT_PATH = -L/usr/local/lib
FFT_LIB = -lfftw
```

If you do not plan to use PPPM, you do not need an FFT library. In this case you can set FFT_INC to −DFFT_NONE and leave the other two FFT variables blank.

Once you have found a correct Makefile.foo file from the available choices and pre-built any other libraries which will be used (e.g., MPI, FFT), all you need to do from the src directory is type one of these two commands:

```
$make foo
$gmake foo
```

where foo represents your platform name. You should get the executable lmp_foo in the src directory when the build is complete. Then you can copy this file to your own directory for your calculation.

Note that on a multi-processor or multi-core platform you can launch a parallel make, by using the -j switch with the make command, which will build LAMMPS more quickly.

If you have not found the compatible Makefile.foo for your own platform, you need to spend some effort to edit a new low-level Makefile.foo file. See the LAMMPS manual and/or its website for more details.

### 10.13.3.3 Running LAMMPS

Computations can occur in the execute directory or in any directory containing the executable lmp_foo, input script files, potential function files, and sometimes other input files. You should then create a separate directory for each single problem. The details for running LAMMPS are described in detail as follows:

1. Create a new folder and rename it whatever one wants.
2. Copy the executable file lmp_foo from src directory into the folder you have just built. Then you need to write an input script file, according to the simulated problem, or copy a simple test script from the examples directory.
3. Run LAMMPS in one's own folder using one of these two commands:

```
$ ./lmp_serial < in.alloy
$ mpirun -np 16 lmp_foo
```

The first command will launch a serial job while the second one launches a parallel job. In the parallel run, the first item is the parallel command, the second and third together will indicate the number of cpu that is used in this simulation, -np is the parameter name, without modification. 16 is the parameter, that is, 16 cpus are used, which can be modified based on the actual circumstances. lmp_foo is the executable file of LAMMPS, produced by the successful compilation. The sign < represents input, while the file in.alloy is the input scripts file that probably needs users to modify it.

### 10.13.3.4 LAMMPS Screen Output and Data Post-processing

As LAMMPS reads an input script, it prints information to both the screen and a log file about significant actions it takes to setup a simulation. Other special LAMMPS commands (such as dump, etc.) also have their own separate output files. For the relevant commands, please refer to the LAMMPS manual.

LAMMPS output data can be visualized through a number of post-processing means, i.e., the high-quality visualization package VMD can easily open the data and help display the image simulation process. The new version of VMD has integrated the visual plug-in for LAMMPS output format.

## 10.13.4   Examples

In the following we will show three examples, namely, "min", "fracture", "nanowire-tension" for using LAMMPS for simulation. In the first 12 sections of this chapter we have introduced in detail MD software such as GULP, DL_POLY, NAMD, and readers should have sufficient training how to run a code and carry on post-processing. Thus, these examples will be used for readers to check their own capability to work independently to understand and run these codes. If they find difficulties they are encouraged to refer to the manual and website of LAMMPS to get the solution, and this is a very important way to learn a new software. Based on this consideration, we will introduce the simulation conditions and the constitution of their input files in the following three subsections. In addition, we have developed their work directories in the computer CSLI and put these input files in those of UNIT13. The path and name of these input files are given as follows:

```
"min": path: ~/UNITS/UNIT13/min; input file: in.minimize
"fracture": path: ~/UNITS/UNIT13/fracture; input file: in.fracture
"nanowire-tension": path: ~/UNITS/UNIT13/nanowire-tension;input file:
in.crack
```

### 10.13.4.1  MD Model for Energy Minimization of α-Fe Crystal Structure

α-Fe is a kind of typical BCC crystal structure which is shown in Figure 3.12 and used in Sections 10.7–10.9. Its space group is IM-3M and unit parameters are given below:

$$a = b = c = 2.8664 \text{ Å}, \; \alpha = \beta = \gamma = 90° \tag{10.11}$$

The position coordinates of atoms in an unit cell for LAMMPS are (0 0 0), (1 0 0), (1 1 0), (0 1 0), (1/2 1/2 1/2), (0 0 1), (1 0 1), (1 1 1), (0 1 1), respectively.

The MD model for energy minimization for α-Fe lattice is set up using LAMMPS. As shown in Figure 10.16, the model is set with $20 \times 20 \times 5$ lattice units with dimensions of 57.328, 57.328 and 14.3325 Å along the x, y and z directions, respectively. The boundary condition is the "sss" command which indicates non-periodic and shrink-wrapped in the x, y and z directions. This is quite different from the "ppp" boundary condition in which periodic conditions are set along the x, y, z directions. The initial temperature of the system is 300 K.

Before running it is necessary to check whether the necessary file, for instance the potential file of Al_jnp.eam is in the directory you are running for the job; otherwise, an error message "cannot open EAM potential file Al_jnp.eam will appear." The command to run a serial calculation is:

```
$./lmp_serial <in.minimize
```

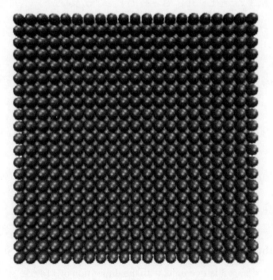

**Figure 10.16**  Initial model for energy minimization

For a long run in the background it is:

```
$ nohup ./path/lmp_serial
```

The commands running that include error messages will appear in the document of nohup.out. The output files include dump.log and dump_minimize.lammpstrj offer basic information on simulation results including atom number, box size, atom type, basic data of temperature, energy, pressure and volume versus simulation steps. The complete input file (in.minimize) is as follows:

```
# energy minimization
units            metal
boundary         s s s
atom_style       atomic
neighbor 2.0 bin
neigh_modify delay 0 every 1 check yes

lattice          bcc 2.8664
region           box block 0 20.0 0 20.0 0 5.0
create_box       1 box
create_atoms     1 box

pair_style       eam/fs
pair_coeff       * * Fe_mm.eam.fs Fe

# temp controllers

compute          new3d all temp

# equilibrate

velocity         all create 300.0 887723 temp new3d
fix              1 all nve
fix              2 all temp/rescale 10 300.0 300.0 10.0 1.0
fix_modify       2 temp new3d

timestep         0.001
thermo           100
thermo_modify    temp new3d
dump             1 all atom 50 dump_minimize.lammpstrj
log              dump.log
run      1000

neigh_modify     delay 0 every 1 check yes
thermo           5
minimize  1.0e-6 1.0e-6 1000 10000
```

## 10.13.4.2 MD Simulation of Crack Propagation for α-Fe Crystal Structure

The model is set with $40 \times 30 \times 3$ lattice units with the dimension of $114.656 \times 85.992 \times 8.5992$ Å. The comparison between the LAMMPS model and the hard sphere crystal structure is given in Figure 10.17.

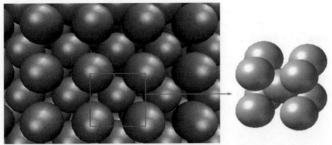

a) Crystal structure based on LAMMPS software   b) Hard sphere model for BCC

**Figure 10.17**   Comparison between LAMMPS model and the hard sphere

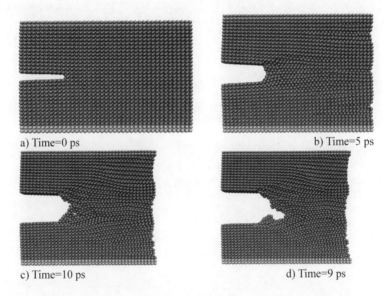

a) Time=0 ps

b) Time=5 ps

c) Time=10 ps

d) Time=9 ps

**Figure 10.18**   The process of crack propagation and deformation

The length of initial crack is 28.864 Å. The boundary condition is the non-periodic boundary condition "sss". The initial temperature of the system is 0.01 K. The loading process is divided into two steps. In the first loading step, the MD model is relaxed with 1000 steps to perform an energy minimization of the system and in the second loading step, the velocity is applied along the y axis until the required load steps are reached. The Finnis-Sinclair (F-S) potential listed in the file Fe_mm.eam.fs is employed to characterize the interatomic interaction. Figure 10.18 shows the fracture pattern by the simulation.

The complete input file (in.fracture) is given as follows:

```
# 3d metal crack simulation
dimension        3        #set dimensions
units            metal     #set the unit in metal
boundary         s s s     #boundary condition
```

```
atom_style        atomic    #set the atom type
neighbor 2.0 bin          #calculate parameters
neigh_modify delay 5     #modify calculate parameters

lattice   bcc 2.8664     #set the lattice structure bcc, a=2.8664
region            box block 0 40.0 0 30.0 0 3.0      #create region of
simulation
create_box        3 box    #create a simulation box
create_atoms      1 box   #create atoms in the simulation box

pair_style        eam/fs   #potential function is F-S
pair_coeff        * * Fe_mm.eam.fs Fe Fe Fe   #potential parameters read
from Fe_mm.eam.fs

region            lower block INF INF INF 1.25 INF INF #define region lower
region            upper block INF INF 28.75 INF INF INF
group             lower region lower       #define group lower by region
group             upper region upper
group             boundary union lower upper
group             mobile subtract all boundary

set     group lower type 2     #set atom type
set     group upper type 3
# crack

region            crack block INF 10 14.5 15.5 INF INF
delete_atoms      region crack   #delete atoms

# temp controllers

compute           new3d mobile temp    #temperature calculation
compute           new2d mobile temp/partial 1 0 1     #define temperature
computation

# equilibrate

velocity          mobile create 0.01 887723 temp new3d #initial velocity
fix               1 all nve   #set ensemble
fix                2 boundary setforce NULL 0.0 NULL  #set force components
fix                3 mobile temp/rescale 10 0.01 0.01 10.0 1.0  #reset the
temperature by explicitly rescaling their velocities
fix_modify        3 temp new3d    #modify parameters

thermo            50       #output frequency of thermodynamics
thermo_modify     temp new3d
timestep 0.001            #time step
run       1000
neigh_modify delay 5          #modify calculate parameters

minimize 1.0e-6 1.0e-6 1000 10000      #energy minimization
```

```
# loading
velocity    upper set 0 0.8 0    #set velocity
velocity    mobile ramp vy 0.0 0.8 y 1.25 28.75 sum yes
unfix                  3
fix       3 mobile temp/rescale 10 0.01 0.01 10.0 1.0
fix_modify        3 temp new2d
thermo            200
thermo_modify     temp new2d
reset_timestep  0        #reset timestep
dump             1 all atom 500 dump.lammpstrj    #output format
log        dump.log        #output log file
run        13000          #running steps
```

Save the file (in.fracture), run a parallel or serial operation simulation using lmp_foo < in.fracture or mpirun -np 2 lmp_foo < in.fracture, get the simulated results in the current folder. The changes of atomic coordinates are in the file named dump.lammpstrj, which can be easily opened by VMD-1.8.7.

### 10.13.4.3 Simulation of Uniaxial Tension Nanowires

The investigation of mechanical properties of nanowires is widely appreciated, the focus of which is the tensile mechanical properties. Tensile loads can be applied in many ways, among which the most convenient one is to give a displacement boundary. Figure 10.19 shows a nanowire stretch. One end is fixed and the other end is subjected to an increment of displacement. This displacement field is linearly distributed between two ends inside the nanowire at the beginning and then relaxed.

The input file in.disp is in the directory of UNITS/UNIT13/nanowire_tension. Enter that directory to find that input file and then run it. Here a brief description for the input file in.disp is given as follows. This file contains the following six parts.

(1)  Set the initial basic conditions:

```
boundary     s s s # Set the boundary conditions for three directions free
units       metal # Set unit
atom_style    atomic # Set atom types
dimension     3 # Set geometric dimensions, 3-D
```

(2)  Create basic geometric models:

```
lattice     fcc 3.986 # Set lattice parameters
region      box block -3 3 -3 3 0 10 # Create a simulation of
                       #the geometric region
```

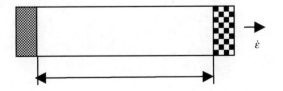

**Figure 10.19**   The nanowire bar

```
create_box        10 box # Create a simulation box
create_atoms      1 region box # Create atoms
```

The following combination of commands are used to divide the simulated atoms into groups:

```
region    left block INF INF INF INF INF 1
region    right block INF INF INF INF 9 INF
group     left region left
group     right region right
group     boundary union left right
group     mobile subtract all boundary
```

(3) Set the initial conditions and parameters of potential functions:

```
neighbor        2.0 bin # Set the calculated parameters
neigh_modify    delay 0 every 10 check yes
mass             * 26.9 # Atomic mass
compute          new3d mobile temp # Set the temperature calculation
pair_style    eam # Set the form of potential function for the eam
potential.
pair_coeff    * * Al_jnp.eam # Set potential function parameters read from the
file.
velocity  mobile create 300.0 87277 mom yes rot yes # Set the initial speed,
the room temperature.
```

(4) Stress calculation and output

```
compute  stress mobile stress/atom # Stress calculated for each atom, as the
form of virial stress
compute  sum_stres mobile reduce sum c_stress[3] # Calculated the stress in
tensile direction
fix    stressout all ave/time 1 500 5000 c_sum_stres file file.stress # Deal with
the stress. It is calculated once for each time step. Output a stress for each
5000 steps, the output stress is the average of the last 500 steps.
```

(5) Thermodynamic output

```
thermo_style custom step temp press pe ke etotal # Set thermodynamic output,
including temperature, pressure kinetic energy, potential energy, total
energy
thermo          500 # Set the output frequencies for thermodynamic
quantities
thermo_modify temp new3d # Modify thermodynamic calculation
timestep        0.001 # Set time step
dump            1 all atom 100 dump.disp # Set the output file format
```

(6) Cyclic tensile loading

```
label           k2 # Set cyclic loading variables
variable        j loop 20 # Set cycle steps
```

```
displace_atoms    right move 0.0 0.0 0.05 # Set the size of each
displacement
displace_atoms mobile ramp z 0.0 0.05 z 1 9+0.05*($j-1) # For the calcula-
tion area, set up a decreasing atomic displacement in advance, to avoid the
destruction of the material microstructure caused by a sudden time dis-
placement

fix               1 mobile nvt 300.0 300.0 0.01 # Ensemble set for
isothermal simulation
fix_modify        1 temp new3d # Ensemble temperature calculation
run               5000 # Set the simulation steps for each load step
write_restart     restart.*.disp # set the state output after each load
balance

next              j # Cyclic loading
jump              in.disp k2 # Skip to cycle markers to start a new cycle
```

Save the input script file (in.disp), run parallel or serial LAMMPS program, the results can be obtained. For this input script, our output files are the following:

- file.stress: The output file is to calculate the stress in the tensile direction, corresponding to the 5000-step relaxation after each load, taking the average for the last 500 steps, in order to approximate simulation of quasi-static situation. By mapping software or programming, output the curve of stress in the function of strain or time.
- dump.disp: This document is based on the changes of atomic coordinates. Through the VMD software, visual observation of the tensile process is realized.
- log.lammps: This output file is the default, belonging to global variable output, such as system temperature, potential energy, total energy, system volume output. They can have related settings.
- restart.*.disp: The actual output of the star will be the actual number of the time step, i.e., in the cyclic loading; load in each cycle and output a state file after the balance relaxation, which records the current system state. If you hit a non-normal interrupt, you can continue the previous simulation, using the output file here as input.

## 10.14    Multiscale Simulation by Generalized Particle (GP) Dynamics Method

Since multiscale analysis is a new area, it is difficult to find codes which can be used in the general public domain. In this section, we offer some GP multiscale methods for readers for training and for developing their own codes, including model development, running simulation, and data processing. They correspond respectively to the three directories of INI_CONF, SIMULATION, and POST_PROC in CSLI UNITS/UNIT14.

### 10.14.1    Multiscale Model Development

In this section, we will introduce model development for a three-scale GP simulation, using the software of "Mater_Multi_2010_4.f90" and input file "model.in" given in UNITS/UNIT14/INI_CONF/GP_ Model. These codes are similar to the code MaterNew_2010.Comb and input file model.in which are described in detail in Appendix 10.F, and interested readers are advised to read that first. The main differences between the source file and input file from the files there are: (1) The code and "model.in" input file are required to produce the imaginary atoms and imaginary particles between different scales as shown in Figure 5.4. (2) They are also required to produce the real atoms and real particles inside the neighbor-link cells of the imaginary atoms and imaginary particles. (3) The output file is Model.MD, however, the first five lines used for the CONFIG file of DL_POLY_2 are no longer used in Model.MD. In the following, we show the input file "model.in" for the three-scale model design.

### 10.14.1.1 Basic Domain Design

To develop a GP multiscale model, the first thing is to design domains with different scales. Now, we would like to take the three-scale copper nanowire in Figure 5.13 to show the design of the model.in file. In general, atomistic scale is the core area where large strain gradients and large deformations may be involved. Figure 10.20 shows a schematic of the three-scale GP model. The most important domain is the middle area where a notch or a crack may be placed. In the figure, all the numerical numbers (i.e., 1, 2, 3 or −1, −2, −3) in the symbols (S$_1$, S$_2$, S$_3$) and (S$_{-1}$, S$_{-2}$, S$_{-3}$) denote the scales with scale 1 the atomistic scale. Scales in the first parenthesis denote real particle domains, scales in the second parenthesis denote imaginary domains which are used for connecting different scales of real particles or atoms (see Section 5.3 for detail).

The full size of the nanowire is 201a × 6a × 6a, where a = 3.62 Å is the copper crystal constant. The origin of the Cartesian coordinates is placed in the centre with the x-axis as a longitudinal axis, along which the loading is applied. The scale domain arrangement is symmetric and Figure 10.20 only shows the right part in which the half length of 100.5a is arranged as follows:

S$_1$: $0 < x < 35.5a$
S$_2$: $35.5a < x < 48.5a$ (each S$_2$ particle is lumped by 8 atoms, lattice constant $= 2a$)
S$_3$: $48.5a < x < 100.5a$ (each S$_3$ particle is lumped by 64 atoms or 8 particles of S$_2$ scale, lattice constant $= 4a$)

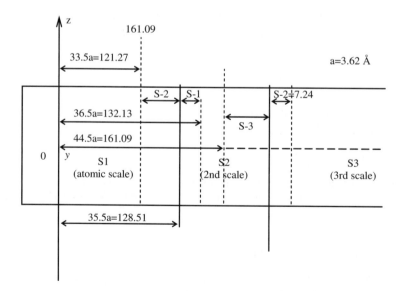

**Figure 10.20** A schematic showing the design of a three-scale model of a copper nanowire with size of $x = \pm100.5a$, $y = \pm3a$, $z = \pm3a$ where $a = 3.62$ Å is the crystal lattice constant

The width of overlapping imaginary area shown in Figure 5.4 is designed to be the same as the lattice length of the scale involved. This can be shown by the following design data:

$S_{-1}$: 35.5a < x < 36.5a (the width is 1a)
$S_{-2}$: 33.5a < x < 35.5a (the width is 2a)
$S_{-2'}$: 48.5a < x < 50.5a (the width is 2a)
$S_{-3}$: 44.5a < x < 48.5a (the width is 4a)

### 10.14.1.2 The Input File "model.in" for 1-scale and 3-scale Copper Nanowire

The design will be realized through the input file "model.in" of the source file "Multi_2010_4.f90". The simplest example of "model.in" is for the model with atomistic scale only, a truly MD simulation, which reads:

```
1
363.81      10.86       10.86
30
1
Y    1        1
-363.81  363.81   -10.86   10.86   -10.86           10.86
s
363.81       -363.81    10.86  -10.86   10.86  -10.86
```

Line 1: denotes the top scale of the model.
Line 2: denotes the half length of the model along the X-, Y-, and Z-directions (Figure 10.20).
Line 3 is irrelevant.
Line 4: denotes the total number of blocks by which the model is formed. A one-scale model only needs one block ($S_1$); a three-scale model needs seven blocks as shown in the above domain division in which three domains (for real atoms $S_1$, $S_2$, $S_3$) are real atoms/particles, four domains (S-1, S-2, S-2', S-3) are imaginary.
Line 5: the symbol denotes the structure type, "Y" is for FCC. The second number denotes whether the block for real or imaginary particles and what is the scale level. For instance, "1" denotes real atoms (scale 1), but "-3" denotes the third-scale imaginary particle. In other words, use positive as real and negative as imaginary. The third number denotes how many cuts will be inside the block. If one needs to cut two edge cracks in the block, then the number is two.
Line 6: The six numbers denote the outside surface location and dimension of the block along the x, y, and z directions. For instance the first two numbers denote the block's left surface is at x = −363.81, the right surface is x = 363.81, and the total length along the x-direction is 727.62.
Line 7: the letter denotes what kind of cutting body (volume and shape) will be cut from the model; "s" denotes a parallelepiped body, "c" denotes a sphere or a cylinder.
Line 8: denotes the location and dimension of the cutting volume. In the first scale model shown above, there actually is nothing to be cut, thus a non-realistic cutting range is given (from positive to negative) so the cut will not occur. However, for the model with two edge notches in the middle shown in Figure 5.13, the design of that block is given as follows:

```
Y        1     2
-125.787     125.787     -14.584     14.584     -58.336     58.336
s
-1.823     1.823     -14.854     14.854     -58.336     -54.69
s
-1.823     1.823     -14.854     14.854     54.69     58.336
```

Lines 5–8 as a whole are used to describe the block. A three-scale model needs to have seven blocks of the same kind, for each block, and needs at least four lines to describe; if it needs to have two cuts then it takes six lines. To have an idea how to design Figure 10.21 by the input file model.in, see the file "model_S3_Cu_in" for copper and "model_S3_square.in" in the UNIT14/INI_CONF/GP_Model for detail.

## 10.14.2 Running "Mater_Multi_2010_4.f90" to Produce the Model.MD for Multiscale Simulation

The Mater_Multi_2010_4.f90 is used to produce the output file Model.MD for simulation. It includes the real and imaginary atoms/particles of different scales as well as the neighbor-link cells, see Model_S3_square.MD in UNIT14/INI_CONF/Model_GP as an example. The construction of the code is not complicated and can be briefly described as follows:

- First, develop different subroutines of cells such as BCC.cell, FCC.cell, Fe4N.cell, based on the structure information given, say, in Section 2.5 and their orientation related to the global X-, Y-, and Z- coordinates (see Appendix 10.F for more detail). These cell subroutines are given at the beginning of the codes after the module CONTROL. They are called in the "Program model" listed in the last part of the code.
- Second, based on the information of input file "model.in", choose the materials, sizes of unit cell for different blocks. The effective volume and location of blocks for atoms/particles are determined by cutting the insertion of parallelepiped and spheres. If the scale is 3, then following the rule (5.3), the generalized lattice constant of the unit cell is four times larger than the crystal lattice constant of the atomistic scale. In each cell, the total number of particles does not change, thus four atoms occupy the volume $a^3$ at scale 1; four particles occupy 64 $a^3$ ($=4a \times 4a \times 4a$) at scale 3. This process is done in the "Input" subroutine of the code.
- The neighbor-link cell is established in subroutine Neighbor. This method is similar to the Link Neighbor algorithm described in Appendix 10.D.1.2 to divide the space into many small cubic cells using three-dimensional grids within a certain control cutoff distance. In MD it uses cutoff distance of the potential, here the atomic lattice distance is used as the grid size for each scale. For scale 1 of copper, gridsize $= 3.62$ Å, for the second scale, it is 7.24 Å. These values are the same as the imaginary domain width as shown above. During the computation, the imaginary atoms (or imaginary particles) first find the real particles (or real atoms) in their own cell and contiguous cells within the cutoff radius and then formulate their neighboring link-cells.

For compiling this code, use the following command for compiling and running:

```
$ gfortran Mater_Multi_2010_4.f90 -o Mater_Multi_2010_4.exe
$ ./Mater_Multi_2010_4.exe
```

The model produced using the data of model_S3_square for a cracked Al specimen is given in Figure 10.21 (Remark: before running a calculation the input file should be changed into the standard name, in this case it should be changed to "model.in".)

## 10.14.3 Running mpi Simulation for Multiscale Analysis and Data Processing

The running of the simulation of a copper nanowire under uniaxial tension along the x-direction is carried out by the GP executable file Multi_ten_0.exe and its input file "Model.MD". Here, the

**Figure 10.21** A three-scale crack specimen model produced by the data in file model_S3_square.in and the source code of Mater_Multi_2010_4.f90

Multi_ten_0.exe file is an executable GP file for nanowire stretching. This run uses a parallel computer cluster for the simulation. The related files are stored in UNITS/UNIT14/SIMULATION directory.

If one cannot obtain the Model.MD by any errors occurred, say, in the input file, one may use the existing Model_S1_Cu.MD and Model_S3_Cu.MD files in the directory of GP_Model for the input file, and then run Multi_ten_S3.exe. Here the files of Model_S1_Cu.MD and Model_S3_Cu.MD are the simulation input file, respectively, for the first scale simulation (i.e., full atomistic or MD simulation) and the three-scale simulation. Needless to say, when running the first scale one should change Model_S1_Cu.MD to Model.MD to run, since the code only recognizes the input file with the name of Model.MD.

The GP simulation code is written for parallel processors. The "mpi" compilation and execution are required with the commands as follows:

```
$ mpif90 Multi_ten0.f90 -o Multi_ten0.exe        ! compiling
$ mpiexec -np 6 $PWD/Multi_ten0.exe </dev/null &   !execution
```

Here, "np 6" denotes the number of computer (node) that will be used in the simulation. After the number 6, the environment variable $PWD denotes the present working directory which makes the absolute path of the executable file.

Since parallel algorithms involve some basic knowledge of computer structure and software, it is not the intention in this book to introduce mpi numerical mechanisms, readers interested may review some basic knowledge of parallel computing in Appendix 10.D.

After running, several output files will be printed in the directory. These files include:

– Conf.xyz file which can be used by VMD for visualization of different loading steps (see Section 10.4.2).

- Check_out.md which can be used by gnuplot to show the variations of parameters.
- xxxxx.MD, the "xxxxx" symbol denotes the step number, this kind of file gives detailed values of the configuration at the given steps. It can be used to develop files such as xxxxx.cfg file for visualization and analysis by AtomEye (see Section 10.4.3).

If one cannot run the Multi_ten_S3.exe for the simulation due to the reason, say, without mpi settings, one may use the existing produced file of Check_out.md_S1 and Check_out.md_S3 (gnuplot) and conf. xyz_S1, Conf.xyz_S3 (vmd and ela_history_2009b.f90) for post-data processing.

# Appendix 10.A    Code Installation Guide

## Prerequisites

Before any programming can be done, the user must obtain a license and appropriate super-user permission to install all necessary software, as well as understand the basics of the UNIX terminal as described in Section 10.1.

## 10.A.1 Introduction

This guide will aid in the installation process of various program codes used in this chapter. Specifically, the following codes will be used for computational simulations throughout UNIT1 through UNIT14:

**vi (or vim):** A powerful text editor that runs directly on the terminal. This program can edit text files and will be used to modify and eventually write computer code.
**gfortran:** A free Fortran compiler.
**gnuplot:** A GNU plotting program used to plot 2D graphs.
**VMD:** Software that creates images of atomic configurations.
**AtomEye:** Free software for data processing of atomistic systems (no license required).

The software listed above has been installed and tested on Linux systems. The installation of the GNU/Linux package has different tools for different distributions. This document mainly provides some hints on how to install the above software on a Debian GNU/Linux system. It is best to install Linux on the computer to be used for all programming listed above. Due to the machine and Linux version variations, the general installation instructions will not be covered here. However, users are encouraged to consult technical support when installing any new programs listed. The most important source of information for the GNU/Linux installation can be found at http://info.ee.surrey.ac.uk/ Teaching/Unix/.

    The KNOPPIX software (or the corresponding CD) can also be used to install the GNU/Linux system as introduced in the next section. This is effective for any system including those running Windows. All Linux software is contained on the KNOPPIX CD, which can be removed as desired to return to the original Windows system.

## 10.A.2 Using the KNOPPIX CD to Install the GNU/Linux System

KNOPPIX is a bootable live operating system that allows the user to run the GNU/Linux software from a CD. Therefore any computer that already has an operating system installed can use the GNU/Linux system. KNOPPIX is free software and can easily be downloaded onto a CD, using the website www. knoppix.org. The method used to run KNOPPIX on personal computers depends on the computer settings. The following steps should cover the basic method:

– Insert the KNOPPIX CD and restart Windows.
– The software will require a short time to boot. After the booting process, a schematic landscape of some mountains should appear. Press "Enter".
– Wait until KDE (the default desktop) is loaded. Mountains should appear as the desktop image as well as a window with some basic KNOPPIX information.
– Close this window and work with the Linux system as needed.

Users should not touch the hard disk icon on the desktop until they are familiar with the file formatting. This icon contains important files that are also used in the computer's normal operating system. If changes are made, the operating system may not restart properly.

To logout, select the first icon in the bottom panel, select logout, and then select restart. The CD-ROM drive tray will open. Remove the CD.

Use the following steps to connect with a remote computer cluster:

• Select the computer monitor icon in the menu bar at the bottom of the screen.
• A UNIX shell should open on the screen. Type the following command:

• $ ssh -X username@machine

Press "Enter". It will then request the user name and password for the remote computer. All the required information such as username, password and machine name should be given by the administrator of the computer cluster. For example, if user name is "douglas" and the computer name is fanlan10@alfred.edu, then the command to connect the current computer to the remote fanlan10@alfred.edu computer is as follows:

```
$ ssh -X douglas@fanlab10.alfred.edu     or     $ ssh -X douglas@fanlab10
```

The second command is effective only if the current computer is local, such as a computer on campus. Note: The "-X" must be a capital letter.

• The system may request that the security type be approved. If this occurs, type "yes".
• The system will ask for the respective password. Type it and then press "Enter". Note: The password is case-sensitive; therefore capital letters and lower-case letters are relevant.
• To exit from the machine, type the command: $ exit or $ logout.

To learn more about GNU/Linux and the free software above, see the website http://www.gnu.org/philosophy/free-sw.html.

## 10.A.3  ssh and scp

In the above, we have used the command "ssh" to connect a remote machine. It is a good position to introduce both "ssh" and "scp" for connecting to a remote machine.

### 10.A.3.1  ssh

The command "ssh" allows us to open a shell on a remote machine by the following command:

```
$ssh username@machine
```

By default the shell is in the home directory on the user machine. The option -X listed in the command of last section allows the establishment of a graphical window.

### 10.A.3.2  scp

The command "scp" allows transfer of a file from (or on) a remote machine. The usage is similar to the "cp" command but the remote machine must be explicitly indicated in the command. To transfer a text file or data from a remote machine to the current one that you are working (or typing), the following command is used:

```
$scp username@machine:path destination path
```

For instance, if the current shell is in UNITS of the computer of fanlab12 and one tries to move a file "crystal_structure.f90" from UNITS/UNIT6/NVE_Ar of computer fanlab15 with the username of fanstud15 to the directory of UNITS/UNIT9/2D_Non_EQUI of the current computer the following command is used:

```
~UNITS$     scp        fanstud15@fanlab15:UNITS/UNIT6/NVE_Ar/crystal_
structure.f90 UNIT9/2D_Non_EQUI/
```

If both paths are correct, the shell will ask you the password of the user "fanstud15" in the remote computer fanlab15. If one wants to move a directory, say, NVE_Ar, not just the file from the remote machine, the symbol "-r" should be used after "scp" as:

```
~UNITS$ scp -r fanstud15@fanlab15:UNITS/UNIT6/NVE_Ar UNIT9/2D_Non_EQUI/
```

To transfer data from the local machine (the one on which the "scp" command is typed) to a remote machine, the command is as follows:

```
~UNITS$scp path username@machine:destination path
```

In both the above cases the path is in reference to the home directory of the user.

## 10.A.4  Fortran and C Compiler

The following will describe how to verify that a Fortran and C compiler are properly installed. Although only Fortran will be used in the chapter, it is suggested that both compilers are installed because the two codes are frequently related and some MD codes such as GROMACS and NAMD are written in C.

### 10.A.4.1  Install gfortran

The software installation procedure will depend on the GNU/Linux distribution, therefore it is not possible to describe a universal procedure. However, for systems similar to Debian, the following steps are required:

– Go to the root of the computer and type "su" on a terminal shell.
– Type the administrator or super-user password.

– Use caution as any changes made here may cause permanent damage to the computer.
– Type the following command for installation:

```
$ apt-get install gfortran
```

– Follow the instructions listed. Typically, it will ask for confirmation of some information related to the computer.
– After the installation is completed, exit the super-user terminal by typing "exit".
– Verify that gfortran works properly using the next section.

### 10.A.4.2  Check that gfortran Has Been Installed Properly

• Open the UNITS/APPENDIX directory.
• Compile test_fortran.f90 program using the command:

```
gfortran test_fortran.f90 -o test_fortran.exe
```

• If an error occurs, such as "gfortran command not found", go to Section A1.2.1 again to reinstall it.
• If an error does not occur, check if the test_fortran.exe file exists (ls -lrt).
• Run it (./test_fortran.exe).
• The program output will appear on the screen (the value of pi/2 and the sin of pi/2).
• If the last step was completed successfully, there is no need to reinstall gfortran.
• If a runtime error is obtained related to the library, reinstall gfortran.

### 10.A.4.3  Installing and Verifying the C Compiler (gcc)

C code is typically part of the default installation on all versions of Linux; the first step is to check whether the C compiler is already installed and working properly. If a compiler has been installed, do not install it again. If one has not, use the same procedure as that used for the gfortran installation described in Section 10.A.4.1. The command is:

```
$ apt-get install gcc
```

Once installed, verify that the C compiler works properly as follows:

• Open the UNITS/APPENDIX directory and find the C code "test_c.c".
• Compile the test_c.c program using the following command:

```
$ gcc -lm test_c.c -o test_c.exe
```

• If an error occurs, such as "gcc command not found", reinstall the compiler.
• If no errors occur, check if the test_c.exe file exists (use command "ls -lrt").
• Run it (./test_c.exe).
• The program output will appear on the screen (the value of pi/2 and the sin of pi/2).
• If the last step was completed successfully, do not reinstall the compiler.
• If a runtime error related to the library occurs, reinstall the compiler.

### 10.A.4.4  Installation of Other Useful Software

Follow the procedure described in 10.A.4.1 to install other useful software such as:

- Vim - Text editor (apt-get install vim).
- Gnuplot - Plotting program (apt-get install gnuplot)
- Gv - PostScript viewer (apt-get install gv)
- Xpdf - PostScript viewer (apt-get install xpdf)
- Image viewing tools (apt-get install imagemagick)
- Make - Program required for compiling several programs such as DL_POLY (apt-get install make)

All software listed above must be installed from a super-user shell. Open the super-user shell using the command "su" as shown in 10.A.4.1. After all installations have been completed exit the super-user shell using the "exit" command.

## 10.A.5  Visual Molecular Dynamics (VMD)

### 10.A.5.1  Introduction

VMD is software designed for molecular visualization and data post-processing. There are several pre-compiled versions of VMD, one for each platform, that can be downloaded from the VMD website http://www.ks.uiuc.edu/Research/vmd/. For a standard GNU/Linux system the two suitable versions are: LINUX OpenGL (32-bit system) and LINUX AMD64 OpenGL (64-bit system).

### 10.A.5.2  VMD Installation

To install VMD, follow the instructions on page 6 of the VMD installation guide (see the vmd_ig.pdf file inside the APPENDIX directory). To install the pre-compiled version of VMD for UNIX, uncompress (extract) the file. Depending on the desktop manager, a tool may be installed to uncompress files; otherwise use the command "tar -xvf " followed by the name of the file to be extracted, for example:

```
tar -xvf vmd-1.10.6.tar.gz
```

Then use the following three steps:

1. Open the vmd-1.10.6 directory (using "cd vmd-1.10.6"). Change the default values for the $install_bin_dir and $install_library_dir directories in the "configure" script file to match the desired directories for the files to be placed.
2. Generate the Makefile based on these configuration variables using the ./configure command.
3. Open the src directory and type "make install". This command will insert the code in the two directories indicated in the "configure" script. After the code has completed, type "vmd" in the terminal to verify that it opens correctly.

Note: Super-user permission is required for the "make install" step. VMD requires several programs to work properly. To verify that the correct files are available on the system, run VMD. If the VMD window appears, the code works properly and all files are available. If VMD does not open, use the error message to troubleshoot the problem.

### 10.A.5.3  Typical Problems with the VMD Installation

A typical problem that occurs when running VMD is that the console window will briefly pop up and then immediately disappear. This is usually caused by either a graphics driver problem or a C++ library problem. To determine the problem, run VMD in text mode using the following command:

```
vmd -dispdev text
```

If VMD fails to run in text mode, this usually indicates a C ++ (or other) dynamic link library problem, which is typically solved by installing the missing libraries. If VMD runs in text mode, but fails to start in graphics mode, run the Linux command "glxinfo", which will access OpenGL and report on the installed hardware, software, and all supported video modes and extensions. If there are fundamental problems with the OpenGL installation, the "glxinfo" output may contain errors or warnings. Additional details regarding this issue can be found at http://www.ks.uiuc.edu/Research/vmd/current/linuxrelnotes .html.

## 10.A.6 Installation of AtomEye

A free version of the AtomEye software as well as a full instruction manual for the installation can be found on the AtomEye website http://mt.seas.upenn.edu/Archive/Archive/Graphics/A/. Right-click on the link and "Save Target As..." Save the file in the desired directory using one of the following systems:

- i686 Linux
- Alpha Linux GLIBC2.1
- Sgi Irix
- Sgi Irix64
- Sun Solaris
- HP UX
- Windows with Cygwin/X (README.txt)
- Alpha Tru64 UNIX
- Mac OS X (v10.4 and before, v10.5 Leopard) with Darwin (see A, B, C for button issues)

After installation, run "chmod 755" on the file. To test the program, run it with a cfg file. Examples of which may be found in /UNIT4/GNUPLOT_ATOMEYE_EXAMPLES or created using CFG_ convrt_cmd.f90.

## Appendix 10.B    Brief Introduction to Fortran 90

This document is intended to provide students with additional knowledge and examples about the Fortran 90 programming language.

## 10.B.1 Program Structure, Write to Terminal and Write to File

Open the UNIT2/F90_CODES directory and open the file example1.f90 using the command:

```
vi example1.f90
```

The file will contain code similar to that listed below:

```
"program example1
! variable declaration
implicit none
character (len=7) : : string          ! line 5
! end variable declaration
string = 'hello 2'                      ! line 8
!write on standard output (terminal)
write(*,*) 'hello 1'
```

```
write(*,*) string
write(*,*) 'hello 1', string
!open file 15
open(15,file="example1.dat")
!write file 15
write(15,*) 'hello 1'
write(15,*) string
write(15,*) 'hello 1', string
!close file 15
close(15)
end program example1"
```

Text after the "!" symbol contains comments for code users and is ignored by the compiler. The program begins with the statement "program example1" and finishes (line 26; on vi use G to move the cursor to the last line) with the instruction "end program example1". These two statements are mandatory.

Lines 3–7 are declaration lines, in which "implicit none" means that any variable must be declared. It is not mandatory; however it is strongly suggested that this command be used.

Line 5 describes the character variable "string". The string is seven spaces long and contains concrete letters. Specifically, on line 8 the value "hello 2" is assigned to the string variable.

Lines 11–13 are statements for printing to standard output, while lines 15–24 are instructions for opening a file entitled "example1.dat" and then writing variables or strings to it. See line 16 for "open" instructions and lines 19–20 for writing instructions.

To understand what the code is designed to acccomplish, compile and run it. Open the UNITS/ UNIT2/F90_CODES directory. Compile the program to produce an example1.exe file using the command:

```
$gfortran example1.f90 -o example1.exe
```

Verify that the file has been created using "ls -lrt". The command should have produced a new file, "example1.exe" in the current directory. Use the following command to run it:

```
$./example1.exe
```

An output on the shell will be obtained. Compare the obtained output file with lines 11–13 of the example1.f90 file. Use "ls -lrt" to determine the new file(s) created. A new file, "example1.dat", has been created. Open it with vi and compare the file's contents with lines 15–24 of the example1.f90 file.

## Exercise 10.17

(a) Running the example1.exe file will produce an output on both the screen and in the output file of example1.dat. Find which three lines write the screen output and which three lines write the output in the example1.dat file.

(b) Cancel the command in line 5 by inserting the symbol "!" at the beginning of that line. Save the file and try to compile it. An error will be obtained. Explain why this occurs.

## 10.B.2  Do Cycle, Formatted Output

Open the "example2.f90" in UNITS/UNIT2/F90_CODES file. The file will contain code similar to that listed below:

```fortran
program example2
! variable declaration
implicit none
integer:: ix, nx
! end variable declaration
nx = 20
!write on standard output (terminal)
write(*,*) 'example program 2'
!open file 15
open(15,file="example2.dat")
do ix = 1,nx
write(*,*) ix, 1./float(ix), sin(float(ix))/float(ix)
write(15,'(i4,2x,2f10.5)') ix, 1./float(ix), sin(float(ix))/float(ix)
enddo !ix = 1,nx
!close file 15
close(15)
write(*,*) 'end example program 2'
end program example2
```

Lines 16–19 contain a do cycle. In this case the do cycle repeats the commands between "do" and "enddo" for each value of the integer variable "ix". This variable ranges in value from 1 to the last value "nx". The last value 20 of ix is assigned in line 8. The do cycle in this code has two functions.

The first function (line 17) writes the value of "ix" on the terminal screen, which includes 1/ix and sin (ix)/ix by introduction of float (ix). Here, "ix" is an integer and the "float" command transforms the integer "ix" to a real number.

The second function (line 18) writes the same quantities in a formatted way in file 15. The string (i4,2x,2f10.5) describes the format in which the data will be plotted. In particular i4 means an integer of four characters, 2x means two blank characters, and 2f10.5 means two real floating point numbers. Each floating number includes five characters after the decimal point. The other five spaces will be occupied by sign, point, and the real number before the decimal point.

The format "i" for integers and "f" for real numbers is frequently used in the "read" and "write" statements. Note that if the symbol * is used instead of a specified format in the write and read statement, as in the previous "write" command, the compiler will decide how to print the data. If the symbol (*,*) is used, such as in lines 11 and 17, it will write the result to the terminal screen.

To understand the program, compile and run it:

- Compile the program using the command:

```
gfortran example2.f90 -o example2.exe
```

- Verify that the output "example2.exe" file exists using "ls -lrt".
- Run the code using the command:

```
./example2.exe
```

- An output will be obtained on the screen. Compare this with the screen output produced by line 17 of the example2.f90 file.
- Check the file system with "ls -lrt". A new file should have been created, example2.dat. Open it with vi and compare the file with line 18 of the example2.f90 file.

## Exercise 10.18

Make a copy of the example2.dat file (for instance cp example2.dat example2_formatted.dat). Change the specified format; for example instead of 2f10.5 write 2f20.10. Compile and run. Check if the file example2.dat was created. Compare the example2_formatted.dat file with the example2.dat file. What are the differences?

### 10.B.3 Arrays and Allocation

This UNIT will first introduce "arrays" using the example3.f90 file found in the UNITS/UNIT2/F90_CODES directory, which contains the same output values of the previous code, but uses the structure "arrays". Open the example3.f90 file. In declaration line 6 the following statement is given:

```
real (kind=8), allocatable :: one_on_x(:), sinx_on_x(:)
```

This statement declares two arrays of real variables, where "allocatable" indicates that the dimension of the vector will be indicated after running the program. "Kind = 8" indicates the number of bytes of memory for each variable. In particular "kind = 8" is often indicated as double precision. The dimension of the two arrays is given so that each array has "nx" elements as shown below:

```
nx=20                                     ! line 9
ALLOCATE (one_on_x(1:nx))                 ! line 11
ALLOCATE (sinx_on_x(1:nx))                ! line 12
```

To identify a specific element of the array it is sufficient to indicate its corresponding number in the array. For instance, the expression "one_on_x(3)" indicates element 3 in the "one_on_x" array. See lines 19, 20, and 27 below:

```
do ix = 1,nx
  one_on_x(ix) = 1./float(ix)                                       ! line 19
  sinx_on_x(ix) = sin(float(ix))/float(ix)                          ! line 20
enddo !ix = 1,nx
.............................................................
do ix = 1,nx
  write(15,'(i4,2x,2f10.5)') ix, one_on_x(ix), sinx_on_x(ix)   ! line 27
enddo !ix = 1,nx
```

Note that the element of the array (e.g., one_on_x) must exist, otherwise runtime errors will result. The type of error will depend on the compiler. Compile and run the code. Verify that the output file, "example3 .dat", contains the desired output data.

## 10.B.4  IF THEN ELSE

Open the example4.f90 file. This program file is very similar to example3.f90. The only difference is in lines 20–24, where an "IF THEN ELSE" command is given as follows:

```
if (sin(float(ix)).gt.0) then               !line 20
   sinx_on_x(ix) = sin(float(ix))/float(ix)
else
   sinx_on_x(ix) = -sin(float(ix))/float(ix)
endif                                       ! line 24
```

After compiling and running the code, compare the output file, "example4.dat", with the example3. dat file. Take note that the values in the third column of the "example4.dat" file are positive, while the corresponding column in the "example3.dat" file are negative. This variation is caused by the **IF THEN ELSE** statement, which requires the program to assign the value sin(float(ix))/ (float(ix)) to the variable sinx_on_x(ix) only if sin(float(ix)) is positive; otherwise change the sign of sin(float(ix))/(float(ix)) and then assign it to the variable sinx_on_x(ix). (Note the minus sign in line 23.)

The general structure of **IF THEN ELSE** block is as follows:

```
IF condition THEN
   block of statements
ELSE
   block of statements
ENDIF
```

The "condition" is a Boolean algebra expression that must have either a "true" or "false" output for the specified condition. In particular, the Boolean expression often contains the following relational operators:

```
.gt.    ! greater than
.eq.    ! equal
.ne.    ! not equal
.lt.    ! less than
```

It is possible to give a more complex Boolean expression by using the operators ".or.", ".and.", and ".not." However, their use with the "if then cycle" in numerical calculation codes is strongly disadvised because they are inefficient.

## Appendix 10.C   Brief Introduction to VIM

### 10.C.1  Introduction

This appendix is designed to introduce the reader to VIM, therefore only a small portion of the vi features are described. Additional information can be found in the websites:

- http://www.vim.org/
- http://www.yolinux.com/TUTORIALS/LinuxTutorialAdvanced_vi.html

Check that VIM is properly installed then open a terminal shell and open vi using the command:

```
$ vi
```

If vi is installed and working properly, a series of text will appear in the middle of the shell window which will include "VIM – Vi IMproved". If this occurs, type ":q" and press Enter to exit from vim and return to the shell. Create a simple file using the "insert mode", then open a terminal shell and create a new vi file using the following prompt:

```
$ vi my_first_vim.txt
```

The string listed after "vi" is the name given to the file to be created and can be changed to any name desired.

Type the letter "i" on the keyboard. The word "INSERT" should appear in the left part of the last line of the vi screen. This indicates that vi is in "INSERT mode", which will allow the file to be altered. The backspace key will cancel a character and the arrow keys will move the cursor within the text. Now press "ESC". The "INSERT" string will disappear. If done correctly, this process has created a txt file using vi. Type ":wq" to save and quit. Use "ls -lrt" to verify that the file was created and saved properly.

## 10.C.2  Simple Commands

Listed below are several common commands for VIM. To illustrate these basic vi commands, the "vimtutorial.txt" file found in the UNITS/APPENDIX directory should be used. Open this directory and make a copy of the vimtutorial.txt file using the following command:

```
$cp vimtutorial.txt vimtutorial_yourname.txt
```

Open the copy using the command:

```
$vi vimtutorial_yourname.txt
```

Save and close a file: To save a file, type ":w". To close the file without saving, type ":q!". To save and close the file simultaneously, type ":wq".

Delete a character (x): Move the cursor to column 7 of line 5. (The row and column number can be found on the right side of the last line of the vi window.) The cursor should be located on the last "d" of the word "alfredd". Type "x". The last d will disappear. Save the file (:w).

Delete a word (dw): Move the cursor to column 8 of line 9. The cursor should be located on the "u" of the word "university". Type the command "dw". The word university will disappear. Save the file (:w).

Delete a line (dd): Move the cursor to line 13. Type the command "dd". The line "alfred university 1" will be deleted. Save the file (:w). Multiple characters and lines can be deleted simultaneously by typing the number of characters or lines to be removed, followed by the value "x" for characters or "dd" for lines. For example, "3x" will delete three characters, and "10dd" will delete 10 lines. Try these commands and close the file without saving (:q!). Open the file again, the "copy" and "paste" commands will be discussed as follows.

Copy a word (yw): Move to the first column of the first line. Type the command "yw". Now move the cursor to the blank space between "This" and "file" (character 1,5). Press "p". A copy of the word will appear. This copy command works without the use of "INSERT" mode, therefore "INSERT" will not appear at the bottom of the window. In vim there are two commands for pasting, including "p" and "P", which produce slightly different results, try both.

Copy a line (yy): Move to line 11. Type the command "yy". Now move the cursor to line 12 and press "p". A copy of the line will appear. Save and close the file. The copy command works much like the delete command; to copy multiple lines type the number of lines to be copied and then type "w" to copy words or "y" to copy lines. For example, the command "3yw" copies three words, while the command "4yy" copies four lines. Try these commands and close the file without saving (:q!).

Move within the file with gg, G and ":n": To move instantly to the first line of the file, type "gg". To move instantly to the last line of the file, type "G". To move the cursor to a given line, type ":" followed by the line number. For example ":3" (no space between ":" and "3") will move the cursor to the third line of the file).

Search for a word using/: To search a word on all lines after the cursor, type ":/" followed by the desired word. For example, if the word ERROR needs to be located within the file, type ":/ERROR". Open the vimtutorial file. Put the cursor on the first line and type "/alfred" (no space between "/" and "alfred"). The cursor will move to the first recurrence of the word "alfred" (i.e., column 34 of line 3). To go on the next recurrence of the word, press "n" on the keyboard. The cursor will move to line 5.

undo "u" and redo "ctrl + shift + r": To undo the last action, type "u". To redo the last action, press "control + shift + r". (Press control, then without releasing the control key press shift + r).

# Appendix 10.D    Basic Knowledge of Numerical Algorithm for Force Calculation

MD simulation requires a great deal of computer resources, so an inefficient program cannot be put into practical use. People have long noticed that in MD simulation about 90% of run time is dedicated to the calculation of force. Therefore, improving the efficiency of force calculation is essential. This session first introduces some highly efficient algorithms for force calculation, and then treats parallel algorithms in a parallel computing environment.

## 10.D.1  Force Calculation in Atomistic Simulation

In MD, the force imposed on a random atom i by an atom j is determined by following equation in classical mechanics:

$$\bar{F}_i = -\frac{\partial U}{\partial \bar{r}_{ij}} \tag{10.D.1}$$

where U is the total potential energy of the system. When one approximates the result using pair potential, U is given by (2.2c), and $\bar{r}_{ij}$ is the position vector starting at i and ending at j. One uses its module $r_{ij}$, which is a scalar, to denote the vector value. Force calculation for atom i is done by combining all the forces imposed on the atom by other atoms. In principle, when calculating the force imposed on atom i (i=1, N), one must take account of every other atom j (j = 1, N, j ≠ i), so the calculation has order of $N^2$ complexity.

In fact, interatomic force decreases dramatically as the distance between atoms increases, until it is equal to zero. So when dealing with the potentials in MD, we introduce the concept of cutoff distance or cutoff radius rcut, implying that when the distance between two atoms exceeds rcut, the interatomic force can be ignored. Thus we need only calculate the forces imposed by adjacent atoms, reducing the complexity order to N. Then in order to get an efficient algorithm of force calculation, we focus on efficient ways of finding the adjacent atoms. A few common methods are listed.

### 10.D.1.1 Verlet Neighbor List Algorithm

Verlet method[16] is to store all the serial numbers of every atom's neighbor atoms in a list, which is dynamically maintained. The force is calculated by going through the entire list. When setting up the neighbor list of an atom, search is conducted among all atoms and the atoms within the distance $r = r_{cut} + \delta$ are stored in the list. Here $r_{cut}$ is the cutoff distance of the potential, and $\delta$ is the extension distance. After each step of integration, the displacement of each atom is judged since the list was last updated. If the displacement is larger than $\delta$, the neighbor list is updated and the displacement of the atom is set to be zero, otherwise the same list is still used in the next step. The value of $\delta$ is adjustable. A larger value increases the length of the neighbor list but reduces the frequency of update, and therefore saves time. So the $\delta$ value needs to be determined according to experience.

### 10.D.1.2 Link Cell Algorithm

As an alternative to Verlet method, which requires a time-consuming search among all atoms, the link cell algorithm was proposed.[17] This method divides the space into many small cubic cells using 3D grids based on the cutoff radius $r_{cut}$ of the potential function. During the computation, the atoms first find out to which cell they belong, and then look for neighbors in their own cells and contiguous cells. As there is no need for the atoms to look for atoms far from their neighbors, this method saves time, especially for the computational models that are far larger than cutoff distance.

### 10.D.1.3 The Mixed Neighbor List/Link Cell Algorithm

Since the lengths of many atomic pairs involved in the computation are larger than the cutoff distance $r_{cut}$, the link cell method is still not efficient enough. Therefore, a mixed algorithm was introduced to deal efficiently with large-scale MD simulations. In this method,[18] one uses the main framework of the Neighbor List Algorithm, and adopts the Link Cell Algorithm in creating and updating the neighbor list to save time. Inevitably, this method needs a more complicated program structure. This mixed method is commonly used in large MD software.

## Appendix 10.E    Basic Knowledge of Parallel Numerical Algorithm

### 10.E.1  General Information

As MD simulation takes so much computer resources, the simulation of millions of atoms on a single computer cannot be accomplished within a reasonable period of time. Therefore, large-scale MD simulations are usually conducted on large parallel computers. In recent years, the Linux cluster has provided researchers with a low-price parallel computing environment. Many cheap PCs are linked together to form a parallel supercomputer using message transmission mechanisms such as MPI,[19] OPENMP,[20] and PVM.[21] There are various free and commercial MD computation software utilizing the parallel environment such as NAMD,[22] AL_CMD,[23] SpaSM,[24] MdynaMix,[25] and CPMD.[26]

The MD method is not as mature as the finite element method, and the widening research scope has posed many new questions. Sometimes available software cannot serve the users' needs well, and this is especially true for frontier scientific research. Therefore, this section discusses the theory of parallel computing in these softwares to provide a reference for those who are interested in developing their own MD programs. The idea of parallel computing in MD is to decompose the calculation of force and the update of velocity/position, and process them on multiple CPUs. The parallel computing of the update of velocity/position can be realized straightforwardly, as we need only the information of each individual atom's motion without considering neighboring atoms. However, the calculation of force requires the positional information of neighbor atoms, so parallel computing involves a complicated message transmission between CPUs. Up to now, there are mainly two methods to realize the parallel computing of force.

The first method is to divide the calculation of force into many subsets, and then allot each of the subsets to a computer. This kind of distribution is kept unchanged throughout the process of calculation. The simplest case is to allot a fixed number of atoms to each CPU. Each CPU then computes the force of the allotted atoms and updates their velocity and coordinate position without considering where these atoms move to. This method is called Atom Decomposition. Another way is to decompose the interatomic forces into a matrix, where each CPU only calculates the interactive forces between specific atom pairs. This is named Force Decomposition. Both Atom Decomposition and Force Decomposition are similar to Lagrangian in mechanics in that the subject computed on each CPU changes in space along with the deformation.

The second method is similar to Eulerian description. The subject of computation for each CPU is fixed in space, while the atoms in the space change as the simulation goes on. This method is called Spatio-Decomposition. The following is a brief introduction to these basic theories. Sample parallel computing programs can be downloaded from the websites given in the references.

## 10.E.2 Atom Decomposition

Assume the total number of atoms is N, and there are P CPUs for the parallel computation. So each CPU is in charge of the force calculation and the update of velocity/position for N/P atoms. In this method, the entire positional information is stored in every computer, so when allotting the atoms to the CPUs, there is no need to consider the positions or the interactive force of the atoms. In each computational step, the local CPUs first calculate the force imposed on their own N/P atoms. Then the velocity/position of these atoms is updated through the integration of the dynamics equation. After the update, the positional information of all atoms is collected. The collected information is then broadcast to all the computers so that it can be used in the next step.

This method is very similar to the common serial computation, and can be achieved by slightly modifying a serial program. Beginners to parallel computation can start with this method to understand the efficiency of parallel computation. This method requires a high memory capacity, and the message transmission between computers is enormous; moreover, it cannot be improved by increasing the number of CPUs. Therefore, this method is commonly used for computation whose scale is not very large. Considering the theory of Newton's third law in this method, nearly half of the time used in force calculation can be saved. As for the interactive force between two remotely separated atoms, we adopt a distribution plan so that two CPUs are allotted equal loads. The mathematical expression of such a plan is:

If $I > J$ and $I + J$ is an odd number, the computation should be done on the local computer, otherwise it should be done on a remote CPU and acquired through mapping.

If $I < J$ and $I + J$ is an even number, the computation should be done on the local computer, otherwise it should be done on a remote CPU and acquired through mapping.

This method reduced the computational burden, but the force imposed on the local atoms can only be acquired after the remote CPU has finished computation. Message transmission between CPUs involves information of position as well as force. Therefore, the improved computational efficiency is obtained at the cost of increased load of communications. In order to reduce the memory consumption, distributed storage is adopted in Atom Decomposition method. The essence is to divide the calculation of the interactive force between local and remote atoms into P–1 steps. In each step, the interactive force between the remote atoms on one CPU and the local atoms is calculated. The positional information of N/P atoms is temporarily stored, and in this way, some memory capacity is spared.

## 10.E.3  Force Decomposition

Force Decomposition is very similar to Atom Decomposition. Each computer has the positional information of all the atoms and is allotted the same amount of atoms. However, it is different from Atom Decomposition in that the force calculated by each CPU is only part of the force imposed on the atom, and a full picture of the forces' calculation relies on all the CPUs.

In fact, when local calculation cannot give a full picture of the forces imposed on local atoms (as in the above case where Newton's third law is taken into consideration), we can view Atom Decomposition as a category in Force Decomposition. A typical method of Force Decomposition is to divide the upper half of the matrix of interactive force into K equal areas, and then the computation of the interactive force within the K areas is assigned to K CPUs. After computation, the results are mapped to the lower half of the matrix according to Newton's third law, so every CPU obtains the integral information on the force of local atoms.

Then the force imposed on local atoms is collected, and the velocity/position information of local atoms is updated. The information is then transmitted to other CPUs so that they can be used in the next step. The defect in both Force Decomposition and Atom Decomposition is that a high memory capacity is required since the positional information of all the atoms must be stored locally.

## 10.E.4  Domain Decomposition

Domain Decomposition is quite different from the above two methods. It is a genuine distributed storage parallel algorithm with a small amount of message transmission. The method is well suited for large-scale MD simulations in a distributed parallel environment. The idea of this method is to divide the domain of the simulated system into multiple subdomains, each one of which is handled by a CPU.

During the process of simulation, the CPU manages the atoms in a dynamic way. When an atom enters the local domain, it becomes a local atom. The local atoms interact with each other and with the interfacial atoms of adjacent domain, so information on the atoms of adjacent domains is also needed.

Only the positional information of these interfacial atom need to be transmitted between CPUs. The decomposition of domain can be one-dimensional, two-dimensional, or three-dimensional. Unlike Atom Decomposition and Force Decomposition, both of which require the transmission of all the atoms' positional information, the Domain Decomposition method transmits positional information within a local domain. So in this method, the cost of message transmission is greatly reduced. For the two-dimensional situation, message transmission is as follows:

Messages are transmitted from left to right between adjacent CPUs. Each CPU transmits the positional information of a selected group of atoms in its subdomain to the CPU on the right that is in charge of the next subdomain. These selected atoms are those involved in the force calculation of the next subdomain on the right. Similarly, each CPU receives positional information from the CPU that is in charge of the subdomain on the left.

Messages are transmitted from right to left between adjacent CPUs. Each CPU transmits the positional information of a selected group of atoms in its subdomain to the CPU that is in charge of the next subdomain on the left. These selected atoms are those involved in the force calculation of the next subdomain on the left. Similarly, each CPU receives positional information from the CPU that is in charge of the subdomain on the right.

Messages are transmitted upward between adjacent CPUs. Each CPU transmits the positional information of a selected group of atoms in its subdomain to the CPU that is in charge of the subdomain above. These selected atoms are those involved in the force calculation of the subdomain above. Similarly, each CPU receives positional information from the CPU that is in charge of the subdomain below.

Messages are transmitted downward between adjacent CPUs. Each CPU transmits the positional information of a selected group of atoms in its subdomain to the CPU that is in charge of the sub-domain below. These selected atoms are those involved in the force calculation of the subdomain below. Similarly, each CPU receives positional information from the CPU that is in charge of the subdomain above.

It merits noting that in the third and fourth steps, the positional information acquired in the first and second steps are also transmitted upward and downward. The one- and three-dimensional situation is similar to the two-dimensional situation discussed in the above passages. From the above algorithm, we can see that each CPU needs only to dynamically store and maintain two data structures, one of the position, velocity, and acceleration of the local atoms, and the other of the positional information of atoms in adjacent domains, which are acquired through message transmission.

This algorithm does not need global but only local transmission, so the amount of information exchanged is only in proportion to the surface area of subdomains. This presents an obvious advantage in large-scale MD simulations. However, the load balance between CPUs requires that each sub-domain has the same amount of atoms, which gives rise to limitations in complicated situations. Still, the improved efficiency brought about by more CPUs is unlikely to be offset if a few CPUs are inadequately loaded.

## Appendix 10.F   Supplemental Materials and Software for Geometric Model Development in Atomistic Simulation

Geometric model development of atomistic systems is important for simulation and is discussed in Section 2.5. Briefly speaking, after obtaining unit cell data from the crystal structure database[27] one knows the unit cell parameters and structure given by the values of crystal constants a, b, c and the angles between crystal axes, $\alpha$, $\beta$, $\gamma$. Table 10F.1 shows this data for $Fe_4O$, $ZnO$, $Al_2O_3$ and $Fe_3N_4$. Databases such as Findit of the National Institute of Standard Technology (NIST) in the USA also provide the number of each type of atom and the SG number of the unit cell. Based on this data, one can use tables for X-ray crystallography[28] to determine the coordinates of each atom related to the crystal unit cell. As we explained in Section 2.5, these coordinates are fractional of the corresponding values of crystal constants a, b, and c.

**Table 10F.1**   Examples for a, b, c values of crystals

| Crystals | Crystallographic parameters a, b, c, $\alpha$, $\beta$, $\gamma$ |
|---|---|
| $Fe_4N$ | 3.795, 3.795, 3.795, 90, 90, 90 <br> SG: 221 |
| $Al_2O_3$ | 4.7607, 4.7607, 12.9947, 90, 90, 120 <br> SG: 167 |
| $Si_3N_4$ | 7.765, 7.765, 5.622, 90, 90, 120 <br> SG: 159 |
| $ZnO$ | 3.2489, 3.2489, 5.2049, 90, 90, 120 <br> SG: 186 |

Since the model coordinates usually use the X, Y, Z Cartesian coordinate system which in general is not coincident with the crystallographic axes a, b, c, the coordinate transformation should be conducted from the crystallographic (a, b, c) coordinates to the Cartesian (X, Y, Z) coordinates. This transformation can determine the Cartesian coordinates for each atom in the model coordinates. Design the box size $L_X$, $L_Y$, and $L_Z$, and then estimate the total atomistic number of the model. Let us take the following cases to show how these models in the Cartesian coordinates can be developed.

## 10.F.1 Model Development for Model Coordinates Coincident with Main Crystal Axes

### 10.F.1.1 Two Cases

There are two cases frequently used in the model development.

**Case 1: BCC and FCC crystals ($a = b = c$, $\alpha = \beta = \gamma = 90$) and the model coordinate axes X, Y, Z are coincident with the crystallographic axes a, b, c**
This is the simplest case which is discussed in Section 10.6 for FCC (e.g., Ar) and Section 10.7 for BCC (e.g., Fe). The software Crystal_M_Simple.f90 in UNIT3 and Crystal_structure.f90 in UNIT6 can be used to produce these models. As shown in these codes, the model sizes are given as follows:

$$L_x = m_x a, \ L_y = m_y b, \ L_z = m_z c \qquad (10F.1a, b, c)$$

where $m_x$, $m_y$, $m_z$ are cell numbers along the model axis X, Y, and Z direction as shown in the fifth line of the input file Model_input in UNITS/UNIT6/NPT_Ar (see Section 10.6.1). The cell number equals the product of $(m_x) \times (m_y) \times (m_z)$; the total atom number, N, of the system equals the cell number times the atom number per primary unit cell in the periodic condition.

**Case 2: Unit cell with a, b, c, $\alpha = \beta = 90$ and $\gamma = 120$**
$Fe_4O$, $ZnO$, $Al_2O_3$, and $Fe_3N_4$ in Table 10F.1 belong to this case. In this case, one may take model axis X, Z, and the XOY plane, respectively, coincident with the crystallographic axis "a", "c", and the plane "aOb" as shown in Figure 10F.1.

### 10.F.1.2 Coordinate Transformation and Directional Cosine

The $X_i$, $Y_i$, $Z_i$ coordinates of any atom "i" along the model axes X, Y, Z can be determined by the crystallographic coordinates $A_i$, $B_i$, $C_i$, with direction cosines as follows:

$$X_i = base(1, 1) \times A_i + base(1, 2) \times B_i + base(1, 3) \times C_i \qquad (10F.2a)$$

$$Y_i = base(2, 1) \times A_i + base(2, 2) \times B_i + base(2, 3) \times C_i \qquad (10F.2b)$$

$$Z_i = base(3, 1) \times A_i + base(3, 2) \times B_i + base(3, 3) \times C_i \qquad (10F.2c)$$

with

$$A_i = a_i \times a_0, \quad B_i = b_i \times b_0, \quad C_i = c_i \times c_0 \qquad (10F.2d-f)$$

where $a_i$, $b_i$, $c_i$ are atomic crystal coordinates; $a_0$, $b_0$, and $c_0$ are lattice constants along the crystallographic axes. The notation base (I, j) is used for the direction cosine in the computer code for the modeling development (see introduction to MaterNew_2010_Comb.f90). The first entry in the parenthesis denotes the model axes, $I = 1, 2, 3$ which correspond, respectively, to the model coordinates X, Y, and

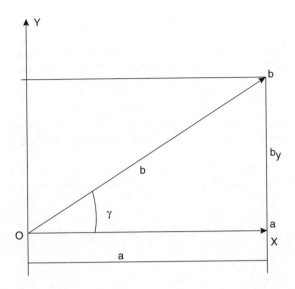

**Figure 10F.1**    An arrangement for the crystallographic axes a, b, c and the model axes X, Y, Z for Case 2 with a, b, c, $\alpha = \beta = 90$ and $\gamma = 120$

Z. The second entry denotes the crystallographic axes, $j = 1, 2, 3$ corresponding, respectively, to the a, b, c axes.

While the above coordinate transformation formulas are generally correct, the directional cosine should be determined based on different orientations of the model axes with respect to the crystallographic axes. For case 2, these directional cosines read:

$$\text{base}(1, 1) = \cos(X, a) = 1.0 \tag{10F.3a}$$

$$\text{base}(1, 2) = \cos(X, b) = \cos(a, b) = \cos\gamma = -0.5 \tag{10F.3b}$$

$$\text{base}(1, 3) = \cos(X, c) = 0 \tag{10F.3c}$$

$$\text{base}(2, 1) = \cos(Y, a) = 0 \tag{10F.3d}$$

$$\text{base}(2, 2) = \cos(Y, b) = \sin\gamma = \sin 120 = 0.8660254038 \tag{10F.3e}$$

$$\text{base}(2, 3) = \cos(Y, c) = 0 \tag{10F.3f}$$

$$\text{base}(3, 1) = \cos(Z, a) = 0 \tag{10F.3g}$$

$$\text{base}(3, 2) = \cos(Z, b) = 0 \tag{10F.3h}$$

$$\text{base}(3, 3) = \cos(Z, c) = 1.0 \tag{10F.3i}$$

### 10.F.1.3  Codes and Input File Used in the Model Development for DL_POLY_2

Here, we take DL_POLY_2 as an example to show how to develop its CONFIG file, which is the geometric model file of DL_POLY_2. All codes and input files used in this section are placed in both the directory of UNITS/UNIT2/F90_CODES/Model_Develop and UNITS/UNIT14/INI_CONF/DLPOLY_Model directory.

The Fortran 90 source code MaterNew_2010_Comb.f90 includes several unit cells for different materials. Besides the BCC and FCC structures, it includes materials of $Al_2O_3$, $Fe_4N$, ZnO, and $Si_3N_4$. The first task of the code is to formulate the model unit cell along the Cartesian coordinates X, Y, and Z. This is done by developing subroutines of cell.BCC, cell.FCC, cell.Al$_2$O$_3$, cell.Fe$_4$N, cell.ZnO, and cell.Si$_3$N$_4$. In these subroutines, the crystallographic coordinates of each atom in the unit cell are transferred into the model coordinates through (10F.2a, b, c). After the model unit cell is developed, the simulation model can be developed by repeating the model unit cell along the crystal axes a, b and c directions.

The file "model.in" is the input file which shows the design of the model. Taking the ZnO/Fe coating system as example, its format can be shown below and then explained line by line as follows:

```
-  0   3                                          ! line 2
-  1 0 0
-  0 1 0
-  0 0 1
-  1                                              ! line 6
-  17.34 13.4933  57.8                            ! line 7
-  2                                              ! line 8
-  F  1  1                                         ! line 9
-  -17.34 17.34 -13.4933  13.4933 -57.8 57.8      ! line 10
-  s                                              ! line 11
-  -17.34 17.34 -13.4933 13.4933 0.0 57.8         ! line 12
-  Z  1  1                                         ! line 13
-  -17.34 17.34 -13.4933 13.4933 0.0 57.8         ! line 14
-  s                                              ! line 15
-  17.34 -17.34 13.4933 -13.493 57.8 0.0          ! line 16
-  0 11                                           ! line 17
-  0 12                                           ! line 18
```

Line 1: The simulation name.

Line 2: The first number 0 is "levcfg" parameter and the second number 3 is IMCON. See Appendix 10.B of DL_POLY_2 manual for more information. For the present case, imcon = 3 indicates that the boundary condition belongs to the parallelepiped periodic boundary condition.

Lines 3–5: These three lines are used to produce the corresponding lines 3–5 in the CONFIG files of DL_POLY, see Section 10.5.6.1.

Line 6: This denotes the scale of the model; for atomistic analysis it is always scale 1.

Line 7: The three values give half dimensions of X, Y, and Z of the model size. If the model size is Lx, Ly, and Lz along the X, Y, and Z direction, the values should be 0.5Lx, 0.5 Ly, and 0.5 Lz.

Line 8: How many blocks the model will need.

Lines 9 and 13: The three values give information of phase, scale, and ncut, respectively; "phase" denotes the material phase, such as F denotes iron in line 9 and Z denotes ZnO in line 15; "scale" is with the same meaning of line 6, but here is for the scale of the phase; "ncut" denotes how many cuts will be carried out in the given block.

Lines 10 and 12: Line 10 defines the outer boundaries of the block. Line 12 defines its inner boundaries which specify the volume of the material to be removed, thus leaving the desired geometry of the block. In this case, the top half of the height is cut with the bottom half of the iron remaining.

Lines 11 and 15: Relate to the type of cutting within the block; "s" means the cutting is a rectangular body; "c" denotes it is a cylindrical hole, perpendicular to the XOZ plane.

Lines 13–16: Denote the ZnO block. The limited values of the outer faces of left, right, back, front, lower, and upper surfaces are given in line 14. The value in line 16 is arranged from a large value to a small value for the X, Y, and Z direction respectively, opposite to a normal cut, thus the cut is not conducted.

The code CONFIG.f90: The output file after running MaterNew_2010_Comb.f90 with model.in is the output file Model.MD. It is not the accurate CONFIG file for DL_POLY_2. It serves as the input file for the code CONFIG.f90 to produce the CONFIG file. The function of this code is to make the produced chemical bonds in the right order, say, two O atoms after two Zn atoms. For different materials, the model needs to have some changes on lines 39 and 51. Where "a" denotes the first chemical element (e.g. Al), "b" denotes the second chemical element (e.g., O), and the number in these two lines equals the chemical element number plus 1. For instance, in the case of $Al_2O_3$, in line 39, the number after "a.eq. " is 3, and in line 51 after "b.eq. " is 4.

### 10.F.1.4 Notable Points

The following points need to be noted in model development.

(1) The numbers $m_x$, $m_y$, $m_z$ of cells along the X, Y, and Z direction can be determined as follows:

$$m_x = L_x/a \quad m_y = L_y/(b\sin\gamma) \quad m_z = L_z/c \qquad (10F.4a, b, c)$$

(2) While it is expected that $m_x$, $m_y$, $m_z$ will be integers, there is a possibility that they are fractional numbers. This situation occurs when layers of materials with different structure and unit size are combined to match each other.

(3) This case may cause the model to have unmatched atoms. Take ZnO as an example to show this situation. To match the dimension of iron substrate, the model size is designed to be $L_x$=34.85 Å, $L_y$=26.8992 Å, and $L_z$=52.049 Å. The X-ray table shows that in each unit cell there are two Zn atoms and two O atoms. From Table 10F.1, it is easily seen that a=3.2489 Å, $b_y$=2.8163 Å, c=5.2049 Å, thus:

$$m_x = 34.85/3.2489 = 10.7267; \quad m_y = L_y/(b\sin\gamma) = 26.8992/2.8163 = 9.5516;$$
$$m_z = 52.049/5.2049 = 10 \qquad (10F.5)$$

The total number M of unit cells and the total number N of atoms in the system can be determined as follows:

$$M = m_x \times m_y \times m_z = 1024.54 \qquad (10F.6)$$

$$N = M \times 4 = 4098 \qquad (10F.7)$$

From running the code of MaterNew_2010_Comb.f90, the total number of 4200 is obtained. This number is larger than the estimated number of 4098.

(4) In general, the unmatched atoms should be dropped to make the unit cells into integer numbers. This will make the FIELD file in DL_POLY easy to develop. In using CONFIG.f90 to formulate the CONFIG file, the unmatched atoms will appear in the last part of the CONFIG file, which must be deleted.

## 10.F.2 Model Development for Model Coordinates not Coincident with Crystal Axes

Frequently, the model coordinates are distributed with different orientations to the crystallographic axes. This arrangement is important to investigate the orientation effects of crystals and loading conditions on crystal behavior; thus it is useful to learn the model development for the case whose coordinates are not coincident with crystal axes.

**BCC and FCC crystals (a=b=c, $\alpha=\beta=\gamma=90$) and the model coordinate axes X, Y, Z are not coincident with the crystallographic axes a, b, c**

Suppose the model coordinate axes are inclined to the crystallographic axes by $X=[1\,1\,0]$, $Y=[-1\,1-2]$, $Z=[-1\,1\,1]$. Here, the numbers in brackets denote the smallest integer values of the crystallographic node that meet the model coordinate axes X, Y, and Z. Since the model axes formulate an orthogonal system, the dot product of any of the two axes should be zero. For instance, the dot product of Y and Z gives $(-1)\times(-1)+(1)\times(1)+(-2)\times(1)=0$. In addition, the lengths of the model box size along the X, Y, and Z direction are, respectively, $\sqrt{2}a$, $\sqrt{6}a$, $\sqrt{3}a$.

The coordinate transformation formulas 10F.2(a, b, c) from the crystallographic $a_i$, $b_i$, $c_i$ to the model axis X, Y, Z are still correct. But the directional cosines of the incline X, Y, Z relative to the crystallographic axis are given as follows:

$$\text{base}(1,1) = \cos(X,a) = 1/\sqrt{2} \tag{10F.8a}$$

$$\text{base}(1,2) = \cos(X,b) = 1/\sqrt{2} \tag{10F.8b}$$

$$\text{base}(1,3) = \cos(X,c) = 0 \tag{10F.8c}$$

$$\text{base}(2,1) = \cos(Y,a) = -1/\sqrt{6} \tag{10F.8d}$$

$$\text{base}(2,2) = \cos(Y,b) = 1/\sqrt{6} \tag{10F.8e}$$

$$\text{base}(2,3) = \cos(Y,c) = -2/\sqrt{6}, \tag{10F.8f}$$

$$\text{base}(3,1) = \cos(Z,a) = -1/\sqrt{3} \tag{10F.8g}$$

$$\text{base}(3,2) = \cos(Z,b) = 1/\sqrt{3} \tag{10F.8h}$$

$$\text{base}(3,3) = \cos(z,c) = 1/\sqrt{3} \tag{10F.8i}$$

The unit model box size $\ell_x$, $\ell_y$, $\ell_z$ and numbers $m_x$, $m_y$, $m_z$ of cells along the model X, Y, and Z directions can be determined, respectively, as follows:

$$\ell_x = \sqrt{2}a, \quad \ell_y = \sqrt{6}a, \quad \ell_z = \sqrt{3}a \tag{10F.9a, b, c}$$

$$m_x = L_x/\ell_x = L_x/\sqrt{2}a; \quad m_y = L_y/\sqrt{6}a, \quad m_z = L_z/\sqrt{3}a \tag{10F.10a, b, c}$$

The volume of the model unit cell along the X, Y, Z directions is

$$V_{\text{model cell}} = \ell_x \times \ell_y \times \ell_z = \sqrt{2}a \times \sqrt{6}a \times \sqrt{3}a = 6a^3 \tag{10F.11}$$

For BCC crystal in periodic boundary conditions, each primary unit cell has two atoms, thus the model unit cell should include 12 atoms; for FCC each cell has four atoms, so each model unit cell has 24 atoms. This is shown in detail for the Al multiscale model in Section 10.14.

**General crystal structure (a, b, c, $\alpha$, $\beta$, $\gamma$) and the model coordinate axes X, Y, Z are not coincident with the crystallographic axes a, b, c**

In this case, formulas (10F.2a, b, c) are still valid. The directional cosine can be determined by the given axis values such as $X=[1\,1\,0]$, $Y=[-1\,1-2]$, $Z=[-1\,1\,1]$ and the angle. The directional cosine will not be as simple as (10F.8a–i) because the crystallographic axes are not perpendicular to each other such that the length of the model box cell along the X, Y, and Z direction cannot be determined by the Pythagorean theorem. The length of model axis, however, can be determined by the sum of projection length of each vector along the crystallographic axis on the corresponding model axis. Taking the unit box length $\ell_x$, along the X direction for example, for $X = [1\,1\,0]$, we have

$$\ell_x = a\cos(X, a) + b\cos(X, b) + 0 \tag{10F.12}$$

After $\ell_x$, $\ell_y$, $\ell_z$ and the model length $L_x$, $L_y$, and $L_z$ are determined, the cell number in each direction and the number of atoms in each model unit cell can be determined, thus the total model unit cells and the total atom number in the system can also be determined as shown above.

# References

[1] https://www.ivec.org/gulp/
[2] Gale, J. D. and Rohl, A. L. (2003) The General Utility Lattice Program (GULP). *Molecular Simulation*, **29**(5), 291.
[3] Li, J. (2003) AtomEye: An efficient atomistic configuration viewer. *Modelling Simul. Mater. Sci. Eng.*, **11**, 173.
[4] Li, J. (2005) Atomistic visualization, in *Handbook of Materials Modeling* (ed. S. Yip), Springer, Berlin, 1051.
[5] Wang, Y. M., Li, J., Hamza, A. V., and Barbee, T. W. (2007) Ductile crystalline-amorphous nanolaminates. *Proc. Nat. Acad. Sci. USA*, **104**, 11155.
[6] Plimpton, S. J. (1995) Fast parallel algorithms for short-range molecular dynamics. *J. Comput. Phys.*, **117**, 1.
[7] Smith, W. and Forester, T. J. (1996) DL_POLY_2.0: A general-purpose parallel molecular dynamics simulation package. *J. Mol. Graphics*, **14**, 136.
[8] Smith, W., Yong, C. W., and Rodger, P. M. (2002) DL_POLY: Application to molecular simulation. *Molecular Simulation*, **28**, 385.
[9] Refson, K. (2000) Moldy: A portable molecular dynamics simulation program for serial and parallel computers. *Computer Physics Communications*, **126**, 310.
[10] Nelson, M. T., Humphrey, W., Gursoy, A., *et al.* (1996) NAMD: A parallel, object oriented molecular dynamics program. *Int. J. High Perf. Computing Appl.*, **10**, 251.
[11] Kale, L., Skell, R., Bhandarkar, M., *et al.* (1999) NAMD2: Greater scalability for parallel molecular dynamics. *J. Comput. Phys.*, **151**, 283.
[12] http://www.itap.physik.uni-stuttgart.de/~imd/download/imd-20010-10-06.tgz
[13] http://www.ks.uiuc.edu/Research/vmd
[14] Lodesh, H., Berk, A., Matsudaira, P., *et al.* (2004) *Molecular Cell Biology*, 5th edn., W. H. Freeman and Company, New York.
[15] Krammer, A., Lu, H., Isralewitz, B., *et al.* (1999) Forced unfolding of the fibronectin type III module reveals a tensile molecular recognition switch. *Proc. Nat. Acad. Sci. USA*, **96**, 1351.
[16] Stoer, J. and Bulirsch, R. (1973) *Einführung in die numerische Mathematik*, II, Springer, Berlin.
[17] Hockney, R. W., Goel, S. P., and Eastwood, J. W. (1974) Quite high-resolution computer models of a plasma. *J. Comput. Phys.*, **14**, 148.
[18] Grest, G. S., Dunweg, B., and Kremer, K. (1989) Vectorized link cell Fortran code for molecular dynamics simulations for a large number of particles. *Computer Physics Communications*, **55**, 269.
[19] http://www-unix.mcs.anl.gov/mpi/mpich/
[20] www.openmp.org/
[21] www.csm.ornl.gov/pvm/pvm_home.html
[22] www.ks.uiuc.edu/Research/namd/
[23] http://cmp.ameslab.gov/cmp/CMP_Theory/cmd/alcmd_source.html
[24] http://bifrost.lanl.gov/MD/MD.html
[25] http://www.fos.su.se/physical/sasha/md_prog.html
[26] http://www.cpmd.org/cpmd_thecode.html
[27] Inorganic Crystal Structure Database, Fachinformationszentrum Karlsruhe, Germany, 2006.
[28] *International Tables for X-Ray Crystallography*. The Kynoch Press, Birmingham, 1965.

# Postface

In the book's website http://multiscale.alfred.edu, there are different simulation softwares; most packages such as GULP, DL_POLY_2, NAMD, VMD and LAMMPS, require registration first (without any financial charge). Readers should register and then download these packages into either their home directory or into their UNITS directory, which should be downloaded from corresponding websites. The book's website also includes answers for all the homework problems and other teaching materials.

This book (in draft) has been used at Alfred University as a textbook for a dual-level course for undergraduate and graduate students in engineering. It was also used for a graduate course at Shanghai University. Approximately 50 teaching hours can cover the basic contents from Chapter 1 to 7 and 10 hours for computer simulation from UNIT1 to UNIT6. In the teaching at Alfred University, two teaching courses each of 80 minutes (or three of 50 minutes) and one 50-minute lab course are arranged every week for a total of 16 weeks. The last five weeks of lab classes are used for course projects, whose content can be something similar to the coating layer simulation described in UNIT7 to UNIT9 of CSLI. Chapters 8 and 9 are written for researchers but can also be used for a course of graduate students such as one covering plasticity and mechanics of biomaterials.

Multiscale analysis is a hot topic at the cutting edge of engineering and science. It can link the knowledge of physics, chemistry, and biology, that students learned in high school and/or in the first year of university, to the leading areas of science and new technology such as nanotechnology and biotechnology. Based on this nature, our experience shows the importance of student reading, oral report, and class questioning and discussions. For instance, Chapter 1 and Chapter 7 can basically be used for reading and discussion (i.e., do not need detailed teaching). Part 10.2 of Chapter 10 can first be used to show the intrinsic relationship between Section 6 and Section 7 through introduction of the connection between Boxes A, B, C and D of Table 10.1 and the directory tree of UNIT7 and UNIT8 in Figure 10.6, and then students can read and do more work themselves. While Chapter 10 includes several softwares, practical teaching must be limited to a selection, for instance one need only teach VMD or AtomEye for the visualization part.

The book has tried not to use complicated mathematics. Simple matrix or vector calculations, however, are necessary for some sections such as Part 6.3 of Chapter 6. Without this knowledge it is difficult to use the Cauchy-Born rule to calculate the interatomic distance and angles between atoms after deformation, or to understand the beauty of atomistic-based continuum theory. A short review of the mathematical content is suggested before teaching it. Chapters 8 and 9 are mainly for graduate students and some undergraduate students with good background. Instructors can also ask students to read papers and make oral presentation in the class. This will greatly inspire students' learning motivation and enhance their capability for critical thinking and questioning.

# Index